Principles of Metamorphic Petrology

Principles of Metamorphic Petrology provides a modern and comprehensive introduction to the latest ideas, techniques and approaches in the study of metamorphic rocks. The book begins with basic concepts, but advances further than most other metamorphic petrology texts. It presents an introduction to the latest chemographic approaches, especially the correct use of pseudosections and the application of X-ray compositional mapping to metamorphic problems. It covers recent advances in thermobarometry and the application of modern geochronological techniques to the absolute timing of tectonometamorphic events. The determination of parent rocks is covered in detail, and there is a modern exposition of partial melting reactions, melt extraction and deformation of migmatites. The book includes a large number of references to lead students into independent investigation. This is an invaluable text for advanced undergraduate and graduate courses in metamorphic petrology. It also provides an up-to-date reference for researchers and exploration geologists.

R. H. Vernon is Emeritus Professor of Geology at Macquarie University and an honorary Research Professor at the University of Southern California. He has taught undergraduate and graduate courses and short courses in Australia, the United States, Italy, Germany, Finland and Mexico and authored several books, including two others with Cambridge University Press: *Beneath Our Feet: The Rocks of Planet Earth* (2000) and *A Practical Guide to Rock Microstructure* (2004). He was an editor of the *Journal of Metamorphic Geology* for 18 years.

G. L. Clarke is a Professor of Geology and the current Head of the School of Geosciences at the University of Sydney, where he has taught geology for more than 15 years. Before teaching, he spent three years as a minerals exploration geologist for Esso Exploration and Production. He is a member of the Editorial Board of the *Journal of Metamorphic Geology* and an editor for the *Journal of Petrology*.

Principles of Metamorphic Petrology

R. H. Vernon
Macquarie University, Sydney

G. L. Clarke
University of Sydney

CAMBRIDGE UNIVERSITY PRESS
Cambridge, New York, Melbourne, Madrid, Cape Town, Singapore, São Paulo, Delhi

Cambridge University Press
32 Avenue of the Americas, New York, NY 10013-2473, USA

www.cambridge.org
Information on this title: www.cambridge.org/9780521871785

First published 2008

Printed in the United States of America

A catalog record for this publication is available from the British Library.

Library of Congress Cataloging in Publication Data

Vernon, R. H. (Ronald Holden)
Principles of metamorphic petrology / R. H. Vernon, G. L. Clarke.
p. cm.
Includes bibliographical references and index.
ISBN 978-0-521-87178-5 (hardback)
1. Rocks, Metamorphic. 2. Petrology. I. Clarke, G. L. (Geoffrey L.), 1958– II. Title.
QE475.V47 2008
552′.4 – dc22 2007045326

ISBN 978-0-521-87178-5 hardback

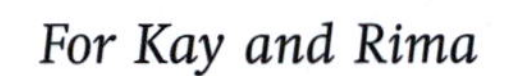

For Kay and Rima

Contents

Colour insert follows page 210.

Preface

We wrote this book for senior undergraduate, honours and beginning graduate students in metamorphic petrology. We start with basic concepts and develop them in terms of modern ideas, techniques and approaches, citing a reasonably large number of references, in the hope that they may lead students into various avenues of independent investigation.

We emphasize the field-based aspects of metamorphic geology because they provide a firm basis for the controlled interpretation of metamorphic and deformation processes. We use as examples rocks and metamorphic terranes on which we have carried out systematic research work (e.g., the Arunta and Musgrave Blocks, Broken Hill and Cooma in Australia, as well as Fiordland in New Zealand, California, Ireland, Antarctica and New Caledonia), in the hope that this may convey to the reader some idea of the spirit of research and the development of ideas in metamorphic petrology.

We also emphasize the recent approach of using pressure-temperature (*P-T*) pseudosections, instead of compositionally simplified *P-T* grids, although these are useful in many circumstances. Pseudosections take into account compositional controls on assemblages in more detail than has been possible previously, thereby permitting more realistic *P-T* estimations and histories for specific rock compositions.

We have also tried to provide some useful information for students intending to enter the mineral exploration field and for professional economic geologists working in metamorphic terranes, in the form of (1) criteria for determining parent rocks (protoliths) for metamorphic rocks, which is basic information for ore search models in such terranes, (2) a review of metasomatism and (3) references to ore deposits formed during regional metamorphism.

The book begins with basic metamorphic processes, emphasizing the control of whole-rock composition on mineral assemblages and showing how modern chemography can be used to elucidate metamorphic conditions and history. After this, we examine the inference of pressure-temperature-deformation history from mineral assemblages, replacement microstructures, porphyroblast-matrix microstructural relationships and geochronology. Then we discuss partial melting in crustal rocks and the influence of fluids and metasomatism in metamorphism. The book continues with a review of the main tectonic structures of metamorphic rocks and concludes with a detailed discussion of criteria for recognizing parent rocks.

Mineral abbreviation symbols (listed in the Appendix) follow Kretz (1983), as extended by Bucher and Frey (1994). Most of the diagrams and photographs are ours; colleagues who kindly supplied illustrations are acknowledged in the figure captions. We thank Nathan Daczko, Joel Fitzherbert, Pascal Phillipot and Richard White for helpful comments on parts of the typescript.

Ron Vernon and Geoff Clarke

Chapter 1

Metamorphic Processes

1.1 Introduction

Metamorphism refers to the mineralogical and structural alteration of rocks in Earth's crust, and excludes alteration at or just beneath the surface, such as weathering and early diagenesis. However, diagenetic changes grade into metamorphic changes in *burial metamorphism*. In fact, metamorphism may occur in several geological environments, because assemblages react to changes in temperature (*T*), pressure (*P*) and activities of mobile chemical components, regardless of the geological processes responsible for the changes. For example, identical or very similar assemblages can be formed by (1) deuteric alteration in an igneous intrusion, (2) wall-rock alteration around a hydrothermal orebody, (3) low-temperature alteration around an igneous intrusion, (4) burial metamorphism of sedimentary/volcanic successions and (5) alteration, both local and regional, in geothermal areas (Vernon, 1976).

Metamorphic geologists try to: (1) infer what the rocks were before metamorphism (the parent or precursor rocks) and (2) work out what has happened to them during metamorphism (the metamorphic-deformation history). Most effort goes into the second aspect (see Chapters 1 to 6), in the form of inferring: (1) compatible metamorphic mineral assemblages; (2) the combinations of pressure (*P*), temperature (*T*) and fluid composition (*X*) implied by those assemblages; (3) changes in *P-T* conditions with time, (4) metamorphic reactions, including partial melting reactions; (5) relating metamorphic minerals and assemblages to deformation events (D) and (6) suggesting *P-T*-time (*t*) paths for the region of Earth's crust under consideration. However, careful determination of parent (source) rocks is also important (see Chapter 7), for (1) providing the necessary field control on laboratory-based interpretations; (2) completing the metamorphic-deformation history of a terrane and (3) identifying pre-metamorphic rock-types, which is necessary for ore search models in metamorphic terranes.

The classification of metamorphic rocks is simple, with the result that metamorphic petrology is not cluttered with a large number of rock-type names. Nine basic terms (Table 1.1) cover nearly everything, apart from a few specialized compositional terms, such as 'eclogite', 'skarn', 'peridotite', 'serpentinite' and 'rodingite'. More detailed naming is achieved by qualifying the nine basic terms with mineral adjectives. For example, various

Table 1.1. Metamorphic rock names

hornfels:	typically granoblastic rock, generally non-foliated ('massive'), occurring in contact metamorphic aureoles; may be foliated where the contact metamorphism overprints earlier regional metamorphism
slate:	fine-grained, low-temperature regional metamorphic rock with a slaty cleavage (see Chapter 6)
phyllite:	similar to slate, but with larger chlorite and white mica grains, resulting in a less perfect cleavage
schist:	regional metamorphic rock, coarser-grained than phyllite, with schistose foliation (see Chapter 6) and commonly (though not necessarily) with porphyroblasts (large metamorphic grains)
gneiss:	coarse-grained, generally high-temperature regional metamorphic rock with a gneissic (commonly lenticular and discontinuous) foliation (see Chapter 6)
granofels:	granoblastic, non-foliated rock, commonly (though not necessarily) high-grade; may be contact or regional metamorphic; useful non-genetic term
amphibolite:	special compositional term for a metamorphic rock (typically regional) composed largely of amphibole (especially hornblende) and plagioclase
marble:	special compositional term for metamorphosed pure or nearly pure limestones or dolomites; metamorphosed impure (clay and/or quartz-bearing) limestones or dolomites are called 'calcsilicate' rocks
quartzite:	special compositional term for metamorphosed pure or nearly pure quartz-rich rocks, such as quartz sandstones (orthoquartzites) or cherts; the term 'metaquartzite' is technically preferable, but not always used

hornfelses and schists may be distinguished by mineral prefixes, such as hornblende hornfels, biotite-andalusite hornfels, garnet-wollastonite hornfels, mica schist, biotite-garnet schist and muscovite-cordierite-andalusite schist. The prefix 'meta-' is also useful (e.g., metasandstone, metapelite, metagabbro, metarhyolite), provided the parent rock is known.

The most direct evidence of metamorphism in Earth's crust is observation of a transition from unaltered to strongly altered rocks. The variation may be due to progressively increasing T and P, which is called *prograde metamorphism* or it may be due to decreasing T and P, which is called *retrograde metamorphism*. The degree of alteration is loosely referred to as the *metamorphic grade* or *grade of metamorphism* (e.g., Tilley, 1924). Rocks are referred to as being of 'low grade', 'medium grade' or 'high grade' informally, with no well-defined P-T boundaries between them. Grade is commonly related mostly to temperature (e.g., in low-P metamorphism), though pressure also has an effect, and dominates over T in high-P metamorphism (Fig. 1.1).

Because chemical reactions proceed faster at higher T, most prograde reactions tend to go to completion, whereas retrograde reactions, are commonly incomplete (showing evidence of partial replacement of high-grade

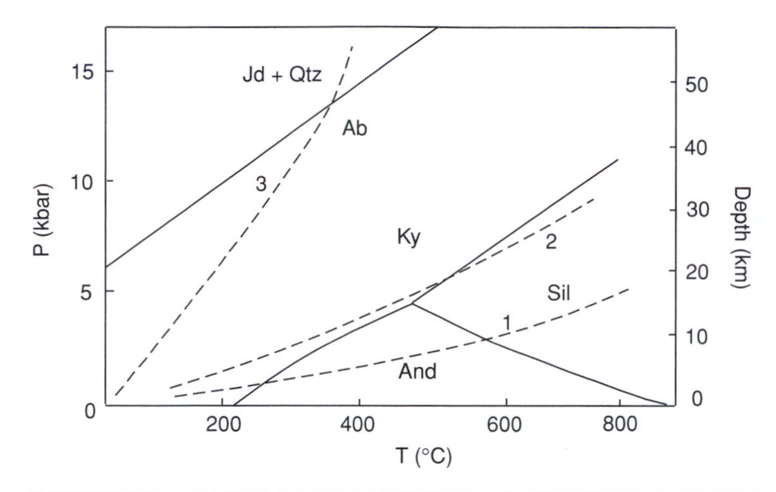

Fig. 1.1 Three 'geotherms' superimposed on a simple phase diagram that shows the approximate *P-T* positions of the Al_2SiO_5 reaction curves and the line for the reaction Ab = Jd + Qtz. Geotherm 1 (characterized by andalusite at lower grades) represents a low-pressure terrane; geotherm 2 (characterized by kyanite at lower grades) represents a medium-pressure terrane; and geotherm 3 (characterized by the assemblage Jd + Qtz at higher *P*) represents a high-pressure terrane.

minerals), owing to decreasing *T* and restricted access of H_2O and/or CO_2 that was lost during earlier prograde metamorphism (e.g., Vernon, 1969; Vernon & Ransom, 1971; Corbett & Phillips, 1981). Exceptions occur in high-strain zones (shear zones, mylonite zones), where the retrograde reactions tend to go to completion, owing to the catalytic effect of strong deformation, which facilitates neocrystallization by inducing plastic strain in the original grains and by opening transient pores for fluid access (Vernon & Ransom, 1971; White & Clarke, 1997; Vernon, 2004). Outside such high-strain zones, rocks may show evidence of both prograde and retrograde metamorphism (Vernon, 1969; Vernon & Ransom, 1971). Both the prograde and retrograde metamorphism can belong to the same metamorphic cycle ('event'), or the retrograde metamorphism can occur much later, the rock then being referred to as *polymetamorphic*. This term is also applied to rocks subjected to more than one period of prograde metamorphism.

Experiments have shown that low-grade metamorphic minerals are stable at about 100°C–300°C and that common high-grade minerals are stable at about 700°C–850°C. Higher temperatures (in excess of 1000°C) occur in exceptionally hot regional metamorphic terranes (e.g., Ellis *et al.*, 1980; Harley, 1989) and around some high-temperature (mafic-ultramafic) intrusions (e.g., Willemse & Viljoen, 1970; Nell, 1985). Pressures range from very low (in high-level metamorphism around igneous intrusions) to in excess of 30 kbar (approximately 100-km depth) in high-pressure metamorphism (Chopin, 2003).

Conventionally, metamorphism is divided into *contact ('thermal') metamorphism*, occurring around igneous intrusions, and *regional metamorphism*, in which metamorphic rocks are produced on a regional scale. However, the same minerals can be formed in both types of metamorphism, and so the distinction has no serious genetic importance, though the terms are widely used for convenience. In fact, the intensity of regional metamorphism at low pressure is commonly related to proximity to granite intrusions (e.g., Vernon *et al.*, 1990; Clarke *et al.*, 1990; Collins & Vernon, 1991; Vernon *et al.*, 1993a; Williams & Karlstrom, 1996; Vassallo & Vernon, 2000), and a useful distinction can be made between *contact aureole granites*, around which metamorphism is very localized, and *regional aureole granites*, around which occur metamorphic aureoles of regional extent (White *et al.*, 1974), as in the Cooma complex, south-east Australia. It is also often useful to refer to *low-P*, *medium-P* and *high-P* metamorphism, based on broad pressure differences (Fig. 1.1), as discussed in more detail in Section 1.7.2.

The minerals in chemically compatible coexistence in a metamorphic event constitute a *mineral assemblage* or *paragenesis*. A problem is to ascertain that the members of an inferred mineral assemblage really did coexist stably (Section 2.3).

The metamorphic process is profoundly asymmetrical. Metamorphic rocks formed at deep crustal or even upper mantle depths are exposed at Earth's surface, and yet their high-grade minerals do not reflect their current physical conditions (1 atmosphere, 25°C). Thus, they are *metastable*, in the sense that they are thermodynamically unstable, but persist because of kinetic barriers to their reaction. The *metastable persistence* of metamorphic rocks at Earth's surface can be explained mainly by the role of a minor but important constituent, H_2O. Sediments being actively deposited have high proportions of hydrous pore fluid (up to 30%); yet their metamorphosed equivalents contain only small amounts of H_2O, which is mostly structurally bound in hydrous minerals (0%–12%). Though hydrous minerals dominate low-grade (low-*T*) metamorphic rocks, only small proportions of hydrous minerals occur in high-grade (high-*T*) metamorphic rocks. One of the main reasons high-grade metamorphic rocks are exposed at Earth's surface without being converted to low-temperature mineral assemblages during cooling is that hydrous fluid lost during the previous prograde metamorphism needs to be returned to the rocks in order to cause retrograde metamorphism, as mentioned earlier.

The mass of rock overlying most metamorphic environments imposes a *confining pressure*, which depends on the density and thickness of the overlying material, and which reduces the porosity. Plastic deformation and consequent *recrystallization* (see Sections 6.2 and 6.3) of minerals during metamorphism, growth of new mineral grains (*neocrystallization*) and movement of grain boundaries to reduce interfacial energy (e.g., Vernon, 1976, 2004) all contribute to porosity reduction. An analogous environment occurs in glacial ice sheets, where the weight of successive snow accumulation compacts underlying snow into nevé (firn) and then ice, with the progressive reduction of extensive porosity by grain-boundary movement, as illustrated by Vernon (2004, fig. 4.43).

These processes ensure that metamorphic rocks generally contain very little pore fluid, probably only up to about 2 weight percent (Yardley & Valley, 1997, 2000). The excess pore water is initially squeezed out, owing to the confining pressure, and H_2O released by prograde metamorphic reactions migrates as supercritical fluid in microfractures formed by local fluid pressure increase and by tectonic deformation. Despite the inability of metamorphic rocks to hold much free fluid, substantial amounts of fluid may pass though rocks undergoing deformation with time, especially in fracture networks and shear zones, in which space for the fluid is made by the opening of transient microfractures and pores caused by the different responses of adjacent minerals to the local stress system, as discussed in Chapter 5.

Pore fluid in most metamorphic environments cannot be observed directly, as most is lost during metamorphism. Vestiges may remain as small primary fluid inclusions in minerals, and indications of fluid composition can be obtained from deep crustal drilling. However, limits to drilling technology restrict sampling to the upper 10 km or so of the crust (i.e., above most regional metamorphic terranes), and fluid inclusions are prone to the partial loss of mobile material (decrepitation). Therefore, much information concerning pore fluid in metamorphic rocks must be gleaned indirectly (see Chapter 5).

Up to 12 weight percent H_2O may be structurally bound as hydroxyl $(OH)^-$ in common hydrous minerals, such as mica, amphibole or lawsonite. These solid H_2O reservoirs convey fluid deep into Earth during subduction, and fluid released consequent to their breakdown during prograde metamorphism may contribute to partial melting and the formation of some arc magmas.

Though most metamorphic rocks experience a dynamic history involving burial, metamorphism and exhumation, they commonly preserve *P-T* evidence of only the maximum or *peak* conditions. Working out the *P-T* history of the burial and exhumation stages (Chapter 2) may be difficult or impossible. However, many terranes partially record evidence of their *P-T* history, and so metamorphic petrologists attempt to suggest plausible models by combining these partial records with *P-T* information obtained on other rocks or from experiments.

On the way to reaching peak conditions, a rock undergoes a series of *devolatilization* (dehydration and/or decarbonation) reactions, with products that are commonly denser and less hydrous than the original minerals (reactants). The inability of rocks to store fluid, owing to restricted or absent porosity, means that H_2O or CO_2 released as fluid by mineral breakdown is lost. Having lost a key potential reactant, the products cannot react to reverse the prograde reactions during cooling, preserving peak assemblages, as mentioned previously.

1.2 Effect of rock composition: The ACF diagram

Most metamorphic processes are considered to be essentially *isochemical*, apart from loss of volatile components, such as H_2O in the metamorphism of pelites and mafic igneous rocks and CO_2 in the metamorphism of carbonate rocks (Harker, 1893; Leake & Skirrow, 1958; Pattison & Tracy, 1991, p. 118; Barton *et al.*, 1991; Greenfield *et al.*, 1998), although some have challenged this generalization (see Section 5.13). However, regardless of the extent of inferred chemical change during prograde metamorphism of metasedimentary rocks, the essential chemical composition (e.g., pelitic character) is preserved. So it is fair to say that the major and trace element composition of most metamorphic rocks largely reflects the *protolith* (parent rock) composition, enabling us to 'see through' the effects of metamorphism (see Chapter 7). Major exceptions to the generalization of approximately isochemical metamorphism involve special circumstances, in which large fluid–rock ratios and chemically complex fluids profoundly influence a rock's chemical composition in a process referred to as *metasomatism* (Chapter 5).

A simple example of at least approximately isochemical metamorphism is an orthoquartzit with ripple cross-bedding delineated by magnetite or hematite: metamorphism produces a metaquartzite by recrystallization of the detrital quartz, magnetite and hematite grains, without chemically modifying them. A more complex example is a low-grade metamorphic rock rich in chlorite and biotite, as well as feldspar and quartz. The chemical elements now forming chlorite and biotite probably were deposited as clay minerals in a mud-rich sediment. Clay minerals are only stable up to a few hundred degrees Celsius, and react to form new minerals, such as chlorite and biotite, which are stable at the elevated temperatures of metamorphism. Large proportions of chlorite tend to occur in low-grade metamorphic rocks, but react with other minerals to form new minerals such as garnet, cordierite and staurolite with biotite, in response to increasing temperature.

In some areas of extensively exposed metamorphic rocks, individual sedimentary layers can be traced through grade variations (e.g., Harte & Hudson, 1979; Stewart, 1981). This illustrates another important principle, namely that minerals in a metamorphic rock reflect both the whole-rock chemical composition *and* the metamorphic grade. This is well exemplified by amphiboles formed in the metamorphism of ocean-floor basalts. At the comparatively low-*T* conditions of subduction metamorphism, metabasalts are dominated by glaucophane, whereas at higher-grade conditions, the same rock-composition is dominated by hornblende. At the highest grade, the metabasalt may lack amphibole, and instead contain the anhydrous chemical equivalent, pyroxene (e.g., Vernon *et al.*, 1990).

Metamorphic rocks exposed at Earth's surface can be subdivided chemically in many ways, but for the purpose of field identification, can be initially divided into a few main compositional types. *Mafic* (*basic*) rocks form the largest proportion of Earth's surface, as basaltic material floors most oceans. Most metamorphosed mafic rocks (amphibolites, greenschists) have basaltic or gabbroic protoliths; they are widespread and define key assemblages for several metamorphic facies (Section 1.7). *Ultramafic* rocks are commonly mantle derived, low-Si, high-Mg rocks, composed chiefly of mafic minerals. *Quartzofeldspathic* (*felsic*) rocks may be of (1) sedimentary origin, namely sandstones (psammites, arenites), such as feldspathic sandstone ('greywacke') or felsic tuff, or (2) igneous origin, namely granitoids or metavolcanic rocks. *Pelitic* rocks (shales, claystones, mudstones) are fine-grained clastic sediments with appreciable clay and quartz; they are commonly carbonate-poor, and their metamorphic equivalents (metapelites) are sensitive indicators of the greenschist, amphibolite and granulite facies (discussed in Section 1.8.1). *Carbonate rocks* are sedimentary rocks formed by organic or inorganic precipitation from aqueous solution of carbonates of calcium, magnesium or iron. *Marls* are sedimentary rocks containing both carbonate and clastic (clay and quartz) components, and may be dominated by either.

Metamorphic minerals are of two types, namely those with fixed composition, such as quartz (SiO_2) or calcite ($CaCO_3$), and those with variable composition, such as garnet or amphibole. Variations in mineral composition are caused by cations of similar charge and size exchanging on sites in the crystal lattices, a process called *solid solution*. For example, garnet is commonly a solid solution of almandine ($Fe_3Al_2Si_3O_{12}$) and pyrope ($Mg_3Al_2Si_3O_{12}$), the ratio of Fe^{2+} to Mg^{2+} being temperature-dependent (for

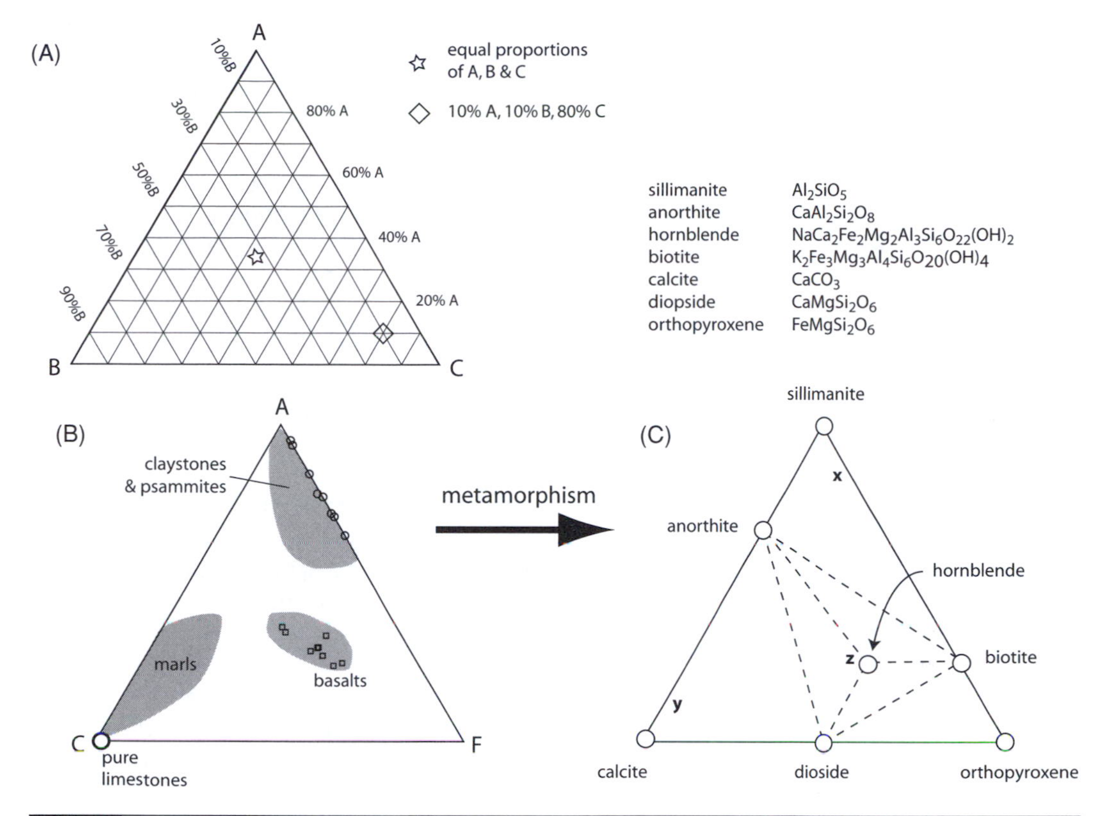

Fig. 1.2 ACF ternary diagram, for which $A = Al_2O_3 + Fe_2O_3 - (Na_2O + K_2O)$, $C = CaO - 3.3\ P_2O_5$, and $F = MgO + FeO + MnO$. (A) Method for plotting on a ternary diagram; the star represents equal proportions of all three components. (B) Common compositional variation shown by a range of claystones, psammites, marls, and basalts. The plotted claystone (pelite) and sandstone (psammite) analyses are from a series of metamorphosed turbidites at Mount Stafford, central Australia (Greenfield *et al.*, 1998). The metabasalt analyses shown in Fig.1.2B are from a series of high-*P* metamorphic rocks in northern New Caledonia (Clarke *et al.*, 1997). (C) Simple mineral compatibility diagram showing the positions of common metamorphic minerals and common whole-rock ('bulk-rock') compositions. X represents the position of a metamorphosed claystone (pelite), plotting in the sillimanite-anorthite-biotite field. Y represents an impure limestone plotting in the calcite-anorthite-diopside field. Z represents a basalt, plotting in the hornblende-biotite-diopside field. As the compositions of most metabasalts are close to that of hornblende, these rocks are dominated by hornblende. Relationships between metamorphic whole-rock compositions and mineral assemblage are discussed in more detail in Sections 1.7 and 2.2.

a given rock composition). Plagioclase shows coupled ($Na^+Si^{4+} = Ca^{2+}Al^{3+}$) substitution, and more complex minerals, such as micas or amphiboles, may show multiple solid solution series.

In rocks dominated by few minerals, the composition of the minerals can provide much useful information concerning variations in metamorphic grade. For example, greenschist facies metabasalts are commonly dominated by the amphiboles actinolite and hornblende, the transition from actinolite to hornblende being a chemical (and crystallographic symmetry) change controlled by temperature (Section 1.7).

Though metamorphosed ocean-floor basalts tend to be uniform in chemical composition, the wide variety of sedimentary environments at Earth's surface generates considerable diversity in the composition of clastic deposits. The variation in compositions is best shown graphically, which is commonly done using ternary diagrams. In Fig. 1.2A, the proportions of hypothetical variables A, B and C are simultaneously plotted, the base of the triangle representing 0% A and the top apex 100% A. Equal proportions of all three variables are represented by the star in the middle of the triangle. For example, variation in common whole-rock (bulk-rock) composition can be simplistically represented on an *ACF diagram* (e.g., Turner, 1981), in which $A = Al_2O_3(AlO_{3/2}) + Fe_2O_3(Fe_{3/2}) - (Na_2O + K_2O)$, $C = CaO - 3.3P_2O_5$ and $F = (FeO + MgO + MnO)$, as shown in Fig. 1.2B. Before plotting rack compositions on this diagram, K_2O and Na_2O are subtracted from Al_2O_3 to remove from consideration the amount of Al_2O_3 tied up in common felspar, and so more accurately reflect excess Al available for other minerals. Similarly, enough P_2O_5 is subtracted from CaO to reflect calcium locked away in apatite. Pure limestones are almost pure $CaCO_3$ and hence plot at the C apex of the ACF diagram (carbon not being considered). Contamination of a pure limestone by variable proportions of clastic material (such as quartz or clay minerals) results in a spread of rock types plotting between the C apex and the clastic contaminant (Fig. 1.2B), the exact position depending on the proportions of the mixture and the composition of the clay minerals. Claystones and sandstones are typically carbonate-poor or carbonate-free, and have a spread of compositions, depending on the proportions of clay and quartz (Fig. 1.2B).

At metamorphic conditions in the mid-amphibolite facies (discussed in Section 1.7.1), namely about 10-km depth and *c.* 550°C, the elements in clay minerals in sedimentary rocks or glass in basalt form metamorphic minerals that broadly reflect the original elemental proportions. Figure 1.2C shows a series of metamorphic minerals common at such temperature conditions plotted on an ACF diagram, their positions depending on the proportions of elements forming them. These minerals subdivide the diagram into a series of chemical domains, diagrammatically shown by joining different minerals by *tie-lines*. The positions of the lines depend on the physical conditions (P,T), as discussed in more detail in Section 1.8. For the chemographic relationships shown in Fig. 1.2C, a sandstone with composition x plots in a ternary sub-field bound by sillimanite, anorthite and biotite, so that this rock has these minerals at the metamorphic conditions considered. A marl with composition y plots in a field defined by calcite, anorthite and diopside; so the metamorphosed equivalent of the marl (a calcsilicate rock) is composed of these minerals. As the composition lies close to where calcite plots, the rock is mostly formed from calcite, with substantially less diopside and anorthite. The compositions of most basalts lie close to the position of hornblende; consequently they are mostly formed from this mineral at amphibolite facies conditions (see Section 1.8). The plotted metabasalt with composition z has the assemblage hornblende, diopside and biotite. At higher temperature conditions, the appearance of orthopyroxene in metabasalt defines the lower limit of the granulite facies (see Section 1.8), and the configuration of the tie-lines in the lower right of the diagram would need to be redrawn with a line between hornblende and hypersthene.

A weakness with the ACF diagram is that most rocks contain large proportions of elements in addition to those explicitly considered in Fig. 1.2, such as SiO_2, Na_2O, K_2O and TiO_2. The adjustment of Al_2O_3 values by subtracting Na_2O and K_2O content primarily addresses the role of plagioclase in metabasalts, and works less well for alkali content in, for example, actinolite, hornblende or biotite. Though the ACF diagram is a good starting point for considering metamorphic changes, it is unable to handle compositional detail.

Some rock compositions are better indicators of metamorphic grade than others. Mafic and pelitic compositions show marked mineralogical changes with changing grade, whereas felsic compositions commonly do not. For example, subdivisions of the blueschist and eclogite facies (Section 1.8.1) are largely based on metabasic mineral assemblages, and subdivisions of the amphibolite and granulite facies (Section 1.8.1) are largely based on metapelitic mineral assemblages. This is because metabasic rocks commonly have *high-variance* equilibria (involving few minerals) that are less useful for subdividing the amphibolite and granulite facies. However, continuous changes in the more calcic high-variance metabasic equilibria can be sensitive indicators of relative pressure variations. Mineral changes in metapelitic equilibria are dominated by *low-variance* dehydration reactions involving many minerals (e.g., Fig. 1.10), which mostly occur over restricted temperature intervals, and so are useful for subdividing the greenschist and amphibolite facies. Therefore, *P-T* analysis of most metamorphic terranes is best done using information from as many rock types as possible.

1.3 Metamorphic phase diagrams

Phase diagrams, such as the ACF diagram (Fig. 1.2B), are graphical representations of the chemical (mineral) and physical (*P* and *T*) variables that control metamorphic reactions. Inferred relationships are based on experimental data, direct field observations and theoretical calculations. Much effort has been exerted in determining phase relationships in metapelitic rocks, as they are the most sensitive indicators of common crustal metamorphic conditions, and can be used to subdivide the amphibolite facies into 'subfacies' (e.g., Harte & Hudson, 1979; Pattison & Harte, 1985). In this section, we address the modelling of metapelitic equilibria in a qualitative sense, with a simple example of how maps of mineral equilibria called *pseudosections* can be used to predict *P-T* conditions for major mineral assemblages (e.g., Hensen, 1988). Chapter 2 covers more complex methods for the quantitative modelling of mineral assemblages.

Single-component systems are the simplest, and H_2O is a good example. At conditions near Earth's surface, H_2O exists as the three phases ice, water and gas, the *P-T* relationships of which are shown in Fig. 1.3A. The three independent variables required to define the system are *P*, *T* and the chemical component H_2O. Increasing *T* by heating at constant *P* can result in a successive change from ice to water to gas, as indicated by the horizontal blue arrow in Fig. 1.3A. Increasing *P* at constant *T* can result in the change from ice to water, as indicated by the vertical blue arrow; this happens at the bottom of large glaciers and enables them to slip on the underlying rocks. An example more relevant to metamorphic petrology is provided by the aluminosilicate phase diagram (Fig. 1.3B). Three solid polymorphs of Al_2SiO_5 occur at common crustal conditions. Kyanite has the greatest relative density and hence occurs at the highest-pressure conditions. If only one phase is stable, there is considerable freedom in choosing *P* or *T* conditions (i.e., the mineral occupies a *P-T* area or *divariant field*, in which two variables, *P* and *T*, can be varied independently).

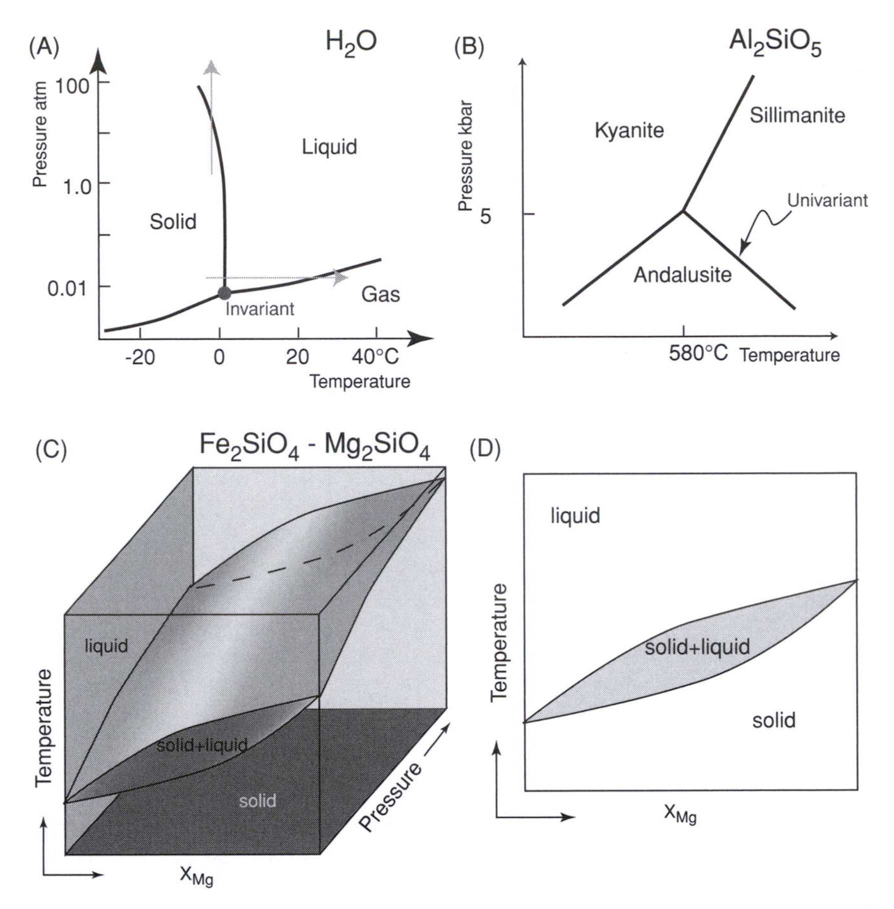

Fig. 1.3 (A) *P-T* diagram for the one-component system, H_2O. The vertical blue arrow indicates a phase transition from solid to liquid with increasing *P*, such as occurs at the base of large glaciers. For pure H_2O, the point at which ice, water, and vapour all coexist is fixed or invariant. (B) *P-T* diagram for the one-component system Al_2SiO_5, showing the relative stability ranges of the three minerals andalusite, sillimanite, and kyanite. The *P-T* conditions at which two of the minerals can coexist are univariant reaction lines, and the *P-T* condition at which all three phases coexist is an invariant point. (C) Diagrammatic three-dimensional illustration showing the three parameters (*P*, *T*, *X*) needed to define solid–liquid relationships in the two component system Fe_2SiO_4-Mg_2SiO_4 (fayalite-forsterite), where X = mole fraction of one of the end-member components. (D) Two-dimensional slice, at constant *P*, through Figure, 1.3C. For a colour version of this figure, please go to the plate section.

If two phases are present in a one-component system, for example, ice and water, or andalusite and sillimanite, the system has less freedom to vary *P* or *T*, so that the two phases can coexist only on a *univariant line* that represents the reaction from one phase to another (Fig. 1.3B). The line is called univariant because it permits only one degree of freedom; once *T* is stipulated, *P* is defined, and vice versa. If three phases coexist, the system has no degrees of freedom, and is fixed or *invariant* (Fig. 1.3A), because the invariant point occurs at only one *P* and *T*, which are the only conditions at which the three phases can coexist. In the system H_2O, the presence of NaCl reduces the freezing point of water to below 0°C, showing that adding more chemical components enables greater freedom in the system being considered. Aluminosilicate minerals can accommodate small proportions of F_2O_3 or Mn_2O_3, displacing the otherwise 'invariant' point for this system over approximately 0.5 kbar, and changing univariant curves to divariant bands (Grambling & Williams, 1985).

Two-component systems are more complex, and require illustration on a three-dimensional diagram. For example, consider melting relationships for olivine, which is composed of a solid solution between fayalite (Fe_2SiO_4) and forsterite (Mg_2SiO_4). The state of olivine, between solid and liquid, depends on P, T and olivine composition, which can be simplified to $X_{Mg} = Mg/(Mg + Fe)$ (Fig. 1.3C). These examples show that the number of axes required to represent any given system is $(v - 1)$, where v is the number of variables. Model systems become unworkable if $v > 4$, as they require diagrams of four or more dimensions.

Most common metapelitic equilibria can be modelled, in the first instance, in a six-component system (e.g., Wei *et al.*, 2004) that also requires P and T to be defined, and so theoretically requires a seven-dimensional diagram. A simplification involves holding one or more variables at fixed values. For example, the three-dimensional complexity of Fig. 1.3C reduces to a more easily comprehensible two-dimensional diagram when a slice (cross section) is drawn at a constant P (Fig. 1.3D).

A simple generalization of all the above relationships is the *phase rule*:

$$F = C + 2 - P$$

where F is the variance, or number of degrees of freedom, as defined by the number of physical variables required to be arbitrarily fixed to define the system, C is the minimum number of chemical components that can be used to express variation in the phases present and P is the number of phases present. The rule provides a useful indication of how variance is increased by adding additional components, such as in altering the freezing point of water by adding NaCl, or decreased by adding additional phases, such as adding kyanite to andalusite + sillimanite in the pure Al_2SiO_5 system, thereby changing a divariant field to an invariant point at which the three aluminosilicate minerals can coexist.

1.4 Metapelitic rocks and the AFM diagram

A commonly used model multi-component system for assessing changes in metapelitic mineral equilibria (Thompson, 1957) involves the chemical components K_2O-FeO-MgO-Al_2O_3-SiO_2-H_2O (KFMASH). K_2O is usually present in muscovite and biotite, or in K-feldspar at higher grade. FeO, MgO and Al_2O_3 occur in many minerals, such as garnet, mica and feldspar. SiO_2 occurs in all silicate minerals, and H_2O may occur in small proportions as a free phase or structurally bound in hydrous minerals. Additional elements, such as CaO, Na_2O or Mn_2O_3 can be important, because they stabilize, for example, plagioclase or garnet. These components are excluded, as a first approximation, in KFMASH, which forms the basis for much of our understanding of greenschist and amphibolite facies metapelitic equilibria (Harte & Hudson, 1979; Pattison & Harte, 1985, 1991).

Six components theoretically require seven axes to define, but the KFMASH system can be simplified by restricting interest to metapelitic assemblages that always have quartz, muscovite and water present, so that SiO_2, K_2O or H_2O can be excluded from being explicitly considered. By fixing P and T, the system reduces to a ternary involving Al_2O_3, FeO and MgO that can be represented in a two-dimensional ternary diagram commonly

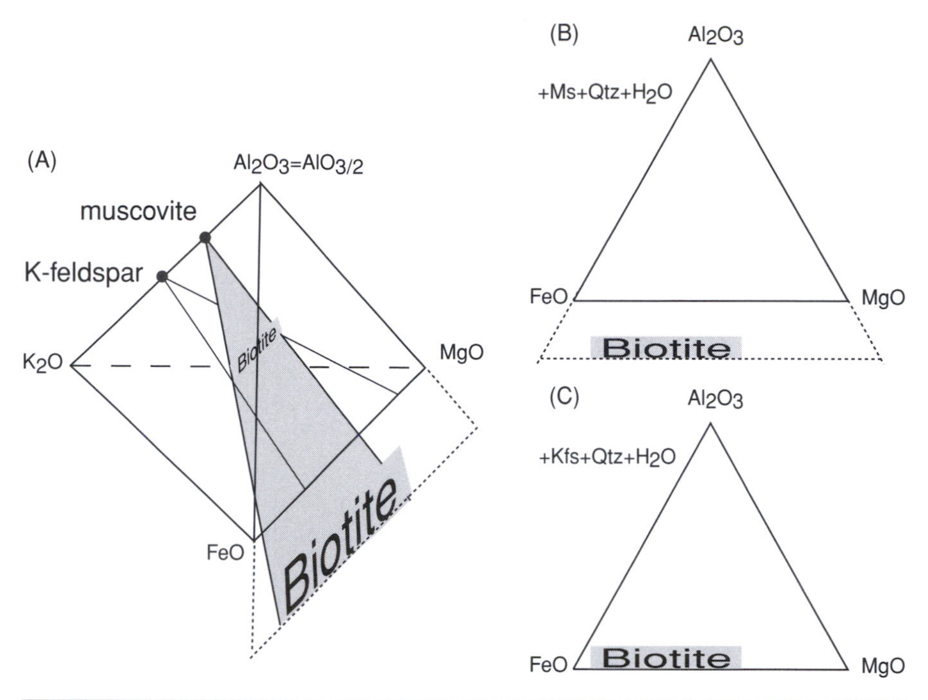

Fig. 1.4 (A) K_2O-Al_2O_3-FeO-MgO tetrahedron illustrating how projecting the position of biotite, which lies in the tetrahedron body, from muscovite onto the Al_2O_3-FeO-MgO face of the diagram causes biotite to plot at negative Al_2O_3. (B) Two-dimensional Al_2O_3-FeO-MgO (AFM) diagram showing the position of biotite projected from muscovite. (C) Two-dimensional Al_2O_3-FeO-MgO (AFM) diagram showing the position of biotite projected from K-felspar. For a colour version of this figure, please go to the plate section.

known as an *AFM projection* in which A = Al_2O_3(=$AlO_{3/2}$), F = FeO and M = MgO. SiO_2 and H_2O occur in pure quartz and water, respectively, but as K_2O occurs only as part of muscovite, biotite or K-feldspar, relationships initially need to be assessed in a K_2O-FeO-MgO-Al_2O_3 tetrahedron (Fig. 1.4A). As using three-dimensional diagrams can be clumsy, it is desirable to reduce the diagram to two dimensions, which is achieved by projecting the chemographic relationships from the main potassic mineral, usually muscovite (K-feldspar at higher metamorphic grade at which muscovite is unstable), onto the AFM face of the tetrahedron (Fig. 1.4A). Biotite plots in the body of the K_2O-FeO-MgO-Al_2O_3 tetrahedron, so that when it is projected from muscovite on to the AFM face, it falls at negative Al_2O_3 values (Fig. 1.4B). Projecting biotite from muscovite or K-feldspar can be calculated as follows, where biotite is expressed in terms of components muscovite, Al_2O_3, FeO, and MgO:

Biotite		Muscovite	$AlO_{3/2}$	FeO	MgO
$K_2Mg_2Fe_2Al_4Si_6O_{20}(OH)_4$	=	$K_2Al_6Si_6O_{20}(OH)_4$	−2	2	2
	=	2 × K-feldspar			
		2 × ($KAlSi_3O_8$)	0	2	2

As muscovite contains proportionately more alumina than biotite, the balancing of components requires a negative Al_2O_3 value on the AFM face. In projections from muscovite, quartz and water, these minerals are no longer explicitly considered on the AFM compatibility diagram, and are assumed to be present in all illustrated assemblages and reactions.

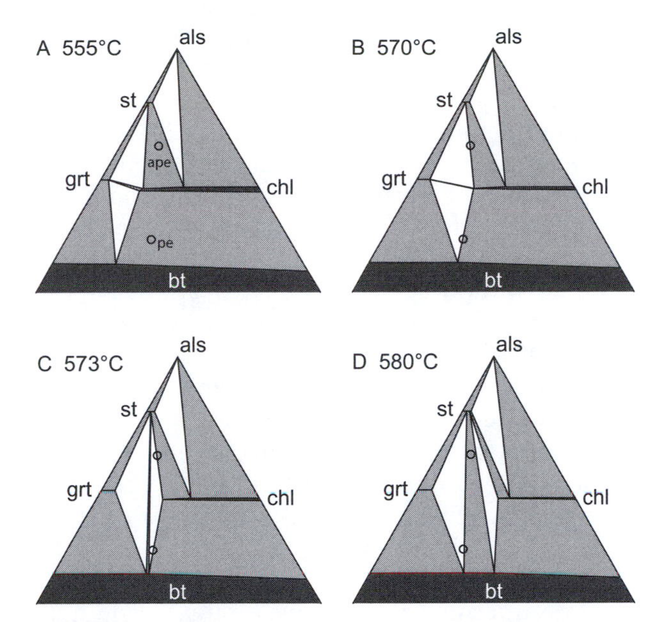

Fig. 1.5 AFM diagrams illustrating the compositional and mineral relationships predicted by phase diagram modelling of metapelitic equilibria in KFMASH using THERMOCALC, for $P = 6$ kbar and $T = 555\,^{\circ}\text{C}$–$580\,^{\circ}\text{C}$, assuming muscovite, water and quartz are in excess; modified from figures in Worley & Powell (1998a). The projected position of a common 'greywacke' (feldspathic sandstone) is shown in all four diagrams as 'pe'; Figure 1.5A additionally shows the projected position of an aluminous metapelite, 'ape'. The white triangles represent divariant assemblages, the light red shading represents trivariant assemblages, and the dark red shading represents quadrivariant assemblages. For a colour version of this figure, please go to the plate section.

Changing the biotite or muscovite compositions results in subtly different positions on the AFM face.

Mineral relationships can be illustrated on an AFM diagram for a lower amphibolite facies schist in the Barrovian garnet zone (Fig. 1.5). Aluminosilicate minerals have a fixed composition and plot as a point at the Al_2O_3 apex of the diagram. At this temperature, garnet is almandine-rich; in the model system it has a single $FeMg^{-1}$ solid solution and plots as a short line between pure almandine and mixed garnet with up to approximately 10% pyrope. Similarly, staurolite plots as a short line at a higher alumina value than garnet. Biotite and chlorite have dual solid solution involving both $FeMg^{-1}$ and $SiAl^{-1}$, so that these minerals plot as quadrivariant fields on the diagram. Figure 1.5A illustrates the chemographic relationships calculated for $T = 555\,^{\circ}\text{C}$ and $P = 6$ kbar in KFMASH, after Worley & Powell (1998a). Garnet, chlorite and biotite (together with muscovite, quartz and water) may occur together as a divariant assemblage in Al-poor rocks. Garnet, staurolite and chlorite or aluminosilicate, staurolite and chlorite form divariant assemblages in rocks with more Al and Mg. Most of the AFM face is covered by trivariant assemblages involving only four minerals (plus water), namely garnet + biotite, chlorite + biotite, garnet + staurolite or staurolite + chlorite, depending on Al-Fe-Mg proportions. These relationships are discussed further in Section 1.6.

The approximate *projected* position of a typical 'greywacke' (feldspathic sandstone) – in terms of *whole-rock* geochemical composition – is shown by the point labelled 'pe' on the diagram (Fig. 1.5A). It plots in the trivariant field involving chlorite, biotite, muscovite, quartz and water. The proportion of chlorite and biotite in the assemblage can be modelled using the relative positions of the whole-rock composition, chlorite and biotite. If the projected whole-rock composition plotted equidistant from chlorite and biotite, it would have equal proportions of the two minerals. However, in Fig. 1.5A, it plots two-thirds of the way towards biotite (relative to chlorite), meaning that the proportion of biotite to chlorite is 2:1. Variations in the proportions of Al_2O_3-FeO-MgO in interlayered metasediments reflect changing proportions of detrital clay minerals, feldspar and quartz in the parent sedimentary rock. Centimetre to metre-scale variations in whole-rock composition are common in sedimentary and metasedimentary rock sequences. For example, a more aluminous metasediment (metapelite) with a similar Fe-Mg ratio to that of the 'greywacke' might fall between staurolite and chlorite (point labelled 'ape' in Fig. 1.5A), or in the divariant field involving staurolite, aluminosilicate and chlorite (plus muscovite, quartz and water). Rocks with intermediate alumina contents would plot in the narrow chlorite-only field, with the quadrivariant assemblage chlorite, muscovite, quartz and water. At upper amphibolite and granulite facies conditions, muscovite is unstable and the main potassic mineral is K-feldspar. K-feldspar has a K:Al ratio different from that of muscovite, which causes biotite to plot on the F-M join instead of at negative Al_2O_3 (Fig. 1.4C).

At $T > 650\,^{\circ}C$ most crustal rocks commence partial melting in the presence of water (e.g., Johannes & Holtz, 1990; Lopez & Castro, 2001), as discussed in Chapter 4. The common evidence that granulite facies terranes experience temperatures up to 750°C and higher, without evidence of extensive water-saturated partial melting (Chapter 4) and melt loss, can only be reconciled by the absence of free water at peak conditions (e.g., Phillips, 1980; Powell, 1983). Accordingly, the assumption that water is 'in excess' is invalid for these rocks. For these situations, projections can be made from a partial melt, into which any water would be mostly partitioned, but relationships on such diagrams cannot be directly compared with those for subsolidus conditions.

1.5 Grade variations in contact metamorphism

Contact metamorphic aureoles have provided the basis for many classic petrological studies (Goldschmidt, 1911; Eskola, 1915; Pitcher & Berger, 1972; Pattison & Tracy, 1991, and additional references therein). The width of most aureoles (commonly 0.5–5 km, as narrow as 20 m or less adjacent to some sills) means that the effects of pressure variation are small or negligible, so that spatial variations in mineral assemblage primarily reflect differences in temperature as the margin of a causal pluton(s) is approached. Owing to limited strain accumulation, contact metamorphic rocks commonly preserve minerals with clear microstructural relationships that can be used for inferring reactions controlling changes at zone boundaries. These inferred key changes potentially have wide application, even in some

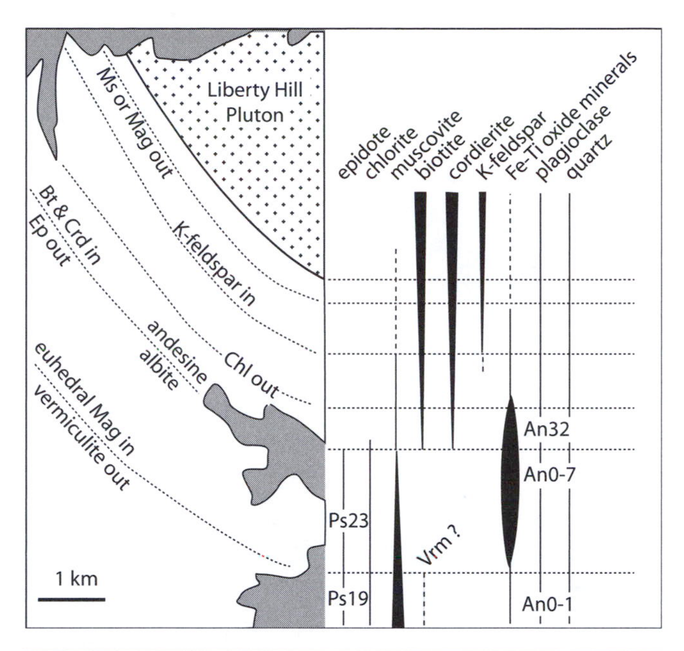

Fig. 1.6 Geological map of part of the Liberty Hill contact metamorphic aureole, South Carolina, USA, and a graphical representation of the major changes in metapelitic mineral assemblages; redrawn from figure 2 of Speer (1981). The chemical compositions of epidote and plagioclase reflect the effects of sliding reactions. An = anorthite component of plagioclase; Ps = pistacite component of epidote.

regional metamorphism, because the *P-T* variation inferred from exposed rocks in a contact metamorphic aureole are similar to those inferred for portions of an isobaric *P-T* path in low-pressure regional metamorphism (Pattison & Tracy, 1991). This is in contrast to the *P-T* variation inferred from exposed rocks in higher-pressure regional metamorphic terranes, which may differ considerably from actual *P-T* paths, as discussed in Section 3.5.3. The distribution of contact metamorphic mineral assemblages and their relative chemical compositions constitute petrological data that, when placed in the context of a model chemical system, provide relative *P-T* information that can be used to construct *petrogenetic grids* (also called *P-T* projections; see Section 2.4.5) of general application. Information needs to be collated from many aureoles, owing to variations in peak temperature and pressure conditions (Dickerson & Holdaway, 1989). In many instances, wall-rock xenoliths in the pluton are used to obtain additional information on the highest-grade conditions. This 'traditional' qualitative approach has direct application to fieldwork-based studies and provides fundamental information for testing modern methods of phase equilibria modelling (Section 2.4.5).

An example of the information obtained from metapelitic rocks in a contact metamorphic aureole adjacent to the Liberty Hill pluton in South Carolina, USA (Speer, 1981) is shown in Fig. 1.6. Key changes in mineral assemblages that occur as the pluton is approached involve: (1) breakdown of vermiculite and appearance of biotite and cordierite; (2) replacement of albite by more calcic plagioclase; (3) elimination of chlorite; (4) appearance

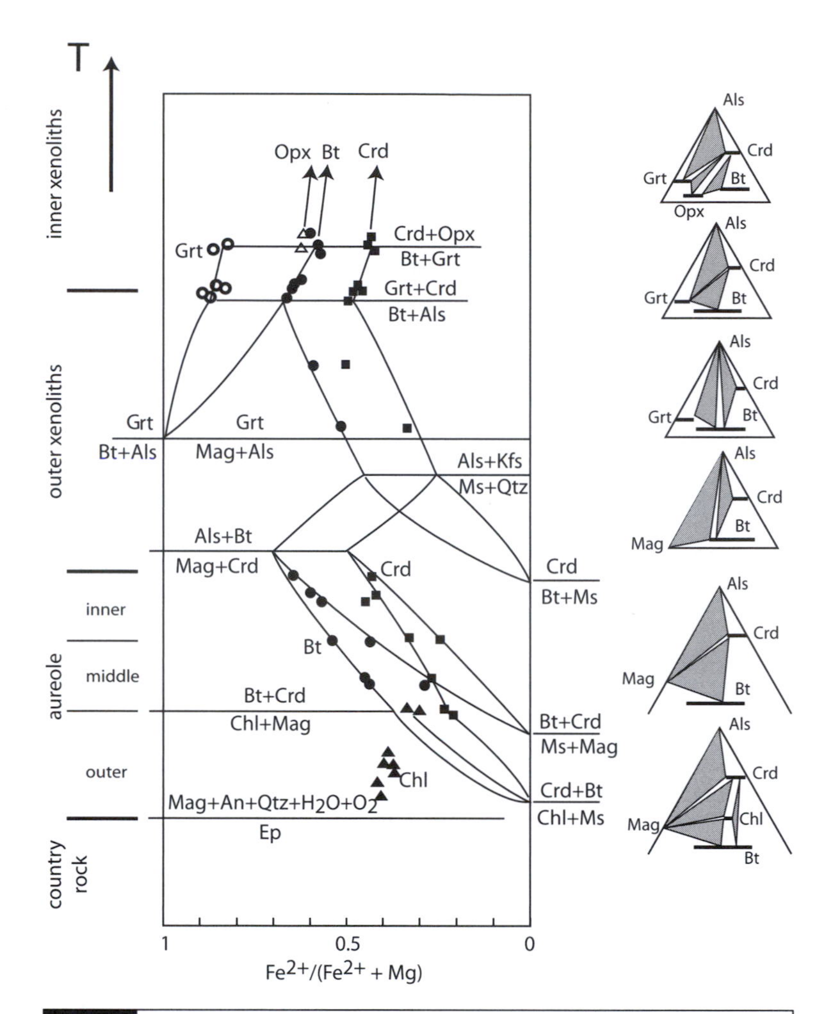

Fig. 1.7 Schematic isobaric T-$X_{Fe\text{-}Mg}$ diagram showing measured mineral compositions in the context of zonation for metapelitic assemblages from the Liberty Hill aureole, South Carolina; redrawn from fig. 3 of Speer (1981). Schematic AFM projections are shown for representative portions of the aureole. The high-*T* part of the profile is compiled from xenoliths in the plution.

of K-feldspar; and (5) elimination of muscovite. Two types of change are represented graphically in Fig. 1.6, namely *discontinuous* changes, such as the appearance of biotite and cordierite or the breakdown of albite, and *continuous* changes, as reflected in the composition of epidote or calcic plagioclase. These two types of change are evaluated further in Section 1.6. Variation in mineral chemical composition with aureole zonation, and schematic AFM diagrams for representative zones are shown in Fig. 1.7, with additional detail obtained from xenoliths in the pluton. Within each zone, mineral compositions change progressively with increasing temperature, and marked changes in assemblages were used by Speer (1981) to infer key reactions. The relationships shown in diagrams such as Fig. 1.7 form a basis for exploring the interdependence of rock composition and mineral assemblage, addressed in more detail in Section 1.6.

On the basis of metapelitic mineral assemblages from a series of well studied contact aureoles, Pattison & Tracy (1991) derived a schematic petrogenetic grid for that part of the KFMASH system for which muscovite,

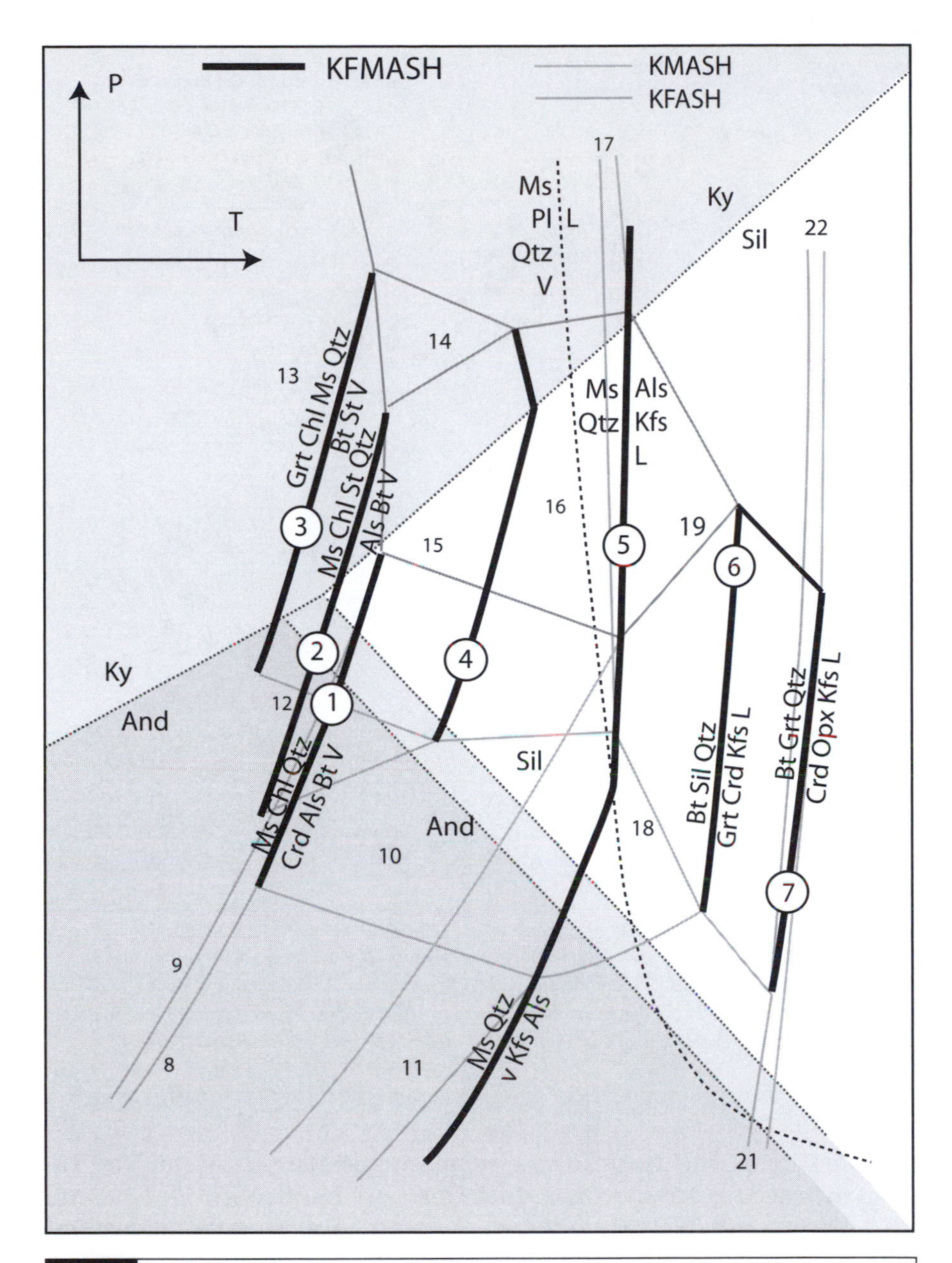

Fig. 1.8 Schematic petrogenetic grid for a portion of the system KFMASH involving the reactions listed in Table 1.2; redrawn from fig. 33 of Pattison & Tracy (1991). Dashed lines represent aluminosilicate relationships, the And = Sil reaction occurring over an interval, which reflects the common interpretation that andalusite metastably persists into the sillimanite field. V = supercritical H_2O fluid; L = silicate liquid (melt). For a colour version of this figure, please go to the plate section.

biotite, quartz and vapour *or* silicate liquid are always present (Fig. 1.8). This grid best reflects subsolidus relationships. The reactions in Fig. 1.8 represent an attempt to account for relative relationships between the majority of natural assemblages in metapelitic rocks of normal composition, on the basis of repeated occurrences of natural assemblages; this is in contrast to model grids based on thermodynamic data for mineral end-members (e.g., Powell & Holland, 1988), which deal with assemblage

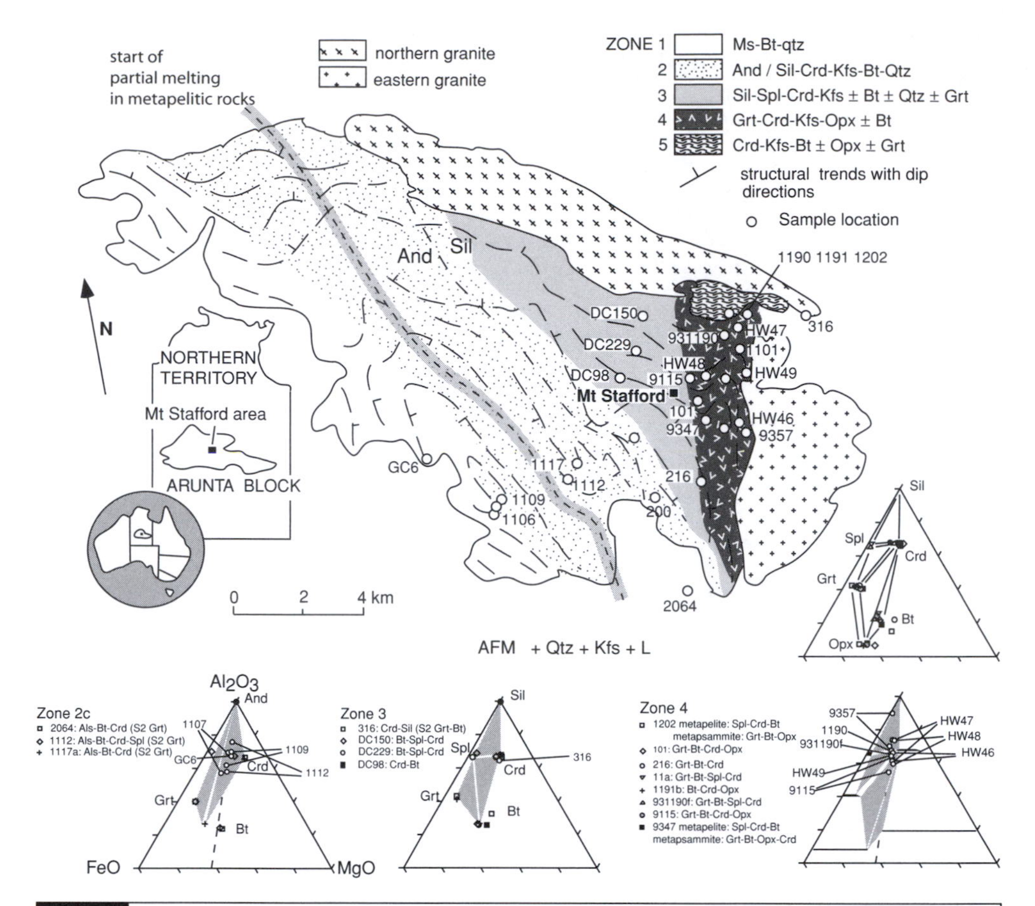

Fig. 1.9 Geological map of the Mount Stafford area, central Australia, showing bedding trends and metamorphic zoning, as well as the position of the And = Sil transition and the onset of partial melting in metapelitic rocks; modified from fig. 1 of Greenfield *et al.* (1998). Schematic AFM diagrams of peak assemblages are shown for Zone 2 to 4 metasediments, illustrating the composition of representative mineral assemblages (numbered) and related whole-rock compositions from XRF data; modified from fig. 4 of Greenfield *et al.* (1998).

variations in specific, realistic rock compositions; see Section 1.6. The key KFMASH reactions are shown as thick lines and labelled in Fig. 1.8. Pattison & Tracy (1991) inferred no invariant points, as most of these reactions have no intersections in KFMASH. The KFMASH reactions are inferred to emerge at low-T conditions from controlling KFASH equilibria and terminate at high-T conditions in controlling KMASH equilibria (see light reaction lines in Fig. 1.8). These reactions, which are divariant in KFMASH, owing to the role of an additional component, are numbered in Fig. 1.8 and summarized in Table 1.2. The discontinuous nature of key univariant equilibria, and the reasons metamorphic rocks rarely record evidence of univariant reactions, are addressed further in Section 1.6.

Most contact aureoles do not reach temperatures high enough to enable partial melting in metasedimentary rocks (reaction 5, Fig. 1.8), or only have a narrow zone of partly melted rocks adjacent to the pluton. This reflects the temperature of emplacement for the causal plutons, which commonly are granite or granodiorite, and means that the highest-grade metamorphic rocks are uncommon in these geological settings. Proterozoic metasedimentary rocks at Mount Stafford, in

Table 1.2. Summary of KFMASH reactions inferred by Pattison & Tracy (1991; table 3) to control sub-solidus metapelitic assemblages

Phases:	Ms, Chl, Qtz, Bt, Crd, Als, St, Grt, Opx, Kfs, V or L (melt) = 11	
Components:	K_2O-FeO-MgO-Al_2O_3-SiO_2-H_2O = 6	
Phases always present:	Bt, Qtz, Ms, V or L	
Exclusions:	Chl+Kfs; St+Kfs; Ms+Crd+St; Crd+Grt+Ms; Opx+Chl; Opx+Ms; Opx+Als; Opx+St.	
Invariant points:	nil	
Univariant curves (invariant points in KMASH or KFASH): 7		
Reaction	Minerals not involved	Number
Ms+Chl+Qtz = Crd+Als +Bt+V	[Opx,Kfs,Grt,St]	1
Ms+Chl+St+Qtz = Als+Bt+V	[Qpx,Kfs,Grt,Crd]	2
Ms+Chl+Grt+Qtz = St+Bt+V	[Opx,Kfs,St,Crd]	3
Ms+St+Qtz = Grt+Als+Bt+V	[Qpx,Kfs,St,Chl]	4
Ms+Qtz = Als+Kfs+V or L	[Opx,Chl,Grt,St,Crd,Bt]	5 degenerate
Bt+Als+Qtz = Grt+Crd+Kfs+L	[Opx,Chl,St,Ms]	6
Bt+(Grt)+Qtz = (Crd)+Opx+Kfs = L	[Chl,Ms,St,Als]	7 degenerate for pure ann-phl
Divariant curves (univariant in KMASH or KFASH): 16		
Ms+Chl+Qtz = Crd+Bt+V	[Opx,Kfs,Grt,St,Als]	8
Ms+Chl = Als+Bt+Qtz+V	[Opx,Kfs,Grt,St,Crd]	9
Ms+Crd = Als+Bt+Qtz+V	[Opx,Kfs,Grt,St,Chl]	10
Ms+Bt+Qtz = Crd+Kfs+V	[Opx,Grt,St,Als,Chl]	11
Ms+Chl = St+Bt+Qtz+V	[Opx,Kfs,Grt,Crd,Als]	12
Ms+St+Qtz = Als+Bt+V	[Opx,Kfs,Grt,Crd,Chl]	13
Ms+Chl+Qtz = Grt+Bt+V	[Opx,Kfs,Crd,Als,St]	14
Bt+St+Qtz = Grt+Ms+V	[Opx,Kfs,Crd,Chl,Als]	15
Ms+Grt = Als+Bt+Qtz	[Opx,Kfs,Crd,Chl,St,V]	16 degenerate
Ms+Bt+Qtz = Grt+Kfs+V or L	[Opx,Crd,St,Als,Chl]	17
Ms+Qtz = Als+Kfs+V or L	[Opx,Crd,Grt,St,Chl,Bt]	5 degenerate
Bt+Als+Gtz = Crd+Kfs+L or V	[Opx,Grt,St,Chl,Ms]	18
Bt+Als+Qtz = Grt+Kfs+L or V	[Opx,Crd,St,Chl,Ms]	19
Bt+Crd+Qtz = Grt+Kfs+L	[Opx,St,Chl,Ms,Als]	20
Bt+Qtz = Opx+(Crd)+Kfs+L	[St,Chl,Ms,Als,Grt]	21
Bt+Qtz = Opx+(Grt)+Kfs+L	[St,Chl,Ms,Als,Crd]	22

the Arunta Block of central Australia, formed in a very high-*T* regional aureole (Section 1.1) involving extensive partial melting at shallow conditions, namely $P = 2.5$–3 kbar (Vernon *et al.*, 1990). The zoned sequence is unusual, because most partially melted rocks occur in regional metamorphic terranes that have more complex geological settings. The metamorphic grade at Mount Stafford varies from low-pressure greenschist facies in the south-west to low-pressure granulite facies (Section 1.8.1) in the north-east, over a lateral distance of only 10 km (Fig. 1.9), indicating a local thermal gradient of about 75°/km. Despite the high metamorphic grade, many sedimentary and igneous structures are well preserved in intricately interlayered metapelitic and metapsammitic rocks (Vernon *et al.*, 1990; Greenfield *et al.*, 1996). The zonal sequence is successively characterized by assemblages involving: (1) muscovite-quartz schist;

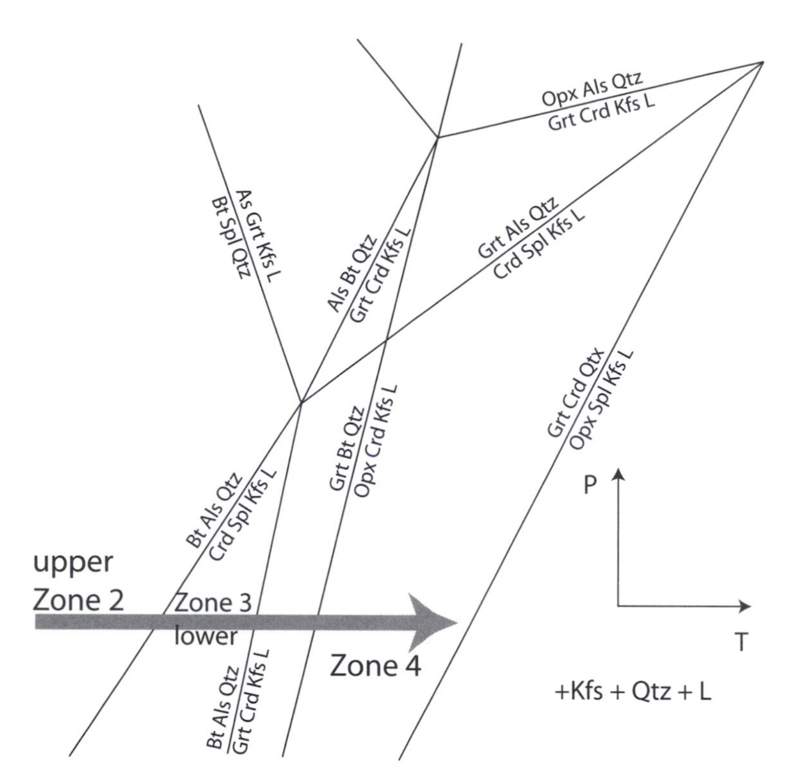

Fig. 1.10 Qualitative KFMASH petrogenetic grid for fluid-absent, but melt-present metapelitic equilibria at Mount Stafford, central Australia (Greenfield *et al.*, 1998, fig. 1.5, with permission of the *Journal of Metamorphic Geology*). The arrow indicates the *P-T* gradient inferred from field exposures, which matches the prograde *P-T* path inferred for Zone 4 rocks.

(2) andalusite/sillimanite-cordierite-K-feldspar granofels with small partial melt segregations; (3) spinel-sillimanite-cordierite-K-feldspar migmatite; (4) garnet-orthopyroxene-cordierite migmatite and minor diatexite; and (5) biotite-cordierite-plagioclase diatexite that shows a transition to granite. The definitions of 'migmatite' and 'diatexite' are presented in Chapter 4. The excellent preservation of sedimentary structures, co-existence of partial melt reactants and products, and similarity in whole-rock chemical composition led Greenfield *et al.* (1996) to infer that partial melting in Zone 2–4 metapelitic assemblages proceeded in situ, without substantial migration of melt. The northern granite (Fig. 1.9) cuts the zonal sequence, and the eastern granite would have provided insufficient heat to generate the observed sequence (Greenfield *et al.*, 1996). However, the northern granite could have been the local heat source, having eventually intruded the metamorphic sequence it produced.

On the basis of field relationships, microstructures and modal data, Greenfield *et al.* (1998) inferred a series of prograde reactions that were influenced by grade, rock-type and fluid availability, and interpreted their conditions of formation using thermobarometry and modelling in KFMASH. Schematic AFM diagrams for Zones 2 to 4 are shown in Fig. 1.9, and key KFMASH reactions inferred for supersolidus conditions are shown in Fig. 1.10. Greenfield *et al.* (1998) inferred that free water was not present at the high-*T* conditions, and used silicate liquid as a fluid proxy in constructing the AFM diagrams. Initial partial melting of metapelitic rocks in Zone 2 was inferred by Greenfield *et al.* (1998; see Fig. 1.9) to reflect vapour-present melting at $T \approx 640°C$, melting having initially been controlled by the congruent breakdown (Chapter 4) of the assemblage Crd-Kfs-Bt-Qtz. At slightly higher temperature, andalusite (occurring in leucosome (Chapter 4) was formed by the reaction: $Kfs + Qtz + Bt + H_2O = And + melt$. The interpretation that partial melting initially occurred in the andalusite field is problematic, but andalusite-bearing pegmatites occur near the andalusite-sillimanite transition shown in Fig. 1.9 (Vernon *et al.*, 1990). In Zone 3, large aluminosilicate aggregates in leucosome are armoured (partially replaced) by spinel-cordierite symplectites, with or without garnet, and all rocks contain appreciably less biotite than in Zone 2. Garnet partially pseudomorphs biotite, cordierite or spinel in high-grade portions of Zone 3. Zone 4 garnet-cordierite-orthopyroxene-bearing metapsammite assemblages and garnet-bearing leucosome reflect $T \approx 800°C$. By considering univariant equilibria in the model KFMASH system, Greenfield *et al.* (1998) inferred that the principal vapour-absent melting step reflected significant changes in mineral proportions (mode) related to the breakdown of the aluminosilicate-biotite tie-line and the establishment of the spinel-cordierite tie-line; the whole-rock chemical composition of most samples straddle the spinel-cordierite tie-line (Fig. 1.9).

We can now improve substantially on qualitative interpretations of metamorphic relationships in most rock types through the quantitative modelling of mineral equilibria using internally consistent thermodynamic datasets. As explained below, when the influences of whole-rock composition are evaluated, we can see that metamorphic rocks rarely undergo univariant reactions. Examples of detailed phase relationships, including melt, modelled for individual rock types at Mount Stafford by White *et al.* (2003; Figs. 1.16B, 1.16C) are discussed below.

1.6 Illustrating regional metamorphic grade variations using the AFM diagram

Barrow (1893, 1912) mapped a series of zones in Caledonian (early Palaeozoic) metasediments in the Scottish Highlands, each characterized by particular minerals. Subsequent work characterizing changes that control the zonation have formed the basis for much modern metamorphic petrology. Pelitic, calcareous, quartzitic and mafic metamorphic rocks form the Neoproterozoic Moine, and Late Neoproterozoic to Ordovician Dalaradian Supergroups. The Great Glen Fault (Fig. 1.11) separates the Northern Highlands on the north-western side, from the Grampian Highlands on the south-eastern side. The Grampian Highlands are divided into two regions: (1) a belt of medium-*P* rocks with kyanite and staurolite, commonly referred to as the Barrovian region; and (2) an area of low-*P* metamorphic rocks with andalusite, sillimanite and cordierite, commonly referred to as the Buchan region (Fig. 1.11).

The Barrovian region exposes a progression through chlorite, biotite, garnet, staurolite, kyanite and sillimanite zones towards the north-east (Fig. 1.11). This change corresponds to increasing metamorphic conditions from lower greenschist to upper amphibolite facies (Section 1.8.1), mostly controlled by increasing temperature. The boundaries of the zones are

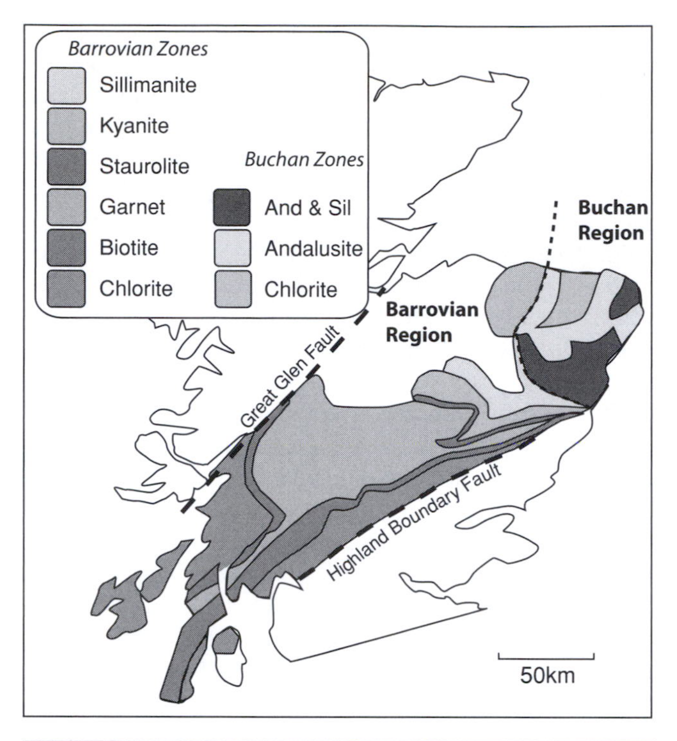

Fig. 1.11 Map of the Grampian Highlands in Scotland, showing the distribution of metamorphic zones, identified on the basis of mineral occurrence, as well as the broad subdivision into medium-*P* metamorphic rocks, commonly referred to as the Barrovian region, and low-*P* metamorphic rocks, commonly referred to as the Buchan region. For a colour version of this figure, please go to the plate section.

not linear, indicating post-metamorphic tectonic disruption. Metapelitic schists in the Barrovian chlorite zone contain chlorite, muscovite and quartz, with or without minor albite and graphite. Equivalent rocks in the biotite zone contain biotite, chlorite, muscovite, albite and quartz, the proportion of biotite increasing towards the garnet zone. Metapelitic schists in the wide garnet zone contain garnet, biotite, muscovite, albite and quartz; the garnet is rich in almandine ($Fe_3Al_2Si_3O_{12}$) with some spessartine ($Mn_3Al_2Si_3O_{12}$) and grossular ($Ca_3Al_2Si_3O_{12}$) components. The narrow staurolite zone has metapelitic schists with almandine-rich garnet, staurolite, biotite, muscovite, plagioclase and quartz. Staurolite and kyanite dominate metapelitic schists in the kyanite zone, which also contain muscovite, plagioclase and quartz. At the highest grade, sillimanite replaces kyanite. Tilley (1924) used the term '*isograd*' to represent a line of equivalent metamorphic grade ideally defined by the appearance of a new mineral in a sequence of new appearances, such as those mapped by Barrow (1893, 1912).

As shown in Figs. 1.6 and 1.9, the AFM diagram can be used to illustrate changes in metamorphic assemblages associated with the appearance of new minerals. Linked to the gross change in metapelitic metamorphic assemblages that defines Barrow's zones are more subtle variations in mineral composition. As a generalization, the ferromagnesian minerals (e.g., garnet, biotite) in high-grade rocks tend to be richer in Mg and poorer in Fe than those in compositionally equivalent lower grade rocks (e.g., Fig. 1.6). For example, garnet in the model KFMASH system is a solid solution of almandine ($Fe_3Al_2Si_3O_{12}$) and pyrope ($Mg_3Al_2Si_3O_{12}$) components. Garnet in the garnet zone is mostly almandine, with only a small proportion of pyrope, but the proportion of pyrope increases subtly with metamorphic grade. A related but more pronounced $FeMg^{-1}$ exchange related to these grade variations occurs in biotite and chlorite, which occur with garnet or staurolite, (Fig. 1.5A–D). As the whole-rock chemical composition is constant, the preservation of mass balance requires that such changes can be achieved only by changes in the proportions of the minerals involved. The changes related to increasing temperature have the effect of moving the plotted positions of minerals, together with the connecting tie lines that define the boundaries of divariant fields, towards the MgO side of the AFM diagram. This is well illustrated by key changes in metapelitic mineral equilibria over the temperature range 500°C–650°C modelled by Worley & Powell (1998a), as seen in movies available at: http://www.earthsci.unimelb.edu.au/tpg/thermocalc/.

Figures 1.5A and 1.5B show changes predicted by the model KFMASH system for a metapelitic rock at $P = 6$ kbar as the garnet-in isograd, which represents the lower boundary of the garnet zone, is approached from the biotite zone. The projected position of the 'greywacke' ('pe' in Fig. 1.5A) initially lies in the chlorite-biotite trivariant field (555°C–570°C). The effect of chlorite and biotite (coexisting with garnet) becoming more magnesian, owing to the increasing temperature, is that at $T = 570$°C the 'greywacke' composition plots in the garnet-biotite-chlorite-bearing divariant field, with the result that garnet begins to crystallize. Such a change is commonly referred to as a *continuous or sliding reaction*, which takes place over a *P-T* interval. It may be marked by the appearance or disappearance of a single mineral, or simply by changes in mode (volume proportion of minerals). Continuous mineral changes present a problem when mapping isograds. Subtle (e.g., 5%) variations in the Fe:Mg ratio of the bulk rock composition would have the effect of displacing the projected plot of the 'greywacke' ('pe') to more Fe- or Mg-rich positions, which would result in garnet appearing at lower (Fe-rich) or higher (Mg-rich) temperatures. Consequently, the garnet-in 'isograd' could be expected to occur over a broad zone (depending on whole-rock composition), and the 'greywacke' represented in Fig. 1.5 would not actually contain garnet until near the upper part of the garnet zone.

The first appearance of staurolite in the 'grewacke' ('pe') in Fig. 1.5C presents a different type of change known as a *discontinuous reaction*. At 573°C, only a few degrees above the first appearance of garnet for the metapelite in Fig. 1.5B, garnet and chlorite react to produce staurolite and biotite (Fig. 1.5C). This represents a *crossing tie-line reaction*, as the garnet-biotite tie-line breaks and is replaced by the staurolite-biotite tie-line (Figs. 1.5C, 1.12A). This corresponds to a reaction that is univariant in KFMASH, namely:

garnet + chlorite = staurolite + biotite (Fig. 1.12B).

This reaction, as written, assumes that quartz, muscovite and water belong to both the reactant and product assemblages. The discontinuous change in phase relations would affect all rocks that lie in the compositional domain defined by staurolite, garnet, biotite and chlorite, namely the yellow diamond shown in Fig. 1.12A. All rocks with bulk-rock compositions

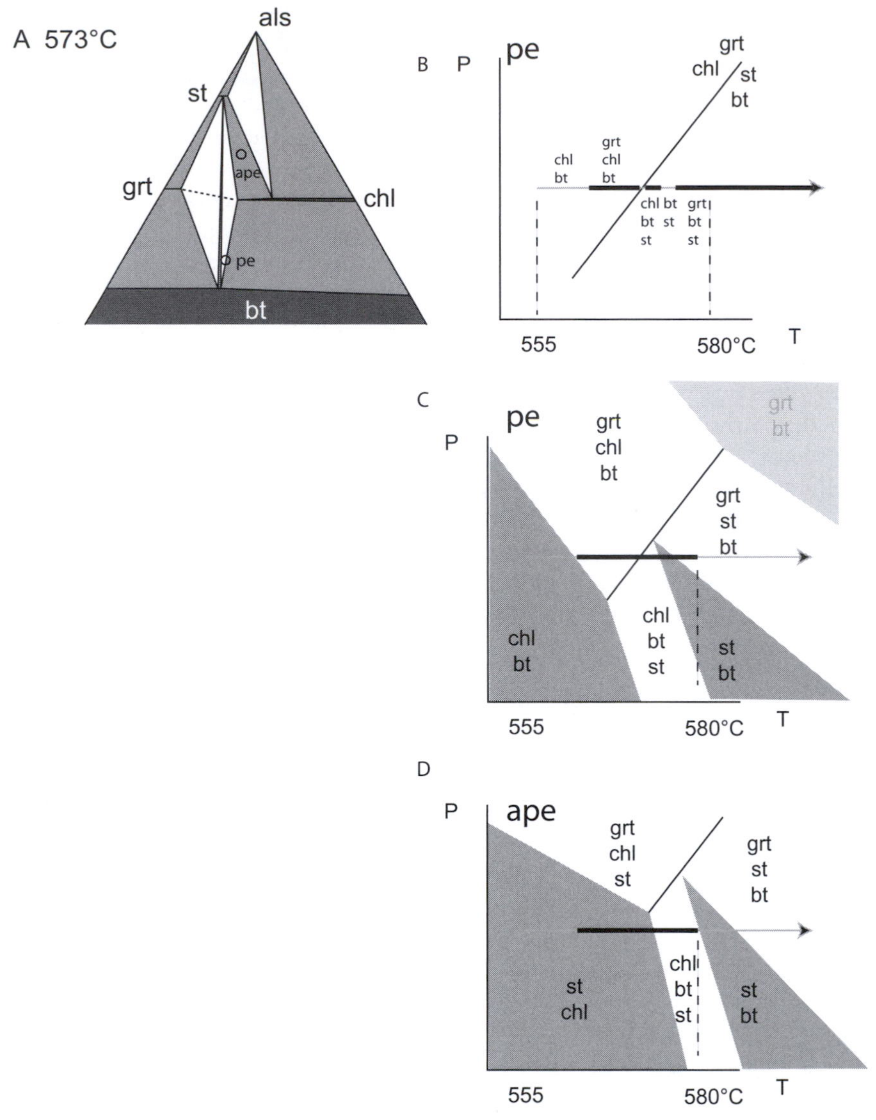

Fig. 1.12 Construction of a qualitative KFMASH pseudosection using the example of isobaric heating of a typical metapelite, 'pe', from $T = 555\,^{\circ}C$–$580\,^{\circ}C$, across the staurolite isograd, as discussed in the text. (A) AFM diagram showing mineral relationships at $P = 6$ kbar and $T = 573\,^{\circ}C$, with muscovite quartz and water in excess. The yellow diamond, defined by the positions of staurolite, garnet, biotite and chlorite, outlines the range of projected rock compositions that would be involved in the univariant reaction. (B) Summary of the mineral assemblages that the 'greywacke' rock composition 'pe' would produce during isobaric heating at $P = 6$ kbar. Divariant equilibria indicated by thick black horizontal line; trivariant equilibria indicated by thin horizontal line. (C) Qualitative *P-T* pseudosection constructed from Fig. 1.5B by extrapolating relationships of the various divariant and trivariant mineral equilibria. The relative geometry of the phase equilibrium boundaries can be calculated by Schreinemaker's analysis (Zen, 1966), using univariant equilibria around an invariant KFASH or KMASH point involving all the minerals present in the KFMASH reaction. (D) Qualitative *P-T* pseudosection constructed for the same heating path as for the aluminous metapelite composition 'ape' in Fig. 1.5A. Though both rock compositions underwent the univariant reaction at the same *T*, they have different starting assemblages and different divariant reaction sequences. For a colour version of this, please go to the plate section.

that project into this domain would experience a change from garnet-chlorite-bearing to staurolite-biotite-bearing. This discontinuous reaction thus produces a far better isograd than the continuous garnet-in isograd, because staurolite appears in a large range of rock compositions at one temperature. However, natural rocks contain elements in addition to those of the simple model system, and the effect of, for example Mn in garnet or Zn in staurolite, add complexity to these simplifications by expanding the stability ranges of these minerals. Rocks plotting outside the compositional domain defined by staurolite, garnet, biotite and chlorite produce no assemblage change in moving from 570°C to 573°C.

The relationship between compositional domains defined by minerals involved in this univariant reaction introduces another technique now commonly employed, namely constructing maps of mineral assemblages on *P-T* diagrams, called *pseudosections* (e.g., Hensen, 1988). Though the yellow diamond defined by the positions of staurolite, garnet, biotite and chlorite represents a large area on the AFM face, a larger portion of the AFM face lies outside this diamond, with plausible rock compositions inappropriate for the univariant reaction. With the exception of small changes in mode, their mineral assemblages would be almost identical below 570°C and above 573°C. For example, the metapelite ('ape') shown in Fig. 1.12A would have the assemblage staurolite and chlorite at 570°C and 573°C, as it lies outside the compositional domain affected by the univariant reaction.

Most univariant equilibria occur in a comparatively small proportion of rock compositions, and even then only over a restricted *P-T* domain. To illustrate this, consider the effect of isobarically heating the 'greywacke' ('pe' in Fig. 1.12A) from 555°C to 580°C, at P = 6 kbar. Between $T = 555$°C and 570°C, the trivariant chlorite-biotite-bearing assemblage is unchanged. At 570°C, the continuous garnet-in isograd changes the 'greywacke' composition to the garnet-biotite-chlorite divariant assemblage, and then the discontinuous staurolite isograd reaction induces another change to the divariant assemblage involving staurolite, chlorite and biotite. As the rock continues to heat towards 580°C, the increasingly magnesian composition of biotite and chlorite stable at these conditions means that the composition initially passes, via continuous changes, into a staurolite-biotite trivariant assemblage and then into a divariant assemblage involving garnet, staurolite and biotite. At some temperature above 580°C, a more magnesian biotite is stable with garnet, and the 'greywacke' ('pe') develops the trivariant assemblage involving garnet and biotite. These changes are summarized in the qualitative pseudosection for the 'normal' greywacke pelite ('pe') in Fig. 1.12C, which is a qualitative *P-T* pseudosection constructed from Fig. 1.12B by extrapolating relationships of the various divariant and trivariant equilibria (see Sections 1.8.1, 1.9, 2.4.5, 4.2.2), as noted in the caption for Fig 1.12C. The univariant reaction related to the staurolite isograd occurs for this rock composition over only a limited *P-T* window; the pseudosection is dominated by the chlorite-biotite-bearing trivariant field at low-grade, and the garnet-biotite-bearing trivariant field at high-grade.

The influence of rock composition on mineral assemblages and reaction sequences can be illustrated by comparing the progression of assemblages that an aluminous metapelite ('ape' in Fig. 1.12A) would experience for the same heating path as that evaluated for the 'greywacke' ('pe') in a pseudosection (Fig. 1.12D). For the first half of the heating path, the aluminous metapelite would have a completely different assemblage evolution: the trivariant assemblage involving staurolite and chlorite would evolve to a divariant assemblage involving garnet, chlorite biotite and staurolite, and then to a trivariant assemblage involving staurolite and biotite. Though the mineral reactions are influenced by the univariant reaction, the rock composition is inappropriate to directly experience the univariant reaction, with the result that the mineral changes are controlled by higher variance assemblages (Fig. 1.12D).

Most prograde reaction sequences involve the progressive dehydration of the rock as more water is released than can be retained, in this example by the breakdown of reactant muscovite and chlorite. In crossing a series of similar reactions on a prograde *P-T* path, the high-*T* products become 'locked-in', owing to there being insufficient water to enable back-reaction during cooling, as previously noted. About 2 weight percent free water that could enable limited retrogression of the peak assemblage could be present. How can we identify such retrograde minerals? As diffusion distances are strongly temperature dependent, retrograde minerals involve diffusion over distances smaller than the grainsize of the peak mineral assemblage, but must form in a coarse-grained aggregate inherited from the metamorphic peak. This leads to characteristic intergrowths of fine-grained retrograde minerals partly or completely pseudomorphing the pre-existing shapes of comparatively coarse-grained peak minerals (see Sections 2.2 and 3.5.1). Extensive retrogression of a peak assemblage may accompany water ingress, commonly along faults or shear zones (e.g., Vernon & Ransom, 1971).

1.7 Extent of equilibrium

Metamorphic petrologists typically apply the equilibrium thermodynamic model to explain observed mineral assemblages and compositions. However, metamorphic reactions occur at non-equilibrium conditions, because no reaction can occur at equilibrium. Because rocks continue to respond to changing conditions until reactions cease (either for either kinetic reasons or because of inadequate supply of reactants), most metamorphic rocks do not achieve chemical equilibrium at the conditions at which reactions stop. However, the general assumption is that the products of inferred metamorphic reactions are those expected at (theoretical) equilibrium, as determined experimentally or by thermodynamic calculation.

Before the advent of the electron probe microanalyser, the chemical compositions of minerals were initially determined using wet chemical methods, following rock crushing and physical mineral separation. The scanning electron microscope enables the imaging of polished rock slabs and thin sections, revealing complex zoning in some common minerals such as garnet. The electron microprobe, developed as a variation of the scanning electron microscope, enables the chemical analysis of very small areas of individual mineral grains. It has several spectrometers used to count secondary X-rays induced in the sample by an electron beam. The frequency of the secondary X-rays indicates which elements are present in the volume affected by the electron beam. By performing matrix corrections that account for the partial absorption of the secondary X-rays (e.g., Bence & Albee, 1968), it is possible to obtain a quantitative analysis of part of a mineral as small as 1–5 microns (10^{-6} metre) across. This development has revealed the following key features of metamorphic rocks: (1) the chemical compositions of different grains of a given mineral are identical or very similar on a scale that approximates the grainsize; (2) the chemical compositions of different grains of a given mineral from adjacent distinct layers of metasediment may vary on a centimetre to metre scale; (3) minerals in high-*T*, for example, granulite facies, rocks are commonly homogeneous; and (4) minerals in low- to medium-grade and high-pressure rocks may be complexly zoned and may contain inclusions of different composition or mineral type from those that occur outside their hosts. These observations have expanded the older idea of an equilibrium view of metamorphism, which is critical to applying equilibrium thermodynamics in interpretations of the development of mineral assemblages, including the application of phase diagrams.

Most petrologists probably would accept that granulite facies assemblages of chemically homogeneous minerals with low-energy grain shapes (see Chapter 3) can be interpreted as being in or close to thermodynamic equilibrium, so providing reliable evidence of *P-T* conditions during a segment of their burial and exhumation, or *P-T* path. The situation is more complex where minerals are zoned, or for rocks with reaction rims between minerals (*coronas*). Such microstructures can be interpreted by considering them in plausible prograde and retrograde contexts that might involve, for example, equilibrium in volumes smaller than the grainsize.

The preservation of compositional heterogeneities, on various scales, in metamorphic rocks is due to limits to the diffusion of the least mobile chemical components at the prevailing metamorphic conditions. These limits enable the preservation of structures, such as bedding and compositional layering, as well as distinct grains of different minerals. In effect, the least mobile components determine the concentrations of the more mobile components that can be fixed in the rock. In compositionally layered rocks, the mineral grains in each layer can be chemically compatible (with equalized chemical potentials of each component); they are said to be in *local equilibrium*. This applies even where compositional differences exist between grains of the same mineral in different layers. In effect, each layer is a separate chemical system, the boundaries of which are determined by the immobile chemical components, while permitting mobile components to move from one layer to the next (see Section 5.10).

The following summary largely follows Powell *et al.* (2005) in presenting a simple macroscopic view of equilibration in a mineral assemblage developed during changing *P-T* (following a '*P-T* path'; see Section 3.5.2). Nucleation and diffusion are key processes in mineral assemblages that experience evolving *P-T* conditions and approach equilibrium at new conditions. In such circumstances, a mineral assemblage may include parts that are in disequilibrium at particular *P-T* conditions, such as where mineral inclusions in a host grain record part of the prograde history, but are not in equilibrium with neocrystallizing matrix minerals. Diffusion controls the development of equilibrium mineral compositions in volumes that depend on temperature, the amount of time available, and grainsize. At most metamorphic conditions, grain-boundary diffusion of chemical components, assisted by the presence of fluid or melt, occurs much faster than intracrystalline ('volume') diffusion. An exponential dependence of diffusion rates on temperature means that diffusion distances are strongly temperature-dependent, a consequence being that minerals in high-*T* metamorphic rocks are usually coarser-grained than minerals in low-*T* metamorphic rocks, though this is also connected with the general tendency for lower nucleation rates with increasing temperatures (e.g., Vernon, 2004). The exponential temperature dependence enables rates of intracrystalline diffusion to approach those of grain-boundary diffusion at the high-*T* extreme of metamorphism.

Factors that influence microstructural development in a neocrystallizing aggregate can be discussed by considering an assemblage with minerals in stable equilibrium at the lowest Gibbs free energy. When the imposed *P-T* conditions change, the mineral assemblage may become metastable if low rates of nucleation prevent the appearance of a new mineral, or if rates of diffusion are too slow to allow the continuous adjustments in mineral compositions required for the new assemblage to be stable. This situation, in which *P-T* conditions evolve more rapidly than equilibrium can be reached or maintained, is known as *reaction overstepping*.

In prograde regional metamorphism, enough time is generally available to allow reactions to proceed to completion with increasing temperature, so that a succession of index minerals or assemblages (as in the Scottish Dalradian) is taken to indicate a succession of equilibrium or near-equilibrium reactions (Fyfe *et al.*, 1958). For example, in the isobaric heating model (discussed above) for a metapelite crossing the staurolite isograd, continuous destruction of the reactants was assumed, stable equilibrium assemblages and mineral compositions being maintained. This inference is especially reasonable for devolatilization reactions, which proceed rapidly in laboratory conditions, though less so for solid–solid reactions, which are commonly sluggish in laboratory conditions. As noted by Ridley & Thompson (1986), overstepping is reduced by high reaction temperatures and by *epitaxial* (epitactic) *nucleation* (growth of a new mineral in an oriented crystallographic relationship with a reactant mineral, which minimizes the interfacial free energy of the new grain).

Equilibrium is generally assumed to be true also of prograde contact metamorphism (e.g., Dipple & Ferry, 1992a; Spear, 1993), though overstepping has been inferred for some contact metamorphic situations (e.g., Naggar & Atherton, 1970; Miller *et al.*, 2004), owing to a relatively short-lived heating duration. Walther & Wood (1984) inferred that, provided fluid is effectively removed from reaction sites by channelled flow, surface reaction rates are too rapid to be rate limiting in metamorphic reactions, and that prograde reactions are not overstepped by more than a few degrees Celsius, except for solid–solid reactions (especially in some low-strain contact metamorphic environments) and/or when CO_2 is dominant in the fluid phase.

However, an example of a relatively large degree of overstepping is provided by the *metastable persistence* of andalusite into the sillimanite zone, for example, in low-pressure metamorphic terranes, such as the Placitas-Juan Tabo area, New Mexico, USA (Vernon, 1987a). If a reaction equilibrium is overstepped for long enough to delay nucleation of the expected equilibrium mineral, the new mineral may have a non-equilibrium chemical composition, or a metastable mineral may crystallize instead (e.g., Hollister, 1969). For example, Miller *et al.* (2004) used detailed microstructural and grain-scale isotopic information to infer a metastable olivine-forming

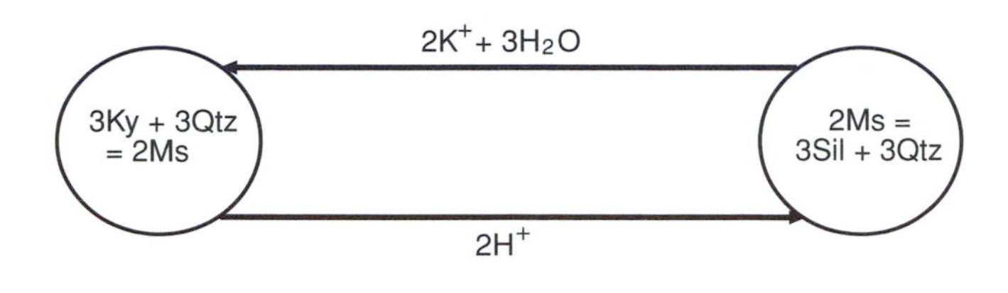

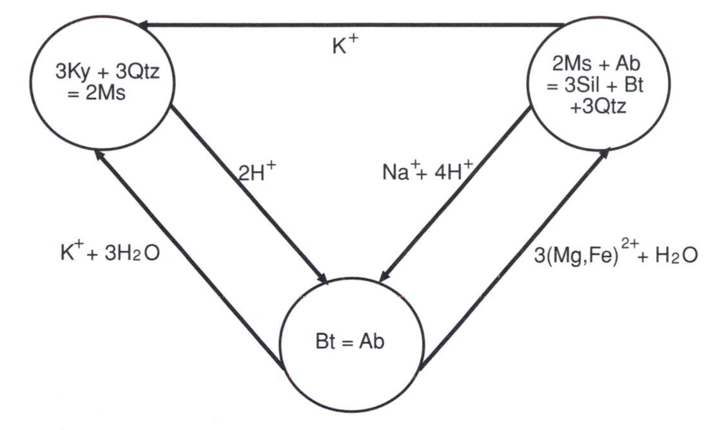

Fig. 1.13 Diagram showing that the net change: kyanite → sillimanite can be the result of several linked, ionic sub-reactions or partial reactions, involving diffusion of chemical components in fluid from one local reaction site to another, the partial reactions combining to form an overall 'cyclic' reaction; after Carmichael (1969).

reaction in calcareous rocks. Presumably the reactions predicted by equilibrium thermodynamics proceeded so slowly that the faster metastable reaction was able to occur (Miller *et al.*, 2004).

An example of reaction overstepping at high temperature is provided by metapelitic rocks at 550°C–560°C subjected to rapid (5–10 km per Ma), near-isothermal decompression, owing to extensional collapse of thickened crust, resulting in (1) coexistence of all three Al_2SiO_5 polymorphs, microstructural disequilibrium, wide compositional variation in minerals in single samples, and unsystematic or abnormal Mg-Fe partitioning among biotite, garnet and staurolite (Garcia-Casco & Torres-Roldán, 1996).

Details of which minerals should be stable at any given *P-T*-fluid conditions also focus attention on how change is achieved in terms of diffusion mechanisms that supply reactants for grain growth and remove unnecessary material. Carmichael (1969) emphasized that even simple reactions can have complex mechanisms, as discussed in Section 3.2, using the example of kyanite breaking down to sillimanite, which marks the sillimanite isograd in the Scottish Dalradian (Section 1.6). If this were a simple pseudomorphous reaction, sillimanite should occur largely around the former kyanite grains. However, most of the sillimanite occurs in biotite and quartz. Carmichael proposed that several local 'ionic' reactions facilitated the change (Fig. 1.13), based on microstructural evidence. For example, the local reaction:

$$3\text{Ky} + 3\text{Qtz} + 2\text{K}^+ + 3\text{H}_2\text{O} = 2\text{Ms} + 2\text{H}^+ \quad (1.1)$$

uses Al and Si from the mineral reactants, as well as K^+ and H_2O supplied by grain-boundary diffusion from the site of the postulated local reaction:

$$2Ms + 2H^+ = 3Sil + 3Qtz + 2K^+ + 3H_2O \quad (1.2)$$

If the two reactions proceeded together in the same volume of rock, the *net reaction* would be the conversion of kyanite to sillimanite. 'Cyclic' ionic reaction patterns have also been inferred by Yardley (1977a, fig. 2) to explain pseudomorphs of sillimanite after garnet, involving movement of Al from staurolite and mica in the matrix to well-separated garnet porphyroblasts. Studies by Foster (1977, 1981, 1983, 1986, 1999), Waters (2001) and Likhanov & Reverdatto (2002) have also revealed the complexity of prograde reactions, showing that net reactions occurring in volumes as small as 1 cm^3 may be the result of several even more local reactions. *Local equilibrium* is also indicated by the growth in cordierite or biotite grains with Ti contents that vary with distance from the nearest Ti source, such as ilmenite (Waters & Charnley, 2002).

Stüwe (1997) discussed a model equilibration volume for a given time and at a given temperature, involving an approximate sphere centred on a point of interest (Fig. 1.14). This model builds conceptually from Blackburn (1968), whose microprobe analyses showed that domains of spatial equilibration of Mg and Fe in garnet range from only a few millimetres to a few centimetres across, and that the domains tend to be elongate parallel to the foliation and lineation of the rock. Blackburn (1968) noted that domains of equilibration become larger with increasing metamorphic grade, and approach hand–specimen scale in a pyroxene granulite. Thus, chemical analyses of aggregates of mineral grains separated from a hand specimen represent average compositions, and usually much more can be learnt from analyses of individual grains or parts of individual grains. The exchange of elements between a point of interest and adjacent grains depends on the relative capacity for a given element to migrate via grain boundary or intracrystalline diffusion, and whether any grain boundary fluid is present.

Variations in the chemical compositions of equilibration volumes (reactive volumes) during retrograde metamorphism can be caused by heterogeneities in the host material and differences in the controlling diffusion mechanism. Consider a nominal area for intracrystalline-controlled diffusion of a given element at peak conditions centred on a garnet grain, indicated by the dashed line in Fig. 1.14. With a small amount of cooling, the diffusion distance would reduce, possibly to the area indicated in yellow. This change would have the effect of reducing the proportion of quartz and feldspar in the equilibration volume, altering the chemical composition of the effective reactive volume from that at peak conditions. Were grain-boundary diffusion to be most important in controlling the elemental exchange of another element, the composition of the effective area/volume could be quite different, as indicated by area 3 in Fig. 1.14. Alternatively, considering two comparable retrograde equilibration volumes in a heterogeneous rock (Fig. 1.14), area 2, containing garnet, would be substantially richer in iron than area 4, which is rich in biotite, quartz and feldspar.

Most prograde reaction sequences involve the progressive dehydration of the rock as more water is released than can be retained, in this example

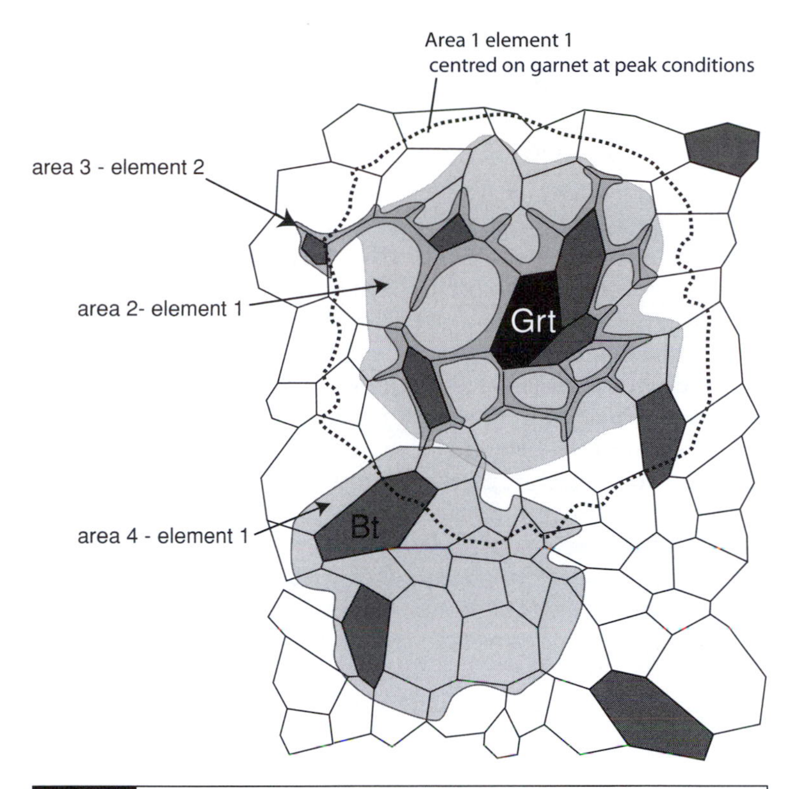

Fig. 1.14 Schematic thin section, modified from Stüwe (1997, fig. 2), consisting of four different minerals, namely garnet (black, Grt), biotite (orange, Bt) and a quartz-feldspar matrix (white). Superimposed are idealized equilibrium areas, (and volumes by extrapolation) for a given time at a given temperature (coloured lobate regions). Areas 1 to 3 overlap, and are centred on a garnet grain; they indicate that the equilibrium volume contracts with cooling and may be quite distinct for different elements. Area 1 indicates the volume for an element for which the ratio of grain boundary diffusion to intracrystalline (volume) diffusion is large at peak conditions; area 2 indicates similar relationships at lower T. Area 3 indicates the volume of another element for which the ratio of intracrystalline to grain boundary diffusion is large. Thus, area 1 extends along grain boundaries and includes only biotite grains that can easily be reached along the grain boundaries. Area 4 is a region, approximately the same size as area 2, centred on quartz and feldspar grains. Each of the four areas would have different chemical compositions. For a colour version of this figure, please go to the plate section.

by the breakdown of reactant muscovite and chlorite. In crossing a series of similar reactions on a prograde path, the high-T products become 'locked-in', because of insufficient water to enable back-reaction during cooling, as previously noted. About 2 weight percent combined water for limited retrograde alteration of the peak assemblage could be present. How can such retrograde minerals be identified? As diffusion distances are strongly temperature-dependent, retrograde minerals involve diffusion over distances smaller than the grainsize of the peak mineral assemblage, but must form in a coarse-grained aggregate inherited from the metamorphic peak. This leads to characteristic intergrowths of fine-grained retrograde minerals mimicking, or pseudomorphing the pre-existing shapes of comparatively coarse-grained peak minerals (Figs. 2.2 and 2.3). Extensive retrogression of a peak assemblage may accompany water ingress, commonly along faults or shear zones (e.g., Vernon & Ransom, 1971).

1.8 Genetic classification of metamorphism

1.8.1 Metamorphic facies

Eskola (1915, 1921) proposed the concept of metamorphic facies before the availability of experimental or thermodynamic data on the stability of metamorphic minerals, recognizing that both temperature and pressure were important in determining the mineral assemblage. Eskola (1915, p. 115) defined a metamorphic facies to include rocks which 'may be supposed to have been metamorphosed under identical conditions. As belonging to a certain facies we regard rocks which, if having an identical chemical composition, are composed of the same minerals'. Eskola (1915, 1921) realized that mineral assemblages formed at high pressures tend to involve denser minerals, and mineral assemblages formed at higher temperatures tend to have progressively lower volatile (H_2O, CO_2) contents. Eskola's original work was based on a series of contact aureoles in Finland and Norway, where he observed reproducible zonations in metamorphic mineral assemblages around different plutons.

The metamorphic facies concept involves a qualitative subdivision of the *P-T* range of metamorphism that is still in use for today. The facies concept is useful in a general way, especially for making metamorphic maps of large regions, because many diagnostic assemblages can be identified by hand-specimen examination. A metamorphic facies includes all mineral assemblages in a set of rock compositions formed under the same broad *P-T* conditions. Most metamorphic facies are named after characteristic metabasic assemblages (greenschist, amphibolite, blueschist, eclogite). However, a metamorphic facies does not refer to a single rock-type, but embraces a number of rock-types, all formed under the same conditions. Several new facies have been added to Eskola's original set (e.g., Turner, 1981), the most commonly accepted distribution of facies fields being shown in Fig. 1.15. Assemblages diagnostic of different facies are listed in Table 1.3.

Applying the facies concept in anything other than a general way encounters several problems, for example, the issues of water activity and composition. Water abundance and composition are not explicitly considered in the facies concept, though assemblages change from being H_2O-present at low-grade to H_2O-absent at high-grade, for example in the granulite facies. The boundaries of many facies *P-T* have slopes broadly parallel to the pressure axis, because they are mostly controlled by dehydration reactions, which are characterized by a large volume change (ΔV). The temperatures at which such reactions occur are highly sensitive to water abundance and composition (see Chapter 5). Hence, all boundaries should be regarded as *P-T* intervals rather than lines. As discussed below, different rock compositions undergo change at different *P-T* conditions, as discussed previously, with regard to the garnet isograd. For example, the boundary between, the greenschist and amphibolite facies occurs at different conditions for metapelitic and mafic compositions. Some rock-types, especially felsic gneisses, register no appreciable mineralogical difference from one facies to another.

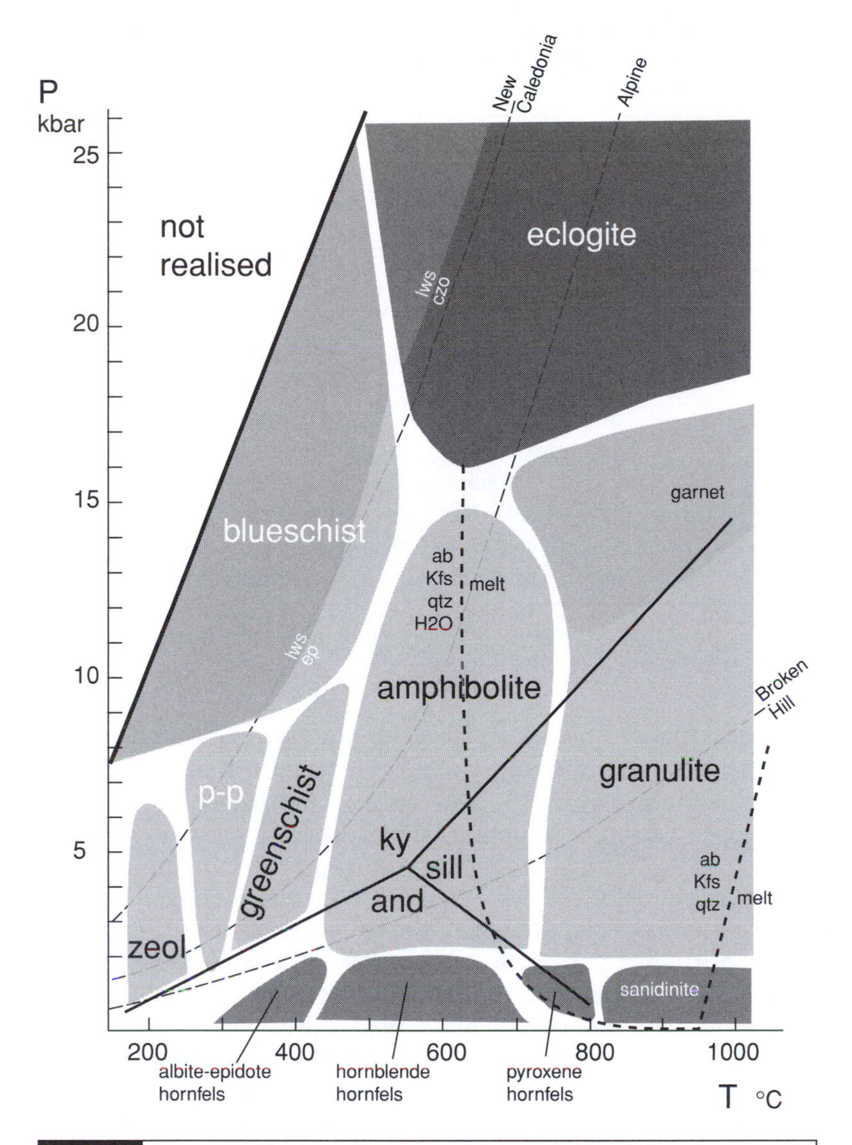

Fig. 1.15 *P-T* diagram showing the generalized distribution of the main metamorphic facies; derived from Eskola (1915, 1921) and Turner (1981). Andalusite is common and almandine garnet is rare in the low-*P* facies, though considerable overlap exists between the low-*P* (typically contact metamorphic) facies and the equivalent intermediate-*P* (regional metamorphic) facies (shown in the same colours). The blueschist and eclogite facies may be subdivided on the basis of the breakdown reaction of lawsonite to form epidote. The low-*P* limit of garnet in the granulite field, delineating the 'garnet granulite subfacies of Green & Ringwood (1967), is sensitive to whole-rock composition, and so is shown as a shaded zone. Representative geotherms are shown for: subduction metamorphism from Eocene rocks in northern New Caledonia; convergent continental metamorphism, such as is responsible for forming the European Alpine metamorphism; and exceptionally high heat flow in low-pressure regional metamorphism, evident at Broken Hill, Australia. Also shown are phase relationships for the pure Al_2SiO_5 system and the water-saturated melting curve for a simplified 'granite' composition. Abbreviations: p-p = prehnite-pumpellyite; zeol = zeolite. For a colour version of this figure, please go to the plate section.

Table 1.3. Diagnostic mineral assemblages for the various metamorphic facies

Facies	Diagnostic minerals or assemblages
zeolite	zeolites, especially laumontite
prehnite-pumpellyite	prehnite-pumpellyite, prehnite-actinolite pumpellyite-actinolite
greenschist	albite-tremolite/actinolite-chlorite-epidote, chloritoid, stilpnomelane
amphibolite	hornblende-plagioclase ($>An_{20}$) staurolite in metapelites
granulite	orthopyroxene-plagioclase, orthopyroxene-K-feldspar, sapphirine (in metapelites) no staurolite or muscovite (in metapelites)
blueschist	glaucophane, lawsonite, jadeite-rich pyroxene, aragonite, Mg-Fe carpholite
eclogite	omphacite-garnet *without* plagioclase

These are mostly for mafic compositions, but some distinctions are also made for metapelitic compositions.

General associations of metamorphic facies in regional metamorphic terranes reflect common geotherms related to tectonic setting (see the three geotherms shown in Fig. 1.15). Blueschist facies (high-*P* low-*T*) conditions are developed during subduction (Miyashiro, 1961; Cloos, 1985), or ephemerally in the early stages of metamorphism related to continental convergence, where blueschist facies assemblage are commonly neocrystallized to greenschist and/or 'epidote amphibolite facies' assemblages (Ernst, 1973; England & Thompson, 1984b). Some facies schemes have an 'epidote amphibolite facies' (characterized by sodic plagioclase, epidote and hornblende) between the greenschist and amphibolite facies, but it is probably best regarded as a *subfacies* delineated in more detailed classifications (Fig. 1.16A). Blueschist and low-*T* eclogite facies metabasites are characterized by lawsonite, commonly in association with glaucophane, chlorite, albite and phengitic white mica. At higher pressure and temperature, the blueschist facies gives way to the eclogite facies (Fig. 1.15), which is characterized by relatively magnesian garnet coexisting with omphacite or another clinopyroxene with high Al and Na content (White, 1964; Carswell, 1989). Lawsonite eclogites have been reported from only a few high-*P*, low-*T* terranes (Zack *et al.*, 2004; Tsujimori *et al.*, 2006); their rarity is probably because lawsonite preservation requires exhumation to be accompanied by cooling (Clarke *et al.*, 2006). Epidote-bearing blueschists (high-*P*) and greenschists (low-*P*) are transitional to the amphibolite facies (Fig. 1.15). Detailed subdivisions of these facies are sensitive to whole-rock composition, and Fig. 1.16A illustrates a detailed subdivision of the high-*P* facies based on mineral equilibria modelled for a mafic eclogite from New Caledonia (Clarke *et al.*, 2006). The positions of many mineral fields change with subtle variations in, for example, Mg and Al contents.

The blueschist, eclogite and granulite fields give way to ultra-high-pressure (UHP) conditions, the lower limit of which can be defined by the reaction: quartz = coesite (Bohlen & Boettcher, 1982; Fig. 1.16A). Early

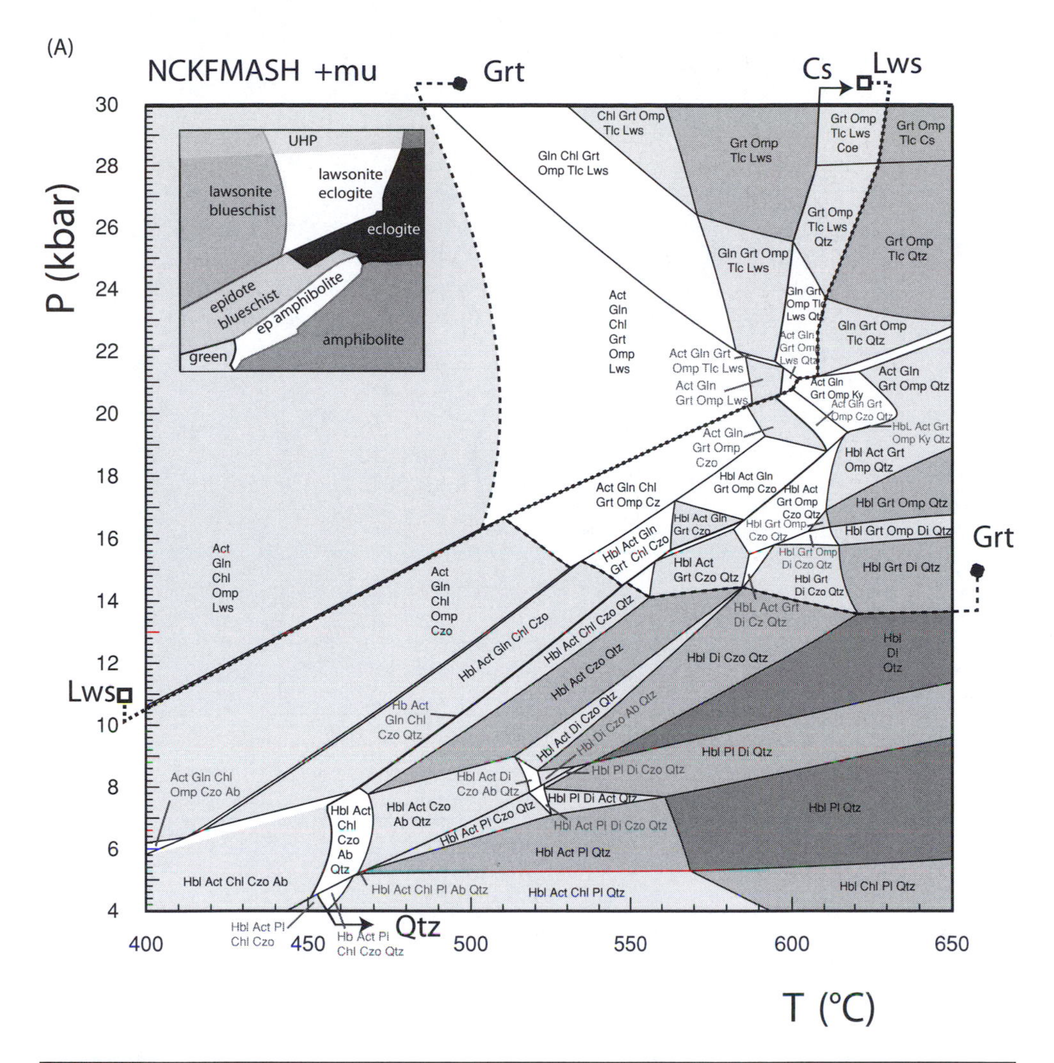

Fig. 1.16 (A) *P-T* pseudosection in NCKFMASH, calculated using THERMOCALC (Section 2.3) with H_2O and muscovite in excess, for a metabasite from New Caledonia inferred to be a subtly altered MOR basalt; modified, with additional material, from Clarke *et al.* (2006, fig. 2a). Univariant reactions experienced by this rock composition are shown as continuous lines. Divariant fields are unshaded six phase fields, trivariant fields are shaded five phase fields and so on. Quartz is present only at high-*T* conditions, and garnet is present at high-*P*/*T* conditions. A narrow band of three amphibole assemblages occurs in the upper epidote blueschist and lower eclogite subfacies conditions, and a glaucophane-hornblende solvus is at intermediate eclogite facies conditions. High-*T* amphibolite facies equilibria are metastable with respect to orthoamphibole-bearing equilibria; so the diagram is less reliable for those conditions. Inset shows the inferred extent of common metamorphic subfacies based on the calculated positions of the mineral equilibria. The lower limit of the ultra high-pressure (UHP) field is delineated by the reaction quartz = coesite. (B, C) *P-T* pseudosections calculated using THERMOCALC for interlayered metapelitic and metapsammitic portions of turbidites at Mount Stafford, central Australia; modified from White *et al.* (2003, figs. 5 and 7). In calculating the mineral equilibria, there is a need to fix water contents in the modelled whole-rock compositions at the solidi, and the water-absent sub-solidus fields are inappropriate for interpreting the pre-melting prograde evolution. The pseudosections also show the calculated mode isopleths in mole precent for several minerals and silicate melt. Insets show details of the mode changes that occur near the solidi, at conditions within the areas covered by the *P-T* boxes. For a colour version of this figure, please go to the plate section.

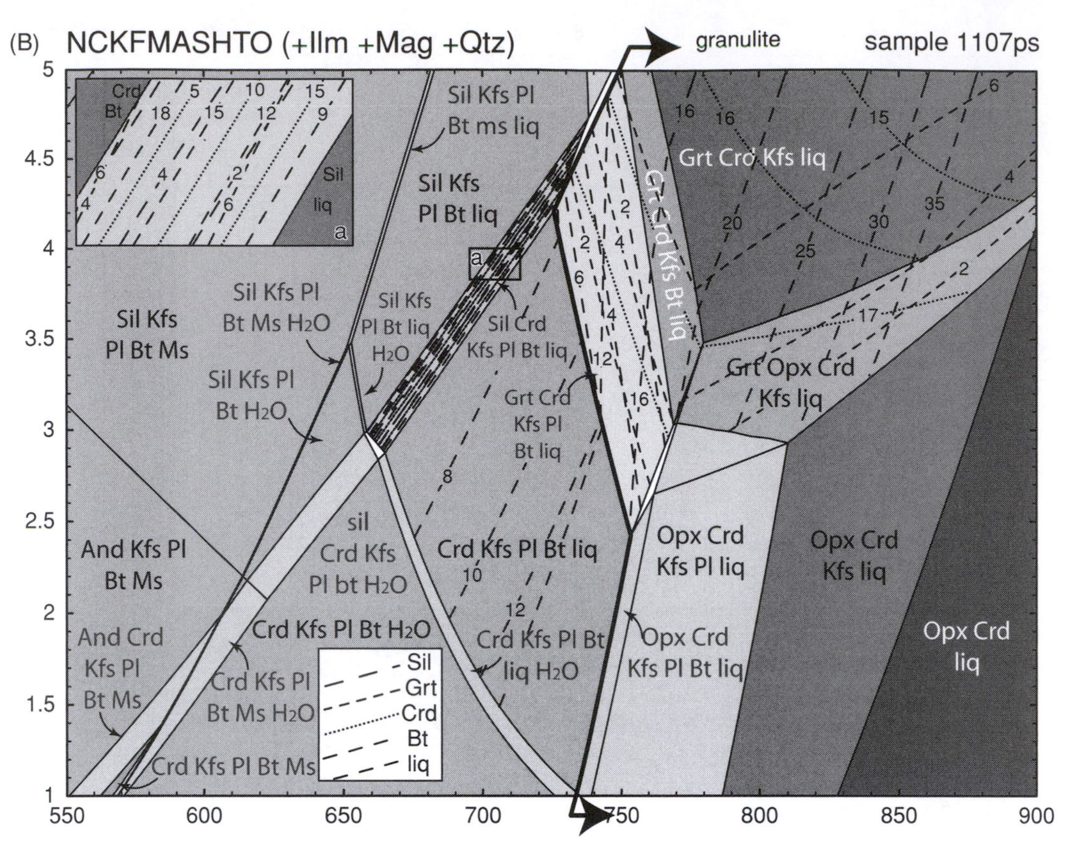

(B)
NCKFMASHTO (+Ilm +Mag +Qtz)
granulite
sample 1107ps
Sil Kfs Pl Bt ms liq
Sil Kfs Pl Bt liq
Grt Crd Kfs liq
Grt Crd Kfs Bt liq
Sil Kfs Pl Bt Ms
Sil Kfs Pl Bt Ms H2O
Sil Kfs Pl Bt liq H2O
Sil Crd Kfs Pl Bt liq
Sil Kfs Pl Bt H2O
Grt Crd Kfs Pl Bt liq
Grt Opx Crd Kfs liq
And Kfs Pl Bt Ms
sil Crd Kfs Pl bt H2O
Crd Kfs Pl Bt liq
Opx Crd Kfs Pl liq
Opx Crd Kfs liq
And Crd Kfs Pl Bt Ms
Crd Kfs Pl Bt H2O
Crd Kfs Pl Bt liq H2O
Opx Crd Kfs Pl Bt liq
Opx Crd liq
Crd Kfs Pl Bt Ms H2O
Crd Kfs Pl Bt Ms
Sil
Grt
Crd
Bt
liq

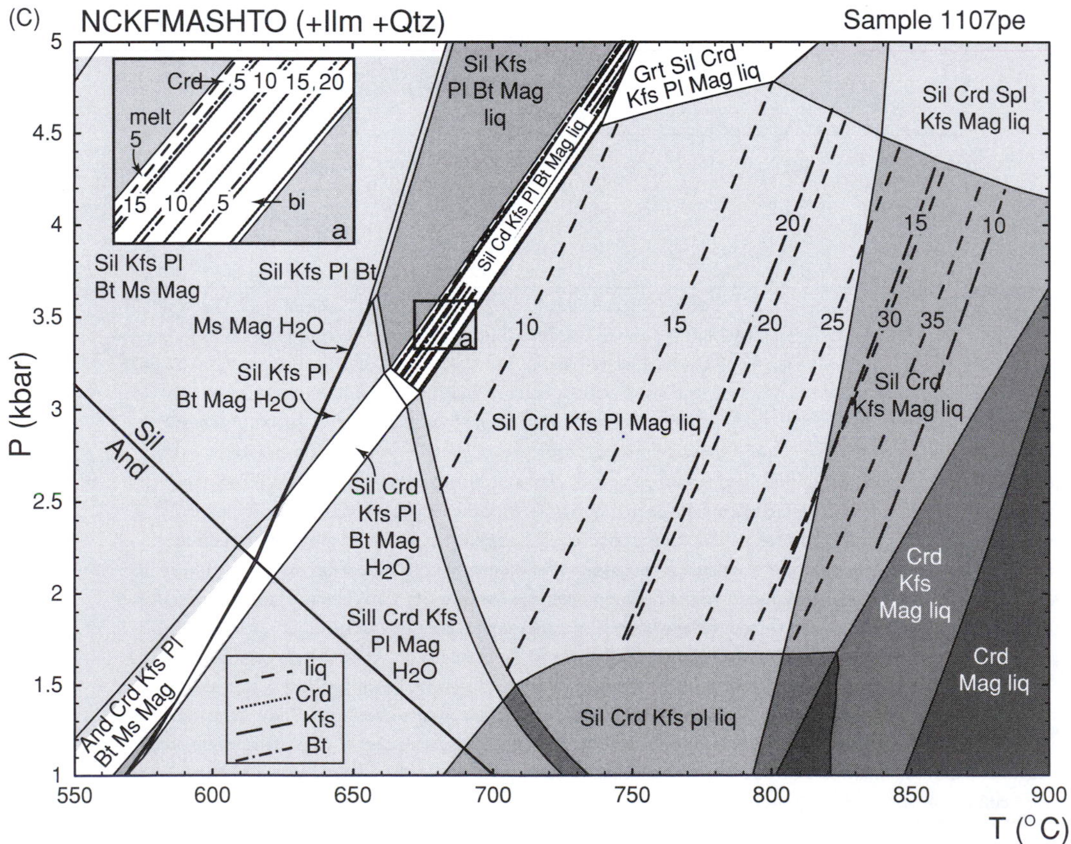

(C)
NCKFMASHTO (+Ilm +Qtz)
Sample 1107pe
Sil Kfs Pl Bt Mag liq
Grt Sil Crd Kfs Pl Mag liq
Sil Crd Spl Kfs Mag liq
Crd
melt
bi
Sil Kfs Pl Bt Ms Mag
Sil Kfs Pl Bt
Sil Cd Kfs Pl Bt Mag liq
Ms Mag H2O
Sil Kfs Pl Bt Mag H2O
Sil Crd Kfs Pl Mag liq
Sil Crd Kfs Mag liq
Sil
And
Sil Crd Kfs Pl Bt Mag H2O
Crd Kfs Mag liq
Sill Crd Kfs Pl Mag H2O
Crd Mag liq
Sil Crd Kfs pl liq
And Crd Kfs Pl Bt Ms Mag
liq
Crd
Kfs
Bt
P (kbar)
T (°C)

Fig. 1.16 (*cont.*).

recognition of such extreme conditions of metamorphism came from the discovery of diamond in eclogite blocks in kimberlite pipes in South Africa (Bonney, 1899; Beck, 1899). Garnet and clinopyroxene in such eclogites are chemically distinct from the equivalent minerals in blueschist and eclogite facies rocks in Alpine-type fold belts (e.g., Coleman *et al.*, 1965), though the nomenclature of garnet-pyroxene rocks in kimberlites is confusing (Dawson & Carswell, 1990). More evidence is progressively being discovered in collisional belts for the burial of crustal rocks to previously unsuspected depths, in excess of 150–200 km (Chopin, 1984; Smith, 1984; Chopin, 2003) and for their subsequent exhumation from such depths. The main UHP mineral indicators are coesite and microdiamond, but these are commonly present only as relics of UHP assemblages that escaped transformation during decompression (e.g., Wain, 1997). In addition, many rocks have compositions inappropriate to form such minerals (Wain *et al.*, 2001; fig. 1.10A). The presence of pyroxene exsolution in garnet and of coesite exsolution in titanite indicate precursor UHP garnet or titanite with six-fold coordinated silicon (Roermund *et al.*, 2000). Though common in UHP garnet, rutile exsolution may reflect high temperature, rather than UHP conditions. The repeated occurrence of UHP metamorphism indicates that it is a common process, and that an unknown portion of subducted continental crust may have escaped exhumation and been recycled into the mantle (Chopin, 2003).

The greenschist and amphibolite facies account for the bulk of metamorphosed continental crust, and involve higher temperature geotherms than subduction metamorphism, as represented by the Alpine geotherm in Fig. 1.15. The Barrovian metamorphic sequence described above presents a well-studied example of the greenschist to amphibolite facies transition (e.g., Harte & Hudson, 1979); the chlorite, biotite and garnet zones belong to the greenschist facies, and the staurolite, kyanite and sillimanite zones belong to the amphibolite facies. Greenschist facies metabasites are characterized by mineral assemblages involving albite, actinolite, chlorite and epidote, with or without quartz and garnet. The transition to the amphibolite facies is marked by the appearance of hornblende (rather than actinolite) and Ca-bearing plagioclase (rather than albite), with or without garnet and clinopyroxene. Amphibolite facies conditions straddle both the aluminosilicate triple point and the beginning of partial melting in the presence of free water in most rock types (Fig. 1.15). Assemblages in metapelitic compositions are highly sensitive to temperature variations in this facies (e.g., Pattison & Harte, 1985).

The granulite facies covers the highest temperature portion of orogenic metamorphism, with temperatures in excess of 900°C commonly referred to as *ultra-high-temperature* (UHT). Granulite facies conditions (Fig. 1.15), such as those recorded in rocks of the of the Mount Stafford (Fig. 1.9) or Broken Hill areas, Australia, reflect heat fluxes that are anomalously high in the context of continental convergence, commonly referred to as *low-pressure/high-temperature* (LPHT) conditions. Such terranes probably require mantle heat. Equivalent amphibolite facies conditions are recorded by andalusite-sillimanite-bearing metapelitic rocks, such as occur in the Buchan zones of the Scottish Dalradian. The amphibolite-granulite transition is marked by the first appearance of orthopyroxene in mafic compositions or by coexisting garnet and cordierite in metapelitic compositions (e.g., Powell & Downes, 1990). In granulite facies metabasites, orthopyroxene and clinopyroxene replace hornblende and the low-*P* part of the granulite facies field is also referred to as the *two-pyroxene granulite* subfacies. At elevated pressures, orthopyroxene is progressively replaced by garnet, delineating a *garnet granulite* subfacies (Green & Ringwood, 1967), as shown in Fig. 1.15; this change is strongly dependent on whole-rock composition.

Water-poor, or even anhydrous, assemblages formed at elevated temperatures present problems for *P-T* diagrams that do not explicitly consider fluid abundance and composition (Figs 1.15, 1.16A). At super-solidus conditions, the small amount of free water is partitioned into melt (Chapter 4). Removal of melt by advection (Chapter 4) substantially dehydrates the residual rock (Powell, 1983), though small proportions of water may remain structurally bound in hydrous minerals, such as biotite, cordierite and hornblende. Mineral equilibria at elevated temperatures thus need to be represented on diagrams that explicitly consider volatile content, and, preferably, the proportion of melt. For example, consider pseudosections (Figs. 1.16B, 1.16C) drawn for intricately interlayered metapelitic and metapasammitic rocks at Mount Stafford, central Australia (Vernon *et al.*, 1990; White *et al.*, 2003). Mineral assemblages in LPHT metapelitic compositions commonly lose muscovite before encountering their solidus (Fyfe *et al.*, 1978; White *et al.*, 2003; Vernon *et al.*, 2003), and partial melting is mostly controlled by biotite breakdown. The appearance of orthopyroxene at low-*P* conditions, or garnet and cordierite at higher pressures, present a low-*T* boundary for the granulite facies in subaluminous metapelitic or metapsammitic compositions (White *et al.*, 2003), as shown in Fig. 1.16B. Interlayered aluminous metapelitic compositions are insensitive to the amphibolite-granulite transition, which is marked only by changes in mode that include progressively higher proportions of melt (Fig. 1.16C).

Owing to the mixture of metamorphic and inherited grains inevitable in low-grade metamorphic rocks, Eskola (1921) did not infer a facies for temperatures below the greenschist facies. The *zeolite facies* was proposed by Fyfe *et al.* (1958) on the basis of assemblages in Triassic metavolcanic rocks in southern New Zealand described by Coombs (1954), and the *prehnite-pumpellyite facies* was proposed by Coombs (1960, 1961), on the basis of Upper Carboniferous-Triassic rocks in south-eastern New Zealand, as well as by Packham & Crook (1960), on the basis of 'diagenetic facies' (effectively burial metamorphic facies) in several areas in New South Wales, Australia. More complex *P-T* sub-divisions of such low-grade rocks have not received widespread acceptance, and care needs to be exercised in applying the facies concept at such low-grade conditions (Banno, 1998).

The boundaries between other facies are complex. Interlayered green and blue amphibole-bearing schists at the blueschist-greenschist transition commonly reflect subtle whole-rock compositional variations. Plagioclase-bearing mafic gneisses with Na-poor clinopyroxene symplectites enclosing omphacite may reflect eclogites that entered the granulite facies, or omphacite granulite assemblages transitional between the eclogite and granulite facies (Scharbert & Carswell, 1983; O'Brien & Rötzler, 2003).

1.8.2 Facies series

The metamorphic facies scheme summarized by Turner (1981) separates contact metamorphic facies from regional metamorphic facies (Fig. 1.15).

Pattison & Tracy (1991) regarded this distinction as arbitrary, and instead favoured the concept of metamorphic *facies series* (Miyashiro, 1961) or *bathozones* (Carmichael, 1978). The following features are diagnostic of the facies series for metapelites suggested by Pattison & Tracy (1991). The scheme predominantly reflects increasingly higher pressure assemblages, but whole-rock composition also has an effect, for example, Fe-rich rocks favouring staurolite and almandine garnet, Mg-rich rocks favouring cordierite, and Mn-rich rocks favouring spessartine garnet.

Facies series 1 is characterized by cordierite and andalusite at low and intermediate grades and by the assemblages And + Kfs or And + Sil + Kfs at high grades, and is equivalent to bathozone 1 and the boundary of bathozone 2 of Carmichael (1978). The following three sub-types can be distinguished. *Type 1a* is characterized by the prevalence of cordierite and absence of andalusite up to the muscovite-quartz breakdown reaction, the appearance of Crd + Kfs at lower grade than the muscovite-quartz breakdown reaction, the stability of And + Kfs, the absence of sillimanite except at high grade, and the prevalence of spinel and hypersthene and absence of garnet in high-grade cordierite-bearing assemblages, all these feature being consistent with low pressure. *Type 1b* is identical, except that at grades above the muscovite-quartz breakdown reaction sillimanite occurs sporadically at the same grade as And-Kfs, and garnet may occur at the highest grade with cordierite, hypersthene and spinel, though Grt + Als assemblages are rare. These assemblages are consistent with slightly higher pressure than for type 1a. *Type 1c* differs from type 1b in that (1) either andalusite or cordierite may occur at low grade, (2) the invariant assemblage: Ms + Chl + Crd + And + Bt + Qtz may occur, (3) the assemblage: Ms + Crd + And + Bt + Qtz is common at intermediate grade up to the muscovite-quartz breakdown reaction, and (4) prograde K-feldspar does not appear except at higher grade than the muscovite-quartz breakdown reaction. High-grade assemblages are similar to those of type 1b, except that Grt + Sil + Crd + Kfs is more common. As in type 1b, And + Kfs ± Sil is common at higher grade than the muscovite-quartz breakdown reaction, suggesting that the pressure ranges of 1b and 1c may partly overlap. Though facies series 1 is common in contact metamorphism, it may also occur in LPHT regional aureole settings, such as the Cooma Complex, south-east Australia, which has assemblages indicative of type 1c (Johnson *et al.*, 1994; Vernon & Johnson, 2000; Vernon *et al.*, 2000, 2003).

Facies series 2 is characterized by various combinations of cordierite, andalusite, staurolite and Mn-rich garnet at low and intermediate grades and by sillimanite at lower grade than the muscovite-quartz breakdown reaction (consistent with higher pressure than facies series 1); high temperature assemblages typically contain the assemblage: Bt-Grt-Sil-Kfs-Crd. Two sub-types can be distinguished, namely *type 2a*, characterized by the prevalence of cordierite at low to intermediate grade and *type 2b*, characterized by St-And at low to intermediate grade, cordierite being rare, except at high grade; garnet is common at all grades.

Facies series 3 is similar to type 2b, except that sillimanite is the only aluminosilicate mineral, and garnet is more prevalent at all grades.

Facies series 4 is similar to facies series 3, except that kyanite, rather than andalusite or sillimanite occurs above the staurolite isograd, and sillimanite appears at lower grade than the muscovite-quartz breakdown reaction. This is typical of Barrovian regional metamorphic sequences, but also occurs rarely in contact metamorphism.

1.9 Determining whole-rock composition and a volume of equilibration

In the context of the microstructural complexities outlined above, involving either zoned minerals or corona reaction microstructures, determining a whole-rock (bulk-rock) composition appropriate for modelling changes in metamorphic phase diagrams may not be a simple task. Though realistic equilibration volumes are impossible to identify exactly, grain-size and the absence or presence of chemically zoned porphyroblasts can be indicative of key prograde and/or retrograde processes. For example, the whole-rock chemical compositions of microstructurally simple upper amphibolite and granulite facies rocks have been successfully modelled using bulk-rock X-ray fluorescence analytical (XRF) data (e.g., White *et al.*, 2002, 2004), with the usual caveat that the size of the sample collected (and analysed) must be representative of the rock and mineral assemblage. A rule of thumb, based on standard statistics, is that the sample should have at least 1,000 grains, to reduce the effect of skewed data resulting from large or uncommon mineral grains (Fig. 1.14). An alternative, for microstructurally simple rocks, involves using assumed average mineral compositions and the mode to obtain a whole-rock composition (Carson *et al.*, 1999; Wei *et al.*, 2004). The application of whole-rock XRF analyses is not generally useful in microstructurally complex rocks because: (1) chemical fractionation of the rock may be related to large or zoned porphyroblasts, (2) compositional layering may be too fine to compositionally separate, and (3) heterogeneous porphyroblast distribution at hand-specimen scale may induce a marked local compositional variation (the 'nugget' effect).

Recently, several methods have become available for estimating the effective rock composition in microstructurally complex rocks (e.g., Marmo *et al.*, 2002; Evans, 2004; Tinkham & Ghent, 2005), involving the application of quantitative element mapping of selected portions of rock samples. Element mapping using an electron microprobe provides a powerful tool for examining the nature of chemical zonation in individual crystals and selected rock areas (Clarke *et al.*, 2001).

Conventional microprobe analyses or profiles of analyses present a random or one-dimensional section that may miss critical zones or minerals present in small proportions. X-ray maps, generated using an electron microprobe, allow fine-scale two-dimensional spatial relationships between minerals to be shown, as well as characteristics of mineral zoning that may be otherwise difficult to detect. Examples of such maps are discussed in Section 2.2. Microprobe analytical data are collected with the diffracting crystal in each wavelength dispersive spectrometer set to one wavelength, to record X-ray intensity as the sample stage is stepped through a chosen area. The number of elements that can be counted simultaneously depends on the number of spectrometers. However, as element data in X-ray intensity maps are not proportional or corrected for partial X-ray re-absorption, they are not suitable for use in petrology. The new generation of fast personal computers offers the possibility of overcoming these limitations by the application of a comparatively simple algorithm for matrix correction to raw intensity maps (Clarke *et al.*, 2001; Tinkham & Ghent, 2005). Electron microprobes now commonly have software, supplied by the manufacturer, for performing this task. Data in the processed maps are much more petrologically informative than maps of X-ray intensity, as they can be used in a variety of routine petrological methods. There is a trade-off between costly data acquisition time and data reproducibility. For example, the maps can be used to calculate rock compositions for small domains that would not be possible using conventional chemical techniques. Alternatively, by applying appropriate thresholds, data relevant to any chosen mineral(s) can be isolated and presented as cation maps, maps of cation ratios, or binary plots of cation proportions (Clarke *et al.*, 2001).

As noted previously, though metamorphic reactions are ultimately driven mainly by *P-T* conditions, the resulting mineral assemblages also depend on whole-rock chemical composition. If the parent rocks are chemically heterogeneous on the outcrop to thin-section scales, local chemical potential gradients may develop between compositional domains, resulting in complex reaction microstructures and dominating the effects of *P* and *T* (Hensen, 1988; Dunkley *et al.*, 1999). Different reactions may occur in different domains at different *P-T* conditions, which complicates the determination of *P-T* histories, but also makes for more realistic determinations, provided the compositional variations are taken in to account (e.g., Halpin *et al.*, 2007). This is best illustrated by *P-T pseudosections*, because they show multivariant mineral assemblage fields, rather than the divariant fields and univariant curves shown on typical *P-T* diagrams. For example, Fig. 1.1 shows aluminosilicate phase relations, which are mostly relevant to aluminous metapelitic rocks, and the breakdown of pure albite to jadeite and quartz, which is mostly relevant to Na-rich metafelsic and metabasic (mafic) rocks. In comparison, the detailed mineral assemblage relations shown in Figs. 1.16A and 1.16B are drawn for realistic single–rock compositions. Pseudosections assist in the interpretation of reaction microstructures that result from growth of minerals as *P* and *T* change across multivariant fields, rather than across univariant curves.

1.10 Open- versus closed-system behaviour during metamorphism

As outlined above, the prograde history of fluid-present mineral assemblages containing hydrous minerals is commonly interpreted as involving progressive dehydration and fluid loss with heating, but only minor retrograde reaction, owing to insufficient water being available to facilitate retrograde reactions during cooling, except in or near high-strain zones (e.g., Vernon, 1969; Vernon & Ransom, 1971; Vernon, 1976; Fyfe *et al.*, 1978). Prograde metamorphism is commonly regarded as involving closed-system behaviour with regard to all components except water, which is not explicitly considered. This idea is supported by (1) distinct layers with preserved original contrasting oxidation states in relatively high-grade metamorphic rocks (Chinner, 1960) and (2) the dominance of relatively low-variance assemblages in prograde metamorphic sequences, in contrast to the typical occurrence of monomineralic or bimineralic zones in systems for which chemical activities are externally controlled (see Chapter 5). However, H_2O structurally bound in minerals appears to play a critical role in determining the preservation of mineral assemblages, even in an inferred closed system, and is thus an important variable in relating calculated phase equilibria (Guiraud *et al.*, 1996; Carson *et al.*, 1999; Guiraud *et al.*, 2001).

Though some metamorphic terranes have been inferred to have involved low fluid:rock ratios (Miller & Cartwright, 2000; Scambelluri & Phillipot, 2001), abundant evidence of high fluid-rock ratios (1 to 5 rock volumes) has been found in some rocks, notably metacarbonate rocks (see Chapter 5), for which water plays an active role during metamorphism at greenschist and amphibolite facies conditions (Wood & Walther, 1986). The water promotes reactions by assisting access of 'nutrient' components and removal of 'waste' components. The effects of added water generally are not as clear in metapelitic and metapsammitic rocks, as H_2O is the only volatile component liberated in the prograde reactions. However, the high water contents of common blueschist facies assemblages indicate that lawsonite-bearing equilibria are difficult to form because they are relatively H_2O-rich, requiring the addition of H_2O at elevated pressure (Clarke *et al.*, 2006). Assessing time-integrated changes in the volume of fluid, either by ingress or egress, is a complex problem, involving stable isotope data (e.g., Miller & Cartwright, 2000) or fluid budget modelling (e.g., Clarke *et al.*, 2006) in the context of field relationships and a knowledge of the protolith(s).

Deformation enables transient openings between grains of different mechanical properties and enhances permeability in, for example, shear zones (Chapter 5). The total volume of fluid at any time must be small, but over time, a lot of fluid can pass through. In many high-grade Precambrian terranes, fluid flow focussed (channelled) along fault structures, either late in a single metamorphic cycle or during a second event, causes narrow zones of extensive retrogression (e.g., Vernon, 1969; Vernon & Ransom, 1971; Corbett & Phillips, 1981; Hobbs *et al.*, 1984). Field relationships and whole-rock chemical data can be used to assess metasomatic changes in such circumstances, as discussed in Chapter 5.

Chapter 2

Metamorphic Conditions: Chemography and Thermobarometry

2.1 Introduction: Intensive factors, extensive factors and rock chemical composition

The protolith of any metamorphic rock determines its chemical make-up, (whole-rock chemical composition) and hence the minerals that can be produced by metamorphism. Chemographic relationships between protoliths and common minerals, such as outlined in ACF diagrams (Section 1.2), provide a context for predicting likely mineral occurrences. For example, aluminosilicate minerals are uncommon in metamorphosed marls because the relatively low Al content of these rocks is usually sequestered in a calcic mineral such as plagioclase.

Extensive metamorphic variables, such as volume, entropy and enthalpy, depend on the nature and amount of the phases (minerals and fluids) present, which are, in the absence of metasomatism (Chapter 5), inherited from the protolith (parent rock). A second set of variables also determines the metamorphic mineral assemblage, in the context of the extensive variables; these are *intensive variables*, such as P and T, which are imposed on the rock system and independent of mass. They include the chemical potential of all components, and at equilibrium conditions, are constant through all phases present.

The phase rule presented in Section 1.3 does not distinguish between intensive and extensive variables, and so fails to separate their influence in determining mineral assemblages. The distinction can become important when representing relationships on phase diagrams (Powell *et al.*, 2005) as discussed below. In metamorphic petrology, the term *paragenesis* refers to mineral assemblages inferred to have grown at the same time, that is, minerals in stable, or metastable, coexistence during the metamorphic event being considered; ideally, though not necessarily, the members of a paragenesis have low-energy mineral grain shapes (Vernon, 1976, 2004), as discussed in Section 3.2. In this chapter, we first discuss a more sophisticated equilibrium model as a basis for inferring metamorphic parageneses and explore the interplay of factors that can result in complex mineral microstructures.

2.2 Chemically zoned porphyroblasts and equilibration

The driving force for any chemical reaction at constant P and T, is the Gibbs free energy change (ΔG). For a spontaneous incremental change at constant T and P, ΔG must be either negative (for an irreversible reaction) or zero (at equilibrium). The model of prograde changes inducing new assemblages that destroy evidence for reactants, owing to maintaining of equilibrium, has widespread, but not complete, acceptance for high-grade metamorphic rocks (Rubie, 1990). Complexity in this model is caused by the common observation of chemically zoned porphyroblasts. For example, garnet grains in eclogite and blueschist facies rocks commonly show extensive chemical zoning (Fig. 2.1; Clarke *et al.*, 1997; Robbo *et al.*, 1999) as do garnet porphyroblast in many greenschist to amphibolite facies terranes (Vance & Holland, 1993; Lanzirotti, 1995; Schumacher *et al.*, 1999; Pyle & Spear, 1999; Willner *et al.*, 2000; Stallard & Hickey, 2002; Yang & Rivers, 2002; Vernon, 2004, pp. 257–68). In such rocks, diffusional processes cannot maintain equilibrium conditions on the grain scale during neocrystallization. The following discussion of the consequences for the equilibrium model is drawn from Marmo *et al.* (2002).

Factors that may influence chemical zonation in porphyroblasts include the comparatively slow rate of intragranular diffusion relative to the growth rate of anhydrous minerals, such as garnet, and the transient nature of high-P/T metamorphism. Prograde zoning is rarely preserved in minerals in granulite facies rocks, as rates of intragranular diffusion at high T are fast enough to maintain chemical equilibration over a volume larger than that of individual porphyroblasts (e.g., Yardley, 1977d). However, coarse-grained anhydrous mineral grains in granulite facies rocks do commonly preserve limited chemical zoning near grain boundaries, as a consequence of limited reaction by diffusional exchange with adjacent minerals during cooling (Woodsworth, 1977; Pattison & Bégin, 1994).

Compositional zoning indicates fractionation of the whole-rock composition, by selectively removing components from the effective reacting system (Schumacher *et al.*, 1999; Marmo *et al.*, 2002; Stallard & Hickey, 2002). Clear examples of this effect in eclogite and blueschist facies rocks involve garnet grains with spessartine and grossular-rich cores formed under lower temperature conditions than pyrope-rich garnet rims. The progressive growth of such porphyroblasts leads to the chemical isolation of the garnet cores, and therefore changes the composition of the matrix. Once chemically isolated, the elements in the garnet cores no longer participate in reactions that involve the matrix.

Analogous fractionation of the whole-rock composition occurs as a consequence of cooling in granulite facies rocks, in the form of *reaction coronas* that enclose comparatively coarse-grained minerals (e.g., Griffin & Heier, 1973; McLelland & Whitney, 1977; Ellis *et al.*, 1980; Mørk, 1985), as shown in Fig. 2.2. The cores of large grains that form under peak (highest-T) conditions remain chemically isolated from retrograde reactions, and the secondary assemblages form at their boundaries, because the peak grainsize is coarser than the scale of retrograde diffusion (Clarke & Powell, 1991; White & Clarke, 1997).

The fractionation of a rock composition means that elements forming the cores of chemically-zoned porphyroblasts do not contribute to the

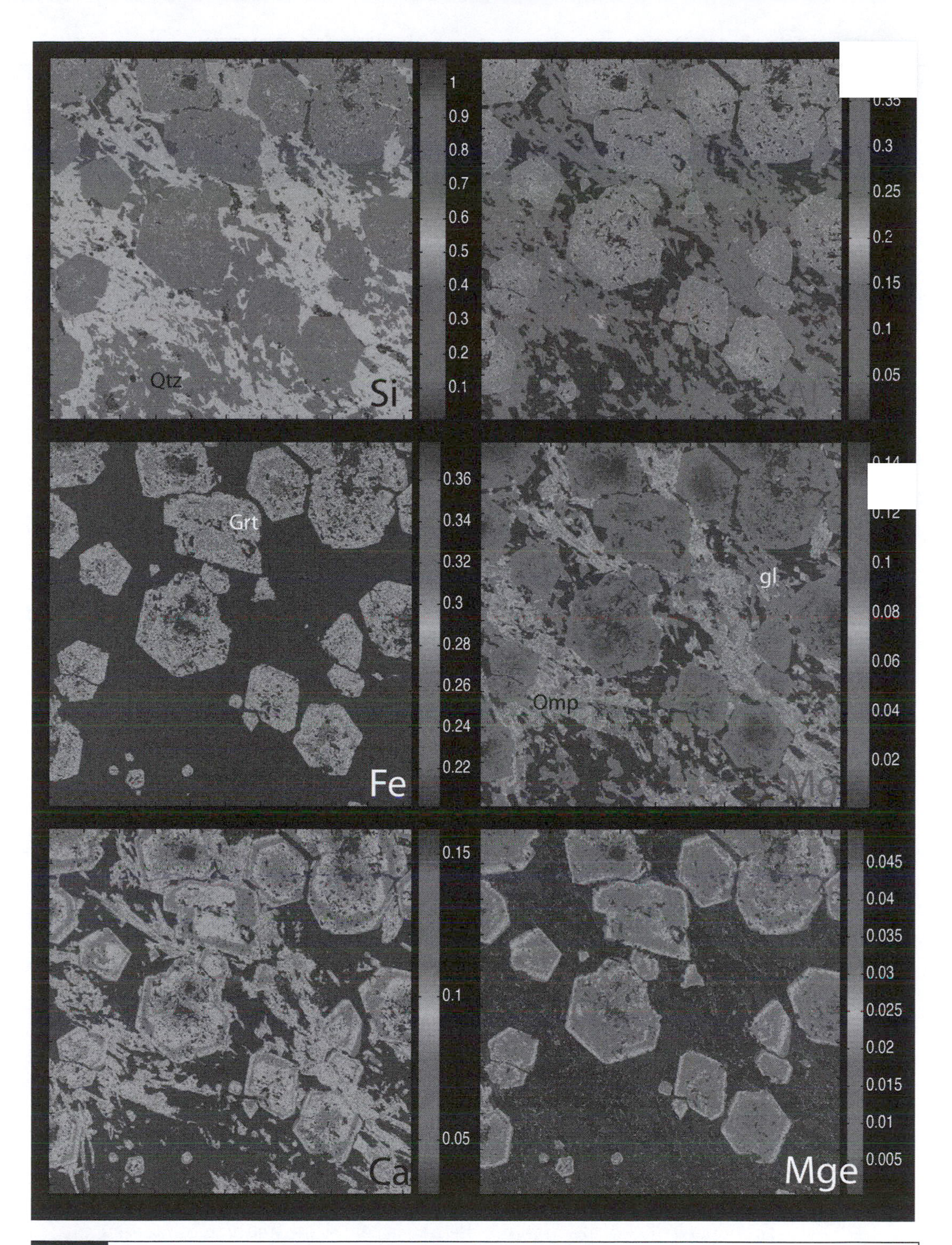

Fig. 2.1 Maps of element concentrations (weight precent; for example, 0.15 = 15%), illustrating complex chemical zoning in garnet from a basaltic eclogite in New Caledonia (Clarke *et al.*, 1997). Garnet growth has been inferred to be related to a change in prograde conditions, from approximately 475 °C to 590 °C and from 15 to 19 kbar (Marmo *et al.*, 2002), with Fe-rich, Mg-poor cores grading to Mg-rich rims (see Mge map with restricted Mg range). Most of the matrix of the rock is composed of random omphacite, glaucophane and quartz grains (see Mg map), omphacite having formed late in the prograde history (Carson *et al.*, 1999). The Ca map shows an unusual spike in garnet Ca content in the outer part of the grains. The imaged area is approximately 1.5 cm across. Maps of X-ray intensity were converted to weight per cent following the method of Clarke *et al.* (2001). A subtle effect from a small focussing error is evident in decreased concentration of iron from the top to the bottom of each image. For a colour version of this figure, please go to the plate section.

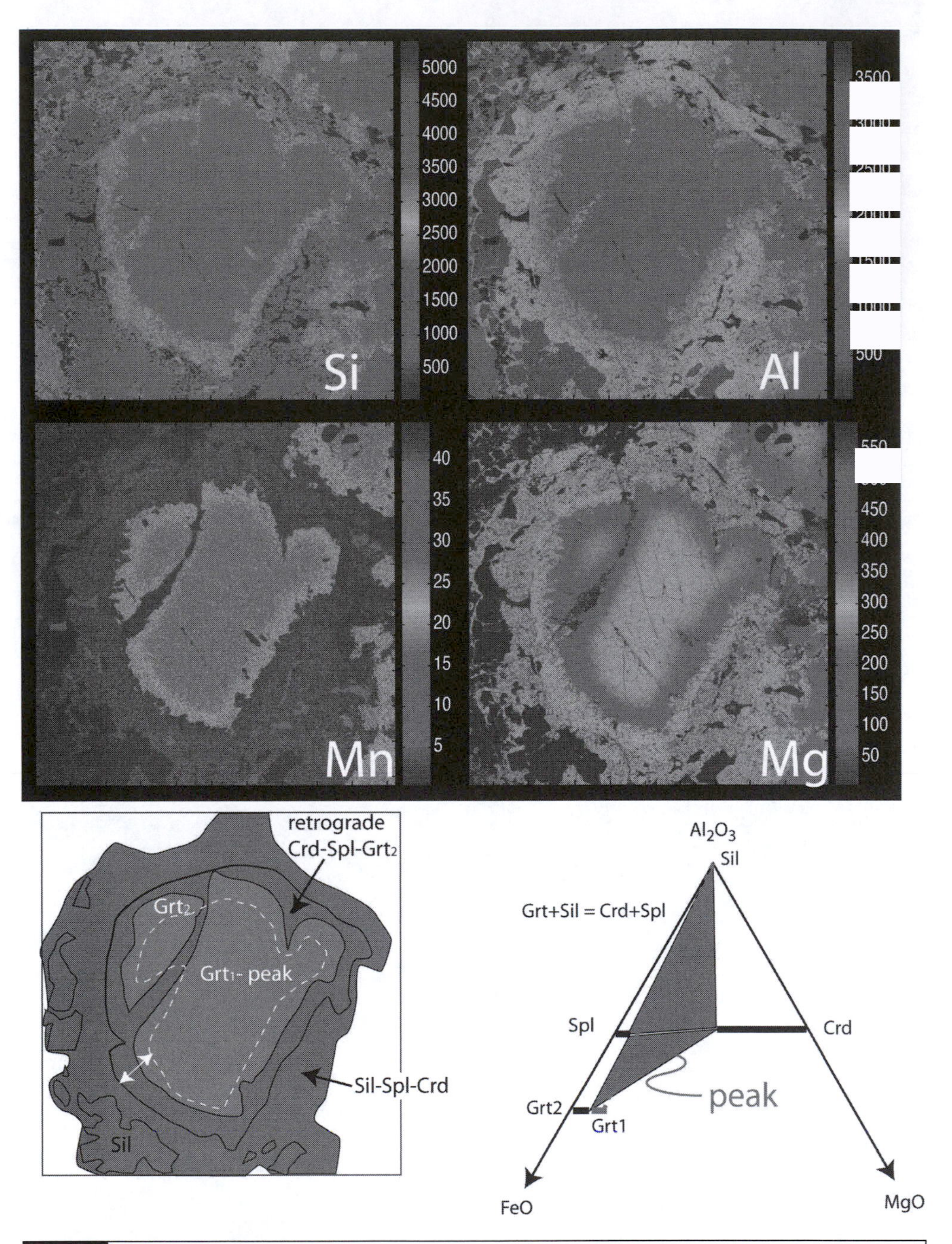

Fig. 2.2 Maps of X-ray counts for Si, Al, Mn and Mg in a granulite facies reaction corona involving spinel and cordierite that separates garnet from sillimanite; Cohn Hill, western Musgrave Block, central Australia. The reaction has been inferred to be the result of decompression from peak metamorphic conditions (Clarke & Powell, 1991). Image approximately 2.5 cm across. For a colour version of this figure, please go to the plate section.

composition of an equilibrium-volume that involves their rims and surrounding matrix. If the zoned porphyroblasts are abundant, and have an average chemical composition different from that of the matrix, the rock may undergo progressive chemical fractionation as a consequence of porphyroblast growth. The mineral assemblage stable in the matrix depends on *P*, *T* and fluid composition (if fluid is present), as well as the chemical composition of its solid equilibrium-volume. The following three examples illustrate the role of chemically zoned porphyroblasts in determining rock major element composition and common prograde and retrograde mineral microstructures. Improvements in analytical procedures are increasingly providing additional information from mineral trace element compositions (Pyle & Spear, 1999; Yang & Rivers, 2002; Clarke *et al.*, 2007).

2.2.1 Equilibration during prograde metamorphism: Garnet growth in New Caledonian eclogites

Northern New Caledonia exposes blueschist and eclogite facies rocks that have been interpreted as reflecting a series of Eocene tectonic events consequent to the attempted subduction of a dispersed Gondwana fragment (e.g., Black, 1977; Aitchison *et al.*, 1995). The high-*P* rocks can be divided into MORB (mid-ocean ridge basalt)-type eclogites of the Pouébo terrane, inferred to represent metamorphosed oceanic crust, and lawsonite- and epidote-blueschists of the Diahot terrane, inferred to represent metamorphosed continental crust (Clarke *et al.*, 1997). The high-*P*, medium-*T* eclogites of the Pouébo terrane include barroisite and glaucophane-bearing eclogites in domains that experienced minor or no post-peak fluid influx. Eclogite assemblages reflect peak conditions of $P \approx 19$ kbar and $T \approx 590$°C (Carson *et al.*, 1999). Extensive garnet-glaucophanite formed at $P \approx 16$ kbar during semi-pervasive fluid influx (Carson *et al.*, 2000). In the eclogites, inclusions of lawsonite, chlorite and albite, which indicate blueschist facies conditions, occur in chemically zoned garnet grains that show low-energy grain-shape relationships with matrix eclogite facies assemblages devoid of these minerals (e.g., Clarke *et al.*, 1997).

Figure 2.1 shows a series of element maps for a portion of one of the eclogites. A high quartz mode can be discerned by comparing the Si and Al maps. Random, late-formed splays of omphacite and glaucophane, intergrown with quartz, can be identified by comparing the Mg and Ca maps (glaucophane having no appreciable Ca). Garnet is strongly zoned from comparatively Mn- and Fe-rich cores to comparatively Mg-rich rims (see the enhanced Mg map labelled 'Mge', which only shows a restricted Mg range). Calcium zoning is complex, with a pronounced maximum concentration concentric with the shapes of the grains, but displaced inward from the grain boundaries. On the basis of the element concentrations, several parageneses can be identified for this rock. The matrix minerals show restricted compositional ranges that are inferred to reflect extensive neocrystallization of these parts of the eclogites, either at peak conditions or when the rocks lost free water during decompression (Carson *et al.*, 2000). The equilibrium volume for these rocks (at peak conditions) was considered by Marmo *et al.* (2002) to consist of the outermost parts of garnet and the comparatively fine-grained matrix of the rock, where chemical equilibrium would have been maintained by grain-boundary diffusion, combined with limited intercrystalline diffusion. Relatively low rates of intercrystalline diffusion and comparatively rapid grain growth enabled the preservation of the prograde history in the garnet grains. As garnet is the main mineral that preserves prograde zoning, the chemical fractionation of these rocks was controlled by the garnet mode and composition. The Mn-rich cores and captured mineral inclusions reflect temperatures some 100°C cooler than peak conditions, on the basis of modelling (Fig. 2.9). The continued growth of garnet porphyroblasts has been interpreted as leading to an ongoing increase in the volume of garnet-forming elements that were isolated from reactions occurring in the matrix of the rock (Clarke *et al.*, 1997; Marmo *et al.*, 2002). Prograde parageneses can be inferred by relating the position of the various mineral inclusions in garnet to the patterns of elemental zoning. For example, quartz, chlorite, glaucophane, titanite and albite occur in garnet cores, but it would be necessary to establish that each of the minerals occurs with chemically similar garnet before inferring that they all coexisted. By selecting core, mantle and rim portions of garnet grains, and either including or excluding them from a calculated 'equilibration' composition, Marmo *et al.* (2002) modelled the effects of composition fractionation on microstructures and assemblages during the prograde history.

2.2.2 Corona reaction microstructures: Cohn Hill, Musgrave Block, central Australia and Rauer Group, Antarctica

Mesoproterozoic granulite facies gneisses of the Musgrave Complex form a latitudinally-trending gravity high with a strike extent of over 700 km, straddling the Western Australian, South Australian and Northern Territory borders, Australia. Outcrops are dominated by granulite facies orthogneiss and charnockite, along with banded gneisses that have been interpreted as having volcanic and/or sedimentary precursors (C. M. Gray, 1971, 1977; Daniels, 1974). Rare metapelitic gneiss preserves spectacular reaction coronas of spinel and cordierite that separate garnet from sillimanite (Clarke & Powell, 1991), as shown in Fig. 2.2.

Figure 2.2 shows Si, Al, Mn and Mg maps of a representative reaction microstructure. The minerals involved are labelled in a sketch. Sillimanite occurs in the outer parts of the corona, but not near garnet. Whereas the cores of garnet grains are homogeneous, the corroded margins are zoned: Mn and Fe (not shown) contents increase and Mg content decreases towards the rim over a short distance from the ragged grain boundary. The microstructure reflects the instability of peak garnet and sillimanite during a retrograde event (which may have shortly followed peak conditions) or a subsequent unrelated event. The original Mg-rich composition of the garnet is preserved in the cores of large grains (garnet 1), and a chemical diffusion front on the garnet rims (garnet 2) was inferred to reflect garnet that chemically adjusted to the retrograde conditions (Clarke & Powell, 1991). The corona has an inner domain of retrograde cordierite and spinel (no sillimanite) inferred to have been in elemental exchange with garnet 2. The outer limit of this domain may represent the original extent of garnet 1. Note the enrichment of Mn in garnet 2, sourced from garnet 1, now consumed by spinel and cordierite. The outer domain of the corona involves spinel and cordierite that partially pseudomorph large sillimanite grains.

Clarke & Powell (1991) inferred that spinel and cordierite became stable with *either* garnet *or* sillimanite, and formed retrograde intergrowths with grainsizes substantially smaller than the peak minerals garnet and sillimanite. The scale of equilibration at peak conditions would have included the garnet cores and the remaining sillimanite grains. During corona formation, equilibration contracted to a scale indicated by the outer (chemically adjusted) garnet and the adjacent spinel and cordierite intergrowths,

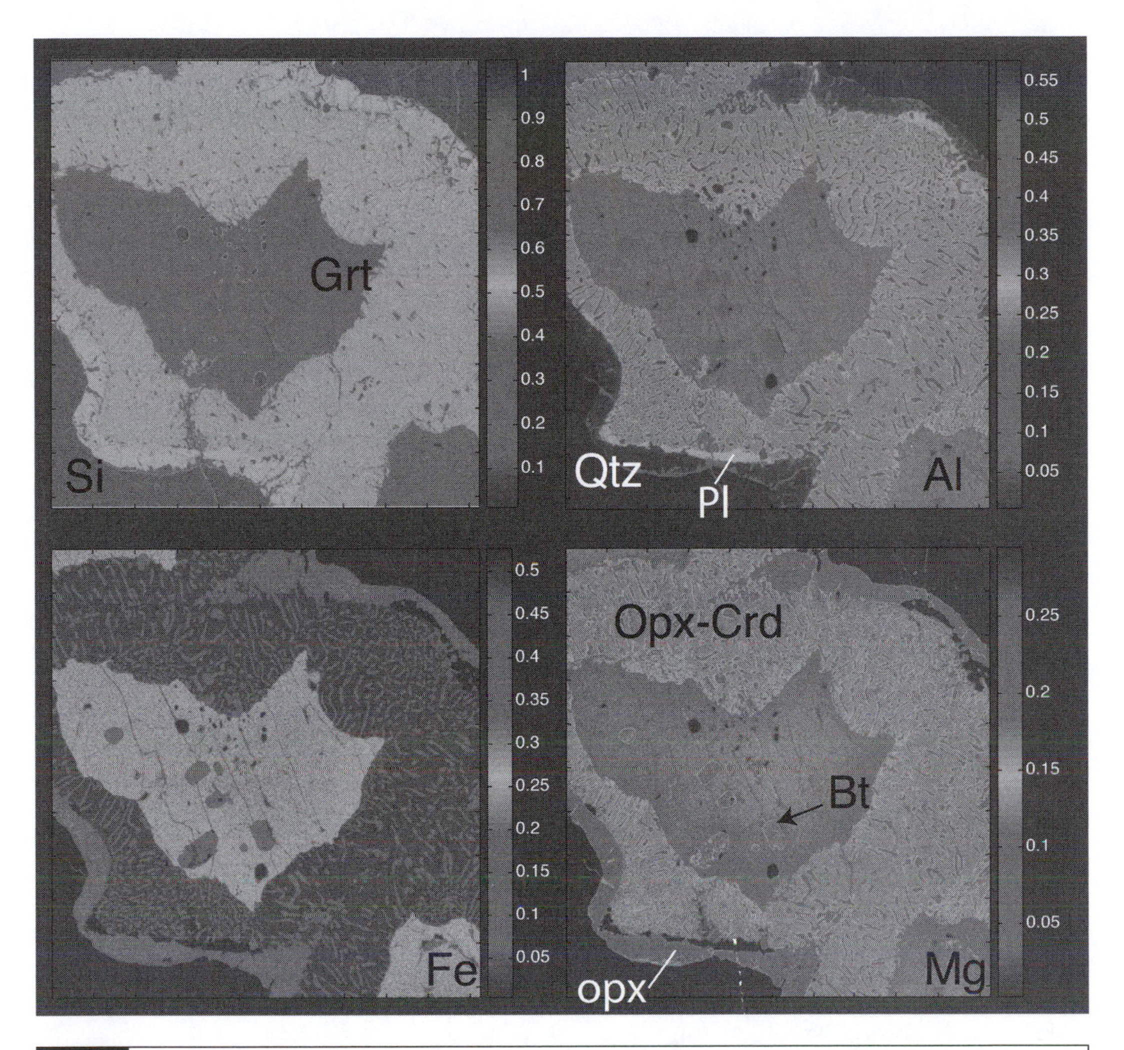

Fig. 2.3 Maps of element concentrations (weight precent), showing a granulite facies reaction corona involving orthopyroxene and cordierite that separates garnet from quartz, the Rauer Group of islands, east Antarctica. The reaction has been inferred to be related to decompression with cooling from peak metamorphic conditions (Kelsey *et al.*, 2003). The biotite inclusions in garnet were inferred to be part of the peak assemblage. Image area approximately 2 cm across. Maps of X-ray intensity were converted to weight percent following the method of Clarke *et al.* (2001). For a colour version of this figure, please go to the plate section.

as indicated by the white arrow on the sketch in Fig. 2.2. Quartz, K-feldspar, biotite and ilmenite form part of most of the peak and retrograde assemblages (Clarke & Powell, 1991). The reaction Grt + Sil = Spl + Crd is univariant in KFMASH, and the structuring of the corona can be explained by a single original bulk rock composition lying along the Grt-Sil tie-line being split into two chemical domains, owing to limited retrograde diffusion. One retrograde composition lies close to garnet (assemblage Grt 2-Spl-Crd) and another lies close to sillimanite (Spl-Crd-Sil). Complexity in the mineral relationships involving the additional minerals biotite (see Mg map), ilmenite and magnetite have been modelled by White *et al.* (2002) by expanding the model system to additionally include TiO_2 and Fe_2O_2.

Another example of a corona reaction microstructure (Fig. 2.3) comes from the Rauer Group, Antarctica. Biotite and garnet are preserved in the core of the garnet grain, chemically isolated from reactions occurring on the grain boundary that formed cordierite-orthopyroxene symplectites. Narrow rinds of plagioclase and orthopyroxene form the outermost parts of the corona.

2.2.3 UHT microstructures and closure temperatures

The poor preservation of prograde features in granulite facies rocks, owing to rapid diffusion rates, presents another problem for high-temperature and ultra-high-temperature (UHT; $T > 900\,^{\circ}C$) assemblages with chemically zoned porphyroblasts. The question is: which part of the retrograde path does chemical zoning in a mineral represent? The preserved mineral chemical information is largely unrepresentative of peak *P-T* conditions, because of retrograde $FeMg^{-1}$ re-equilibration caused by intragranular diffusion at the elevated temperature conditions (e.g., Pattison & Bégin, 1994). Further complexity can be expected, owing to variability in: (1) the temperature dependency of rates of intracrystalline diffusion for different elements; and (2) rates of intracrystalline diffusion of a given element in different minerals. These complexities can be considered in the context of 'closure temperatures' (Lasaga, 1983) for the chemical re-equilibration of different minerals on common geological time scales (for instance, 5–20 Ma). Anhydrous minerals, such as garnet, orthoproxene and clinopyroxene, tend to have higher closure temperatures (*c.* 700°C–850°C) than hydrous minerals, such as amphibole and biotite (*c.* 600°C–750°C), for the intracrystalline diffusion of major elements. As a consequence, most rocks probably do not accurately record chemical evidence for UHT conditions, owing to diffusive re-equilibration (Sandiford & Powell, 1986). A partial record may be provided by elements inferred to diffuse comparatively slowly, such as Al exchanged between orthopyroxene and garnet (Baba, 1999). The most reliable evidence for UHT conditions comes from key mineral assemblages, such as sapphirine, quartz and osumilite in metapelitic granulite facies rocks (Ellis, 1980) or evidence of inverted metamorphic pigeonite (Sandiford & Powell, 1986).

Hollis *et al.* (2006) and Baba (1998, 1999) described evidence for prograde metamorphic changes preserved in UHT granulites from South Harris, Scotland. A 1–2 km-wide belt of granulite facies metasedimentary/metavolcanic gneiss occurs adjacent to a large igneous complex consisting of anorthositic, tonalitic and gabbroic components. Baba (1998) identified the following features indicative of increasing pressure during development of the high-grade metapelitic assemblages: (1) sillimanite occurring as inclusions in aluminous orthopyroxene that coexists with kyanite; and (2) garnet grains with biotite and quartz inclusions in their cores, sillimanite inclusions near their rims, and grossular-rich rim compositions inferred to be stable with matrix kyanite. Fig. 2.4 shows Ca and Mg maps of a garnet grain from these metapelitic rocks. The garnet contains oriented inclusions of sillimanite and biotite, and kyanite occurs in the quartz-rich matrix. Two distinct stages of garnet growth are evident from the Ca map (Hollis *et al.*, 2006). The first involved the growth of subhedral garnet (core and rim 1) with zonation from a high-Ca core to an inclusion-poor, low-Ca rim 1; this has been interpreted as growth zoning, rather than diffusional re-equilibration, on the basis of a preserved 'bell-curve' type of chemical profile and because the zonation is centred on the inclusion-rich core, which is off-centre with respect to the grain shape. The

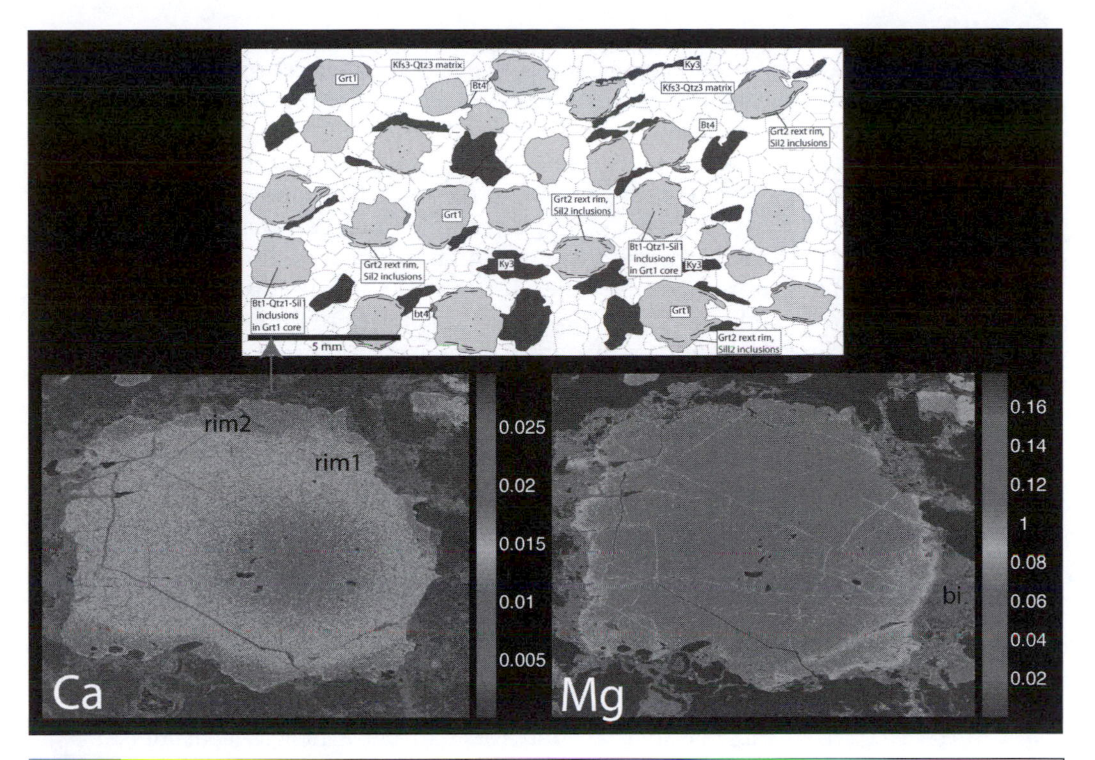

Fig. 2.4 Ca and Mg maps showing patterns of complex chemical zoning in garnet that have been inferred to reflect differences in diffusion rates of these elements at high-grade conditions. The sample is a UHT metapelitic gneiss from South Harris, Scotland (Baba, 1999); thin section sketch modified, with additional material, from Hollis *et al.* (2006, fig. 3). For a colour version of this figure, please go to the plate section.

second stage (rim 2) involved garnet overgrowths preferentially developed along foliation-parallel grain boundaries, the overgrowths capturing sillimanite. The transition from rim 1 to rim 2 is sharp. The zonation of Mg and Fe is approximately concentric to the whole grain, without evidence of the garnet overgrowths evident in the Ca map. These features are consistent with the destruction of prograde Mg (and Fe) zonation in garnet at peak conditions by comparatively rapid intracrystalline diffusion of these elements; the comparatively slow rates of Ca diffusion enabled the preservation of prograde zoning in this element. The pattern of Mg zoning was inferred to reflect the effects of retrograde re-equilibration (Hollis *et al.*, 2006).

2.3 Recognition of metamorphic parageneses

The examples discussed above address issues concerning the *formation* of the various mineral microstructures, but also draw attention to concerns regarding their *preservation*. In terms of the model closed system developed in Section 1.7, the prograde history of fluid–present mineral assemblages containing hydrous minerals involves low instantaneous fluid–rock ratios, progressive dehydration and fluid loss with heating, but only minor retrograde reaction, owing to insufficient water being available to facilitate retrograde reactions during cooling. The amount of H_2O structurally-bound in the minerals, in addition to *P*, *T* and kinetic aspects of mineral reaction, plays a critical role in considering the preservation of mineral assemblages in such a closed system (Guiraud *et al.*, 1996; Carson *et al.*, 1999). Mineral assemblages can only evolve to new assemblages that involve lower water contents, and the reaction ceases at the point where the *P-T* path moves to assemblages involving higher water contents (Guiraud *et al.*, 2001; Proyer, 2003; Clarke *et al.*, 2006). Though reactions may occur in dry rocks, especially if deformation is sufficiently intense, the typical equilibrium model of metamorphism involves the idea that the formation, destruction, new formation and preservation of minerals in different stages of a metamorphic history depend largely on the presence, absence or availability of H_2O and/or melt (Guiraud *et al.*, 2001; Brown, 2002; White & Powell, 2002), and also on the presence or absence of deformation, owing to its ability to deform and recrystallize pre-existing mineral grains, and thus promote the re-equilibration of mineral assemblages (e.g., Stipska & Powell, 2005).

Vernon (1976, pp. 40–43) listed four criteria to identify a mineral paragenesis: (1) common grain boundaries between all minerals, especially smooth, 'clean' contacts; (2) lack of evidence for one mineral replacing another, such as fine-grained aggregates in veinlets or along grain boundaries; (3) the presence of low-energy shapes of grains and inclusions (Section 3.2); and (4) the presence of a relatively small number of minerals, in the context of the rock-type and an appropriate model system. The recognition of such assemblages can be difficult, especially in polymetamorphic rocks. In a heterogeneous rock, the observations should be repeated for each compositionally distinct part, because mineral assemblages vary with whole-rock composition.

Criteria 1–4 are reasonable indicators, especially when taken together. Criteria 2 can be unreliable because minerals rimming each other can reflect incomplete reaction and can mean that one of the reactants became exhausted, so that the reaction stopped, leaving a partial replacement microstructure. In this situation, the product minerals can be compatible, but the reactant mineral would not be part of the new paragenesis. Incomplete reaction may also reflect comparatively limited diffusion distances, such as retrograde coronas (Section 2.2.2), as a result of which, more than one paragenesis may be present, owing to changes in composition. Alternatively, one mineral may partially pseudomorph another, as a consequence of a change in mode, owing to a continuous reaction, such as that involved in crossing the Barrovian garnet isograd (Section 1.6). Criteria 3 and 4, though important, are not sufficient indicators of equilibrium. For example, growing mineral grains are in chemical equilibrium as soon as they impinge, after which adoption of stable grain shapes occur, so that absence of low-energy shapes is not necessarily indicative of chemical disequilibrium. The application of criterion 4 needs to be made in the context of rock composition and metamorphic facies. For example, metabasites commonly have high-variance assemblages over large *P-T* domains, whereas metapelitic rocks can have low-variance assemblages that are stable for only small sections of the amphibolite facies (Sections 1.4, 1.7).

The following are the best criteria for chemical equilibrium compatibility (after Vernon, 1976): (1) all minerals in contact with each other, thereby excluding minerals 'shielded' from reaction, such as inclusions in another mineral; (2) a similar composition for the edges of different grains of the same mineral (Section 2.2.1); (3) similar distribution of components between groups of minerals in different parts of the rock being evaluated; these components might be major or trace elements, or ratios of elements sensitive to grade variations; and (4) low-energy shapes of grains and inclusions of the minerals concerned, as discussed above.

As pointed out by Vernon (1996a), care is needed with questionable criteria that are often taken to infer lack of compatibility of minerals and assemblages, especially the inferences that: (1) minerals occurring as inclusions are necessarily older than the host mineral, and (2) minerals occurring in folia that wrap around porphyroblasts are necessarily younger than the porphyroblast. These problems are discussed in Chapter 3.

2.4 Equilibrium thermodynamics in petrology

An extension of the equilibrium model is the application of equilibrium thermodynamics to infer, in a quantitative manner, the nature of the extensive variables imposed on rocks during formation of their mineral assemblage(s). The following sections concentrate on application of the computer program THERMOCALC (Powell *et al.*, 1998) to calculate mineral equilibria for the representative microstructures discussed above, though alternative programs are Perple_X (Connolly & Kerrick, 1987; Connolly, 1990) and GEOCALC (Berman, 1988). Parts of the following sections draw on information presented by Powell (1978), Spear (1993 chapter 7) and Elmer (2004). Websites that cover related material, and from which parts of the information below are drawn, include the following:

- http://www.esc.cam.ac.uk/astaff/holland/thermocalc.html
- http://www.earthsci.unimelb.edu.au/tpg/thermocalc
- http://www.earth.ox.ac.uk/~davewa/pt/pt-start.html
- http://www.perplex.ethz.ch

Phase equilibria calculations depend on the assumption that chemical equilibrium is attained among the phases involved. The state of chemical equilibrium involves coexisting phases – minerals with or without fluid and/or melt – reaching the lowest free energy configuration for a given set of imposed *P-T-X* conditions. As outlined in the discussions of the equilibrium model (Section 1.7), we assume that equilibrium is maintained during a series of changes until, at some point, an assemblage is locked in and the rock returns to Earth's surface. Chemical equilibrium may be attained on only a restricted scale, and, as discussed above, that scale may change.

2.4.1 Chemical systems, components and end-members

Metamorphic changes can be modelled in a chemical system composed of an independent set of N components that satisfactorily account for the major chemical variations in the system, including any fluid or melt present on grain boundaries. For example, many of the changes in the metapelitic rock examples discussed above can be modelled using the chemical components K_2O-FeO-MgO-Al_2O_3-SiO_2-H_2O (KFMASH). This system adequately describes the major ion-exchange and/or net transfer reactions responsible for subsolidus changes in mineral composition and assemblages in such rocks. In detail, larger systems incorporating MnO, Fe_2O_3, TiO_2, Na_2O and CaO are required, but the KFMASH system is suitable for evaluating many metamorphic situations. Minerals at higher and lower grades than the observed assemblage should also be calculated, so that the peak conditions can be inferred.

Before considering thermodynamic variables and calculations for such model systems, we initially review the main solid solution mechanisms in a few common silicate minerals that enable quantitative estimates of *P-T* conditions to be calculated. Spear (1993, chapter 4) has described the crystal chemistry of a range of mineral groups.

Aspects that need to be considered include the sites in each crystal structure on which the principal substitutions take place and the typical ranges of observed compositional variation. For example, garnet has the basic formula: $X_3Y_2Si_3O_{12}$, where X is an 8 coordinated site containing, in the KFMASH system, divalent cations (Fe^{2+}, Mg^{2+}, Ca^{2+}) and Y is an octahedral site containing trivalent cations (Al^{3+}). Expansions of the model KFMASH system might, for example, additionally incorporate Mn^{2+} on the X site and/or Fe^{3+} on the Y site.

A more complex example involves the mica group, where minerals consist of composite sheets in which a layer of octahedrally-coordinated Y cations is sandwiched between two identical layers of linked $(Si,Al)O_4$ tetrahedra (Deer *et al.*, 1992, p. 279). Apices of the SiO_4^{4-} tetrahedra connect to the octahedra of the central layer. Between each sandwich layer are interlayer sites that can contain large cations. The coordination of the octahedra is completed by OH^-anions. The general formula of the mica group minerals can be written as: $XY_{2\text{-}3}Z_4O_{10}(OH)_2$, where X represents elements in the interlayer site, Y represents elements in the octahedral sites and Z represents elements in the tetrahedral sites. The octahedral sheet can be made dominantly from: (1) divalent cations such as Mg and Fe, in which case all three available sites are filled (trioctahedral mica); or (2) trivalent cations such as Al, in which case one of the three sites is vacant (trioctahedral micas). If the tetrahedra are occupied solely by Si, the sandwich has a neutral charge; common examples are talc (trioctahedral) and pyrophyllite (trioctahedral). Appreciable substitution by Al^{3+} for Si^{4+} in the tetrahedral sites occurs in more common micas, requiring that charge balance be maintained by K^+, Na^+, or Ca^{2+} in the interlayer site. Common rock-forming micas are the trioctahedral phlogopite-biotite series and the dioctahedral 'white' micas such as muscovite, paragonite and margarite. Important substitutions and site preference are:

- tetrahedral: Si^{4+}, Al^{3+}
- octahedral: Al^{3+}, Cr^{3+}, Fe^{3+}, Ti^{3+}, Fe^{2+}, Mg^{2+}, Mn^{2+}
- interlayer site: K^+, Na^+, Ca^{2+}
- hydroxyl site: OH^-, F^-, Cl^-, O^{2-}

In the model KFMASH system, Cr^{3+}, Fe^{3+}, Ti^{3+}, Mn^{2+}, Na^+, Ca^{2+}, F^-, Cl^- and O^{2-} are ignored.

Though most minerals occur with combinations of elements on the various sites, *end-members* are compounds where each site is occupied by one (or a fixed number of) element(s) that have the stoichiometry of the mineral into which they substitute. Thermodynamic data have been obtained for a range of naturally occurring end-members, enabling thermodynamic calculations for rocks involving minerals with these end-members. For example, the common end-members of garnet are pyrope

($Mg_3Al_2Si_3O_{12}$), almandine ($Fe_3Al_2Si_3O_{12}$), spessartine ($Mn_3Al_2Si_3O_{12}$), grossular ($Ca_3Al_2Si_3O_{12}$) and andradite ($Ca_3Fe^{3+}{}_2Si_3O_{12}$). For the model KFMASH system, only pyrope and almandine are considered, and variation between them is modelled by $FeMg^{-1}$ exchange.

The mica group minerals involve multiple solid solution series. Thermodynamic data are available for the biotite end-members phlogopite ($KMg_3[AlSi_3]O_{10}(OH)_2$), annite ($KFe_3[AlSi_3]O_{10}(OH)_2$), eastonite ($K[Mg_2Al][Al_2Si_2]O_{10}(OH)_2$) and Na-phlogopite ($NaMg_3[AlSi_3]O_{10}(OH)_2$). Variation in natural micas can be accounted for by $FeMg^{-1}$ substitution in octahedral sites (phlogopite-annite), $Al_2Mg^{-1}Si^{-1}$ or Tschermak substitution (phlogopite-eastonite), and NaK^{-1} substitution in the interlayer site. Important end-members in 'white' mica are muscovite ($KAl_2[AlSi_3]O_{10}(OH)_2$), paragonite ($NaAl_2[AlSi_3]O_{10}(OH)_2$), margarite ($CaAl_2[Al_2Si_2]O_{10}(OH)_2$), Mg-Al-celadonite ($K[MgAl][Si_4]O_{10}(OH)_2$) and Fe-Al-celadonite ($K[FeAl][Si_4]O_{10}(OH)_2$).

Though the thermodynamic properties of simple substances, such as pure end-members, can be measured or estimated, natural minerals are mostly complex solid solutions and vary greatly in composition. This demands an understanding of the parameters that control the mixing of various mineral end-members, as discussed further below.

2.4.2 Thermodynamic variables and calculations

As outlined previously (Section 2.1), all phases at equilibrium in a model system have equal values of intensive variables (P, T and chemical potential, μ). Values of the extensive variables (V, S and number of moles, n_i) depend on the amount of material present. The mole fraction of i in a given phase may be defined by $x_i = n_i/(\text{sum}(n_i)$, where n_i is the number of moles of end-member i. The application of equilibrium thermodynamics to minerals involves a mathematical expression that includes all variables necessary to express minimization of Gibbs free energy. The resulting expression enables the determination of the stable mineral assemblage in a given chemical system, with respect to the chosen intensive or extensive variables.

Assuming that equilibrium conditions were attained, one or a series of balanced chemical equations involving relevant end-members can be written for element exchange between two or more minerals in the chosen system. Ion-exchange and net-transfer reactions modify the mineral assemblage and chemistry in response to changing imposed variables (P, T and chemical potential, μ). Thermodynamic data for the relevant end-members, combined with information concerning the way that the end-members dissolve in each phase, is required to perform the calculations. Once the relevant chemical reactions are written, the defined thermodynamic relationships are used to undertake thermodynamic calculations. The equilibrium relation for any balanced reaction, written between end-members of the phases, can be expressed in terms of chemical potential (Gibbs, 1961). At equilibrium

$$\Delta\mu = 0 = \sum_i v_i n_i \tag{2.1}$$

where v_i is the reaction coefficient of end-member i in a balanced reaction. For example, at equilibrium, $\Delta CO_2 = CO + 1/2O_2$ can be expressed as

$$\Delta\mu = 0 = \mu_{CO} + 1/2\mu_{O_2} - \mu_{CO_2} \tag{2.2}$$

Chemical potential is a function of T, P and composition. As most natural minerals are solid solutions of end-members and vary in composition by substitutions on sites, the chemical potential can be split into a composition-independent term (u°_{iA}) and a composition-dependent term ($RT \ln a_{iA}$):

$$\mu_{iA} = u^\circ_{iA} + RT \ln a_{iA} \quad (2.3)$$

where u°_{iA} is the standard chemical potential of end-member i in phase A at 1 bar and 298K. The most convenient standard state for solids is the pure end-member at the P and T of interest; this part of the chemical potential can be easily measured and looked up in a table of thermodynamic variables. The composition-dependent term describes deviations from the standard chemical potential due to solid solution, and is the part that needs to be measured using mineral analysis obtained from an electron microprobe. R is the gas constant (0.0083144 kJ K^{-1}), T is temperature in K, and a_{iA} is the activity of end-member i in phase A. Substituting equation 2.3 into equation 2.1 gives

$$\Delta\mu = 0 = \sum_i v_{iA}(u^\circ_{iA} + RT \ln a_{iA}) \quad (2.4)$$

Generally, this equation is written as the fundamental thermodynamic equation:

$$\Delta\mu = 0 = \Delta G^\circ + RT \ln K \quad (2.5)$$

where K is the equilibrium constant and ΔG° is the change in Gibbs free energy between the end-members involved in the reaction. Equation 2.5 can be applied to each balanced end-member reaction in the relevant chemical system. The Gibbs free energy of a phase can be measured, and depends on T and P through equations involving enthalpy, H (energy association with heat), entropy, S (a measure of disorder), volume, V and heat capacity, C_P. The Gibbs energy of a reaction at a specified P and T is determined by the change in enthalpy, entropy, volume and heat capacity in the relation:

$$\Delta G_{P,T} = \Delta H_{1,298} - T\,\Delta S_{1,298} + \int_{298}^{T} \Delta C_P \, dT - T \int_{298}^{T} (\Delta C_P/T)\, dT + \int_{1}^{P} \Delta V \, dP. \quad (2.6)$$

The general form of the heat capacity equation (for each phase) is

$$C_P = a + bT + c/T^2 \quad (2.7)$$

where a, b and c are experimentally-determined constants.

For reactions involving only solids, it is commonly assumed that heat capacity and volumes of minerals are independent of P and T. As $\Delta V_{reaction}$ is a constant, it can be moved outside the integral component of equation 2.6, such that

$$\int_{1}^{P} \Delta V \, dP = \Delta V \int_{1}^{P} dP = (P - 1)\Delta V \quad (2.8)$$

Making these assumptions, most solid–solid reactions have a linear relationship to P and T such as

$$\Delta G^\circ = a + bT + C(P - 1) \tag{2.9}$$

where

$$a = \Delta H_{1,298}$$
$$b = -\Delta S_{1,298}$$
$$c = \Delta V_{1,298}$$

The slope of any given reaction with respect to P and T is thus given by the Clausius-Clapeyron equation:

$$\frac{\delta P}{\delta T} = \frac{\Delta S}{\Delta V} \tag{2.10}$$

The derivation of these equations has been discussed in more detail by Wood & Fraser (1976), Powell (1978) and Spear (1993).

2.4.3 Activity-composition relations and non-ideal mixing

Solid solution contributes to Gibbs free energy, as related by equation 2.5, with the relationship expressed as

$$G = \sum_i p_i \mu_i = \sum_i p_i(\mu_i^\circ + RT \ln a_i) \tag{2.11}$$

where p_i is the proportion of end-member i in the phase (also referred to as the mole fraction). It is convenient to express the Gibbs energy of natural minerals as

$$G = G^{mechanical} + G^{mix} \tag{2.12}$$

where $G^{mechanical}$ is a function of P and T, analogous to equation 2.6. G^{mix} is controlled by how the various elements are distributed in crystal sites. This can be complex, depending on the mineral and elements involved, and is described in thermodynamic equations by activity-composition relationships. The G^{mix} of a phase is commonly described in two parts:

$$G^{mix} = G^{configurational}_{ideal} + G^{excess}_{non\text{-}ideal} \tag{2.13}$$

or

$$G^{mix} = p_i \mu_i = \sum_i p_i RT \ln a^{ideal}_{iA} + \sum_i p_i RT \ln \gamma_{iA} \tag{2.14}$$

where a^{ideal}_{iA} is the ideal mixing activity of end-member i in phase A and γ_{iA}, the activity coefficient, accounts for departures from random mixing. For ideal mixing, $\gamma = 1$. For each end-member of a solid solution series, activity-composition relations must be determined involving both the ideal and non-ideal contributions.

Thermodynamic datasets have been compiled for a range of common mineral end-members (Holland & Powell, 1985; Berman, 1988; Holland & Powell, 1998; Gottschalk, 1997) through a combination of reliable calorimetric data and experimentally-determined phase equilibria. The different datasets use different approaches to generate internally-consistent data, though most computer program applications of these datasets now use the Holland & Powell (1998) dataset. However, this standard state dataset is used in substantially different ways in thermodynamic calculations involving real minerals, mostly owing to differences in activity-composition models. The issue has been addressed by Spear (1993; chapter 7), and Powell & Holland (1993) have presented 'symmetric formulism' as a macroscopic model to simplify the writing of activity-composition relations. This model combines microscopic interaction parameters, w_{ij}, (related to substitution on sites) linearly to form a macroscopic interaction energy parameter W_{ij} for end-members i and j in a system. This limits the number of mixing parameters and simplifies the mathematic calculations of the activity-composition relations. Holland & Powell (1996a, b) developed order–disorder models to address cation ordering on sites via macroscopic mixing parameters, as reflected in common asymmetric solvi, such as in the calcite-dolomite-magnesite system, which involves exchange of different sized cations (Ca^{2+}, Mg^{2+}). Considerable care needs to be used in the application of much petrological software to ensure an understanding of activity-composition relations.

2.4.4 Thermobarometry

The mineral compositions that can be measured were locked in at some time; *P-T* conditions and, sometimes, fluid composition data corresponding to that time can be obtained by using either thermodynamic data or the application of 'directly calibrated' thermometers or barometers. In this section we initially address the application of experimentally-determined exchange equilibria to recover *P-T* estimates, and then outline an alternative forward modelling approach using phase diagrams.

The formal thermodynamic relationship between *P*, *T* and mineral composition can be calculated by combining equations 2.5 and 2.9. Knowing that at equilibrium, the free energy of reaction must be zero and accepting the simplifications presented above in terms of heat capacity and volume expansion data,

$$\Delta G_{P,T}^{reaction} = 0 = \Delta H_{1,298} - T\,\Delta S_{1,298} + (P-1)\Delta V + RT\ln K \tag{2.15}$$

For pure phases, $K = 1$, so $\ln K = 0$ as involved in determining calorimetric data. Equation 2.15 can be rearranged:

$$\Delta H_{1,298} - T\,\Delta S_{1,298} + (P-1)\Delta V = -RT\ln K$$

or

$$\ln K = -\frac{\Delta H_{1,298} - T\,\Delta S_{1,298} + (P-1)\Delta V}{RT} \tag{2.16}$$

Net transfer reactions between minerals with appreciable solid solution are very useful for thermobarometry, as appropriate continuous or sliding equilibria can occur over a wide *P-T* range. The equilibrium constant for any exchange reaction is calculated in the standard way using activity coefficients of reactants and products (e.g., equation 2.19). Equations 2.16 and 2.17 can be considered to involve two parts: (1) that involving the thermodynamic data, on the right hand side of the equation, which remains fixed for a given reaction and standard state thermodynamic data, so there is a linear relationship between *P* and *T*; and (2) that involving the equilibrium constant, on the left hand side of the equation, with activity coefficients of reactants and products calculated from microprobe data and knowledge of the activity-composition relationships. The first part determines the slope

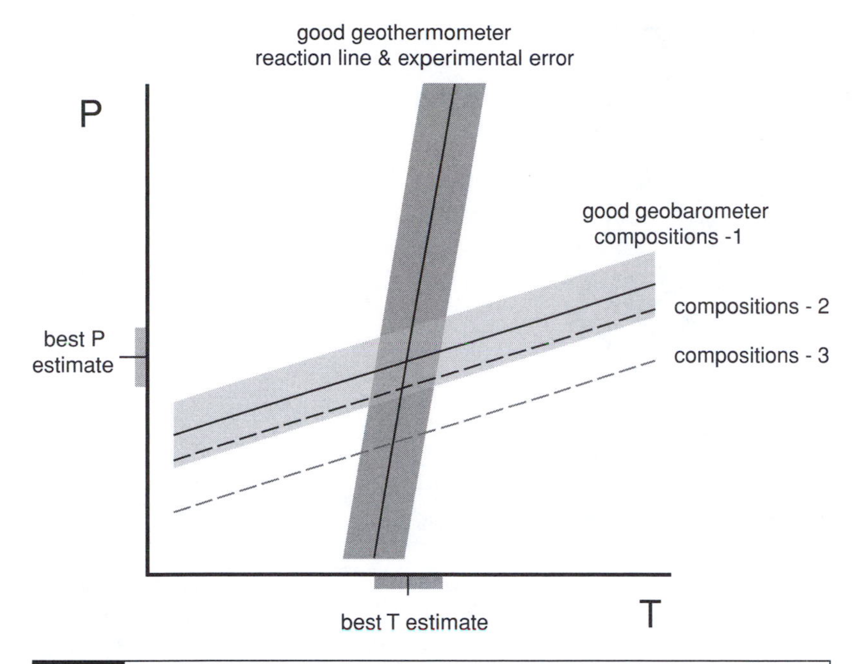

Fig. 2.5 *P-T* diagram indicating the position of two theoretical exchange reactions, the slope of the reaction being strongly influenced by $\Delta V_{reaction}$. The slope determines whether it is more suitable as a geothermometer or geobarometer. By using minerals with exchange reactions suitable as thermometers and barometers, a more robust estimate of *P* and *T* conditions can be obtained from the intersection of two (or more) reactions. The effect of mineral composition differences, in terms of displacing an exchange reaction to different *P-T* conditions, is shown for the geobarometer; subtle differences may result in displacements that lie within experimental error, as shown for compositions 1 and 2. For a colour version of this figure, please go to the plate section.

of the reaction on a *P-T* diagram (Clausius-Clapeyron equation; 2.12), with a precise geothermometer needing a small ΔV to have a slope similar to the pressure axis, and a precise geobarometer needing to have a large ΔV to have a slope similar to the temperature axis (Fig. 2.5). Variations in the second, composition-dependent, part of equation 2.16, due to solid solution, displace the reaction line of fixed slope to different *P-T* conditions. Activity-composition relations for minerals involved in the exchange reaction need to be known or modelled; complexity in nature means that this is usually the most contentious part, as there may not complete agreement on, for example, the site distribution of elements in a given mineral for different concentrations or the influence of changes in crystal structural state on solid solution. In addition, the simplifications involved in deriving equation 2.16 are not valid for dehydration reactions, in which the volume of one of the phases (e.g., H_2O) has a strong dependence on *P* and *T*.

Information from several exchange reactions *may* provide an indication of technique reliability; clustered results could reflect minimal or extensive diffusive re-equilibration. Alternatively, by combining information for two exchange reactions, one being a good thermometer and the other a good barometer, estimates for both *P* and *T* can be obtained (Fig. 2.5). The effect from minerals of different composition – for example, garnet with

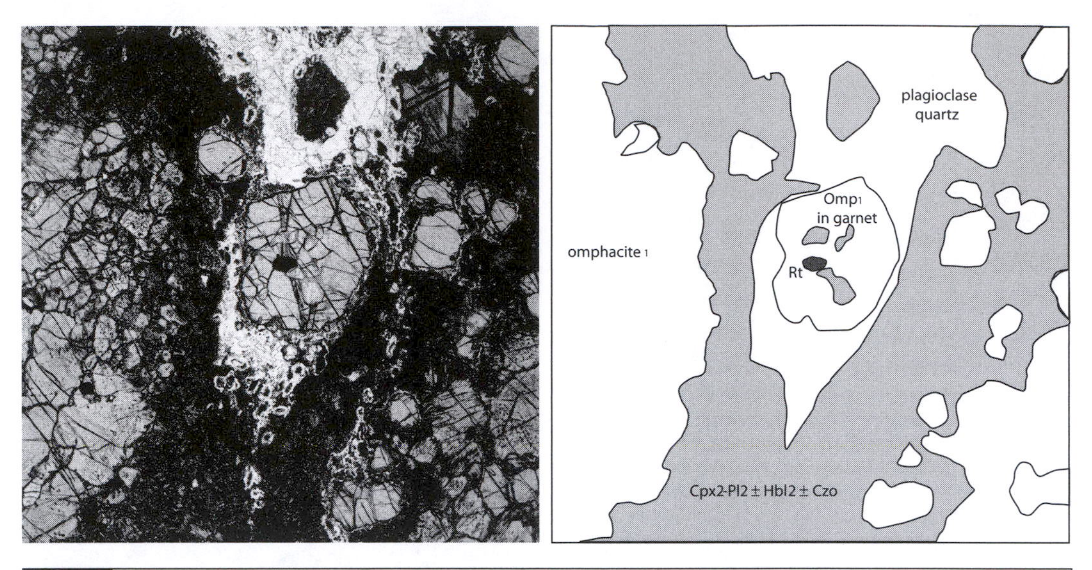

Fig. 2.6 Plane-polarized light photomicrograph and sketch of part of sample 0669, a garnet granulite from Fiordland, New Zealand. An early paragenesis is inferred to comprise the large garnet, plagioclase, omphacite and quartz grains, with a secondary paragenesis formed from comparatively fine-grained symplectites of sodic diopside, plagioclase 2, and quartz. Rims of garnet grains surrounded by symplectite have compositions distinct from those of grain cores, and are inferred to be part of a secondary paragenesis. Field of view 6 mm across. For a colour version of this figure, please go to the plate section.

different Fe/(Fe+Mg) or plagioclase with different Ca/(Ca+Na) values – in displacing the *P-T* conditions of an exchange reaction is shown in Fig. 2.5. Subtle reaction displacement (compositions 1 and 2 on the barometer, Fig. 2.5) could reflect whole-rock compositional differences and lie entirely within experimental error.

As an example, consider sample 0669, a garnet granulite from Fiordland New Zealand (Clarke *et al.*, 2000) with a granoblastic microstructure involving coarse-grained (grains 5–10 mm across) garnet, omphacite, hornblende, plagioclase, quartz, rutile and ilmenite (Fig. 2.6). Omphacite and plagioclase are separated by fine-grained (grains 0.5–2 mm across) symplectites of sodic diopside and plagioclase, and garnet is chemically zoned. Table 2.1 shows representative microprobe analyses of two key parageneses: (1) the early coarse-grained assemblage involving garnet cores (garnet$_1$), omphacite, plagioclase$_1$, hornblende$_1$ and quartz; and (2) garnet rims (garnet$_2$) with sodic-diopside, plagioclase$_2$ and quartz. Analyses of minerals in the first paragenesis were taken from garnet inclusions, to minimise the effects of diffusive re-equilibration; analyses of minerals in the second paragenesis were made on symplectites and the rim of an adjacent garnet grain.

$FeMg^{-1}$ exchange between garnet and clinopyroxene is strongly temperature dependent and forms an important thermometer for high-grade rocks. The *P-T* dependence of the $K_{D_{\mathrm{Fe-Mg}}}^{\mathrm{Grt-Cpx}}$ distribution coefficient (commonly designated simply as K_D) was first experimentally calibrated by Råheim & Green (1974) for a series of basaltic rock compositions crystallized to eclogite at 20–40 kbar and 600°C–1400°C. Further experimentation and data re-evaluation have arisen, owing to disagreement concerning the influence of garnet composition, the composition and structure of the

Table 2.1. Representative microprobe analyses of minerals in a garnet granulite from Fiordland, New Zealand, used for the thermobarometric results shown in Fig. 2.7

	garnet$_1$ Core	garnet$_2$ Rim	Cpx$_1$ Omph	Cpx$_2$ di sympl	Plag$_1$ in garnet$_1$	Plag$_2$ sympl	Hb$_1$ in g$_1$
SiO_2	39.64	39.14	52.05	50.65	64.81	62.73	42.44
TiO_2	0.17	0.05	0.52	0.62	0.00	0.02	1.46
Al_2O_3	21.49	21.79	10.45	4.38	22.49	22.48	12.47
Cr_2O_3	0.00	0.02	0.00	0.00	0.00	0.00	0.00
FeO	22.28	21.39	8.14	12.41	0.09	0.24	12.38
MnO	0.38	0.46	0.06	0.06	0.03	0.00	0.03
MgO	9.69	7.56	8.22	15.29	0.00	0.04	12.96
CaO	7.04	10.29	15.36	16.08	2.25	4.70	11.87
Na_2O	0.04	0.00	5.17	0.72	9.86	8.35	1.77
K_2O	0.00	0.01	0.00	0.08	0.17	0.65	1.40
Total	100.72	100.71	99.96	100.29	100.71	100.20	96.79
Number of Oxygen ions	12	12	6	6	8	8	23
Si	2.001	2.982	1.887	1.872	2.839	2.775	6.332
Ti	0.010	0.003	0.014	0.017	0.000	0.001	0.164
Al	1.918	1.958	0.447	0.191	1.161	1.225	2.193
Cr	0.000	0.001	0.000	0.000	0.000	0.000	0.000
Fe^{2+}	1.411	1.363	0.134	0.299	0.003	0.009	1.545
Fe3			0.113	0.084			
Mn	0.025	0.029	0.002	0.002	0.001	0.000	0.004
Mg	1.093	0.859	0.444	0.842	0.000	0.002	2.881
Ca	0.571	0.840	0.597	0.637	0.153	0.223	1.898
Na	0.006	0.000	0.363	0.051	0.837	0.716	0.511
K	0.000	0.001	0.000	0.004	0.010	0.036	0.267
Sum	8.033	8.036	4.000	4.000	5.004	4.987	15.796

clinopyroxene, how Fe^{2+} and Mg are distributed on sites, and how the high-T experimental results should be extrapolated to lower T conditions. This has resulted in a number of revised calibrations (e.g., Ellis & Green, 1979; Powell, 1985; Krogh, 1988) that mostly reduce the dependence of K_D on P to that inferred by Råheim & Green (1974).

The exchange of Fe^{2+} and Mg between garnet and clinopyroxene can be expressed by the reaction:

$$\underset{\text{pyrope}}{\frac{1}{3}\mathrm{Mg_3Al_2Si_3O_{12}}} + \underset{\text{hedenbergite}}{\mathrm{CaFeSi_2O_6}} = \underset{\text{almandine}}{\frac{1}{3}\mathrm{Fe_3Al_2Si_3O_{12}}} + \underset{\text{diopside}}{\mathrm{CaMgSi_2O_6}} \tag{2.17}$$

The equilibrium constant, K, for the reaction can be written as:

$$K = \frac{\left(a_{\mathrm{Alm}}^{\mathrm{Grt}}\right)^{1/3}\left(a_{\mathrm{Di}}^{\mathrm{Cpx}}\right)}{\left(a_{\mathrm{Prp}}^{\mathrm{Grt}}\right)^{1/3}\left(a_{\mathrm{Hd}}^{\mathrm{Cpx}}\right)} = \frac{(\mathrm{Fe^{2+}/Mg})_{\mathrm{Grt}}}{(\mathrm{Fe^{2+}/Mg})_{\mathrm{Cpx}}} = K_D \tag{2.18}$$

assuming that all iron is ferrous and there is equi-partitioning of Fe^{2+} and Mg on sites. On the basis of further experimental data, Pattison & Newton (1989) inferred asymmetric, compositionally dependent partitioning of Fe^{2+} and Mg.

Ellis & Green (1979) presented one of the most widely cited calibrations of garnet – clinopyroxene thermometry according to the following equation:

$$T\,^{\circ}C_{\text{grt-cpx}} = \frac{3104X_{\text{Ca}}^{\text{grt}} + 3030 + 10.86P\,(\text{kbar})}{\ln K_D + 1.9034} - 273 \tag{2.19}$$

where

$$K_D = \frac{\text{Fe}}{\text{Mg}}\text{grt} \Big/ \frac{\text{Fe}}{\text{Mg}}\text{cpx} \quad \text{and} \quad X_{\text{Ca}}^{\text{grt}} = \frac{\text{Ca}}{(\text{Ca} + \text{Mg} + \text{Fe})}$$

Graham & Powell (1984) calibrated the following equation, which is independent of P, for $FeMg^{-1}$ exchange between garnet and hornblende in amphibolite and granulite facies assemblages, on the basis of experimentation and natural garnet-clinopyroxene-hornblende assemblages:

$$T\,^{\circ}C_{\text{grt-hbl}} = \frac{2880 + 3280X_{\text{Ca}}^{\text{grt}}}{\ln K_D + 2.426} - 273 \tag{2.20}$$

where

$$K_D = \frac{\text{Fe}}{\text{Mg}}\text{grt} \Big/ \frac{\text{Fe}}{\text{Mg}}\text{hbl} \quad \text{and} \quad X_{\text{Ca}}^{\text{grt}} = \frac{\text{Ca}}{(\text{Ca} + \text{Mg} + \text{Fe})}$$

The exchange of Fe^{2+}, Mg and between garnet, clinopyroxene and plagioclase can be expressed by the reaction (Newton & Perkins, 1982):

$$\underset{\text{anorthite}}{CaAl_2Si_2O_8} + \underset{\text{diopside}}{CaMgSi_2O_6} = \underset{\text{grossular}}{2/3Ca_3Al_2Si_3O_{12}} + \underset{\text{pyrope}}{1/3Mg_3Al_2Si_3O_{12}} + \underset{\text{quartz}}{SiO_2} \tag{2.21}$$

This reaction has a comparatively shallow slope (with respect to the P axis) and is stable over a wide P-T range. Eckert *et al.* (1991) reformulated the Newton & Perkins (1982) calibration of this exchange reaction to derive the formula:

$$P = 2.60 + 0.0171T + 0.003596T \ln K \tag{2.22}$$

where

$$K = \frac{a_{Py}.a_{Gr}^2}{a_{An}a_{Di}}$$

assuming that quartz is present. The activity of grossular is squared, as it has a reaction coefficient of 2/3, and activities of the garnet end-members are cubed because solid solution involves sharing on 3 sites. This is analogous to defining garnet components as 1/3 of the formula units shown in the reaction (Eckert *et al.*, 1991). The reaction has a nominal error of ±1.9 kbar propagated from the calorimetric data, and draws attention to non-ideal behaviour in the solid solution of common silicate minerals. Though equi-partitioning of Fe^{2+} and Mg is commonly assumed for high-T garnet, there is no agreement on the influence of calcium and manganese components. For clinopyroxene, equi-partitioning of Fe^{2+} and Mg on a two-site basis (M1, M2) that has all calcium on the larger M2 site. The activity model for plagioclase is markedly non-ideal, owing to the paired

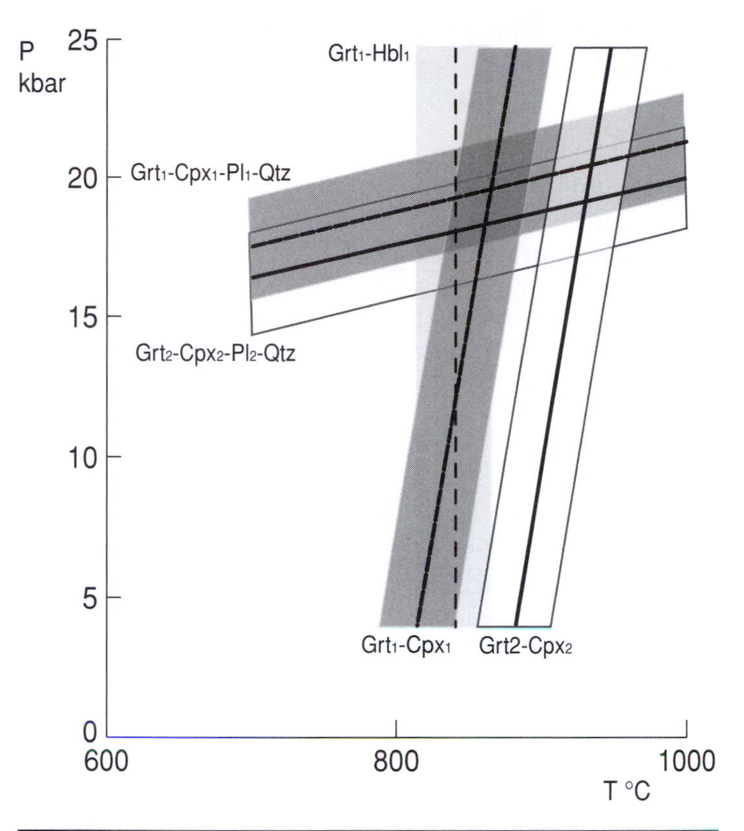

Fig. 2.7 *P-T* diagram showing the position of various exchange equilibria for the two parageneses discussed in the text for sample 0669, a garnet granulite from Fiordland, New Zealand (see Fig. 2.6). Nominal error ranges of 50°C and 1.9 kbar are shown for the positions of the exchange equilibria used as thermometers and barometers, respectively. The *P-T* evolution of the reaction microstructure can be related to minor decompression with heating, but the *P* estimates lie within the error ranges. For a colour version of this figure, please go to the plate section.

$Na^+Si^{4+} = Ca^{2+}Al^{3+}$ exchange. Newton & Perkins (1982) and Eckert *et al.* (1991) discussed these in detail and suggested:

$$\textit{Plagioclase:}\quad a_{an} = \frac{X_{an}(1 + X_{an})^2}{4} \exp\left(\frac{(1 - X_{an})^2(2050 + 9329X_{an})}{TR}\right)$$

for $X_{an} = Ca/(Ca+Na) \leq 0.25$ T is in Kelvin and $R = 1.9872$ cal K^{-1}.

Pyroxene: Assuming equi-partitioningof $Fe^{2+}$and Mg (Wood & Banno, 1973), $a_{Di} = X_{Ca}^{M2}.X_{Mg}^{M1}$where M2 denotes the larger clinopyroxene lattice sitecontaining divalent and monovalent cations, and M1 the smaller site.

Garnet:

$$RT \ln \gamma_{Ca} = (330 - 1.5T)(X_{Mg}^2 + X_{Mg}X_{Fe})$$

$$RT \ln \gamma_{Mg} = (330 - 1.5T)(X_{Ca}^2 + X_{Ca}X_{Fe})$$

where X is the relevant atomic fraction as a ternary (Ca, Mg, Fe) garnet.

P-T estimates for the two parageneses from Fiordland, New Zealand, made using garnet-clinopyroxene and garnet-hornblende thermometry, as well as garnet-clinopyroxene-plagioclase-quartz barometry, are shown in Fig. 2.7, with nominal error ranges of 25°C (thermometer) and 1.9 kbar (barometer) for the positions of the exchange equilibria. The intersection of the equilibria involving the garnet, clinopyroxene and plagioclase compositions for the chosen thermometer and barometer yields *P-T* estimates of 870 ± 25°C and 18.9 ± 2.8 kbar for the early paragenesis, and 920 ± 25°C and 18.2 ± 2.8 kbar for the later paragenesis. The quoted uncertainties are based on calibration and experimentation, and do not consider, for example, uncertainties in mineral analyses or activity models involved in calculating composition parameters. A result is that the quoted uncertainties, for example 25°C in the application of the thermometer, are probably unrealistically small. The composition of the hornblende inclusion in garnet indicates a slightly lower temperature estimate of 840°C for the early paragenesis. These results are consistent with development of the microstructures during terrane heating accompanied by exhumation, though pressure estimates for the two parageneses are the same, within error (Fig. 2.7).

The calculations raise issues about the assumptions common to much thermobarometry. The clinopyroxene in the first paragenesis is omphacite, though all experimental data for the two exchange reactions involving clinopyroxene are based on clinopyroxenes involving diopside-hedenbergite solid solution. Though there appears to be no obvious influence of the jadeite content of clinopyroxene on the K_D for garnet-clinopyroxene thermometry of common metabasites (Krogh, 1988), K_D in more jadeitic clinopyroxenes that are common in quartzofeldspathic eclogites appears to vary inversely with jadeite content (Koons, 1984). The influence of the omphacite content of clinopyroxene on the position of the exchange reaction used as a barometer is not known. In addition, a serious problem exists for the application of all thermometers based on Fe^{2+}–Mg exchange, because the $Fe^{2+}/(Fe^{2+}+Fe^{3+})$ ratio is unknown for minerals analysed by microprobe techniques. The assumption that all Fe is Fe^{2+} in garnet and clinopyroxene generally results in a minimum K_D value, which in turn gives a maximum *T* estimate. In the Fiordland example, the $Fe^{2+}/(Fe^{2+}+Fe^{3+})$ ratio in clinopyroxene has been calculated on the basis of overall charge balance (after Morimoto *et al.*, 1988) and the garnet is assumed to have no andradite component. All such assumptions should be stated when presenting *P-T* estimates.

Table 2.2 lists some exchange reactions that are useful for thermobarometry, and the facies in which the relevant assemblage is common. Key relevant references are given for each technique; for many, additional publications address recalibrations and or compositional influences not evaluated during experimentation.

2.4.5 Thermodynamic calculations for phase diagrams

Phase diagrams are graphical representations of mineral equilibria in a given model chemical system and, from a petrological viewpoint, are constructed for interparting microstructures and mineral compositions. As thermodynamic data for an increasing number of mineral end-members have become available, phase diagram calculations have been facilitated in chemical systems that increasingly approach natural assemblages (Spear & Cheney, 1989; Connolly, 1990). Software such as THERMOCALC (Powell & Holland, 1988), GEOCALC (Berman, 1988) and Perple_X (Connolly & Kerrick, 1987, Connolly, 1990) use internally consistent thermodynamic datasets to calculate phase equilibria that involve mineral solid solution reactions

Table 2.2. Some equilibrium exchange reactions useful for estimating conditions of formation of mineral assemblages in common rock types

Mineral Assemblage	Facies	Exchange Reaction	T/B	Reference
Grt-Chl	blueschist to amphibolite	5Prp+3Fe-Chl = 5Alm+3Mg-Chl	T	Grambling (1990)
Phengite	blueschist, eclogite	Phel = Phe2-Ksp+Phl+Qtz+2H2O	B	Massonne & Schreyer (1987)
Cpx-Pl-Qtz	blueschist, eclogite, granulite	Jd+Qtz = Ab	B	Holland (1980)
				Carswell & Harley (1989)
Ol-Opx	eclogite, granulite	Fo+Fs = Fa+En	T	Carswell & Harley (1989)
Opx-Cpx	eclogite, granulite	En+2Hd = Fs+2Di	T	Lindsay (1983)
Grt-Opx	eclogite	Prp = En + MgTs	B	Carswell & Harley (1989)
	granulite	2Prp+3Fs = 2Alm+3En	T	Harley (1984)
Grt-Ol	eclogite, granulite	2Prp+3Fa = 2Alm+ 3Fo	T	O'Neill & Wood (1979)
Grt-Cpx	eclogite, granulite	Prp+Hed = Alm+Di	T	Ellis & Green (1979)
				Pattison & Newton (1989)
Ttn-Ky-Pl-Rt	eclogite	Ttn+Ky = An+Rt	B	Manning & Bohlen (1991)
Grt-Hbl	amphibolite, eclogite, granulite	4Prp+3Fprg = 4Alm+3Prg	T	Graham & Powell (1984)
Grt-Bt	greenschist to granulite	Prp+Ann = Alm-Phl	T	Ferry & Spear (1978)
Grt-Crd	amphibolite, granulite	2Prp+3Fe-Crd = 2Alm+3Mg-Crd	T	Bhattacharya et al. (1988)
Grt-Hbl-Pl-Qtz	amphibolite, granulite	2Grs+Prp+Prg+18Qtz = 6An+3Ab+3Tr	B	Kohn & Spear (1989)
Crt-Rt-Als-Ilm-Qtz	amphibolite, granulite	Alm+3Rt = 3Ilm+Als+Qtz	B	Bohlen et al. (1983)
Grt-Rt-Pl-Ilm-Qtz	amphibolite, granulite	Grs+2Alm+6Rt = 3An+6Ilm+3Qtz	B	Bohlen & Liotta (1986)
Grt-Als-Pl-Qtz	amphibolite, granulite	Grs+2Als+Qtz = 3An	B	Koziol & Newton (1988)
				McKenna & Hodges (1988)
Grt-Opx-Pl-Qtz	granulite	Grs+2Alm+3Qtz = 3An+3Fs	B	Faulhaber & Raith (1991)
Grt-Opx-Pl-Qtz	granulite	2Grs+Prp+3Qtz = 3An+3Di	B	Eckert et al. (1991)
Grt-Opx-Cpx-Pl-Qtz	granulite	Alm+Hd+Qtz = 2Fs+An	B	Paria et al. (1988)
Grt-Spl-Sil-Qtz	granulite	Alm+Sil = 3Hc+5Qtz	B	Bohlen et al. (1986)

(Worley & Powell, 1998a, b; White *et al.*, 2002; Wei *et al.*, 2004). Several methods have been used for calculating mineral equilibria involving solid solutions. Spear & Cheney (1989) used the Gibbs method, THERMOCALC solves for a set of simultaneous non-linear equations and Perple_X focusses on defining the minimum free energy surface.

Powell *et al.* (1998) presented a tutorial regarding the use of THERMOCALC, which is an excellent starting point for using this approach and has been adapted below. The calculation of phase diagrams involves the following steps.

(1) *Choose the model system in which to perform the calculations.* The model chemical system is usually specified in terms of oxides, in terms of which the equilibria to be calculated can be represented. This specifies which phases, and their solid solutions, will be able to be considered. The phases can only involve end-members that occur in the thermodynamic dataset (or linear combinations of them). As outlined in Chapter 2, the model chemical system normally used to consider metapelitic equilibria is K_2O-FeO-MgO-Al_2O_3-SiO_2-H_2O (KFMASH; Thompson, 1957) allowing most of the critical minerals and the $FeMg^{-1}$ and (Fe,Mg)Si $Al^{-1}Al^{-1}$ (Tschermak's) substitution to be considered. An expansion of this system to include the effect of MnO was completed by Wei *et al.* (2004) and the effects of Na_2O, CaO, Fe_2O_3 and TiO_2 on phase relations are currently being evaluated (White *et al.*, 2007).

(2) *Formulate the thermodynamics of the phases in the system.* The activity-composition (a-x) relationships of the phases are needed in algebraic form, in terms of the compositional variables, to perform the calculations. The first stage of this is matching the required substitutions in each phase with end-members in the thermodynamic dataset, and assigning compositional variables to the substitutions. For the example of KFMASH, two variables that account for the range of $FeMg^{-1}$ and (Fe,Mg)Si $Al^{-1}Al^{-1}$ are used. A mixing model needs to be chosen for each phase.

(3) *Decide on which phase diagrams are to be calculated*, which depend on problem being addressed. The following important types of diagrams are employed.

  (a) *P-T projections* show the stable invariant points and univariant reactions for all the bulk compositions in the system. A *P-T* projection for KFMASH for both muscovite and melt-bearing equilibria, is shown in Fig. 2.8. Such projections form the basis for many other diagrams and are commonly referred to as petrogenetic grids. Many such grids have been drawn in a qualitative manner using petrographic experience (e.g., Harte & Hudson, 1979), as shown in Fig. 1.10, as well as using experimentally based data, coupled with petrographic experience (Fig. 1.8), which may be inadequate to reveal accurate phase equilibria. For example, THERMOCALC is unable to distinguish stable from metastable equilibria, and skill in Schreinemakers analysis (Niggli, 1954; Korzhinskii, 1959; Zen, 1966; Yardley, 1989) is critical in using this software to construct phase diagrams.

  (b) *Compatibility diagrams* show the mineral assemblages and ranges of solid solutions at specified *P-T* conditions, for all of the bulk compositions in the model system. AFM compatibility diagrams for a

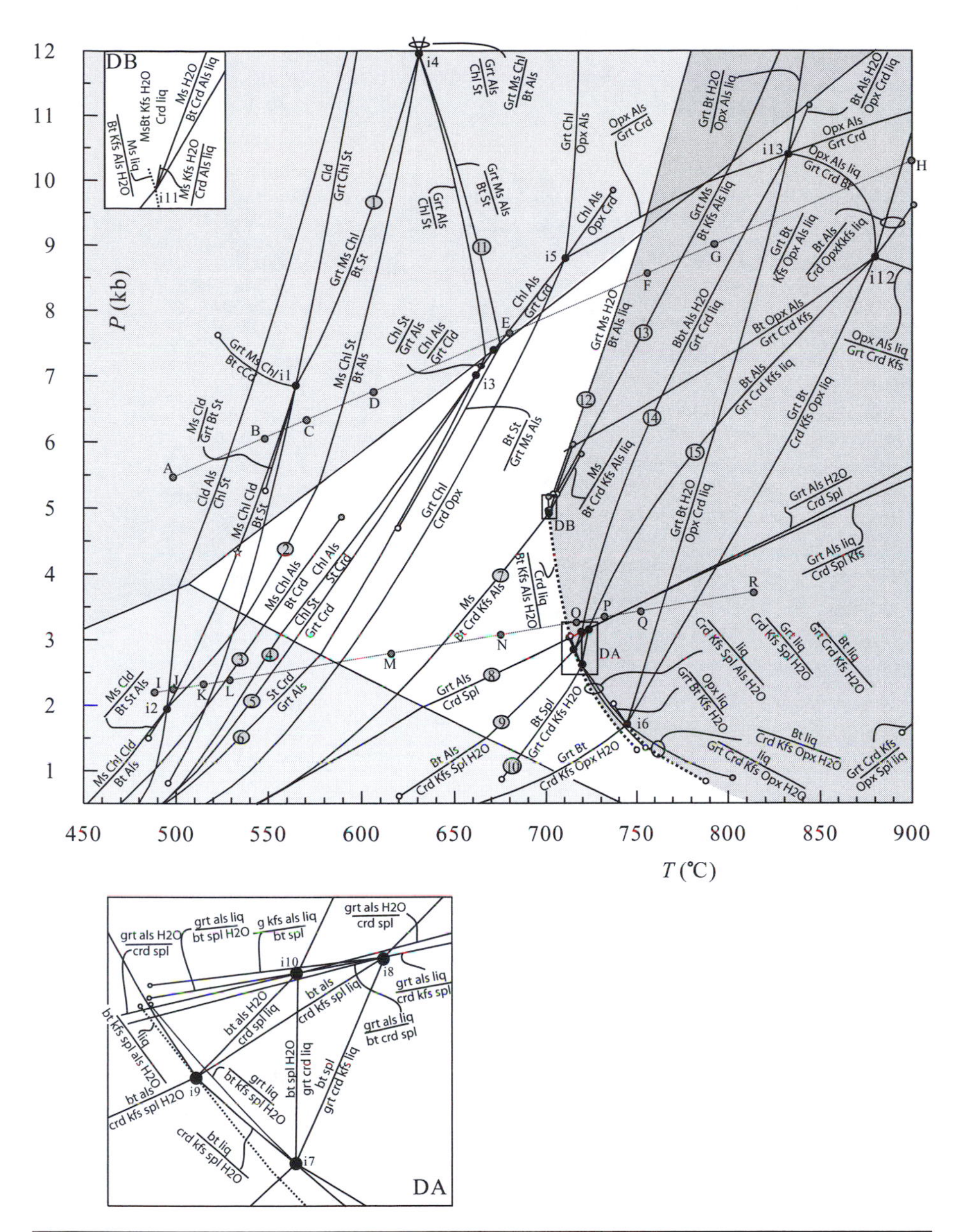

Fig. 2.8 *P-T* projection for the system KFMASH, with quartz and H_2O in excess for subsolidus conditions, and for just quartz in excess for supersolidus conditions (pale red); modified from Wei *et al.* (2004, fig. 2.1). Solid dots with labels i1 to i13 refer to invariant points, and grey-shaded circles identify reactions discussed by Wei *et al.* (2004). Unfilled stars indicate the location of singularities, where one or more phases swap sides of the reaction. Smaller open dots refer to points invariant in KFASH, and smaller grey dots to points invariant in KMASH. Heavy dotted lines represent the H_2O-saturated solidus in KFMASH. Larger grey dots with letters A-R are *P-T* locations of the compatibility diagrams shown in Fig. 2.9. Inset DA shows detail around invariant points i7 to i10, and inset DB shows detail around invariant point i11. For the thermodynamic data used to construct the diagram, the andalusite field (yellow) does not intersect the solidus, but the andalusite-sillimanite transition is now generally considered to plot at higher pressure, such that partial melting of metapelitic rocks can occur in the andalusite field (Vernon *et al.*, 1990). For a colour version of this figure, please go to the plate section.

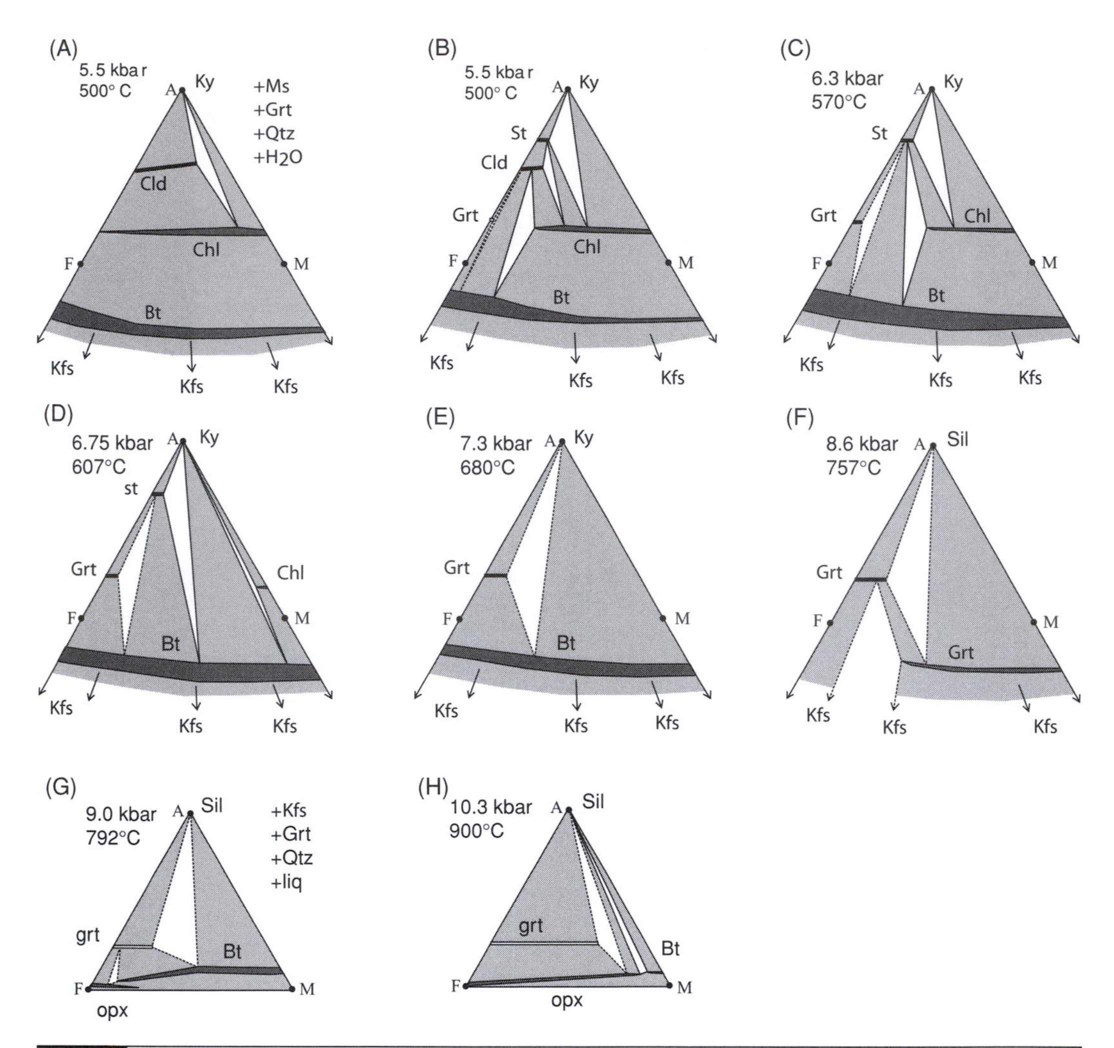

Fig. 2.9 AFM compatibility diagrams calculated for the locations A to H labelled in Fig. 2.8 and their *P-T* conditions; modified from Wei *et al.* (2004, fig. 2.3). A–F are projected from muscovite, quartz and H_2O, whereas melt is substituted for H_2O for G and H (supersolidus conditions), and K-feldspar is substituted for muscovite at the stage when muscovite + quartz have reacted out.

succession of *P-T* points approximating a Barrovian-style geotherm on Fig. 2.8 are shown in Fig. 2.9 (from Wei *et al.*, 2004).

(c) *P-T pseudosections* show the phase relationships for a particular whole-rock composition. Examples of *P-T* pseudosections are shown in Chapter 1 for metamorphosed mid-ocean ridge basalts (Fig. 1.16A) and metapelitic and metapsammitic rocks (Figs. 1.16B, C).

(d) *T-X or P-X pseudosections* show the phase relationships for a particular bulk composition line at specified *P* or *T*, respectively. Fitzherbert *et al.* (2003) modelled the effect of variations in basaltic protolith composition on phase equilibria in the lawsonite-blueschist to eclogite transition in New Caledonia (Fig. 2.10).

Pseudosections provide an excellent basis for evaluating mineral equilibria and microstructures because they represent reactions, including any variations in solid solution, relevant to the chosen

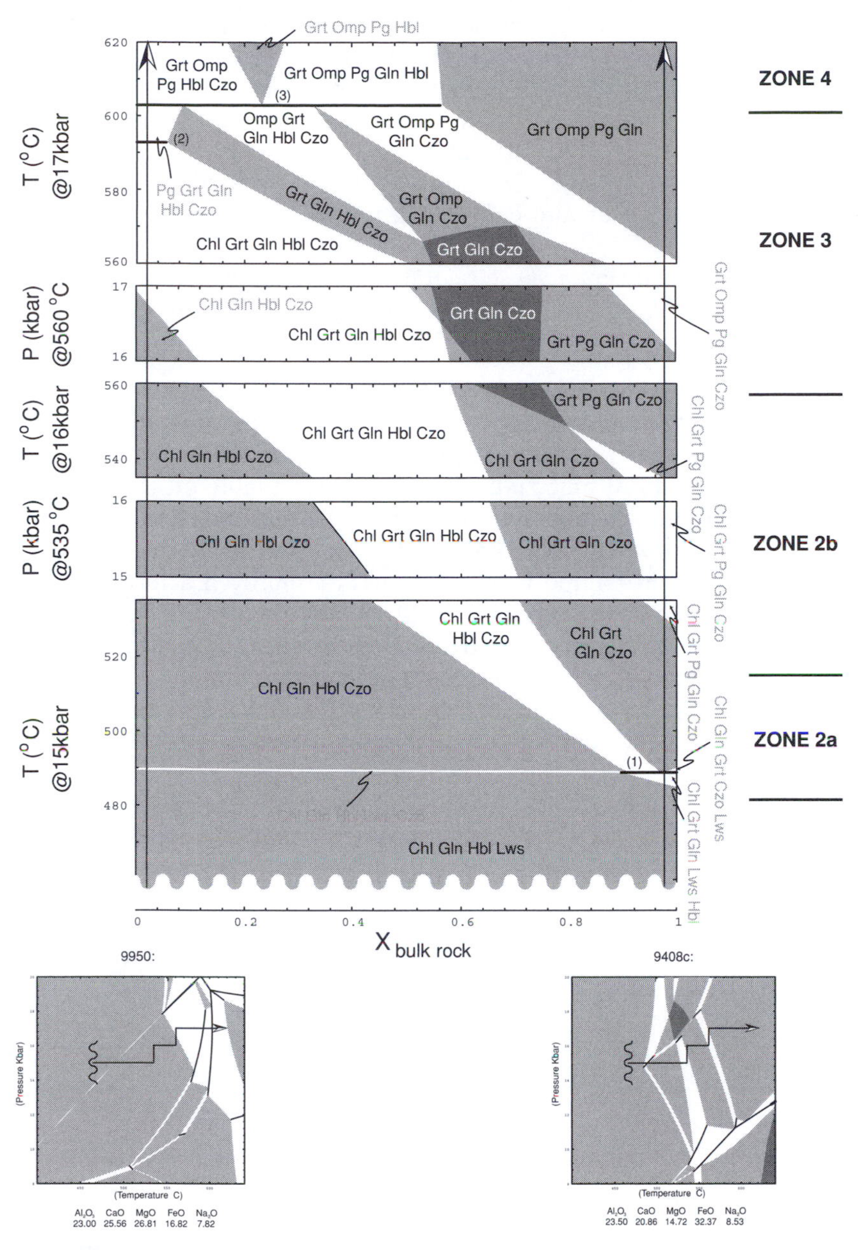

Fig. 2.10 Composite P-$X_{\text{bulk rock}}$ and T-$X_{\text{bulk rock}}$ pseudosection in CNFMASH, with H_2O in excess, for the spectrum of bulk-rock compositions between two metabasalts (9950 and 9408c) from New Caledonia; after Fitzherbert *et al.* (2003, fig. 11), with permission of Oxford University Press. The pseudosection is drawn for the stepwise trajectory shown in the lower diagrams. At $X_{\text{bulk}} = 0$, the bulk-rock composition is appropriate to that of sample 9950, and at $X_{\text{bulk}} = 1$ the bulk-rock composition is that of sample 9408c. Univariant reactions in this compositional range are shown as heavy solid lines, and include: (1) Grt + Hbl + Lws = Chl + Gln + Czo, (2) Chl + Gln + Czo = Grt + Pg + Hbl and (3) Gln + Czo = Grt + Omp + Pg + Hbl. Divariant fields are unshaded five-phase regions, trivariant fields are shaded four-phase fields, and quadrivariant fields are deeply shaded three-phase regions.

whole-rock composition. *P-T* conditions related to mineral assemblage formation can commonly be relatively precisely estimated for low-variance assemblages. For high-variance assemblages that commonly cover broad *P-T* domains, information can be combined from the modelling of several rock compositions (e.g., Kelsey *et al.*, 2003). Further refinements can be made by plotting or model mineral compositional data in broad *P-T* fields and comparing modelled results with observational data.

(4) *Build up the phase diagram via calculations of the equilibria involved.* Generating a phase diagram usually involves many computationally-intensive calculations that are performed using computer software.

Becoming familiar with the use of the specialist software required to produce metamorphic phase diagrams is best done by downloading the appropriate software using the links given at the start of this section, reading the relevant tutorial material, which is also available on the website, and then completing the example problems. The methods are constantly being updated, as new experimental results improve thermodynamic data and models for activity-composition relations are improved. Thermobarometry and calculated phase diagrams need to be applied with a clear understanding of field, petrographic and mineral chemical relationships; all methods should be used with caution, common sense and an understanding of technique limitations and uncertainties. It may be possible to attend training workshops that are held occasionally, where these issues can be explored more fully.

Chapter 3

Tectonothermal History of Metamorphic Terranes

3.1 Introduction and precautions

Among the main aims of metamorphic geology are: (1) to infer time sequences of mineral assemblages ('metamorphic events', 'M'), and (2) to relate the time of growth of metamorphic mineral assemblages to the time of formation of a specific structure, such as a foliation (S), lineation (L) or fold set (F), with implications for a deformation event (D). These aims are discussed in Sections 3.2 and 3.3, respectively. The relative chronology inferred from assemblage and structural considerations may be made absolute by applying microscale radiometric techniques (Section 3.7).

Two microstructural approaches used for inferring time sequences of mineral assemblages are: (1) inferring a set of minerals that grew simultaneously (a 'mineral assemblage' or 'paragenesis') and (2) inferring metamorphic reactions. Inferring a mineral assemblage partly depends on an appreciation of low-energy grain shapes and configurations, as discussed in the next section. Criteria for inferring a mineral assemblage, as well as problems involved, are discussed in Section 2.3.

3.2 Metamorphic grain shapes

3.2.1 Low-energy grain shapes

During metamorphism, grains of new minerals grow as products of metamorphic reactions, in an attempt to minimize the *chemical free energy* of the system. The first step is nucleation of new (lower energy or more stable) minerals. This involves (1) assembly of the required atoms by diffusion inside the old minerals, (2) probable change of the atomic arrangement into one or more unstable intermediate structures ('activated complexes'), and (3) nucleation of a new stable or metastable mineral. The production of a viable nucleus is followed by growth, forming new grains (neocrystallization), which involves (a) diffusion of the required chemical components through the old minerals and through any fluid between them, (b) transfer of these components across the boundary into the new mineral, (c) simultaneous diffusion of unwanted components from the reaction site.

Nucleation may be *epitaxial* ('epitactic'), which means that the atomic structure of the new mineral is closely related to the atomic structure of a pre-existing mineral. An example is andalusite replaced by sillimanite.

Fig. 3.1 Polygonal aggregate of albite, Broken Hill, Australia. From Vernon (2000b, fig. 122) and Vernon (2004, fig. 3.4). Crossed polars; base of photo 3.4 mm.

However, most metamorphic reactions are more complicated, not obviously pseudomorphous, and involve several minerals.

Once the grains of stable minerals impinge on one another, filling the volume of rock with the new minerals, the chemical free energy of the system is reduced to a minimum. At this stage, the grainsize and grain shapes (but not the minerals) change, so as to minimize the *interfacial free energy* (γ) of the system, the process being known as *grain growth*. Though the interfacial free energy is much smaller than the chemical free energy, it nevertheless controls the final shapes of the grains. The process involves either: (1) reduction of the *interface area* to a minimum, which reduces the total proportion of atoms in high-energy sites (namely in grain boundaries, as opposed to those in lower energy sites in the more regular, internal parts of grains), producing polygonal (*granoblastic*) aggregates (Figs. 3.1, 3.2), or (2) the formation of *low-energy crystal faces* (*idioblastic* crystals), as shown in Figs. 3.3 and 3.4. Process (1) occurs mainly in minerals that have 'relatively isotropic' crystal structures (though not completely structurally isotropic and not necessarily optically isotropic), such as quartz, feldspar, calcite, scapolite, olivine, chromite, magnetite, galena, pyrite and ice, whereas process (2) occurs in minerals with strongly anisotropic structures, such as the sheet silicates (e.g., mica, chlorite), hematite, molybdenite, graphite and sillimanite. For example, {001} in mica and {110} in sillimanite commonly dominate the microstructure in metapelitic rocks (Fig. 3.3). The microstructure of aggregates of structurally anisotropic minerals appears to be the result of mutual *impingement*, rather than mutual *adjustment* of interfaces (Vernon, 1976), so that local dominance of low-energy faces of one grain or the other depends on accidents of nucleation and impingement. The predominance of crystal faces implies their greater stability, relative to random boundaries, in minerals of this type. However, some minerals (e.g., garnet, staurolite and andalusite) develop crystal faces in many rocks, but not in others, indicating that crystal structure is not the

Fig. 3.2 Polygonal aggregate of plagioclase, clinopyroxene and orthopyroxene in a granulite facies metabasite. Broken Hill, Australia. From Vernon (2004, fig. 3.23). Crossed polars; base of photo 4 mm.

only factor controlling the development of crystal faces in metamorphic rocks. The processes involved in attaining minimum-energy grain shapes have been discussed by Voll (1960), Kretz (1966a) and Vernon (1968, 1976, 1999, 2004).

Growth during deformation can give rise to grain-boundary irregularities, elongate shapes and strong preferred orientations, though the interfacial angles nevertheless may tend towards low-energy configurations. Minimum-energy grain shapes are best shown by high-grade metamorphic

Fig. 3.3 Aggregate of muscovite dominated by low-energy {001} crystal faces. Broken Hill area, Australia. From Vernon (2000b, fig. 125) and Vernon (2004, fig. 3.13). Crossed polars; base of photo 1.75 mm.

Fig. 3.4 Aggregate of quartz and muscovite in amphibolite facies rock, Grenville Province, Ontario, Canada. The microstructure is dominated by low-energy {001} crystal faces of the muscovite. Though the quartz appears to be 'interstitial' to the muscovite, both minerals grew simultaneously. From Vernon (2004, fig. 3.27). Crossed polars; base of photo 2 mm.

rocks, because the process is favoured by prolonged heating. At low grades of metamorphism, these ideal shapes may not be achieved.

3.2.2 Low-energy shapes of inclusions

Inclusions typically belong to the same mineral assemblage as the host mineral, and if so, the inclusion and host mineral are chemically compatible. Given sufficient time and heat, the interfacial energy of the inclusion-host boundary tends to a minimum, as it does in the surrounding aggregates of independent grains, because the interface between an inclusion and its host mineral is a true grain boundary. If both inclusion and host are structurally isotropic, the minimum-energy shape is a sphere (Kretz, 1966a, 1994; Vernon, 1968, 1970, 1976, 2004), which has the minimum surface area per unit volume, and if both minerals are not too structurally anisotropic, spherical or elliptical shapes are approached (Fig. 3.5). Even where the inclusions are elongate and aligned (forming 'inclusion trails'), they tend towards rounded shapes, as discussed in Section 3.3.2. Common examples are quartz inclusions in feldspar or cordierite. Inclusions develop crystal faces and are elongate if the included mineral has a strongly anisotropic crystal structure, for example, sillimanite or biotite in K-feldspar (e.g., Vernon, 1968, 1976, 2004), as shown in Fig. 3.5.

3.3 Porphyroblasts

3.3.1 Nucleation and growth of porphyroblasts

As with igneous rocks, the size of isolated crystals in metamorphic rocks depends on the ratio of nucleation rate (N) to growth rate (G). If abundant

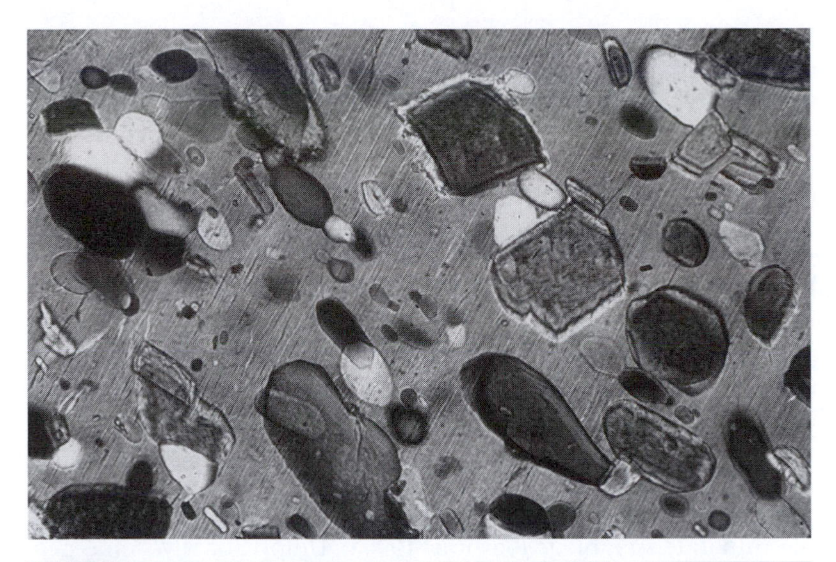

Fig. 3.5 Inclusions of quartz and biotite in K-feldspar, Cooma Complex, south-eastern Australia. The quartz inclusions tend towards spherical shapes, and quartz–feldspar interfaces show dihedral angles where they meet quartz–quartz interfaces. Sections of biotite inclusions with cleavage show {001} crystal faces (commonly with rounded corners), whereas sections of biotite inclusions without cleavage have rounded shapes. From Vernon (2004, fig. 3.30). Crossed polars; base of photo 1.5 mm.

viable nuclei form, many grains are produced, whereas if nucleation is difficult, few grains grow and well separated, large grains (*porphyroblasts*) are formed. The nucleation rate depends on temperature, as well as on minerals available to act as nucleating agents. For example, nucleation may be assisted by crystallographic ('coaxial' or 'epitaxial') relationships between new and old grains (e.g., Bosworth, 1910; Chinner, 1961; Spry, 1969; Brearley, 1987; Vernon, 1987a; Worden *et al.*, 1991; Putnis, 1992), as illustrated in Figs. 3.2 and 3.3.

Growth of porphyroblasts requires diffusion of chemical components over larger distances than the average grainsize. Crystal size distributions and compositional zoning patterns indicate that nucleation and growth rates of garnet porphyroblasts are controlled by rates of transport of chemical components through intergranular fluid (Denison & Carlson, 1997; Hirsch *et al.*, 2000). Detailed analyses of compositional zoning, grain shape, grainsize and spatial distribution have shown that garnet crystals nucleate continuously and randomly, as the linear growth rate of all the garnet crystals (regardless of size) remains constant (Kretz, 1993, 2006; Daniel & Spear, 1999). With increasing temperature, the garnet nucleation rate tends to increase slowly, then stays nearly constant, and then declines (Kretz, 2006).

Porphyroblasts generally grow at the same time as minerals in the surrounding matrix, as products of the same prograde metamorphic reaction. No time difference is implied by a grainsize difference, as explained in detail by Vernon (1977). Porphyroblasts develop especially in metapelites. During the progressive metamorphism of shale to slate, phyllite, schist and gneiss, many nuclei are available for quartz and mica, which consequently form a relatively fine-grained matrix (high ratio of nucleation rate to growth rate). However, when minerals such as cordierite, andalusite, garnet and staurolite are produced by prograde reactions, few or no suitable minerals to act as nuclei are present, with the result that only a few nuclei can form, and therefore these minerals typically grow as porphyroblasts.

At the same time as the porphyroblasts grow, mica and quartz in the matrix may continue to grow, as required by the relevant prograde metamorphic reactions, but remain as much smaller grains, owing to an abundance of suitable nuclei. Even at the highest metamorphic grades, readily nucleated minerals such as quartz, biotite and sillimanite may remain relatively fine-grained, commonly occurring as inclusions in, or as folia anastomosing around large grains of the other minerals, such as garnet, cordierite and K-feldspar (Section 3.4.1). One of the reasons for the fine grainsize of sillimanite may be that it appears to nucleate easily in mica (Chinner, 1961; Vernon, 1979, 1987a).

Both fine-grained and coarse-grained minerals can belong to the same compatible mineral assemblage (Vernon, 1996a). Unfortunately, some people have made the unjustified assumption that the minerals in finer-grained folia form later than the porphyroblasts around which the folia anastomose, whereas commonly certain minerals (e.g., sillimanite and mica) grow simultaneously in the folia because of their ability to nucleate readily and survive under conditions of strong deformation. This is an example of *deformation partitioning* (Section 6.15), as emphasized by Bell & Rubenach (1983).

The common preservation in porphyroblasts of *inclusion trails* outlining pre-existing microstructures, such as microfolds and bedding (Section 3.3.2) indicates that porphyroblasts replace an equal volume of old matrix minerals. This is also implied by uncommon porphyroblasts that preserve the shapes of former matrix mica grains, as discussed in relation to prefoliation porphyroblasts in Section 3.5.3. Of course, this does not mean that the entire metamorphic reaction is restricted to the volume now occupied by the porphyroblast, as chemical components are exchanged with other grains that are dissolving and growing throughout the rock. The growth of matrix minerals may or may not involve equal-volume replacement of old minerals, depending on local mass transfer and mineral density changes. This may apply also to porphyroblasts without inclusion trails.

3.3.2 Inclusions in porphyroblasts

Owing to their relatively low ratios of nucleation rate to growth rate, porphyroblasts overtake and enclose smaller grains (*inclusions*) of matrix minerals (Fig. 3.5). Some inclusions may be relics inherited from lower grades of metamorphism, but generally most belong to the same metamorphic assemblage as the porphyroblast (Fig. 3.5). That is, the inclusions generally are either products of, or excess reactants in the prograde reaction that produced the porphyroblast. Therefore, many or most of them grow at the same time as (but at a much slower rate than) the porphyroblast that engulfs them. Consequently, the inclusions do not react chemically with the porphyroblast, but tend to change their shapes to minimize the interfacial free energy of their boundaries with the porphyroblast (Kretz, 1966a, 1994; Vernon, 1968, 1970, 1976, 2004), as mentioned previously. Evidence of this mutual grain-boundary adjustment confirms that the two minerals were chemically compatible under the prevailing metamorphic conditions.

Inclusions of minerals inherited from a previous assemblage may be either (1) stable or metastable relics (e.g., quartz or graphite) or (2) minerals from an unstable earlier assemblage (absent from the matrix) trapped in the porphyroblast and hence isolated from further reaction with matrix minerals. It is important to distinguish between these alternatives. A reliable interpretation of inclusions as unstable or metastable relics can be made only if the minerals occurring as inclusions do not occur anywhere in the matrix, and if they are compatible with paragenetic interpretations made on the basis of other evidence, such as chemography based on experimentally controlled data (Chapter 2). Such inclusions have been used to unravel the metamorphic history (Krogh, 1982; St-Onge, 1987). Radiometric dating of the inclusions (e.g., using monazite; see Section 3.7.5) compared with the dating of matrix minerals, would provide strong supporting evidence.

3.4 Inferring sequences of mineral assemblages ('metamorphic events')

3.4.1 Concept of 'events'

The first problem is to decide what a 'metamorphic event' (typically labelled 'M') really is. Generally it is thought of as a single heating cooling episode, and if clear evidence of two such episodes, well separated in time, can be recognized, they are generally labelled 'M1' and 'M2'. The question is how to ensure that these 'events', as well as an inferred sequence of deformation 'events' (Section 3.5), could not equally well be regarded as progressive stages in a single tectonometamorphic episode. A mineral assemblage that now delineates a foliation did not necessarily grow when that foliation was initiated, and so minerals delineating 'S2' (for example) should not automatically be assigned to a metamorphism 'M2', inferred to be distinct from 'M1', which is a common practice. In addition, evidence of a microstructural change should not automatically be related to a new metamorphic event. Radiometric methods (Section 3.7) can help separate distinct heating events, as well as delineating smaller stages in the evolution of an inferred single event.

3.4.2 Criteria for inferring metamorphic reactions

Microstructural evidence is potentially useful for inferring time sequences of minerals and for inferring metamorphic reactions, but care needs to be taken, especially for the more complex reactions. Generally the approach is to identify minerals that appear to have replaced other minerals, analysing the minerals with the electron probe microanalyser, and then writing a balanced chemical equation on the basis of the volumes and chemical compositions of the minerals and volatile components (such as H_2O or CO_2) inferred to have been involved.

Incomplete metamorphic reactions enable recognition of both solid reactants and solid products, and so assist in the writing of realistic metamorphic reaction equations. This situation is relatively uncommon in prograde metamorphism, because progressive heating and continued deformation

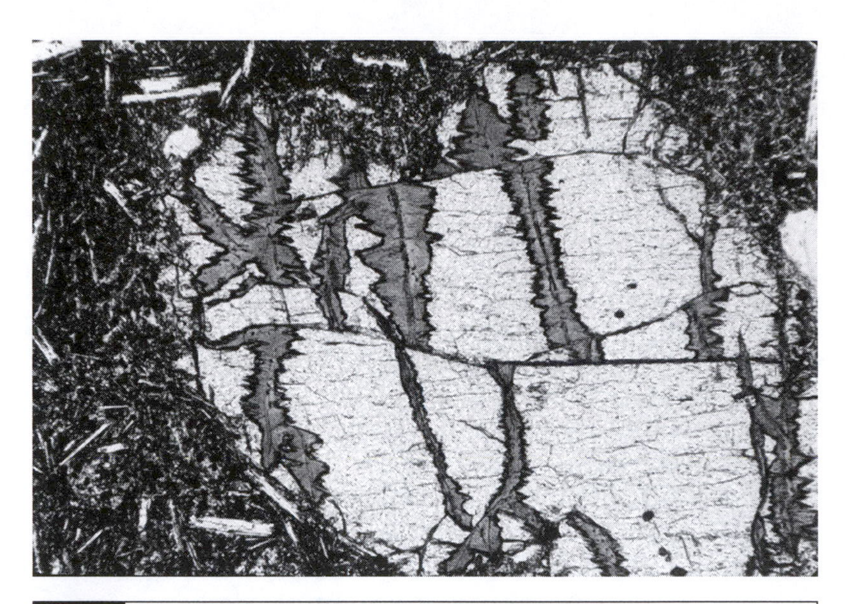

Fig. 3.6 Olivine partly replaced by chlorite along fractures in altered basalt, Mount Wilson, New South Wales, Australia. From Vernon (2004, fig. 3.70). Plane-polarized light; base of photo 1.3 mm.

tend to produce new, lower-energy grain shapes and so obliterate reaction microstructures. However, they can be preserved in prograde successions that have not been subjected to strong deformation, as in the Mount Stafford area, central Australia (Vernon *et al.*, 1990; Greenfield *et al.*, 1998). Preservation of reaction microstructures is most likely to be encountered in retrograde metamorphism. This is because falling temperature tends to reduce reaction rates, and also because most retrograde reactions need addition of H_2O or CO_2, which have to be returned to a rock that was devolatilized during prograde metamorphism, and fluid access generally requires fracture systems. This situation also applies to the low-grade metamorphism of igneous rocks, which requires the addition of H_2O to form hydrous amphibole, epidote and chlorite from anhydrous pyroxene, olivine and plagioclase, and may also involve addition of CO_2 to form carbonate minerals.

For these reasons, retrograde metamorphism is commonly incomplete and patchy. However, in high-strain deformation zones (mylonites, shear zones, retrograde schist zones; see Section 6.6), transient grain-scale openings constitute low-pressure sites that suck in fluid, and so promote retrograde reactions. Consequently, the reactions tend to go to completion in these zones, in contrast to typically incomplete reactions in adjacent, less deformed rocks (Vernon & Ransom, 1971; Corbett & Phillips, 1981; Clarke & Powell, 1991; White & Clarke, 1997; Keller et al., 2004).

Reliable microstructural criteria

The following metamorphic microstructural criteria are generally reliable as indicators of metamorphic reactions.

(1) *Partial pseudomorphism* (constant-volume replacement) is the most reliable criterion of a reaction (Vernon, 1996a), because the replaced mineral and the solid products can be observed (e.g., Guidotti & Johnson, 2002). The replacing minerals may project inwards from fractures (Fig. 3.6)

Fig. 3.7 Clinopyroxene partly replaced around its margins by hornblende in a metabasic rock, Wellington, New South Wales, Australia. From Vernon (2004, fig. 3.71). Plane-polarized light; base of photo 1.3 mm.

or from grain boundaries (Fig. 3.7), but replacement may also begin internally (Fig. 3.8). Radiating aggregates of the new mineral inside porphyroblasts of the old mineral (Fig. 3.8) are relatively reliable indicators of a reaction, because these are high-energy grain shapes that are not typical of true inclusions. Though pseudomorphs are especially common in retrograde metamorphic rocks (e.g., Ferry, 2001), they may also occur in prograde metamorphism, especially if the deformation is weak (e.g., Guidotti & Johnson, 2002).

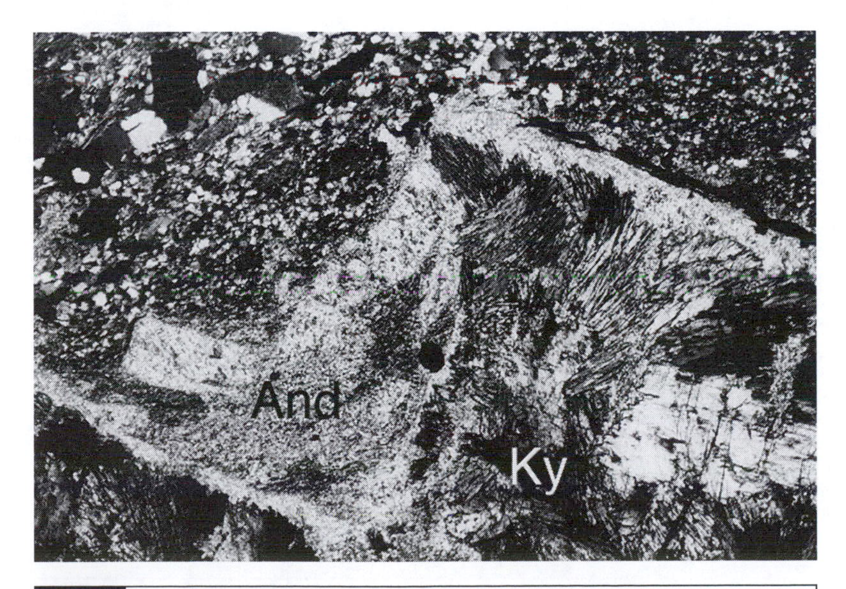

Fig. 3.8 Porphyroblast of andalusite (And) partly replaced by radiating aggregates of kyanite (Ky) in a schist, Fiordland, New Zealand. From Vernon (2004, fig. 3.72). Crossed polars; base of photo 3.4 mm.

Fig. 3.9 Former polygonal grains of olivine that have been completely replaced (pseudomorphed) by serpentine in an Archaean serpentinized peridotite. Western Australia. From Vernon (2004, fig. 3.73). Crossed polars; base of photo 1.3 mm.

Even where replacement is complete, former grain shapes may be preserved (Fig. 3.9), as indicated by (1) shapes of former euhedral phenocrysts, (2) preservation of inclusions with boundaries that show evidence of solid-state microstructural adjustment with the former host mineral (Kretz, 1966a; Vernon, 1968, 1976, 2004), now completely replaced by finer-grained aggregates that could not have equilibrated microstructurally with the much larger inclusions (Vernon, 1978b, fig. 1B; Gibson, 1992, fig. 4d), as shown in Fig. 3.10, (3) shapes of former polygonal grains now completely replaced by much finer-grained aggregates, which could not have produced such coarser-grained polygonal shapes by interface adjustment, because their individual grains are far too small (Fig. 3.9), and

Fig. 3.10 Cordierite replaced by fine-grained, retrograde mica-chlorite aggregate, leaving the original shapes of quartz and biotite inclusions intact. Plane-polarized light; base of photo 1.75 mm.

Fig. 3.11 Sillimanite (Sil) partly pseudomorphing andalusite (And), Chiwaukum Schist, Cascade Range, Washington, USA. All the sillimanite sections are in the same crystallographic orientation (possibly even parts of a single dendritic grain), and the relationship between both minerals is coaxial (epitaxial), as discussed by Vernon (2004, pp. 174–5).

(4) crystallographically controlled ('epitaxial', 'coaxial') replacement of one mineral by another, such as biotite by sillimanite (Chinner, 1961) and andalusite by sillimanite (Bosworth, 1910; Vernon, 1987a; Gibson, 1992, fig. 5e), as shown in Fig. 3.11.

Other examples of pseudomorphous reactions are (1) aggregates of kyanite replacing andalusite (Ward, 1984; Brown, 1996, fig. 7), shown in Fig. 3.8, (2) aggregates of kyanite and staurolite replacing andalusite (Evans & Berti, 1986), and (3) aggregates of epidote + margarite + quartz + plagioclase replacing lawsonite (Droop, 1985; Selverstone, 1993), all of which have been used to infer a pressure increase. Pseudomorphs of muscovite after staurolite and of biotite + sillimanite replacing garnet have been described by Guidotti & Johnson (2002). Another common example is the serpentinization of olivine in hydrated peridotites (Fig. 3.9). Aggregates of wollastonite replacing aggregates of calcite, by reaction with quartz, may preserve the curved shapes of fossil shells (Fig. 3.12).

(2) *Coronas* (reaction rims) provide excellent examples of incomplete replacement of earlier minerals (Figs. 2.3, 3.13, 3.14, 3.15, 3.16). The incomplete reaction could be due to (a) slower reaction rates with lowering of temperature, (b) the thickening coronas limiting volume diffusion, thereby 'armouring' the residual cores from further reaction, and (c) scarcity of fluid for access by diffusion of nutrient components and removal of waste components in the reaction. Factors (a) and (b) are kinetic, whereas factor (c) concerns the control of the chemical composition of the local reacting system. For example, if H_2O is a reactant, apparently incomplete reactions may actually have gone to completion in terms of the composition of the local system, if that system contains insufficient fluid; in other words, the reaction need not have been stopped for kinetic reasons.

Fig. 3.12 Aggregates of elongate wollastonite grains that have pseudomorphed calcite in fossil shells, preserving the curved outlines of the shells, in a hornfels, Hartley, New South Wales, Australia. From Vernon (2000b, fig. 120) and Vernon (2004, fig. 3.38). Crossed polars; base of photo 1.8 cm.

Partly replaced minerals in the cores of these structures are taken to represent part of a primary paragenesis, and minerals in the corona are taken to represent all or part of a later paragenesis. The implication is that equilibration volumes during the reaction are smaller than the original grainsize (Clarke & Powell, 1991). Many coronas are composed of symplectic intergrowths (Figs 3.14), which result from simultaneous growth of

Fig. 3.13 'Colonies' of quartz-plagioclase symplectite ('myrmekite') replacing a porphyroclast (former phenocryst) of microcline from its margins inwards, forming a type of 'corona' structure, in a felsic mylonite (strongly deformed granite), near Lago Maggiore, northern Italy. From Vernon (2004, fig. 3.57). Crossed polars; base of photo 3.4 mm.

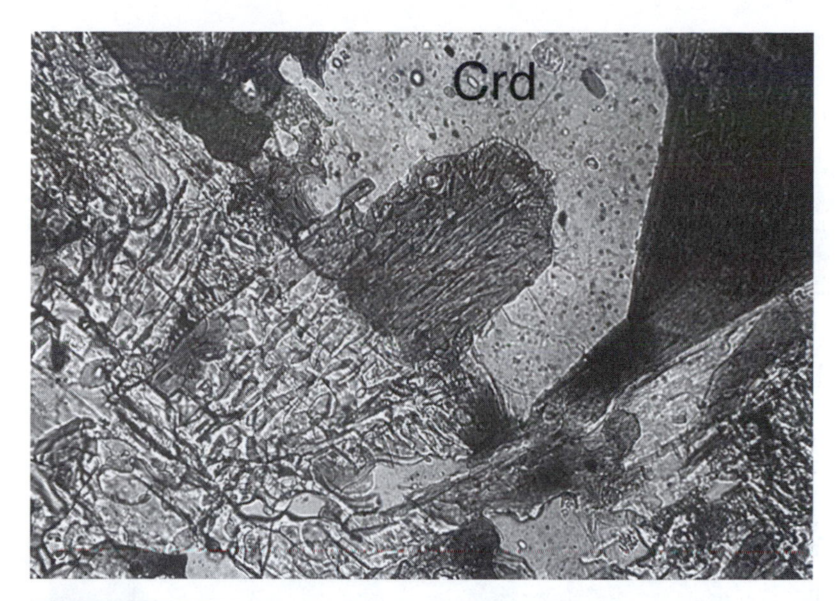

Fig. 3.14 Symplectic intergrowths of quartz-andalusite (bottom left) and quartz-biotite (centre) that formed as a result of the reaction: cordierite + K-feldspar = quartz + andalusite + biotite, in an amphibolite facies metapelite, Cooma Complex, south-eastern Australia. The quartz rods pass continuously from the andalusite into the biotite. The symplectic aggregates replaced both reactant minerals, though this photo shows clearly only the local replacement of cordierite (Crd). From Vernon (2004, fig. 3.58). Crossed polars; base of photo 3.4 mm.

two or more minerals (Fig. 3.15). Some coronas replace an earlier mineral (Figs. 2.3, 3.13), whereas others replace two minerals as they react along their grain boundaries (Figs. 3.15, 3.16).

Corona structures are preserved best in volumes of rock that do not undergo deformation during the reaction, or at least, develop in areas of a deforming rock that are protected from deformation. Therefore, coronas may provide information about reactions in adjacent rocks, from which deformation has removed the evidence of incomplete reactions (e.g., Clarke & Powell, 1991). Even so, deformation may assist transfer of chemical components to and from replacive symplectite and corona reaction sites (e.g., Simpson & Wintsch, 1989; Vernon, 1991b, 2004).

Coronas are potentially useful for inferring metamorphic reactions, because both the original mineral(s) and solid product minerals are observable, from which reactions and *P-T* histories have been inferred (Ellis *et al.*, 1980; Droop & Bucher-Nurminen, 1984; Clarke & Powell, 1991). However, care should be taken in the interpretation of reaction coronas, for the following reasons. (1) Inferring that the minerals of the cores and coronas were the only ones involved in the reaction can be justified only by ensuring that other minerals in the rock (if any) were not involved (Passchier & Trouw, 1996, p. 194). (2) Many, if not most, symplectic coronas reflect disequilibrium. However, some coronas reflect changes in the volumes of minerals as *P-T* conditions change within the stability field of a single mineral assemblage; i.e., all the minerals remain chemically compatible (White *et al.*, 2002). (3) Zoning in cores conceivably could complicate the reaction, especially if the zoning follows the outline of the relic, though

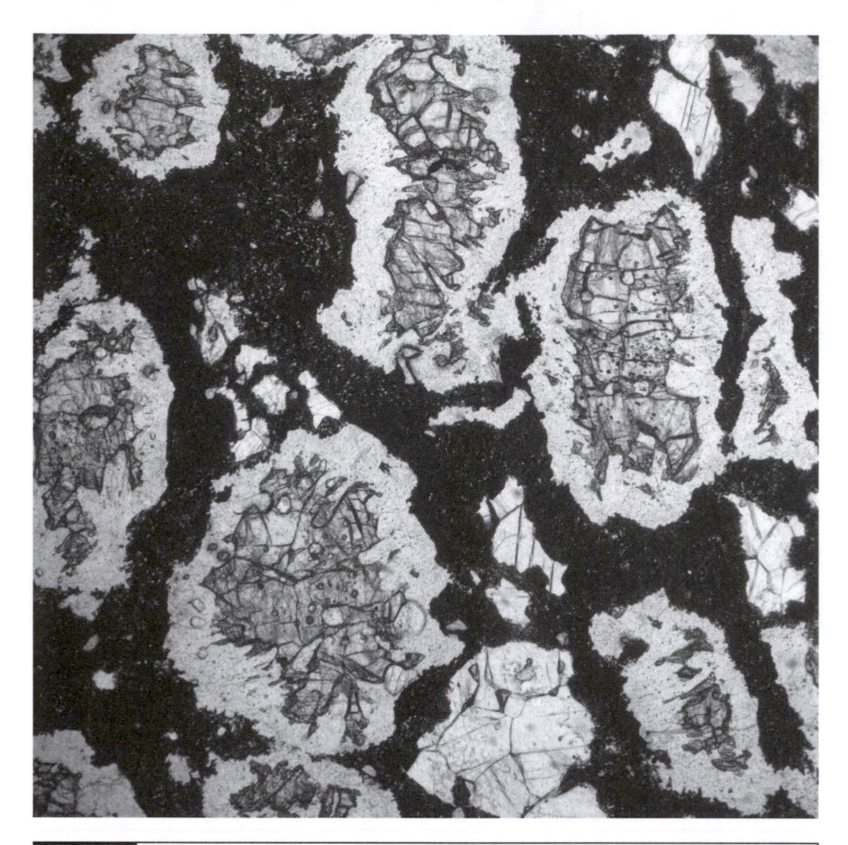

Fig. 3.15 Reaction coronas between partly replaced garnet relics in a granulite facies rock at Cohn Hill, western Musgrave Ranges, central Australia. The corona developed a result of the inferred reaction: Grt + Sil + Bt = Crd + Spl + Ilm, which occurred in response to decompression resulting from uprooting of the terrane from peak metamorphic conditions of 8 kbar to 4 kbar (Clarke & Powell, 1991). See Fig. 2.2 for X-ray intensity maps of these corona structures. Plane-polarized light; base of photo 3.5 mm.

if the zoning is truncated by the edge of the relic, it may be inferred to be primary (Passchier & Trouw, 1996, p. 194). (4) Factors such as strain, nucleation and growth rates, and changing volumes of equilibrium, may control the proportions of minerals produced in the intergrowths (White *et al.*, 2002).

Some coronas are double (Fig. 3.13) or multiple. Their microstructures are controlled by: (1) the chemical reaction(s) involved and the consequent chemical potential gradients developed, (2) the rate(s) of the reaction(s), (3) the relative rates of intergranular and intragranular diffusion of the chemical components involved in the reaction(s), (4) whether or not components are added or removed from outside the volume occupied by the obvious reactants, and (5) the temperature and pressure and their variation during growth of the corona (White & Clarke, 1997). For example, complex multiple coronas may develop from an apparently simple reaction between two minerals, owing to differing rates of diffusion of migrating chemical elements (Grant, 1988; Johnson & Carlson, 1990). Moreover, in multiple coronas, the layers may not all grow at the same time, because changes

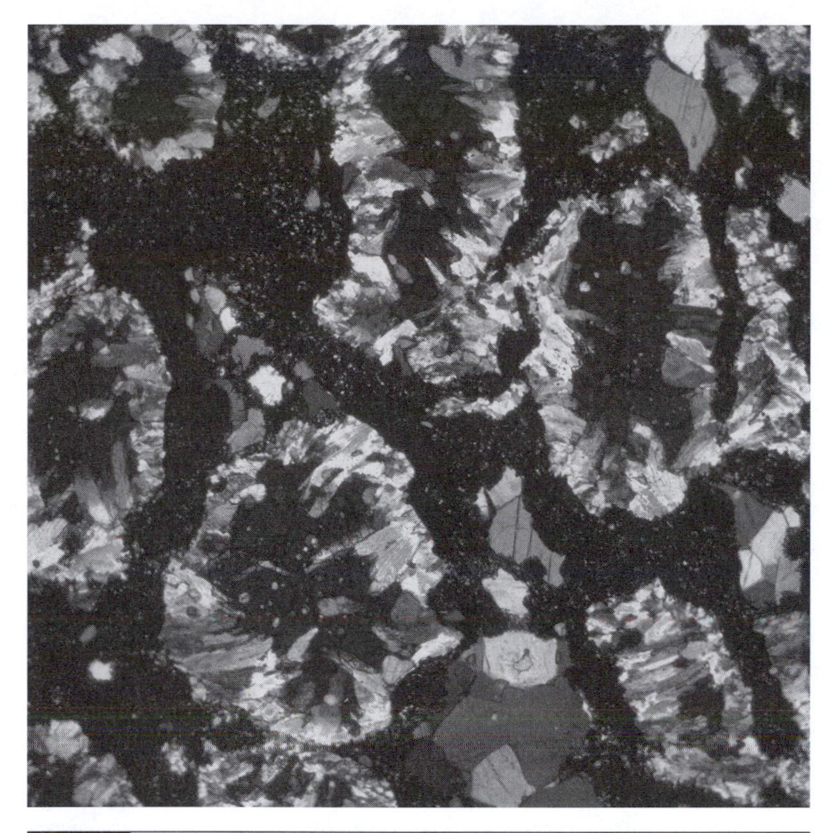

Fig. 3.16 Same field of view as shown in Fig. 3.15. Crossed polars; base of photo 3.5 mm.

in temperature and pressure can cause reaction between two early layers to form a younger intermediate layer (Griffin, 1972; Indares, 1993).

A few prograde reactions involve pseudomorphism of one mineral by another or by an aggregate of minerals of the same composition, suggesting a possible reaction, for example, replacement of andalusite by sillimanite. However, though we write equations and refer to them as 'reactions', the reaction mechanisms in most prograde metamorphic changes are harder to infer, even where pseudomorphism is involved, because components external to the pseudomorph are commonly also involved (Vernon & Pooley, 1981; Likhanov & Reverdatto, 2002; Guidotti & Johnson, 2002).

A net change from one mineral to another (e.g., kyanite → sillimanite) can be the result of several linked, ionic sub-reactions or partial reactions, involving diffusion of chemical components in fluid from one local reaction site to another, the partial reactions combining to form an overall 'cyclic' reaction (Carmichael, 1969; Yardley, 1977a), as discussed in Section 1.7. Detailed studies by Foster (1977, 1981, 1983, 1986, 1999), Waters (2001) and Likhanov & Reverdatto (2002) have shown that net reactions occurring in volumes as small as 1 cm^3 may be the result of several even more local reactions.

A distinction should be made between when a mineral becomes stable and when a particular grain of that mineral grows, because a mineral stable over a relatively large *P-T* range may nucleate new grains every time a prograde reaction that produces more of the mineral takes place. The new grains need not grow on existing grains of that mineral; for example sillimanite that forms prismatic grains at the metamorphic peak may grow as 'beards' on cordierite during cooling, while still in the sillimanite stability field (R. W. White, personal communication). Thus, a mineral becomes stable at a single point along a *P-T* path, but particular grains of that mineral may nucleate at several points along the *P-T* path while the mineral remains stable. This situation has been modelled for garnet, which nucleates continuously with increasing temperature, though the nucleation rate may fluctuate and may also vary from place to place, in response to differences in whole-rock composition, strain, diffusion rates, or ability of water to escape from reaction sites; see Vernon, (2004, pp. 206–7) for a summary. In addition, Carlson (2002) suggested that local disequilibrium with respect to some chemical components, but not others ('partial disequilibrium') may be common in prograde metamorphism, and Waters & Lovegrove (2002) found that the inferred order of appearance of porphyroblastic minerals in metapelites in the aureole of the Bushveld Complex, South Africa, is different from the order predicted from calculated equilibrium phase relationships; this emphasizes the potential importance of kinetic factors, such as nucleation rates, in determining actual metamorphic histories.

Unreliable microstructural criteria

Microstructures that generally do *not* indicate metamorphic reactions (Vernon, 1996a) include the following.

(1) *Inclusions and partial inclusions* are commonly used to infer orders of commencement of crystallization, both for metamorphic and igneous rocks. Assumptions generally made are that: (a) inclusions begin to crystallize earlier than the host mineral (e.g., Boger & Hansen, 2004), and (b) a mineral begins to crystallize later than a mineral on which it is 'moulded' (i.e., which it partly surrounds). However, the only logical inference than can be made from these simple geometrical relationships is that an included or partly surrounded mineral finished growing before the host mineral, and only for that particular grain of the host mineral. Elsewhere in the same rock, the included mineral may still be growing, if the same minerals also occur outside the host grain.

Even where one mineral is consistently enclosed or partly enclosed in another, nothing can be inferred about the relative beginning of crystallization of each mineral (Fig. 3.17). The two minerals may nucleate at the same time, one growing faster and outlasting the growth period of the other (Vernon, 1977). The host mineral may even nucleate before the included mineral (Vernon, 2004), as shown in Fig. 3.18. Inclusion-host microstructural relationships depend on variations in the N/G ratios of the two minerals involved (Vernon, 2004).

A clear example is provided by ophitic microstructure in dolerites or gabbros, in which elongate grains of plagioclase are enclosed in much larger grains of clinopyroxene. Both minerals crystallize together at a eutectic for most of the cooling history, as shown experimentally. Therefore, the relative nucleation times of each mineral cannot be determined by inspection of the microstructure (Mathison, 1987; Flood & Vernon, 1988; McBirney & Hunter, 1995; Vernon, 2004). It cannot even be inferred that all the plagioclase finished crystallizing before the pyroxene, because they

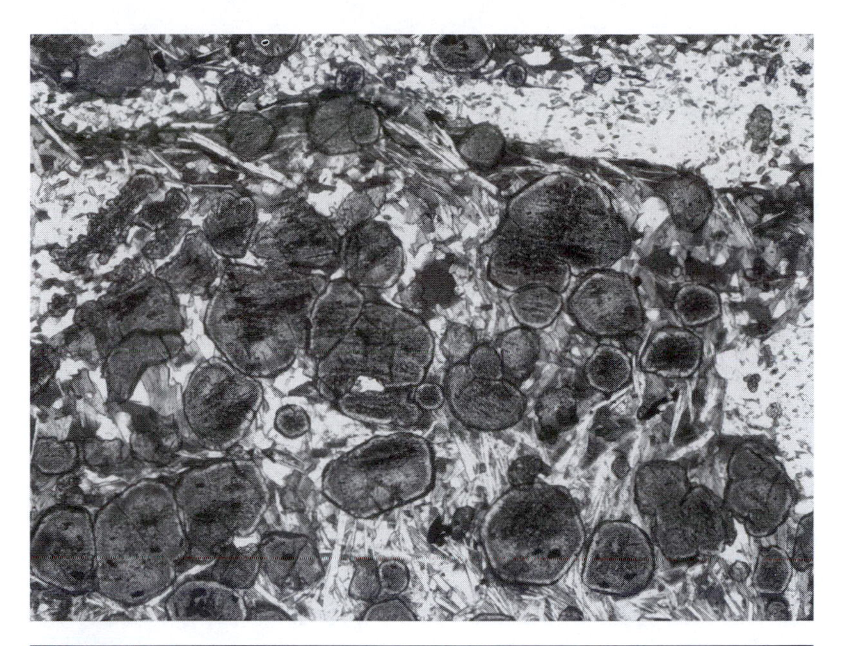

Fig. 3.17 Garnet grains that impinged on and moulded around each other during growth from separate nuclei. No time difference in nucleation should be inferred from the moulding relationship (Vernon, 1977). Graphite inclusion trails are continuous across several garnet grains, suggesting growth after the formation of a matrix foliation, but before coarsening of the matrix by neocrystallization (which eliminated evidence of the foliation from the matrix). Specimen by courtesy of the late Charles Guidotti. From Vernon (2004, fig. 3.33). Plane-polarized light; base of photo 3 mm.

both crystallize together right to the solidus. In other words, though an individual plagioclase grain may have finished crystallizing before the pyroxene in one place, this cannot apply everywhere in the rock. Ophitic microstructure is the result of more abundant nucleation of plagioclase and less abundant nucleation of pyroxene (Wager, 1961). In this context, nucleation refers to the production of viable nuclei, taking into account

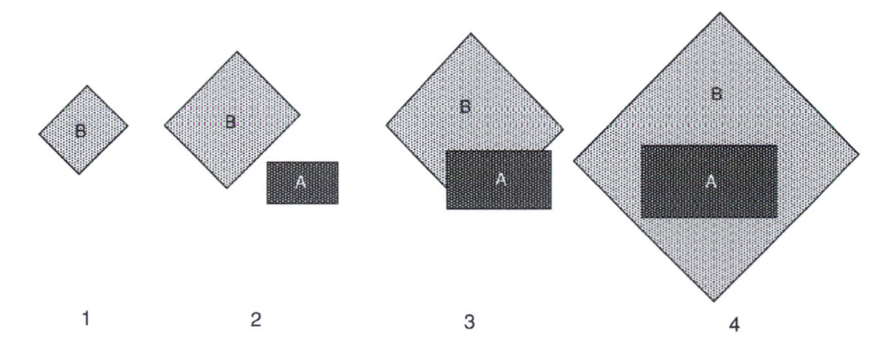

Fig. 3.18 Sketches showing a crystal of mineral B nucleating before, but growing faster than, a crystal of mineral A, with the result that A becomes included in B. This shows that the order of nucleation of two minerals, one included in the other, cannot be inferred from inspection of inclusion-host mineral relationships. This reasoning applies to both igneous and metamorphic rocks. From Vernon (2004, fig. 3.47).

the possibility of 'Ostwald ripening' at the immediate post-nucleation stage (Vernon, 2004).

The same principle applies to inclusions in metamorphic rocks. Inclusions and partial inclusions should not be used to infer an order of crystallization or sequences of mineral assemblages in the absence of other evidence. If mineral A is enclosed or partly enclosed in mineral B, but the reverse situation never occurs in the same rock, it may be argued that A nucleated before B (e.g., Gibson, 1992, pp. 639–40). However, this is not necessarily so, because if A is always finer-grained than B, it has a good chance of always being enclosed in B.

If an included mineral is absent from the matrix everywhere (i.e., on the scale of a large, thin section), it may be a remnant of an older assemblage, and so inference of a reaction may be justified (e.g., Thompson *et al.*, 1977; Krogh, 1982; St-Onge, 1987).

If inclusions and partial inclusions form inclusion trails (Sections 3.3.2, 3.4), more reliable interpretations may be made in some circumstances. For example, Karlstrom & Williams (1995, fig. 14d) illustrated an inclusion of garnet in staurolite, both minerals showing slightly curved inclusion trails that are continuous from one mineral to the other. This indicates that both minerals grew when one planar foliation was undergoing the very earliest stages of crenulation, implying effectively simultaneous growth of both minerals, at this specific place. However, this does not imply that all garnet and staurolite grains in the rock have the same relationship. In fact, most staurolite grains have inclusion trails outlining a more advanced stage of crenulation (Section 3.5.4) than shown by inclusion trails in the garnet (Karlstrom & Williams, 1995, p. 74), which implies that the growth of staurolite outlasted the growth of garnet.

Another example is provided by an inclusion of garnet in andalusite, which contains ilmenite inclusion trails that are deflected around the inclusion, implying that a former matrix was deformed around the garnet grain before it was enclosed by the andalusite (Gibson, 1992, fig. 4a, p. 646). This indicates that this particular grain of garnet finished growing before it was engulfed by this particular grain of andalusite, but it does not indicate that the garnet nucleated before the andalusite grain; the reverse could be true. However, examination of many grains has shown that most andalusite grains in the rock have inclusion trails outlining a more advanced stage of crenulation (Section 3.5.4) than shown by inclusion trails in garnet (Gibson, 1992, p. 646), which implies that the growth of andalusite outlasted the growth of garnet.

(2) *The 'wrapping' of porphyroblasts by minerals in folia* is often used as an indication that the foliated assemblage is younger than the mineral around which it is deflected. An example is illustrated in Fig. 3.19, which shows a folium rich in biotite and sillimanite deflected around a K-feldspar porphyroblast. The temptation may be to infer that the biotite and sillimanite crystallized later than the K-feldspar, but this is unjustified (Vernon, 1996a), because the relationship commonly results from simultaneous growth of all the minerals. Some minerals (for example, mica, sillimanite) can grow and survive in high-strain (folial) zones, whereas others (for example, garnet, cordierite, K-feldspar) can grow only in lower-strain (interfolial) zones. The folial zones are compressed against, and consequently anastomose around, the stronger pods rich in cordierite, garnet or K-feldspar that grow at the same time. (Fig. 3.19). This process is often

Fig. 3.19 Folium of fibrous sillimanite (Sil) deflected around a porphyroblast of K-feldspar (Kfs) with rounded inclusions of quartz and minor biotite, in a granulite facies metapelite, Broken Hill, Australia. The minerals in the porphyroblast and folium belong to the same mineral assemblage (paragenesis). From Vernon (2004, fig. 3.31). Crossed polars; base of photo 3.4 mm.

referred to as 'deformation partitioning', and has been described by Cobbold (1977), Bell, (1981), Bell & Rubenach (1983) and Vernon (1987b, 1996a, 2004). All the minerals grow simultaneously as members of the same mineral assemblage (Vernon, 1996a, 2004). Therefore, it should not be assumed that minerals in anastomosing folia form later than minerals (such as porphyroblasts) in low-strain pods.

On the other hand, if the mineral in the resistant pods is strongly deformed, recrystallized or partly replaced by new minerals that are compatible with a lower-temperature assemblage in the matrix, as in many retrograde schists and mylonitic rocks, the inference of a reaction is justified. For example, in retrograde metamorphic rocks and many deformed igneous rocks, the timing of reactions relative to deformation events can be related to the development of new assemblages in foliations, because development of the new foliations is attended by (1) a reduction of grainsize (commonly with partly replaced remnants of the initial minerals) and (2) the development of a new mineral assemblage at different *P-T* conditions from those of the original assemblage, the interfolial minerals being metastable relics.

Collins & van Kranendonk (1999, fig. 3) presented photomicrographs of kyanite schist in the Pilbara region of Western Australia, in which the kyanite occurs as: (1) porphyroblasts with inclusion trails oblique to the main foliation, (2) porphyroblasts elongate parallel to the main foliation, with inclusion trails outlining this foliation, and (3) random, partly radiating aggregates. This microstructural evidence does not imply several discrete periods of kyanite growth, but suggests that kyanite grew, in several local deformation environments, during the development of the schistosity, and continued after it had been completed. In other words, kyanite could have been growing during all stages of a single metamorphism-deformation event. This is a particularly clear example, because only one porphyroblastic mineral is present. But the same principles apply where two or more porphyroblastic minerals are involved.

3.4.3 Using inclusions and folia wrapping porphyroblasts to infer metamorphic 'non-events'

Example 1: Metapelitic gneisses at Round Hill, north of Broken Hill, western New South Wales, Australia, contain the lower granulite facies peak assemblage: Qtz + Kfs + Pl + Bt + Crd + Grt + Sil. They have undergone partial melting with the formation of leucosomes containing coarse-grained garnet (White *et al.*, 2004). Forbes *et al.* (2005) inferred an early metamorphic event, based on inclusions in large grains of K-feldspar, garnet and cordierite, namely an 'inclusion assemblage' consisting of quartz, biotite, sillimanite, plagioclase and muscovite. The same minerals occur in the matrix (with biotite of similar composition to that occurring as inclusions), except for muscovite. However, inferring muscovite as a primary inclusion is contentious, as it is typically retrograde in these lower granulite facies rocks. In lower grade metapelites at Broken Hill, muscovite was removed in andalusite-forming and partial melting reactions, and spectacular pseudomorphs of andalusite by sillimanite are common in the high-grade metapelites. So it is logical to infer that the rock contains a single peak assemblage, namely Qtz-Bt-Sil-Pl-Kfs-Grt-Crd – the normal Broken Hill high-grade metapelitic assemblage, which embodies both host minerals and inclusions, except for the contentious muscovite (Vernon *et al.*, in press). The fact that the quartz and biotite inclusions in coarse-grained K-feldspar, cordierite and garnet in these rocks show evidence of solid-state adjustment of their boundaries towards low-energy shapes (Fig. 3.19) is evidence of their chemical compatibility with the host minerals (Vernon, 1968, 1999, 2004). This example emphasizes that, as stated previously, inclusions do not necessarily belong to a different metamorphic paragenesis from that of the host mineral.

The 'inclusion assemblage' occurs in trails (Section 3.5.2) oblique to the main foliation ('S_2'), which can be taken to indicate a pre-existing foliation ('S_1'). However, if so, it does not mean that the minerals now delineating that foliation were there before the main metamorphism. Moreover, the obliquity could represent a previous orientation of the same foliation that was transposed into a new orientation in the same deformation event. In the absence of detailed accurate radiometric dating of mineral growth in specific foliations, there is no logical reason why one or more progressive deformation stages (so-called 'events') cannot occur during the same thermal (metamorphic) 'event'. Therefore, the inference by Forbes *et al.* (2005) of a separate tectonometamorphic event from an 'inclusion assemblage' is doubtful (Vernon *et al.*, in press).

Example 2: Hand *et al.* (1992) inferred two metamorphic events from inclusion and foliation relationships in granulite facies gneisses of the Weldon terrane, Arunta Block, central Australia. They inferred a reaction between garnet and cordierite to form biotite and sillimanite, on the basis of biotite-sillimanite folia wrapping around and truncating cordierite and garnet porphyroblasts. Vernon (1996a) showed that some of the biotite and sillimanite inclusions in the garnet and cordierite porphyroblasts are in contact (without any sign of reaction), and so presumably these minerals were chemically compatible with garnet and cordierite during the metamorphic event responsible for the porphyroblasts, as confirmed by the tendency of the inclusions to adjust to low-energy shapes. Furthermore, as noted previously, wrapping of porphyroblasts by biotite-sillimanite folia is typical of high-grade metapelitic rocks, and simply reflects a tendency for some minerals (e.g., garnet, cordierite) to form few nuclei and grow to large, strong porphyroblasts, which consequently constitute low-strain lenses or pods, whereas minerals such as sillimanite and biotite nucleate abundantly and concentrate in high-strain folia, where they are aligned and can slip on cleavages and crystal faces during high shear strain. The folia anastomose because the porphyroblasts are stronger, not because they grew earlier than the folia (Vernon, 1996a, p. 145). Therefore, only one metamorphic event is needed to explain the microstructural relationships.

3.4.4 *P-T-t* paths

Much current effort in metamorphic geology goes into the determination of pressure (*P*) – temperature (*T*) conditions from mineral assemblages and compositions, using techniques described in Chapter 2. Attempts are also made to determine how these inferred sets of *P-T* conditions vary with time (*t*) in a metamorphic terrane, known as the *P-T-t* path. This is done by using structural data, either (1) microstructural (for example, by noting the replacement of one mineral or assemblage by another mineral or assemblage, or by using porphyroblast-matrix relationships, as discussed in Section 3.5) or (2) mesostructural (for example, by observing assemblages in shear zones that cut earlier assemblages).

Ideally, several rock-types from the same area should be investigated in the same way, so that different reactions or assemblage changes occurring along different parts of a *P-T-t* path can be evaluated; in this way, a relatively complete *P-T-t* path may be inferred, at least in principle. This information is typically used to make further interpretations of the regional tectonic-metamorphic (*tectonometamorphic*) history (Section 3.6). However, as pointed out by Passchier & Trouw (1996, p. 6), an inferred *P-T-t* path is likely to apply to a limited volume of rock (perhaps at most a few cubic kilometres), and so care has to be applied in extending this inference to a metamorphic terrane. Commonly different *P-T-t* paths can be inferred for different major units in a terrane, increasing the complexity of tectonometamorphic interpretations (e.g., Spear *et al.*, 1984).

The approach encounters many potential difficulties (Vernon, 1996a). For example, the inferred intersection of a single reaction line or *P-T* reaction band cannot give any information about the relevant *P-T* vector (Fig. 3.20). Intersection of two reaction lines improves the situation. For instance, the following two reactions can be inferred from microstructural relationships involving partial pseudomorphism in the Chiwaukum Schist, Washington, USA (Evans & Berti, 1986): (1) andalusite + biotite + H_2O = staurolite + quartz (symplectite) + muscovite (Fig. 3.21) and (2) andalusite = kyanite (Fig. 3.22). Plotting of just one of the *P-T* lines inferred for these reactions gives little information on the *P-T* vector, the situation being identical to that shown in Fig. 3.20. For example, reaction (1) has a positive *P-T* slope, and so is sensitive to both *P* and *T* changes, so that inference of a *P* increase cannot be made on the basis of this inferred replacement reaction alone. Even reaction (2), which is invariably taken to imply an increase in *P*, can be equally well interpreted as isobaric or even involving a small *P* decrease (Fig. 3.20). However, the problem can be

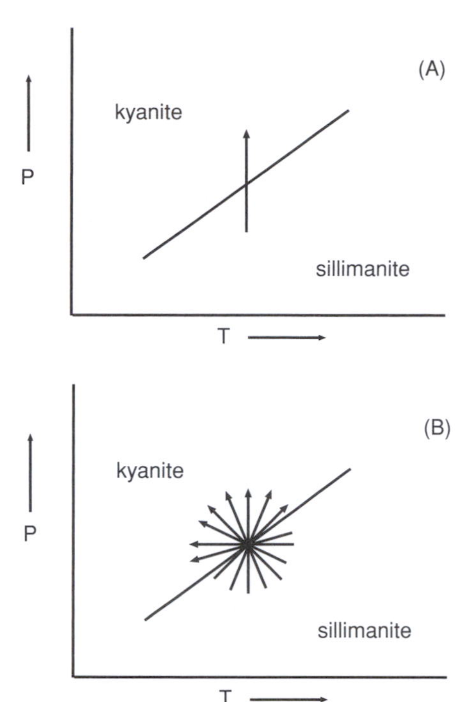

Fig. 3.20 Schematic line for the reaction: kyanite = sillimanite, showing the usual interpretation involving isothermal *P* increase (A) and possible *P-T* vectors (B).

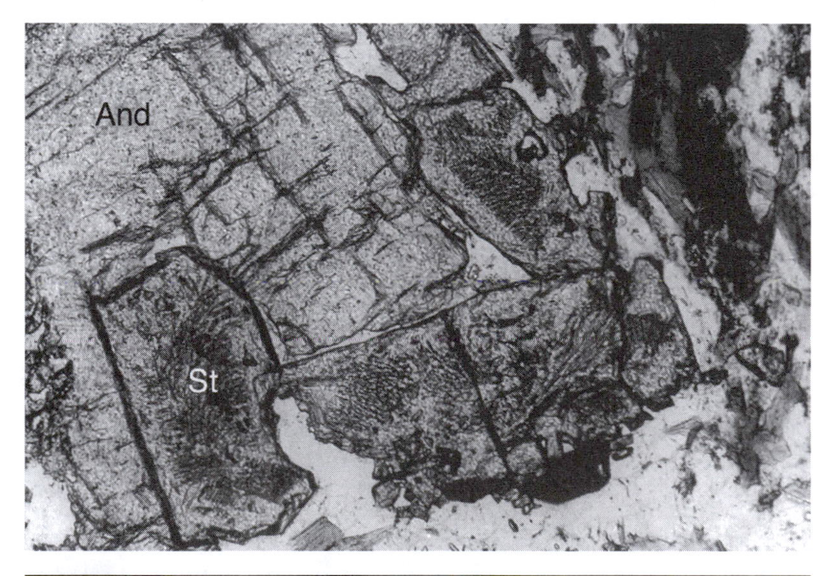

Fig. 3.21 Idioblastic crystals of staurolite (St), symplectically intergrown with quartz, which have partly replaced andalusite (And) from the margins inwards, Chiwaukum Schist, Cascade Range, Washington, USA. Plane-polarized light; base of photo 1.5 mm.

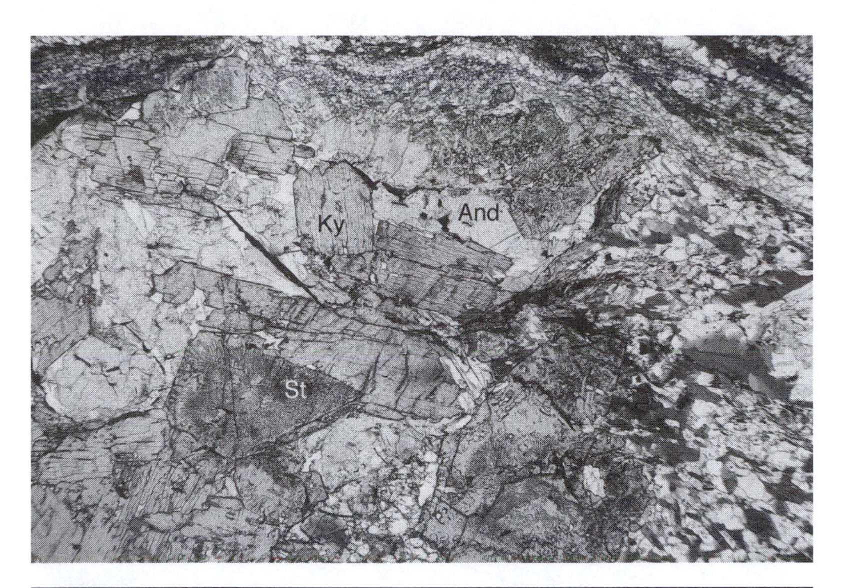

Fig. 3.22 Andalusite (And) partly replaced by staurolite (St) and kyanite (Ky), Chiwaukum Schist, Cascade Range, Washington, USA. The staurolite shows crystal faces against both andalusite (top-right of centre) and kyanite (top-left). This relationship is consistent with later growth of the kyanite, because the same staurolite shapes are present in other andalusite grains in the rock, from which kyanite is absent (Fig. 3.21), inferred that the kyanite grew later than the staurolite, in response to a progressive increase in pressure. Plane-polarized light; base of photo 4 mm.

resolved by inferring the relative timing of the reactions. Kyanite appears to have grown later than the staurolite-quartz symplectite, because (1) some andalusite replaced by the symplectite has not been replaced by kyanite (Fig. 2.1), and (2) where andalusite has been replaced by kyanite, the symplectic staurolite crystals have the same shapes and relationships as where kyanite is absent (Fig. 2.2). This evidence implies that the two reactions occurred in a consistent sequence, which makes for a more confident interpretation of at least a general P increase (Vernon, 1996a). The implication is that the staurolite-forming reaction proceeded at a lower P than that of the kyanite-forming reaction, which is consistent with previous inferences of a regional P increase based on P-T determinations.

Several sequential crossings of reaction lines and/or several thermobarometric P-T determinations made from inferred sequential assemblages give a more reliable vector, but even then, alternative paths may be available, in the absence of other evidence. For example, though smooth, single-episode paths are typically inferred from such data, detailed geochronological evidence is needed before these paths can be accepted with confidence, because some detailed studies have revealed unsuspected evidence of multiple heating events (Vernon, 1996a). The more recent use of P-T pseudosections (see Chapters 1 and 2) involving whole-rock compositions of actual rocks and, in particular, the integration of several such sections for different rocks in a terrane, can give a more complete and realistic P-T path.

P-T-t paths are of two main types: 'clockwise' (England & Richardson, 1977; Holland & Richardson, 1979; Hollister *et al.*, 1979; Earle, 1980;

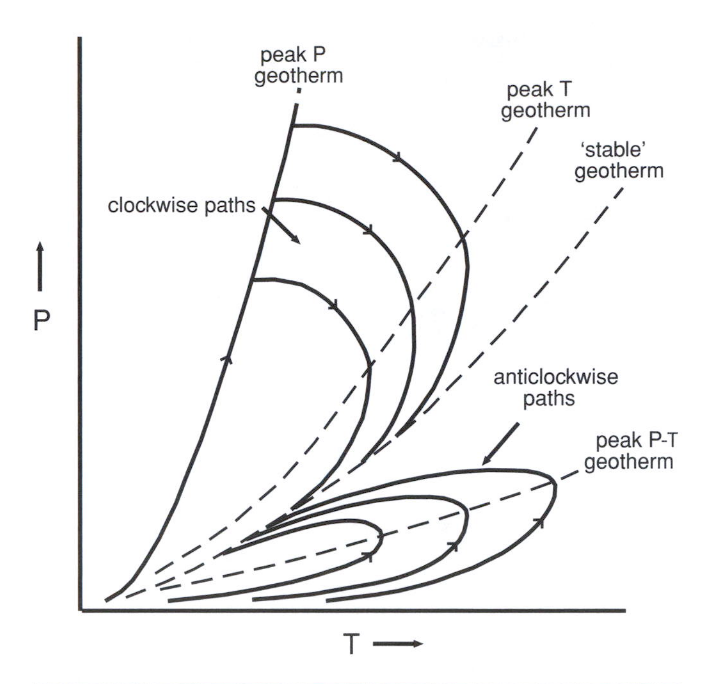

Fig. 3.23 Sketch showing the general shapes of clockwise and anticlockwise *P-T-t* paths, as well as some representative geotherms (dashed lines) and random field gradients (dotted lines).

England & Thompson, 1984a) and 'anticlockwise' (Vernon, 1982; England & Thompson, 1984b; Schumacher *et al.*, 1990), as shown in Fig. 3.23. *Clockwise paths* are produced by continental collision, during which rocks on a 'normal' continental geotherm undergo relatively rapid *P* increase, after which conduction and radioactive decay slowly heat the rocks (over *c.* 20–80 Ma) during exhumation, producing a curved *P-T* path for each rock in the terrane, until a 'steady-state' continental geotherm is reached and metamorphism effectively ceases. Thus, the geotherm is initially perturbed (Richardson, 1970) with respect to *P* and later with respect to *T*, so that the *P* maximum is attained before the *T* peak for each rock (Fig. 3.23). *Anticlockwise paths* are produced in relatively low-*P* terranes undergoing mainly magmatic heating. As the rock sequence undergoes thickening by folding during heating, *P* rises with *T* until the peak *P* and *T* are reached together, after which cooling at or near the maximum *P* takes place, until a 'steady-state' geotherm is reached (Fig. 3.23). More complex *P-T-t* paths, including 'loops', have been proposed for some areas (Williams & Karlstrom, 1996; Dumond *et al.*, 2007). For example, in the Proterozoic of the southwestern United States, looping *P-T* paths record evidence of heating with crustal thickening, followed by long-term residence of the terrane in the middle crust, which resulted in isobaric cooling (Williams & Karlstrom, 1996).

An important problem is that the *P-T* curves shown in Fig. 3.23 may be intersected randomly by an erosion surface, with the result that rocks showing evidence of the maximum *P* and/or *T* may not be exposed in that segment of crust (e.g., Thompson, 1978). In continental collision zones, true *P-T* paths commonly intersect at high angles the *P-T* progression inferred by examination of a metamorphic field area, with the result that the field progression gives no indication of the *P-T* history of individual exposed rocks (Fig. 3.23). Evidence of the maximum *P* cannot be detected in rocks with a clockwise *P-T* history, unless highest-*P* mineral relics are preserved as inclusions, and exposure of rocks with evidence of peak *T* would be fortuitous. Parts of the *P-T* history may be inferred from mineral inclusions (e.g., Wells, 1979), fluid inclusions (Hollister, 1969; Hollister *et al.*, 1979; Selverstone *et al.*, 1984) or mineral compositional zoning (Holland & Richardson, 1979; Spear *et al.*, 1984; Selverstone *et al.*, 1984). The chances of finding exposed rocks with peak *P-T* assemblages are higher for terranes with anticlockwise histories, because the *P* is relatively low throughout the history, so that the curves are typically rather flat and erosion surfaces would be likely to intersect assemblages relatively close to peak conditions, unless the terrane is markedly tilted after metamorphism.

3.5 Inferring time of growth of metamorphic mineral assemblages relative to time of formation of specific structures

3.5.1 Introduction

Relating the time of growth of metamorphic mineral assemblages to the time of formation of a specific structure, such as a foliation (S), lineation (L) or fold set (F), with implications for a deformation event *(D)*, depends on microstructural interpretations. Several problems attend this approach (Vernon, 1977, 1978a, 1989, 2004), some of which are discussed in the next sections.

The approach depends on the ability to reliably relate growth of an individual mineral or a mineral assemblage (paragenesis) to the formation or deformation of a tectonic structure, on the basis of microstructural interpretation. One question is: because a mineral assemblage delineates a particular foliation, can it be confidently inferred that the assemblage grew as the foliation initiated, or is it just as valid to infer that the assemblage grew in an existing foliation ('mimetic crystallization')?

If a deformation event does not produce a structure on the scale of observation (for example, an outcrop or thin section), the growth of a mineral assemblage cannot be related to that deformation event. Therefore, it is not sufficient to say simply that a particular mineral is 'predeformation' or 'prekinematic', without referring to a specific deformation structure. Instead, we need to use expressions like 'pre-foliation' or 'pre-folding' (Vernon *et al.*, 1993a), referring to a specific structure.

3.5.2 Inclusion trails in porphyroblasts

As noted previously, growing porphyroblasts overtake and enclose smaller growing grains of matrix minerals, trapping them as inclusions (Figs. 3.5, 3.19, 3.24). Some inclusions may be relics inherited from lower grades of metamorphism, but generally inclusions belong to the same metamorphic assemblage as the porphyroblast.

Care needs to be taken to ensure that inferred inclusions are true inclusions and not products of exsolution or coaxial growth (Ingerson, 1938; Vernon, 1978a). The following criteria (Ferguson & Harte, 1975, p. 474) can

Fig. 3.24 Elongate, aligned inclusions (with rounded corners, owing to solid-state grain boundary adjustment) of microcline delineating inclusion trails in a porphyroblast of andalusite, Mount Stafford area, central Australia. From Vernon (2004, fig. 3.30B). Crossed polars; base of photo 1.5 mm.

be used to argue against an exsolution or replacement origin of inclusions: (1) inclusion trails being continuous right across the porphyroblast, (2) most trails being unrelated to the crystallographic orientation of the host porphyroblast and not deflected across twin interfaces in the host, and (3) inclusion trails passing continuously from the host mineral, into and through relatively large inclusions of other minerals, where present.

Special caution is needed where later alteration of inclusions is suspected, owing to fluid penetrating along cracks in the host mineral. For example, Perchuk *et al.* (2005) found that high-pressure heating of hydrous inclusions (e.g., clinozoisite) in garnet may liberate fluid and cause local melting, thereby changing the inclusion assemblage. They inferred that inclusions and former melt patches in ultrahigh-pressure garnet can be due to heating after the main metamorphic episode.

Though many elongate inclusions tend to develop curved corners (Vernon, 1968, 1976, 2004), their original shapes may be preserved as a dimensional (shape) preferred orientation of elongate inclusions, forming inclusion trails (Fig. 3.24). Because alignment of elongate minerals can result from growth in a tectonic foliation, inclusion trails potentially permit the timing of growth of porphyroblastic minerals relative to a foliation-forming deformation event, or an earlier stage in the progressive development of the same foliation-forming deformation event.

Though inclusion trails reveal the existence of former foliations, the minerals currently delineating the trails were not necessarily present prior to growth of the porphyroblast. Therefore, to confidently infer that an 'inclusion assemblage' crystallized in a previous deformation or metamorphic event, it is necessary to ensure that the minerals forming the foliation actually grew during that event, and not mimetically in the earlier foliation, as products of the reaction responsible for the enclosing porphyroblast. Earlier syndeformational growth is unlikely if the minerals involved occur both inside and outside the porphyroblast, unless the inclusions are of a mineral stable throughout all or much of the metamorphic evolution (e.g., quartz).

3.5.3 Using porphyroblasts to determine the relative timing of metamorphic reactions and foliations (*P-T-D-t* paths)

If suitable inclusion trails are present, porphyroblast-matrix microstructural relationships may be used to infer the timing of metamorphic reactions relative to foliation-forming deformation events, and so elucidate the 'tectonometamorphic history'. A major aim of metamorphic geology is to integrate inferred *P-T-t* paths with an inferred sequence of foliation-forming deformation events (the *D* path) – provided time differences between these 'events' can be successfully determined – thereby elucidating *P-T-D-t* paths (e.g., Johnson & Vernon, 1995a).

This is a complex problem, not only because of difficulties of microstructural interpretation (Johnson & Vernon, 1995a,b; Alias *et al.*, 2002), but also because patterns of cyclic deformation and flow instabilities may fluctuate with changes in temperature and cooling rate, as emphasized by Knipe (1989). For example, the duration, *T* and stress experienced by a rock between deformation events affects the frequency of events and the instabilities developed. Also, the stress and *T* paths at the end of a deformation event influence the preservation of microstructures, to the extent that that some microstructures produced during prograde metamorphism may be obliterated.

Considering the growth of porphyroblasts alone, without considering growth of matrix minerals, is inadvisable. For example, if porphyroblasts of albite are inferred to have grown independently of neocrystallization of matrix minerals, the implication is that sodium metasomatism has occurred, and chemical evidence is required to substantiate this. Generally, all or nearly all minerals in a rock are involved in prograde metamorphic reactions, including those that produce porphyroblastic minerals, which is relevant for attempts to relate the timing of the growth of porphyroblasts to deformation.

Metamorphic minerals can grow before, during or after a particular foliation-forming or foliation-deforming event. Therefore, a porphyroblast may be classified as being *pre-foliation/pre-folding*, *syn-foliation/syn-folding* or *post-foliation/pre-folding*. Alternative terms commonly used are '*prekinematic*', '*synkinematic*' and '*postkinematic*', but to them must be added: 'with respect to the formation or deformation of...' a particular tectonic structure, such as: (1) the initiation and development of an S-surface (assuming that the minerals defining the S-surface grew while it was developing and not mimetically) or (2) the flattening or folding of an existing S-surface.

The general approach is shown in Fig. 3.25. Zwart (1960a,b, 1962) suggested a basic set of microstructural criteria (Fig. 3.26), based on geometrical relationships between S_i (the 'internal' foliation delineated by inclusion trails in the porphyroblast) and S_e (the 'external' foliation in the matrix). Subsequent work has shown that though these criteria are broadly useful for the inference of relative time relationships between growth of porphyroblasts and foliation development in regional metamorphic rocks, many porphyroblast-matrix microstructural relationships are ambiguous, and so only clear, critical relationships should be used (Vernon, 1978a).

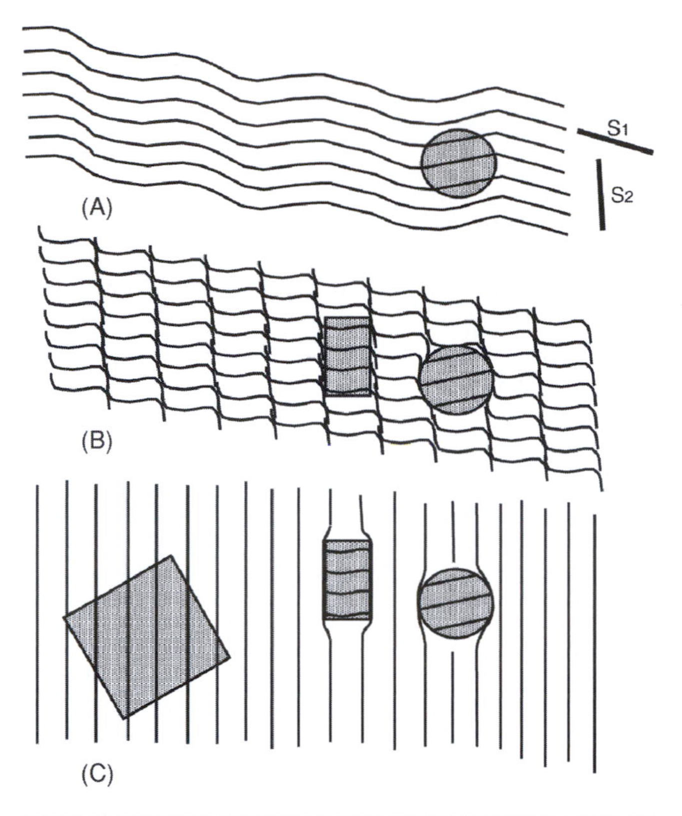

Fig. 3.25 Sketches showing how different minerals growing as porphyroblasts may preserve inclusion trails that reveal evidence of different stages of foliation development. The spherical porphyroblast grew during the early stages of development of a crenulation cleavage (labelled S_2), the vertical rectangular porphyroblast grew during further development of S_2, and the square porphyroblast grew later, during or after the final stages of development of S_2, which produced a planar foliation devoid of crenulations. After Johnson (1999b, fig. 4), with permission of the *American Mineralogist.*

Jamieson (1988) has provided a useful summary of the approach. The best rocks for determining time relationships between deformation and prograde metamorphism are medium-grade metapelitic schists, which commonly contain porphyroblasts and deform relatively easily. Most low-grade rocks lack porphyroblasts, which makes timing difficult. Some high-grade metamorphic rocks with porphyroblasts (Fig. 3.27) are suitable (Vernon, 1989), but coarsening of the matrix commonly removes evidence of earlier foliations and microfolds (Fig. 3.18). On the other hand, high-grade rocks are suitable for inferring retrograde metamorphic histories, because reactions are commonly incomplete and the grade contrast is obvious. Some medium-grade rocks may also be suitable for this purpose.

Reliable inferences from porphyroblast-matrix microstructural relationships depend not only on careful and detailed observation, but also the use of critically oriented sections. Fagan (1979) emphasized the importance of observing the right section for the correct identification and interpretation of porphyroblast-matrix relationships in folded, crenulated rocks. For

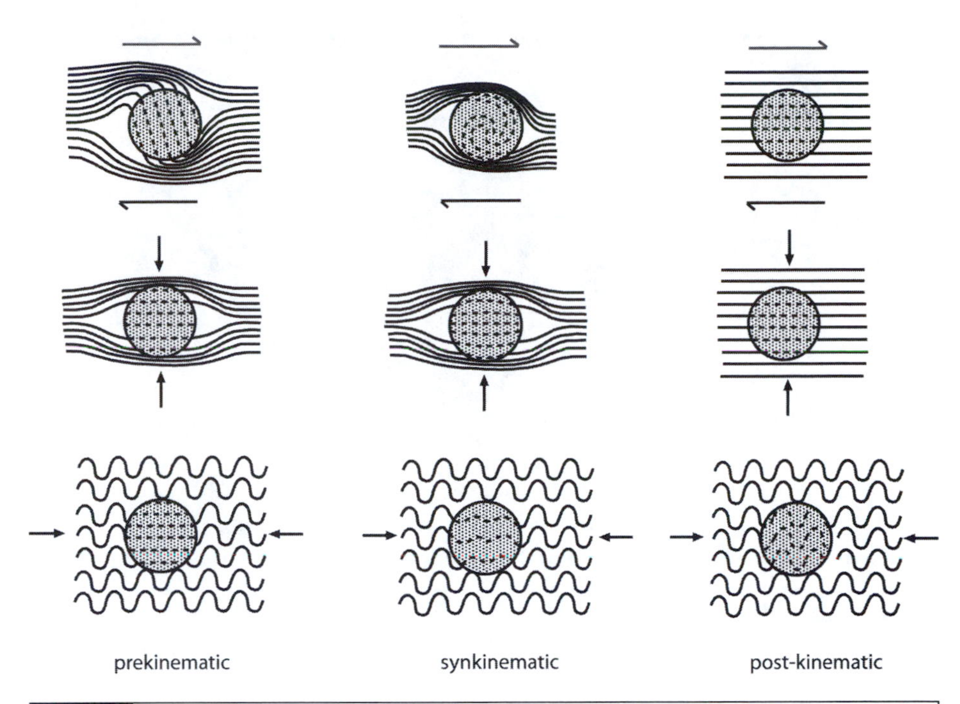

Fig. 3.26 Idealized porphyroblast-matrix microstructural relationships suggested by Zwart (1962) as indicators of growth of porphyroblasts pre-, syn- and post the deformation of a matrix foliation, depending on the type of strain (indicated by arrows).

example, a section cut parallel to the axes of crenulations shows only straight inclusion trails that pass into parallel folia in the matrix (Fig. 3.28). This may suggest that the porphyroblasts grew after the development of the present orientation of the matrix foliation. However, a section cut perpendicular to the axes of the crenulations reveals not only microfolds in the matrix, but also the curvature of inclusion trails in porphyroblasts (Fig. 3.28). Thus, evidence of syndeformational growth may be obtained.

General precautions that should be taken when using porphyroblast-matrix microstructural relationships (Bell & Rubenach, 1983; Vernon, 1988a) include: (1) labelling S surfaces in the field and being careful to refer matrix foliations and inclusion trails to these labels, (2) making thin sections in several orientations, (3) observing as many oriented sections as are necessary to reveal the critical evidence, (4) examining sections showing sedimentary bedding (S_0) wherever possible, to increase the reliability of inferring the sequence of S surfaces, (5) cutting sections through the hinge areas of the earliest folds, so that a clear obliquity between S_0 and S_1 can be seen, (6) using only porphyroblasts with clear inclusion trails, and (7) taking care that inclusion trails are not crystallographically controlled, for example, by exsolution or coaxial growth. Other complications have been pointed out by Johnson & Vernon (1995a,b) and Cihan (2002).

The importance of local strain heterogeneity should also be emphasized when using porphyroblast-matrix microstructural relationships. Different parts of a rock may follow different strain paths, producing different structures, and may deform at different rates (Gibson, 1992). The result is that

Fig. 3.27 Garnet porphyroblast (Grt) with inclusion trails of sillimanite delineating folded shapes in a granulite facies metapelite, Anmatjira Range, central Australia. The sillimanite grew during or after the formation of a crenulation foliation, after which all evidence of the crenulations was removed from the matrix by deformation and neocrystallization. The garnet could have grown during or after the sillimanite, but simultaneous growth is more likely, in view of the abundant sillimanite in the matrix. From Vernon (2004, fig. 5.80). Crossed polars; base of photo 12 mm.

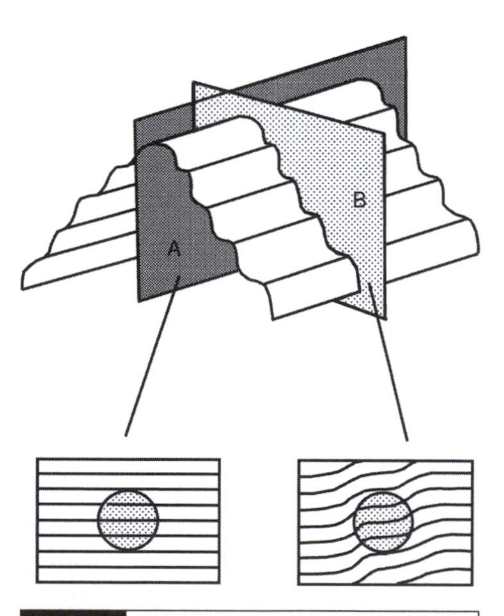

Fig. 3.28 Sketch showing how a section (A) cut parallel to axes of microfolds fails to indicate syn-crenulation porphyroblast growth, whereas a section (B) cut perpendicular to the microfold axis gives the information. From Vernon (2004, fig. 5.82).

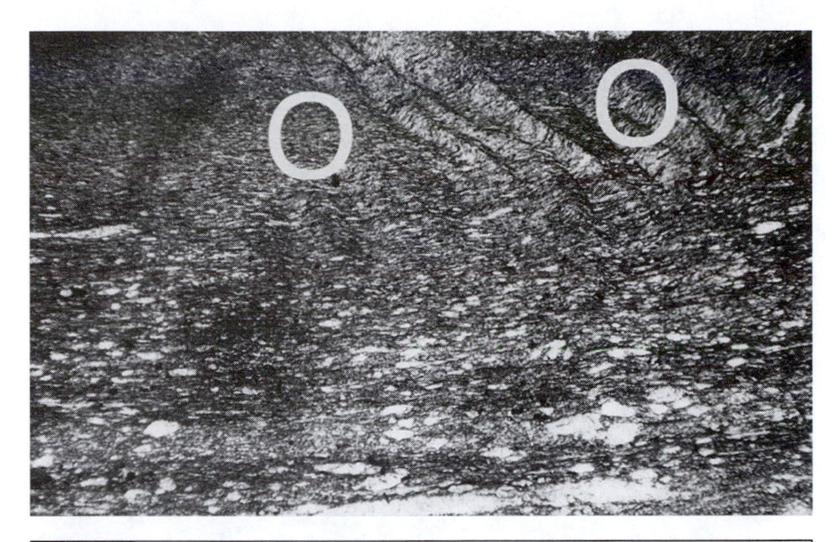

Fig. 3.29 Crenulation cleavage formed very locally in a mica-rich layer in slate, Hill End, area, New South Wales, Australia. Hypothetical porphyroblasts are shown by circles. The porphyroblast at left would show no evidence of crenulations in its inclusion trails, whereas the porphyroblast at right would show crenulations. This suggests caution when viewing only a small number of porphyroblasts in the one rock. Plan-polarized light; base of photo 11 mm.

two or more porphyroblasts with different inclusion trail patterns do not necessarily form at different times, and conversely, two or more porphyroblasts with the same inclusion trail pattern do not necessarily form at the same time. For example, as shown in Fig. 3.29, crenulations may be very patchy, especially in the earlier stages of their development, with the result that porphyroblasts can grow simultaneously in crenulated and non-crenulated areas of the same rock (on the thin-section scale), giving the false impression of different timing of growth relative to the development of the crenulations.

3.5.4 Criteria

Regional metamorphic rocks typically show sequences of foliations (labelled S_0 for sedimentary bedding, and S_1, S_2, etc. for tectonic foliations) as discussed in structural geology textbooks (Ramsay, 1967; Hobbs *et al.*, 1976; Ramsay & Huber, 1987; Passchier & Trouw, 1996). The aim is to relate growth of metamorphic assemblages to one or more inferred S-surfaces or stages of folding. In the absence of detailed geochronological measurements to date the growth of porphyroblasts and/or development of foliations, it cannot be safely assumed that each foliation results from a tectonic event well separated in time or, conversely, that all inferred foliations are formed in the same tectonic event.

The following general criteria are commonly used to indicate that porphyroblasts grow before, during or after the development of a foliation. Though observations are consistent with a syn-foliation origin for many or most porphyroblasts during prograde metamorphism (Bell *et al.*, 1986), other observations suggest that some porphyroblasts grow before foliation development or after foliation deformation, at the scale of observation.

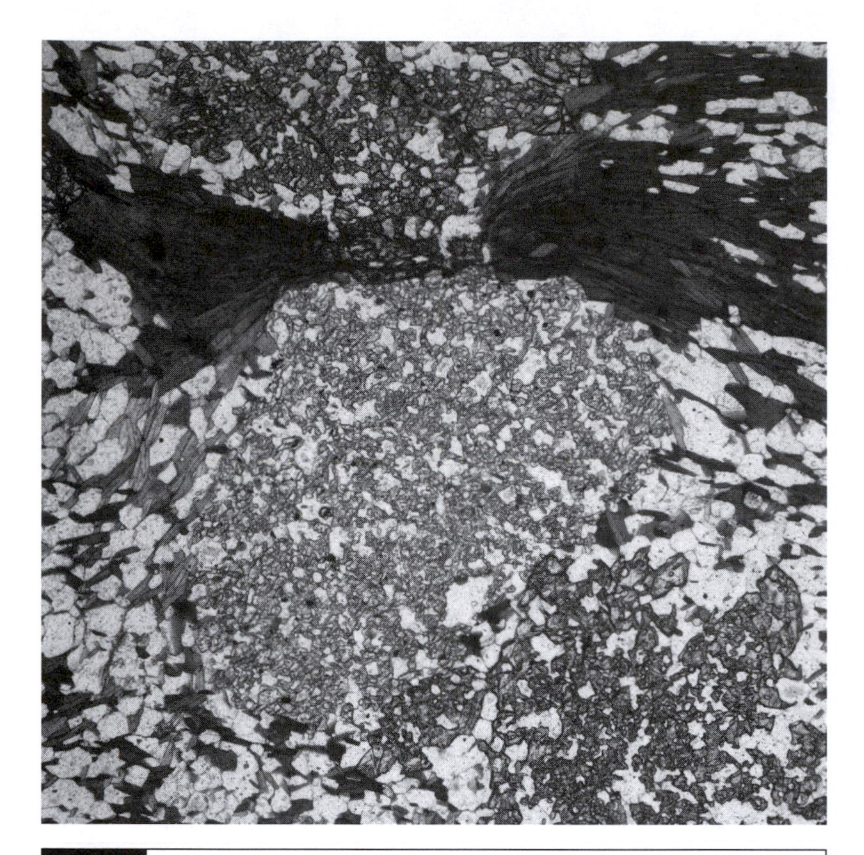

Fig. 3.30 Porphyroblasts of andalusite (centre) and staurolite (top and bottom, with higher relief) containing small, random inclusions of quartz and opaque material, set in a foliated, much coarser-grained matrix, in a schist from Harrogate, Mount Lofty Ranges, South Australia. This indicates that the porphyroblasts grew before a foliation developed in the matrix (Fleming & Offler, 1968) and that the matrix underwent deformation and coarsening after the porphyroblasts grew. Plane-polarized light; base of photo 3.5 mm.

Pre-foliation porphyroblasts

The best indicators of porphyroblasts that grow before the development of a foliation are: (1) random inclusions that are very small, compared with the average grainsize of the matrix minerals (Zwart, 1960b; Johnson, 1962; Fleming & Offler, 1968; Vernon *et al.*, 1993a,b), the porphyroblasts apparently having grown before foliation development or matrix coarsening (Figs. 3.30, 3.31A) and (2) porphyroblasts that have overgrown and preserve the shapes of former random mica grains (Fig. 3.31B).

Syn-foliation porphyroblasts

The following microstructural features (unfortunately relatively uncommon) are indicative of porphyroblast growth during the folding of a slaty cleavage to form a crenulation cleavage: (1) progressive increase towards the edges of porphyroblasts in the curvature of inclusion trails formed from matrix folia deflected around the growing porphyroblasts (Zwart, 1962; Zwart & Calon, 1977), as shown in Fig. 3.32A; (2) progressive increase in the intensity of microfolding of inclusion trails towards the edges of porphyroblasts (e.g., Vernon, 1989), as shown in Fig. 3.32B; (3) 'paracrystalline microboudinage' (Misch, 1969, 1970; Vernon, 1976; Vernon & Flood, 1979),

(A)

(B)

Fig. 3.31 (A) Porphyroblasts of cordierite with small inclusions of quartz and biotite, with some fine-grained opaque material, Chiwaukum schist, Cascade Range, Washington, USA. The quartz layers (top edge and middle of porphyroblast) may be deformed former quartz veins. The very faint tendency to alignment or layered concentration of the inclusions may suggest growth of the porphyroblast during the very earliest stages of foliation development, or may represent former sedimentary bedding. The matrix foliation is oblique to the elongation of the porphyroblast, and the matrix has undergone post-porphyroblast coarsening. Plane-polarized light. Base of photo 4 mm. (B) Porphyroblast of andalusite with random inclusions of quartz and biotite in a schist, Snowy Mountains, New South Wales, Australia. The shapes of andalusite areas between the inclusions suggest the former presence of random mica flakes, the components of which were used by the andalusite during its growth. They also suggest that the andalusite grew before a foliation had formed, after which a foliation developed in the matrix, which has coarsened by neocrystallization. From Vernon (2004, fig. 5.78A). Crossed polars; base of photo 1.75 mm.

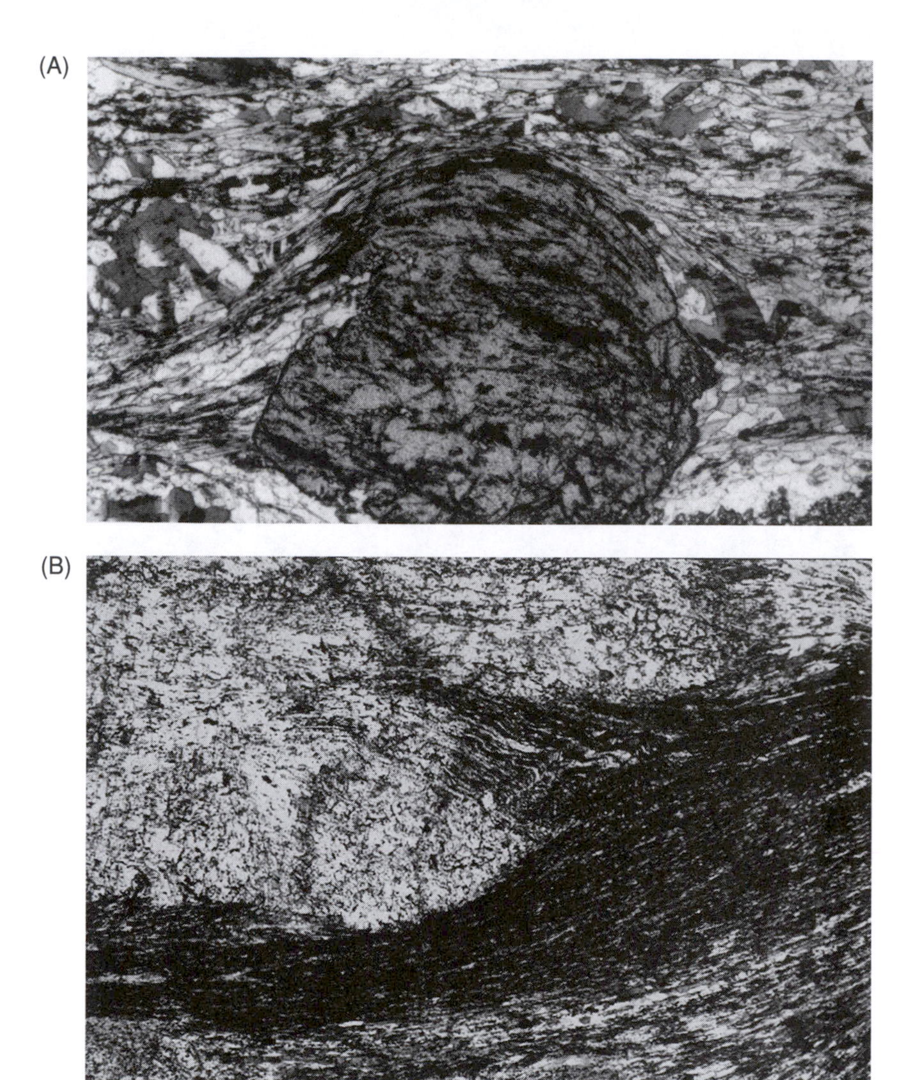

Fig. 3.32 (A) Porphyroblast of garnet showing straight graphite inclusion trails in the centre changing to progressively more curved trails at the edge, which indicates flattening of the matrix foliation around the growing porphyroblast. The bottom part of the porphyroblast has been truncated during intense matrix deformation. Chiwaukum Schist, Cascade Range, Washington, USA. Plane-polarized light; base of photo 4.4 mm. (B) Porphyroblast of cordierite (top left half of photo), showing inclusion trails that are straight in the central parts of the porphyroblast and show progressively tighter microfolding towards the edge of the porphyroblast, in a schist from Cascade Bay, western Gulf of Alaska. These relationships indicate that the porphyroblast grew while the crenulations were developing. The microfolds preserved in the inclusion trails have been almost completely obliterated from the matrix by intense deformation/neocrystallization continuing after the growth of the porphyroblast. From Vernon (2004, fig. 5.85). Plane-polarized light; base of photo 2.7 mm.

as shown in Fig. 3.33; (4) 'millipede' microstructure (Bell & Rubenach, 1980; Johnson, 1993a,b; Johnson & Moore, 1996; Johnson & Bell, 1996), shown in Fig. 3.34; and (5) inclusion spirals, which occur mainly in garnet porphyroblasts (Rosenfeld, 1968, 1970; Schoneveld, 1977, 1979; Powell & Vernon, 1979; Johnson, 1993a; Johnson & Moore, 1996), but which have also been

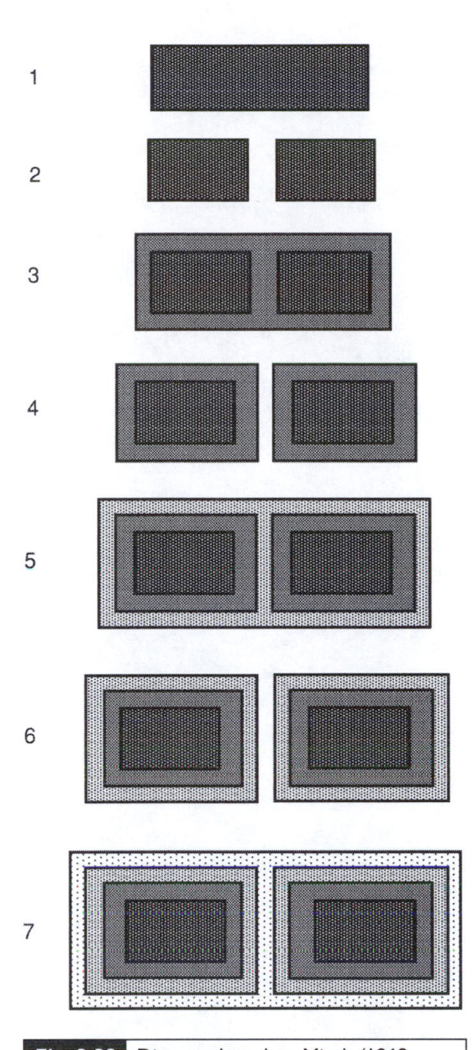

Fig. 3.33 Diagram, based on Misch (1969, p. 48), showing growth-deformation sequence involved in 'paracrystalline microboudinage' of chemically zoned grains with a single site of rupture. Though the process is represented as finite steps, both stretching and mineral growth may occur simultaneously. From Vernon (2004, fig. 5.86).

observed in pyrite porphyroblasts (Craig, 1990; Craig *et al.*, 1991). These 'snowball' porphyroblasts (Fig. 3.35A,B) are generally considered to have rotated (with respect to geographic co-ordinates) while they grew, though this interpretation has been challenged by Bell & Johnson (1989). The controversy has been reviewed by Johnson (1993b, 1999a,b). This difference in interpretation does not alter the inference that the 'snowball' porphyroblasts are syn-foliation development, provided their inclusion trails are continuous with the matrix foliation. Trails discontinuous with the matrix foliation could indicate that the porphyroblasts overgrew an earlier microfold or grew during an earlier deformation event.

(A)

(B)

Fig. 3.34 Porphyroblast of staurolite with straight quartz inclusion trails that curve near the edge of the porphyroblast into the matrix foliation without truncation, indicating growth of the porphyroblast during folding of the matrix foliation. The sense of curvature of the foliation at one end of the porphyroblast is opposite to that at the other end ('millipede structure'). From Vernon (2004, fig. 5.76). Plane-polarized light; base of photo 3.4 mm.

The most common microstructural evidence of syn-foliation porphyroblasts is the presence of curved inclusion trails that are continuous with the matrix foliation (Bell & Rubenach, 1983; Bell *et al.*, 1986; Jamieson & Vernon, 1987; Prior, 1987; Vernon, 1987a, 1988a, 1989; Vernon *et al.*, 1993b; Johnson & Vernon, 1995a,b), as shown in Figs. 3.34, 3.36, 3.37 and 3.38. The inclusion trails should curve inside the porphyroblast, as well as in the matrix.

Porphyroblasts with dominantly straight inclusion trails that curve into the matrix foliation at the edges of the porphyroblasts are generally interpreted as having grown in the early stages of development of a crenulation

(A)

(B)

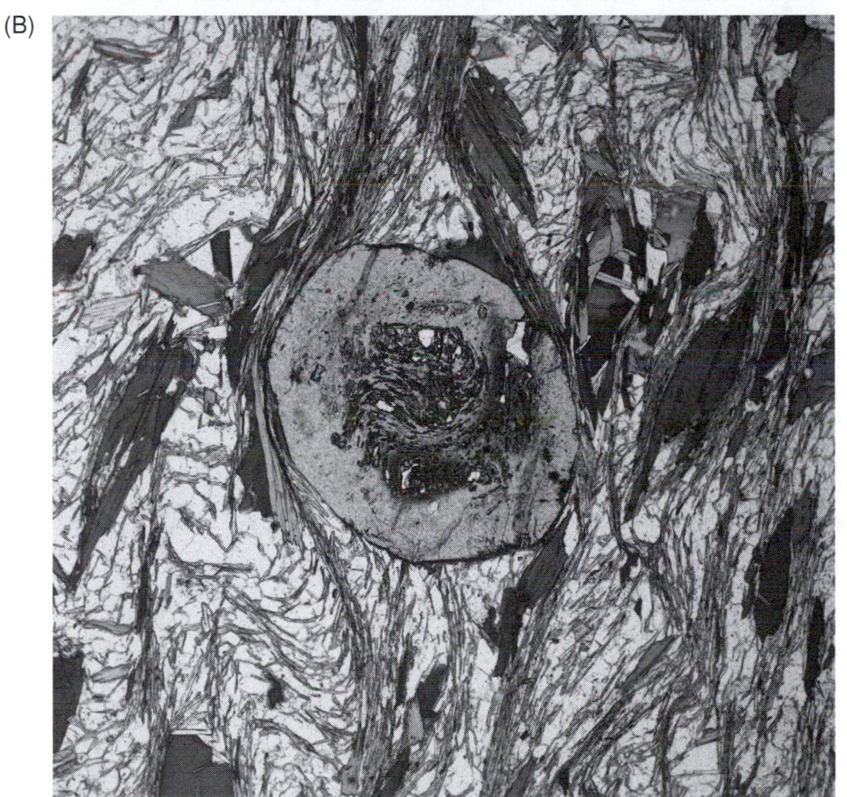

Fig. 3.35 (A) Garnet porphyroblast ('snowball garnet') with spirals of graphite (opaque) and quartz (clear) inclusions in a schist from the Himalayas. From Vernon (2000b, fig. 159; 2004, fig. 5.87). Plane-polarized light; base of photo 3.4 mm. (B) Garnet porphyroblast with a 'snowball' core, formed by syndeformational growth, and an inclusion-free rim, formed later during a period of relatively slow growth when the porphyroblast cleared itself of potential inclusions, in a schist form Cordillera Darwin, Patagonia. The crenulated matrix has been deflected around the porphyroblast during later deformation. Plane-polarized light; base of photo 3.5 mm.

Fig. 3.36 Garnet porphyroblasts with graphite-quartz inclusion trails ('spirals') that mainly curve continuously into the matrix foliation, indicating growth during microfolding (crenulation) of the foliation, Chiwaukum Schist, Cascade Range, Washington, USA. From Vernon (2004, fig. 5.74). Plane-polarized light; base of photo 4 mm.

Fig. 3.37 Crenulated schist, Lukmanier area, Swiss Alps, showing a porphyroblast of staurolite (St) with straight inclusion trails and a porphyroblast of plagioclase (Pl) with curved inclusion trails that are continuous with the matrix foliation. The staurolite evidently grew before the initiation of the crenulations (or possibly in the very earliest stages of their formation), and the plagioclase grew during the development of the crenulations. Although these relationships suggest a clear difference between the times of growth of the staurolite and plagioclase, this could be the result of local variations in nucleation and growth conditions, and so may not apply everywhere in the rock, in which case both minerals could belong to the same metamorphic mineral assemblage. More examples would need to be investigated. From Vernon (2004, fig. 5.75). Crossed polars; base of photo 4 mm.

Fig. 3.38 Porphyroblasts of andalusite with curved inclusion trails passing continuously into the matrix foliation, in a schist from the contact metamorphic aureole of the Ardara pluton, Donegal, Ireland. The porphyroblasts grew when the crenulations were well developed, after which deformation/neocrystallization of the matrix removed evidence of the crenulations. The continuity of the inclusion trails with the matrix foliation suggests that the matrix deformation was continuous with the crenulation event represented by the inclusion trails. Compare with the situation shown in Fig. 3.39. Crossed polars; base of photo 2 cm.

foliation (Bell *et al.*, 1986; Bell & Hayward, 1991), as shown in Fig. 3.37. Straight or slightly to moderately curved inclusion trails continuous with a more tightly folded or planar matrix foliation (Figs. 3.32, 3.37, 3.38) indicate post-porphyroblast deformation of the matrix that was probably a continuation of the deformation responsible for the curved inclusion trails. Straight or curved inclusion trails truncated by the matrix foliation (Figs. 3.39–44) also indicate post-porphyroblast deformation of the matrix, but the deformation may have occurred well after the deformation responsible for the curved inclusion trails.

Only porphyroblasts for which the curved inclusion trails are continuous with the matrix foliation (Figs. 3.34, 3.36–38) can be used to reliably time porphyroblast growth relative to the development of a matrix foliation (Johnson & Vernon, 1995b), as shown in Fig. 3.45.

'Decrenulation' may occur in the matrix by either (1) intense strain and neocrystallization at the culmination of the crenulation-forming event, causing transposition of crenulations into a single composite foliation (Bell & Rubenach, 1983) or (2) reactivation of matrix foliations during a later deformation event (Bell, 1986; Vernon, 1989). The result of these processes may be a porphyroblast with curved or crenulated inclusion trails, in a matrix from which evidence of former crenulations has been obliterated, at least at first sight (Figs. 3.28, 3.32, 3.35, 3.38, 3.39, 3.42–44). However, close inspection of strain shadows adjacent to porphyroblasts or of low-strain areas elsewhere in the matrix commonly reveals residual crenulations (Bell & Rubenach, 1983; Meneilly, 1983; Bell, 1985; Vernon, 1989), the orientation of which can be related to those preserved as inclusion trails in the porphyroblasts (Bell, 1985; Vernon, 1989), as shown in Figs. 3.39 and 3.40. However, even this information may be removed by very intense deformation (Fig. 3.41). Moreover, several cycles of crenulation, transposition

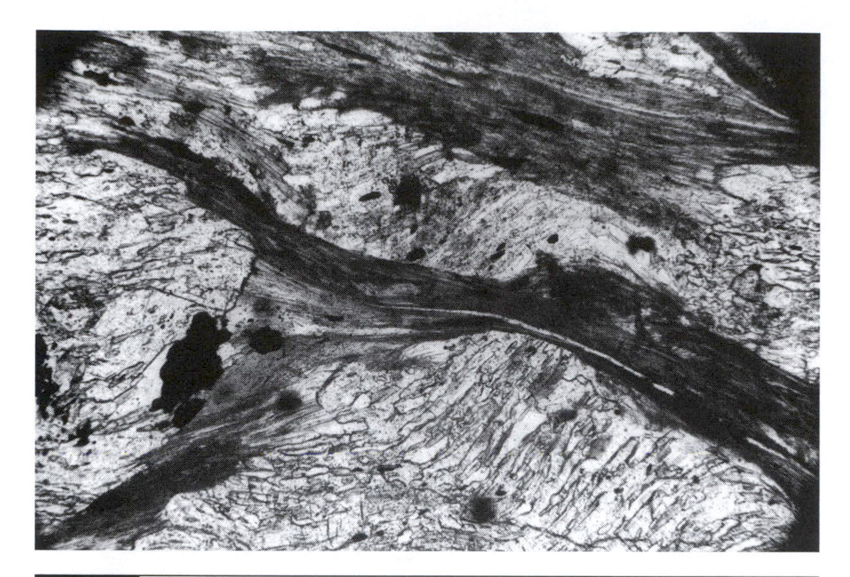

Fig. 3.39 Porphyroblasts of andalusite with curved (crenulated) inclusion trails markedly truncated by the sillimanite-rich matrix foliation, in a schist from the contact metamorphic aureole of the Ardara pluton, Donegal, Ireland. This relationship indicates strong deformation of the matrix after growth of the porphyroblast, which could have occurred some time after the growth of the porphyroblast. Compare with the situation shown in Fig. 3.38.

and reactivation can occur in the same deformation event, producing composite foliations (Tobisch & Paterson, 1988; Gibson, 1992).

Schists commonly show evidence of deformation and metamorphism continuing after porphyroblast growth, confirming, at least in a general

Fig. 3.40 Intensely deformed matrix with residual tight isoclinal folds and isolated fold hinges, resulting from 'decrenulation' in response to intense deformation, Chiwaukum Schist, Cascades, Washington, USA. More open crenulations developed at a less strongly deformed stage are preserved in the strain-shadow of an andalusite (And) porphyroblast that was protected from the intense deformation. Plane-polarized light; base of photo 8 mm.

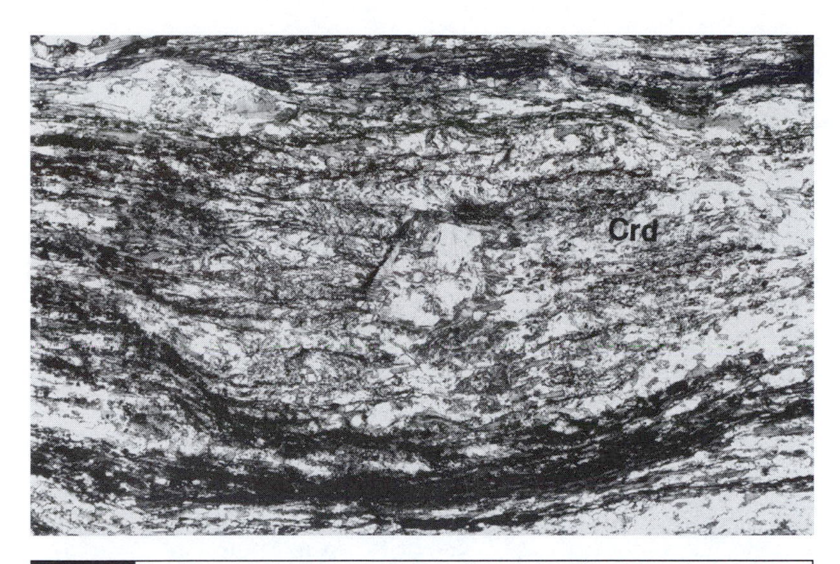

Fig. 3.41 Porphyroblast of cordierite (Crd) with mica-graphite inclusion trails that outline a crenulation cleavage, evidence of which has been obliterated from the matrix by either continued or later deformation, Chiwaukum Schist, Cascade Range, Washington, USA. Even in the relatively protected strain-shadow at the left end of the porphyroblast, little or no evidence of the former crenulations remains. The cordierite contains an inclusion of chiastolite (centre). Plane-polarized light; base of photo 4 mm.

sense, that the porphyroblasts grew during a deformation episode or series of episodes. For example, matrix folia are commonly deflected around and crowded against porphyroblasts (Figs. 3.31B, 3.34, 3.35, 3.38, 3.41). Relatively large strain accumulations ('strain caps') adjacent to some porphyroblasts result in concentrations of mica and/or opaque material, probably due to solution of quartz (Figs 3.32A,B, 3.39).

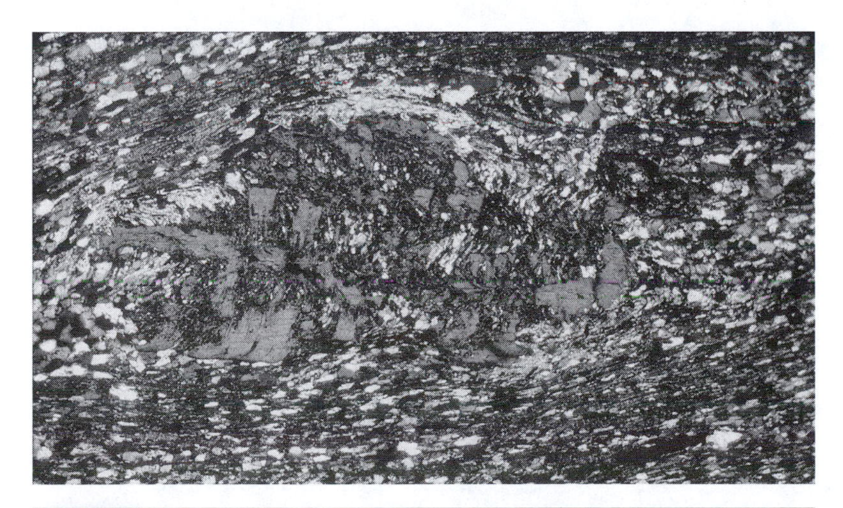

Fig. 3.42 Dendritic porphyroblast of andalusite with inclusion trails delineating former crenulations that since have been obliterated from the matrix by either continued or later deformation, except in the strain shadow to the right of the porphyroblast, Chiwaukum Schist, Cascade Range, Washington, USA. The deforming matrix foliation has been forced to deflect around the much stronger porphyroblast. Crossed polars; base of photo 4 mm.

Fig. 3.43 Porphyroblast of staurolite with graphite inclusion trails delineating tight microfolds that have been obliterated from the matrix by either continued or later deformation, Chiwaukum Schist, Cascade Range, Washington, USA. From Vernon (2000b, fig. 149; 2004, fig. 5.72B). Plane-polarized light; base of photo 4.4 mm.

The presence of metamorphic minerals, consistent with the grade of metamorphism of the rest of the rock, in the 'necks' between microboudins (Vernon & Flood, 1979; Vernon *et al.*, 1993b) indicates that the microboudinage occurred during the prograde metamorphism, suggesting

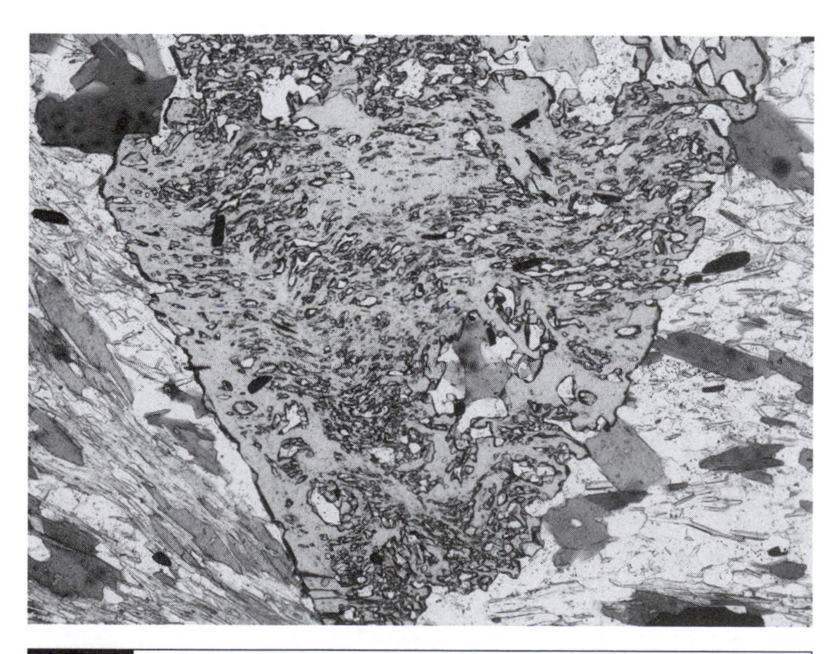

Fig. 3.44 Porphyroblast of staurolite with inclusion trails that preserve outlines of crenulations that have since been removed from the matrix by either continued or later deformation, Rangeley area, Maine. Sample by courtesy of Charles Guidotti and Scott Johnson. Plane-polarized light; base of photo 3 mm.

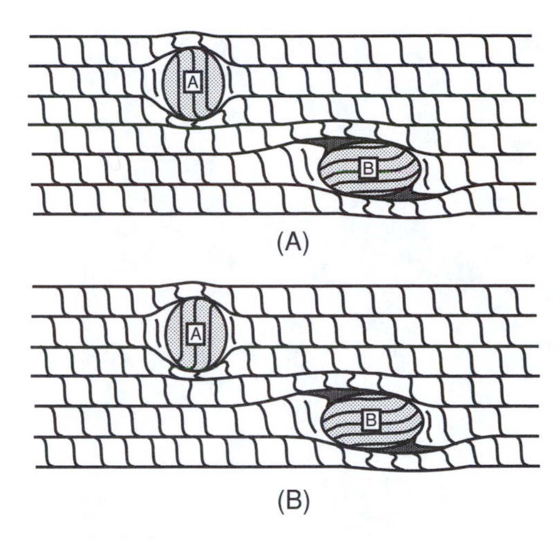

Fig. 3.45 Diagrams showing the importance of continuous versus discontinuous inclusion trails, deformed strain-shadows and asymmetry of inclusion trail curvatures in determining relative time sequences of porphyroblast growth. (A) Porphyroblast A has inclusion trails with the same asymmetry as crenulations in the matrix, and the inclusion trails are continuous with folia in the matrix. These features are reliable evidence for inferring growth during development of the crenulation foliation. Strain shadows around porphyroblast B are folded into the crenulation foliation, and so B pre-dates both A and the matrix foliation. If the inclusion trails in A were not continuous with the matrix foliation, this correlation would be less certain. (B) The inclusion trails in A have the opposite asymmetry to that of the matrix crenulations, and lack continuity with the matrix folia. Therefore, A did not grow during the development of the crenulation foliation, and the rock shows no reliable evidence of its timing with respect to B. After Johnson & Vernon (1995b, fig. 3); copyright *Journal of Structural Geology*; published with permission.

that the porphyroblasts were formed and deformed as part of the prograde metamorphic history. Retrograde aggregates in the 'necks' indicate that the microboudinage occurred during cooling or a later, lower-grade metamorphic event.

Post-foliation porphyroblasts

Some rocks contain porphyroblasts that have overgrown and preserved the shapes of P (mica-rich) and Q (quartz ± feldspar-rich) domains of a crenulation foliation, with identical crenulations in the matrix, the inclusion trails being continuous with the matrix foliation (e.g., Vernon *et al.*, 1993b), as shown in Fig. 3.46. This implies that the porphyroblasts grew after the deformation responsible for the microfolds, at least at the scale of the thin section. This is sometimes referred to as 'static' growth. The

(A)

(B)

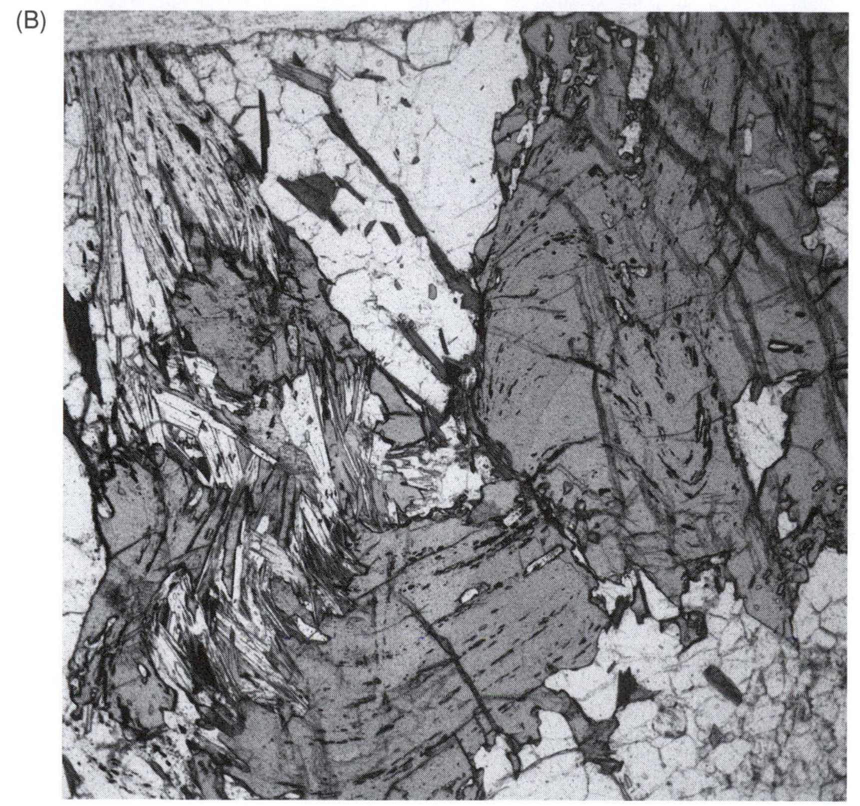

Fig. 3.46 (A) Graphite inclusion trails outlining open crenulations, passing continuously from staurolite (St), through biotite (Bt) to quartz (Qtz). The identical, continuous microfolds indicate that all three minerals grew (as part of the same assemblage) after the deformation responsible for the microfolds. Chiwaukum Schist, Cascade Range, Washington, USA. Plane-polarized light; base of photo 1.5 mm. (B) Staurolite porphyroblasts in a schist from Cordillera Darwin, Patagonia, showing opaque mineral and quartz inclusion trails that delineate former crenulations and pass continuously into folded folia in the muscovite-quartz matrix. This relationship is consistent with growth of the staurolite after the crenulations had been formed in the matrix. Plane-polarized light; base of photo 3.5 mm.

porphyroblast growth model of Bell *et al.* (1986) requires that nutrient components are supplied from actively deforming, P domains outside the growing porphyroblast. These domains must be close enough to the growing porphyroblast for diffusion of nutrient components to be effective, and because these domains are inferred to be active, the matrix should contain tighter microfolds than those preserved as inclusion trails in the porphyroblast. Therefore, if this is not observed (as in Fig. 3.46), the most logical conclusion is that the porphyroblasts effectively grew after the deformation responsible for the microfolds.

Other examples of post-deformational porphyroblasts are random crystals and aggregates of white mica and chlorite that preserve evidence of matrix folia as inclusion trails (e.g., Vernon *et al.*, 1993b). The crystals typically are not deformed, and some occur in radiating aggregates, which would be unlikely if they grew during or before deformation. Presumably they grew during retrograde cooling after deformation had ceased, on the scale of observation. However, radiating aggregates of minerals lying in a foliation, not transecting it, (e.g., radiating sillimanite in high-grade gneisses at Broken Hill, Australia) can be formed during deformation, resulting in shortening perpendicular to the foliation.

Care must be taken with the interpretation of radiating aggregates of fibrous sillimanite in Q domains between P domains that consist of abundant, aligned sillimanite in high-grade metapelitic schists (Vernon 1987a,b). A superficial interpretation could be that the aligned sillimanite is syn-foliation deformation and that the radiating sillimanite is post-foliation deformation, whereas both may grow simultaneously in zones of contrasting strain accumulation; i.e., both can be syn-foliation deformation (Vernon 1987a,b).

Inferences of post-deformation porphyroblasts or aggregates do not preclude simultaneous deformation of other rocks in the same area. For example, the observed rock could have been in a mesoscopic or macroscopic low-strain zone surrounded by rocks that are inferred to have been actively deforming at the time the aggregates grew. However, if deformation zones spaced closer than the length of a large thin section or hand specimen are not inferred in the field, they probably would be too far away to contribute nutrient components to the growing porphyroblasts, as required by the porphyroblast growth model of Bell *et al.* (1986).

Fig. 3.47 shows a staurolite porphyroblast that has overgrown a single foliation (a slaty cleavage or a 'decrenulated' cleavage). Consequently, the porphyroblast is interpreted as being post-foliation. The foliation itself was overgrown statically by non-deformed mica grains oblique to the foliation, and both the foliation and mica grain shapes are preserved in the staurolite.

Fig. 3.48 shows a garnet porphyroblast that is both syndeformational and post-deformational, with regard to different structures. The outer parts of the porphyroblast grew during the flattening and consequent deflection of the matrix foliation around the porphyroblast. Part of the internal foliation is continuous with the matrix foliation, and part has been truncated by continuing deformation of the matrix. The porphyroblast has a core of earlier-formed garnet with inclusion trails perpendicular to the later trails. The core also shows sector arrangement of some of the inclusions. A reasonable interpretation of these microstructural features is that the core grew during or after the development of the first foliation

Fig. 3.47 Porphyroblast of staurolite with inclusions delineating a single foliation (slaty cleavage or 'decrenulated' cleavage). Consequently, the porphyroblast is interpreted as having grown post-foliation. The foliation itself was overgrown statically by non-deformed mica grains oblique to the foliation, and both the foliation and mica grain shapes are preserved in the staurolite. Staurolite-mica schist, Appleton Ridge, Maine, USA. Specimen by courtesy of Charles Guidotti and Scott Johnson. Crossed polars; base of photo 3 mm.

Fig. 3.48 Garnet porphyroblast that is both syndeformational and post-deformational, with regard to different structures. The outer parts of the porphyroblast grew during the flattening and consequent deflection of the matrix foliation around the porphyroblast. Part of the internal foliation is continuous with the matrix foliation, and part has been truncated by continuing deformation of the matrix. The porphyroblast has a core of earlier-formed garnet with inclusion trails perpendicular to the later trails. The core also shows sector arrangement of some of the inclusions. A reasonable interpretation of these microstructural features is that the core grew during or after the development of the first foliation and the rim grew during the deformation of the second foliation. No large time gap is implied. Specimen kindly provided by Margaret Macfarlane. Plane-polarized light; base of photo 3.4 mm.

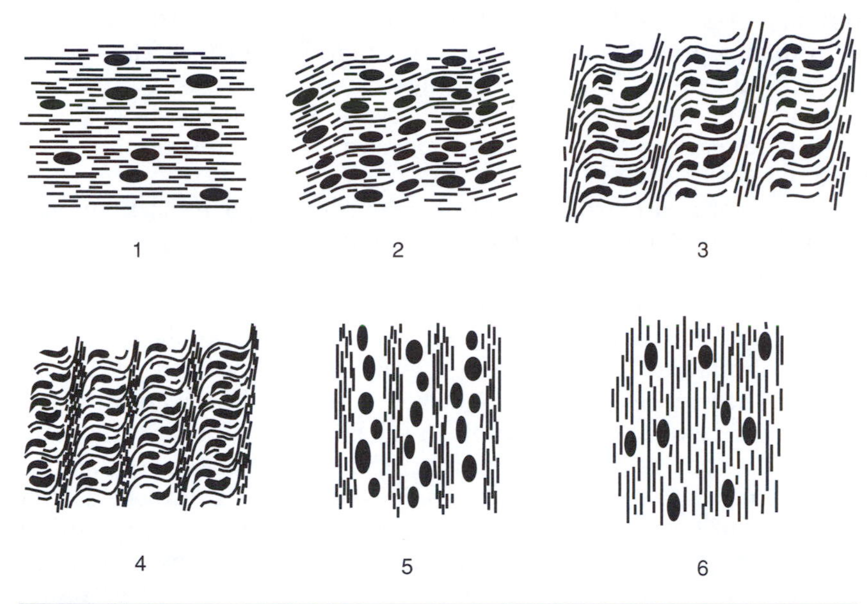

Fig. 3.49 Six stages of progressive development of a crenulation foliation (crenulation cleavage) suggested by Bell & Rubenach (1983). Sketches show (1) original foliation, (2) initiation of crenulations, (3) initiation of a compositional layering (by metamorphic differentiation) in the crenulation limbs (P or M domains), (4) growth of new mica in P domains, (5) destruction of crenulations in fold hinges (Q domains), and (6) homogeneous new foliation in a new orientation. After Johnson (1999b, fig. 2), with permission of the Mineralogical Society of America, and Vernon (2004, fig. 5.57).

and the rim grew during the deformation of the second foliation. No large time gap is implied.

Rocks that undergo metamorphic peak heating late in their deformation history (possibly continuing after deformation effectively ceases) show (1) generally undeformed grains, (2) evidence of grain coarsening without reorientation of previously aligned minerals, and (3) occurrence of only the latest crenulation cleavage as inclusion trails in the grain margins (Williams, 1991).

3.5.5 Examples of the approach

By using porphyroblast-matrix relationships, Bell & Rubenach (1983) inferred a sequence of crystallization of porphyroblastic minerals, and made inferences about the metamorphic history of the Robertson River Formation, north Queensland, Australia. They proposed a constant sequence of porphyroblastic mineral development, regardless of the local stage of development of crenulations (Fig. 3.49) that were forming during the metamorphism. For example, crenulations preserved as inclusion trails in andalusite porphyroblasts are always at an earlier stage of crenulation development than those preserved in garnet porphyroblasts in the same rock, even though the stage of development of crenulations preserved in andalusite varies form place to place. This variation depends on the time of nucleation of andalusite relative to the local stage of development of the crenulations.

Vernon (1988a) used porphyroblast-matrix microstructural relationships to infer a prograde metamorphic reaction in schists of the Cooma Complex, south-eastern Australia. Porphyroblasts of andalusite grew during the development of a crenulation cleavage (labeled 'S_3'), in response to a prograde reaction that consumed cordierite formed during the development of 'S_2'.

Inferences of relationships between growth of porphyroblastic minerals to progressive deformation, temperature and time are exemplified by prograde metamorphic mineral sequences in the Robertson River Formation and the Corella Formation, north Queensland, Australia. Reinhardt & Rubenach (1989) compared inclusion trails in porphyroblasts of each mineral with matrix foliations, from the first appearance of the mineral through its occurrences in higher-grade zones. They found that most of the foliations preserved as inclusion trails in the porphyroblasts are at a more advanced stage of crenulation development in lower-grade zones than for the same minerals in higher-grade zones. This result is to be expected from the tendency of higher-grade parts of a metamorphic terrane to begin to heat up earlier than lower-grade parts (den Tex, 1963). Therefore, porphyroblasts of a particular mineral should grow during the earlier stages of a particular crenulation-forming event at higher grades than at lower grades, assuming approximate contemporaneity of the crenulation event across the terrane.

However, this assumption may or may not be justified, and conceivably both deformation and metamorphism could progress through a volume of rock at different and even spasmodic and locally variable rates (Gibson, 1992; Johnson, 1999b, pp. 1713–14). Problems involved in correlating structures in multiply deformed areas have been discussed by Means (1963), Park (1969) and P. F. Williams (1985). In addition, M. L. Williams (1991) has shown that in terranes that have been deformed heterogeneously, the structural histories of refolded and repeatedly sheared regions are only of local importance, and can be misleading on a regional scale. Moreover, Alias *et al.* (2002) suggested that the development of a foliation (labelled 'S_2' may be diachronous across the Adelaide Fold Belt, South Australia, reflecting variations in the thermal state of the lithosphere caused by heterogeneous production and ascent of magmas.

Many studies of *P-T-t* paths assume a single heating-cooling event, implying that all foliations present formed during that event. This interpretation is supported by geochronological measurements on multiply foliated schists of the classical Barrovian sequence of Scotland (Baxter *et al.*, 2002), which indicate no age difference (within the error of measurement) between the development of garnet zone rocks (500°C–550°C) during deformation D_2 and the development of sillimanite zone rocks (*c.* 660°C) during deformation D_3.

On the other hand, evidence for repeated heating events in some areas has been obtained (e.g., Vernon, 1996a). For example, Crowley *et al.* (2000) used U-(Th)-Pb dating to infer multiple periods of metamorphism and deformation in a Barrovian sequence in the Canadian Cordillera. They found that the thermal peak of metamorphism occurred at different times (roughly 10–50 Ma apart) in three kilometre-scale domains, and that the peak events coincided with secondary events in the other domains. Either the domains were tectonically assembled along unrecognized shear zones, or the events must have been relatively local (Crowley *et al.*, 2000), indicating potential problems in making evidence based on structural evidence alone.

This situation applies especially in low-pressure/high-temperature (LPHT) metamorphism (both contact and regional), for which local magmatic thermal pulses may control metamorphic reactions (e.g., Collins & Vernon, 1991, 1992; Collins *et al.*, 1991; Vernon *et al.*, 1993a; Stüwe *et al.*, 1993; Williams, 1994; Alias *et al.*, 2002); the same sequence of mineral growth and foliation development may occur in different areas at different times. For example, Williams (1994), investigated three metamorphic terranes (two contact metamorphic, one regional) in Arizona, USA and New Mexico that show the same sequence of porphyroblast-foliation development, despite being of different ages. Garnet, biotite, staurolite and andalusite porphyroblasts have inclusion trails indicating overgrowth of progressive stages in the transformation of a slaty cleavage (S_1) into a crenulation cleavage (S_2). Garnet and biotite have trails indicating crenulation stage 2 or 3 of Bell & Rubenach (1983), and staurolite and andalusite trails indicate stages 3 to 6. The transition from greenschist to amphibolite facies conditions coincides with the change from S_1 to S_2, and the transition from chlorite-dominated to biotite-dominated mineral assemblages occurs over a small temperature interval (400 °C–450 °C). The rapid production of abundant water as the chlorite dehydrated may have weakened the rocks and so contributed to the evolving crenulation cleavage, during which time the porphyroblasts grew rapidly. Thus, metamorphism and deformation were intimately linked, producing growth-foliation sequences that are similar to those observed in many areas (Williams, 1994, p. 19).

These investigations show that careful interpretation of porphyroblast-matrix microstructural relationships can provide relatively clear evidence of sequential porphyroblast growth and prograde metamorphic reactions, especially with regard to the development of foliations in a single rock. However, correlations between rocks of different areas are often based on the assumption of contemporaneity of foliation-forming deformation across the region concerned, which may not be justified.

Another outcome of the investigations of Williams (1994) is that porphyroblasts may grow relatively rapidly, especially in LPHT metamorphic terranes, in which relatively short magmatic heat pulses may control the metamorphism. He suggested that the growth durations of porphyroblasts may be no more than a few hundred thousand years, which supports an earlier estimate of 300–300,000 years by Paterson & Tobisch (1992). Microstructural evidence consistent with this inference includes (1) porphyroblasts with many small inclusions, implying rapid growth (Vernon, 1976), (2) porphyroblasts with small inclusions and no inclusion trails, implying growth before foliations had time to develop (Fleming & Offler, 1968; Vernon *et al.*, 1993b), and (3) porphyroblasts with either straight inclusion trails curving slightly into the matrix foliation at their edges or with weakly crenulated inclusion trails, implying growth before crenulation of an existing foliation had time to develop.

Even where reliable evidence for *P-T-D-t* paths is available, detailed geochronological evidence on the microscopic scale is needed before the foliation-forming events can be separated in time (Section 3.7.5).

3.6 Using inclusion trails in porphyroblasts to infer tectonometamorphic events

3.6.1 Use of orthogonal inclusion trails in porphyroblasts

Porphyroblasts-matrix microstructural relationships have been used to infer orogenic histories, such as multiple orthogonal orogenic movements and changing plate motions (e.g., Bell & Johnson, 1989; Johnson, 1990, 1992, 1999a,b; Bell *et al.*, 1995, 1998; Bell & Hickey, 1999; Bell & Mares, 1999). For example, Bell & Johnson (1989) suggested that successive inclusion trail patterns (crenulation cleavages) develop statistically orthogonal to each other, and that evidence of these orthogonal cleavages may be preserved as inclusion trails in garnet porphyroblasts. Their interpretation was that vertical and horizontal foliations represent alternating stages of compression and collapse of a mountain belt (Bell & Johnson, 1989). An alternative model could be alternating periods of orogenic extension and contraction (Collins, 2002). However, as emphasized by Johnson (1999a, p. 1186), many of the published examples of inferred orthogonal foliations are based entirely on microstructural evidence, and so 'it may be a big jump to infer orogen-scale processes from such small-scale observations.'

Moreover, not all statistical distributions of inclusion trails indicate orthogonal foliations (Paterson & Vernon, 2001), and some folia may be very oblique to each other (Barker, 1994; Johnson, 1999b, p. 1721; Paterson & Vernon, 2001). Another important point is that strongly foliated materials, such as schists, invariably tend to crenulate or kink at high angles to the foliation, in response to a relatively wide range of local stress distribution.

Even if it is accepted that orthogonal foliations do represent regional deformation events driven by successive horizontal and vertical compression and, moreover, are in their original positions, a major difficulty remains with accepting the use of porphyroblast-matrix microstructural relationships to infer orogenic histories. The problem is that evidence (e.g., orthogonal crenulation-foliations) for these inferred events in the matrix between porphyroblasts is generally lacking. This is generally explained by inferring obliteration of the foliations during subsequent deformation, except in strain shadows adjacent to porphyroblasts. To avoid this problem, Bell & Hickey (1999) suggested that the putative orthogonal foliations (crenulations) produced in response to orogenic movements form only right against the porphyroblast. This interpretation implies that the obliteration process is remarkably efficient throughout a metamorphic terrane. Paterson & Vernon (2001) found excellent examples of local crenulations against porphyroblasts in schists of the Foothills terrane, central Sierra Nevada, California, USA, though the axial surfaces of the crenulations are not necessarily orthogonal to the previous foliation. However, the question arises as to why major orogenic movements would affect only the matrix adjacent to porphyroblasts, when orogenic movements typically produce widespread foliations. Some evidence of these foliations outside porphyroblasts would be expected. Therefore, rather than inferring regional deformation events, it may be better to consider local strain heterogeneities as an explanation of the observations (Paterson & Vernon, 2001). Similarly, the observation of Stallard & Hickey (2001) that different porphyroblasts in the same outcrop may show different growth histories, with reference to successive foliations, suggests that porphyroblast growth may tend to reflect local strain heterogeneities, rather than regional deformation. As mentioned previously (Section 3.5.3), the local nature of crenulations in their earlier stages of development (Fig. 3.29) could lead to growth of porphyroblasts with straight inclusion trails in non-crenulated areas and curved trails in crenulated areas of the same thin section, thereby causing misleading timing relationships.

3.6.2 Foliation inflection/intersection axes (FIAs)

The position of a fold axis, even an early one in a multiply deformed rock, can be determined by using the asymmetry of crenulations preserved as inclusion trails in porphyroblasts, provided they are inferred or assumed to remain fixed, with regard to geographical co-ordinates, during subsequent deformation (e.g., Hayward, 1990). The basis for this approach is that crenulations commonly change asymmetry across crenulation microfolds. Though originally referred to simply as a 'fold axis' (Hayward, 1990), the line is now called a '*foliation inflection/intersection axis*' or 'FIA'.

The technique of estimating the orientation of a FIA has been described by Hayward (1990, 1992) and Bell *et al.* (1995, 1998, 2004). It involves recording the orientation of the line separating opposing asymmetries of curved inclusion trails in porphyroblasts in vertical thin sections of varying orientation. Only vertical sections are employed; no other section planes are considered, because the proponents of this technique attempt to ascertain the directions of horizontal shortening. Initially, three to six vertical sections are cut at every 60° or 30° interval from true north, and then closer sections are cut, in an attempt to locate the FIA to within 10° (Fig. 3.50). The technique may be used for single or overprinting (multiple) crenulation inclusion trails (Fig. 3.50). Individual FIAs have been used to infer bulk compressive movement directions, and sequences of FIAs (inferred from rocks with multiple crenulation inclusion trails) have been used to correlate different early orogenic 'events' in a multiply deformed terrane (e.g., Hayward, 1990).

Potential measurement errors include the initial orientation of the rock sample in the field, the re-orientation of the sample for sectioning, and the orientation of the section slices. The total accumulated error in determining the trend or plunge of an FIA has been estimated at ±8° (Bell & Hickey, 1999). However, both clockwise and anticlockwise asymmetries commonly occur in thin sections oriented close to the FIA, owing to the curvilinear geometry of FIAs in individual porphyroblasts (Hayward, 1990) and variation in orientation between porphyroblasts (Stallard *et al.*, 2003). Thus, the FIA orientation is best reported as a mid-point value plus the interval over which the asymmetry switches (Stallard *et al.*, 2003), for example 070 ± 25°.

Other sources of variation in FIA orientation (Stallard *et al.*, 2003) are the heterogeneous rheology of rocks, the anastomosing of foliation around heterogeneities such as granite intrusions and rotation of porphyroblasts relative to other porphyroblasts in the same rock (see references mentioned below). A detailed review by Stallard *et al.* (2003) has shown that in a set of temporally related FIAs, the typical spread of orientations occupies a 60° range, though outliers occur at other orientations, including near-normal to the peak distribution. In fact, considerable variation in FIA

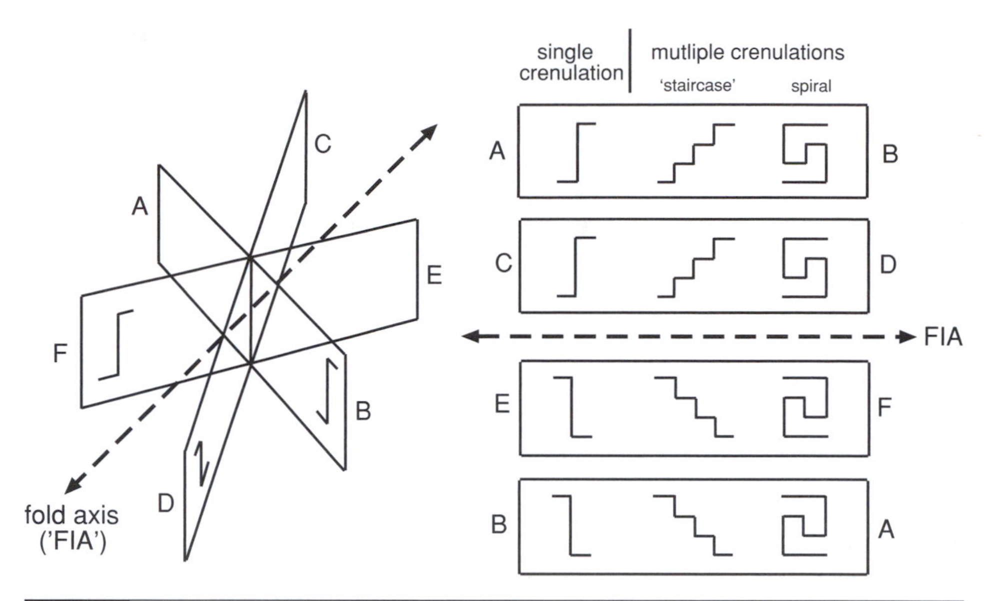

Fig. 3.50 Location of an approximate fold axis ('FIA') orientation using at least three vertical thin sections orientated at 60° to each other. The sections are viewed in the same consistent sense, in this instance clockwise, starting with section AB. A change in crenulation asymmetry occurs between sections CD and EF, which gives an initial location of the FIA. This can be located more precisely by using more closely spaced sections. Section BA repeats section AB, but is viewed from the opposite direction. The styles of crenulation are schematic. After Hayward (1990, fig. 2); copyright Elsevier; published with permission.

orientation may occur in the same rock, at least in some areas. For example, Stallard *et al.* (2003) showed that temporally related FIAs may have a large range of orientations (up to 147° in the Canton Schist, Georgia, USA), making them unreliable for correlation purposes. Stallard *et al.* (2003) suggested that FIAs are best used as semi-quantitative indicators of general trends, rather than as exact measurements for quantitative analyses.

FIAs have been used to infer regional horizontal shortening (e.g., plate motion) events, separated by 'tectonic breaks' (e.g., Sayab, 2006). However, even if the geometrical distinction can be made (i.e., assuming the technique is accurate enough), more evidence would be needed to show that each FIA is separated widely in time, or even that a particular FIA orientation was preserved during later intense deformation/metamorphism.

P-T pseudosections (Chapters 1 and 2) and FIAs have been used to infer and characterize multiple periods of deformation and metamorphism (Bell & Kim, 2004; Kim & Bell, 2005; Sayab, 2006). Provided the FIAs are taken to represent growth of different generations of porphyroblasts, reactions inferred to be responsible for the porphyroblastic minerals can be modelled for each generation using the pseudosection technique and so a tectonometamorphic history can be inferred.

The degree of confidence in such a history depends on methods used to infer the reactions and the sources of variation in orientation referred to previously, as well as acceptance of the general validity of the FIA approach. This acceptance itself requires acceptance of the hypothesis that porphyroblasts do not rotate, with respect to external geographic co-ordinates, during subsequent deformation, so that they preserve their original orientation of their inclusion trails. This proposition, though reasonable for compositionally differentiated schists (Ramsay, 1962; Fyson, 1975, 1980; Fyson & Frith, 1979; Vernon, 1987a, 1988b; Steinhardt, 1988), may not be universally applicable; see Johnson (1993b, 1999b) and Vernon (2004, pp. 446–451) for reviews of the problem.

Quite apart from these problems, the inference that well established differences in FIA orientation necessarily indicate different metamorphic or deformation events is itself questionable, requiring accurate dating (e.g., with reliably inferred monazite inclusions). Conceivably, these changes could represent preserved stages of a single evolving metamorphic-deformation 'event'.

3.7 Absolute dating (geochronology) of mineral growth and foliation-forming events

3.7.1 Introduction

Radiogenic isotopes can be used to determine the age of rocks and minerals, and several publications deal with this subject in considerable detail (e.g., Faure, 1986; DePaolo, 1988; Rollinson, 1993). The application of radiogenic isotopes in igneous petrogenesis has been addressed by Wilson (1989) and Rollinson (1993). Despite their low modal abundance in crustal rocks, minor ('accessory') minerals commonly contain measurable concentrations of elements of geological and geochronological interest. Rare earth elements (REE), including the important heat-producing elements, U and Th, can be key tracers for many geological processes (Kelly *et al.*, 2006). As accessory minerals may contain a large percentage of a rock's trace element content, they may exert an important influence over the composition of reaction products and melt composition. Minerals containing radio-isotopes, such as zircon, monazite, titanite and xenotime, can provide data used to infer the timing and duration of orogenic events.

As most geochronology involves analysing isotopic ratios of elements present in small proportions in a rock, or ratios of elements present in moderate proportions in accessory minerals, traditional techniques involved rock crushing and/or mineral separation, which destroy the microstructural context for age results. However, recent improvements in analytical procedures that involve in situ trace element composition of minerals (in a thin section or grain mount) enable integration of isotopic studies with metamorphic petrology. By carefully determining mineral chemistry in metamorphic assemblages, it is now possible to confidently relate some ages to their host metamorphic assemblages, and obtain more information on diffusion processes using trace elements, which tend to have higher closure temperatures (Chapter 2) than major elements. The temperature-dependent nature of daughter-isotope diffusion (and hence retention) always needs to be considered when evaluating data, and can enable the resolution of thermal history of metamorphic belts (e.g., Zeitler, 1989).

This section addresses the basic principles of geochronology, and then describes some common methods used in dating metamorphic assemblages.

3.7.2 Radioactive decay principles

Rutherford & Soddy (1903) established the foundations of modern geochronology by showing that the process of radioactive decay is exponential and independent of chemical or physical conditions. The rate of decay of an unstable parent atom is proportional to the number of atoms (N) present at any time (t):

$$-\frac{dN}{dt} \propto N \tag{3.1}$$

where dN/dt is the rate of change of the parent atoms and the minus sign indicates that the rate decreases with time. This is converted into an equation by a constant (λ), called the *decay constant*.

$$-\frac{dN}{dt} = \lambda N \tag{3.2}$$

Rearranging equation 3.2 and integrating:

$$-\frac{dN}{N} = \lambda\, dt \tag{3.3}$$

$$-\int \frac{dN}{N} = \lambda \int dt \tag{3.4}$$

$$\text{leads to} \quad -\ln N = \lambda t + C \tag{3.5}$$

where C is a constant. If at $t = 0$, $N = N_0$, then $C = -\ln N_0$. Therefore:

$$-\ln N = \lambda t - \ln N_0 \tag{3.6}$$

Substituting equation 3.6 into equation 3.5 gives:

$$\ln N - \ln N_0 = -\lambda t \qquad \text{or} \qquad \ln \frac{N}{N_0} = -\lambda t \tag{3.7}$$

$$\frac{N}{N_0} = e^{-\lambda t} \qquad \text{or} \qquad N = N_{0e}^{-\lambda t} \qquad \text{or} \qquad N_0 = Ne^{\lambda t} \tag{3.8}$$

Assume that the decay of a radioactive parent produces a stable radiogenic daughter, and the number of daughter atoms at $t = 0$ is zero. For a closed system (i.e., no loss or gain of daughter or parent atoms), the number of daughter atoms (D^*) produced by decay of the parent at any time t is given by:

$$D^* = N_0 - N \tag{3.9}$$

Substitution of equation 3.8 into equation 3.9 gives:

$$D^* = N_0(1 - e^{-\lambda t}) \tag{3.10}$$

This equation relates the number of stable radiogenic daughter atoms (D^*) at any time t formed by the decay of a radioactive parent, the initial number of which at $t = 0$ was N_0.

For most decay schemes relevant to geological problems, it is not possible to directly measure the initial number of atoms. However, the number of radiogenic daughter atoms and the number of parent atoms remaining can be measured using mass spectrometry. It is thus more convenient to relate the number of radiogenic daughters (D^*) to the number of parent atoms remaining (N), rather than to N_0. Substituting for N_0 (from equation 3.8) in equation 3.10, gives the equation:

$$D^* = Ne^{\lambda t} - N = N(e^{\lambda t} - 1) \tag{3.11}$$

The total number of daughter atoms in a system in which decay is occurring can be expressed as:

$$D_{total} = D_0 + D^* \tag{3.12}$$

where D_0 is the initial number of daughter atoms (at $t = 0$) and D^* is the number of radiogenic daughter atoms produced by decay of the parent. Combining equations 3.12 and 3.11:

$$D = D_0 + N(e^{\lambda t} - 1) \tag{3.13}$$

This is the basic equation from which age determinations are made in many geological systems, based on the decay of a radioactive parent to a stable daughter. D and N can be measured, whereas D_0 can be calculated from the data or assumed. Once these parameters are determined, the equation can be solved for t, which is the age of the 'system'. This assumes: (1) the isotopic system remained 'closed' with respect to loss or gain of both parent and daughter atoms; (2) the decay constant* (λ) is known accurately; and (3) D and N are measured accurately.

The decay constant can be easily visualized in terms of a related quantity called the *half-life*, which is the time required for half of any given amount of an isotope to decay to its daughter products. It is a relatively easy quantity to measure and provides an estimate of the decay constant λ. If $t = T_{1/2}$, the time to reduce the initial number of parent atoms by 50%, $N = 0.5N_0$. Substituting into 3.8 gives:

$$\frac{1}{2}N_0 = N_0 e^{-\lambda T_{1/2}} \tag{3.14}$$

Solving for $T_{1/2}$:

$$T_{1/2} = \frac{\ln 2}{\lambda} = \frac{0.693}{\lambda} \tag{3.15}$$

This provides a convenient relationship between the decay constant and the measurable half-life of a radioactive isotope.

3.7.3 Isochron calculations

An isochron diagram is a bivariant plot of measured parent-daughter isotope ratios for a suite of isotopically-related samples or minerals. The diagram is commonly used in igneous petrology to relate cogenetic samples of magma, with variations in isotopic ratios inherent from compositional variations. The technique can also be used to date metamorphic assemblages, as different minerals in an isotopically-equilibrated assemblage have distinct isotopic ratios. The main decay schemes of interest to metamorphic petrology have been the Rb-Sr, Sm-Nd and K-Ar systems, the decay constants of which are shown in Table 3.1. The example below describes the application of the Rb-Sr isochron diagram to date a metamorphic rock (see also Rollinson, 1993), but the method is similar for the other techniques. As the geochemical behaviour of Rb and Sr is similar to that of the alkali elements, the Rb-Sr system is most appropriate to the study of felsic metamorphic rocks.

From equations given in Section 3.2, the total number of ^{87}Sr atoms in a rock or mineral that has been closed for t years is given by the equation:

$$^{87}Sr_{now} = {}^{87}Sr_0 + {}^{87}Rb_{now}(e^{\lambda t} - 1) \tag{3.16}$$

Table 3.1. Decay constants for the Rb-Sr (Steiger & Jäger, 1977), Sm-Nd (Lugmair & Marti, 1978) and K-Ar isotopic systems (Steiger & Jäger, 1977).

			Ratios plotted on the isochron diagram	
	Decay scheme	Decay constant	X axis	Y axis
Rb-Sr	$^{87}Rb \rightarrow {}^{87}Sr+B$	1.42×10^{-11}	$^{87}Rb/^{86}Sr$	$^{87}Sr/^{86}Sr$
Sm-Nd	$^{147}Sm \rightarrow {}^{143}Nd+He$	6.54×10^{-12}	$^{147}Sm/^{144}Nd$	$^{143}Nd/^{144}Nd$
K-Ar	$^{40}K \rightarrow {}^{40}Ar - B$	0.581×10^{-10}	$^{40}K/^{36}Ar$	$^{40}Ar/^{36}Ar$

where $^{87}Sr_0$ is the number of ^{87}Sr atoms present when the rock or mineral formed. The precise measurement of absolute isotopic concentrations is difficult; so isotopic ratios are measured. A related isotope not involved in the radioactive decay scheme is used in the ratio; for the Rb-Sr scheme ^{86}Sr is used.

By dividing equation 3.16 through by ^{86}Sr:

$$\left(\frac{^{87}Sr}{^{86}Sr}\right)_{now} = \left(\frac{^{87}Sr}{^{86}Sr}\right)_0 + \left(\frac{^{87}Rb}{^{86}Sr}\right)_{now} (e^{\lambda t} - 1) \quad (3.17)$$

This equation is the basis for age determinations using the Rb-Sr method. The $^{87}Sr/^{86}Sr$ and $^{87}Rb/^{86}Sr$ ratios of a rock or mineral can be determined by mass spectrometry, leaving $(^{87}Sr/^{86}Sr)_0$ and t unknown. As equation 3.17 is a straight line, the age and initial ratio can be calculated using a plot of measured $^{87}Sr/^{86}Sr$ and $^{87}Rb/^{86}Sr$ ratios for a suite of isotopically-related rocks or minerals, if they have sufficient variation in $^{87}Sr/^{86}Sr$ ratios. The method is well illustrated using an example shown in Fig. 3.51. The age of the suite of isotopically-related rock or mineral samples is given by the slope of the line according to:

$$t = \frac{1}{\lambda} \ln(slope + 1) \quad (3.18)$$

As an example of the application of the technique (Clarke, 1988), consider a granulite facies terrane in Kemp Land, east Antarctica with interlayered charnockitic orthogneiss (Stillwell Gneiss) and garnet-sillimanite metapelitic gneiss (Colbeck Gneiss). Several large rock samples were collected of each rock type over restricted areas, after identifying related samples that had subtle modal variations, suggesting distinct $^{87}Sr/^{86}Sr$ ratios. At the time of sampling, Archaean rocks of the Napier Complex had been identified further west, and extensive Neoproterozoic rocks had been identified further east, but the ages of the two selected rock types were unknown. The orthogneiss samples showed limited variation in whole-rock $^{87}Sr/^{86}Sr$ ratios, but gave an Archaean Rb-Sr whole-rock Rb-Sr isochron age of 2692 ± 48 Ma (Fig. 3.52A). The metapelitic gneiss samples showed far greater variation in whole-rock $^{87}Sr/^{86}Sr$ ratios, reflecting larger changes in mineral assemblage than for the orthogneiss, and this gave a precise Proterozoic Rb-Sr whole-rock Rb-Sr isochron age of 1256 ± 17 Ma (Fig. 3.52B).

The following two geological interpretations of these data are possible (Clarke, 1988): (1) Proterozoic protoliths for the Colbeck Gneiss were

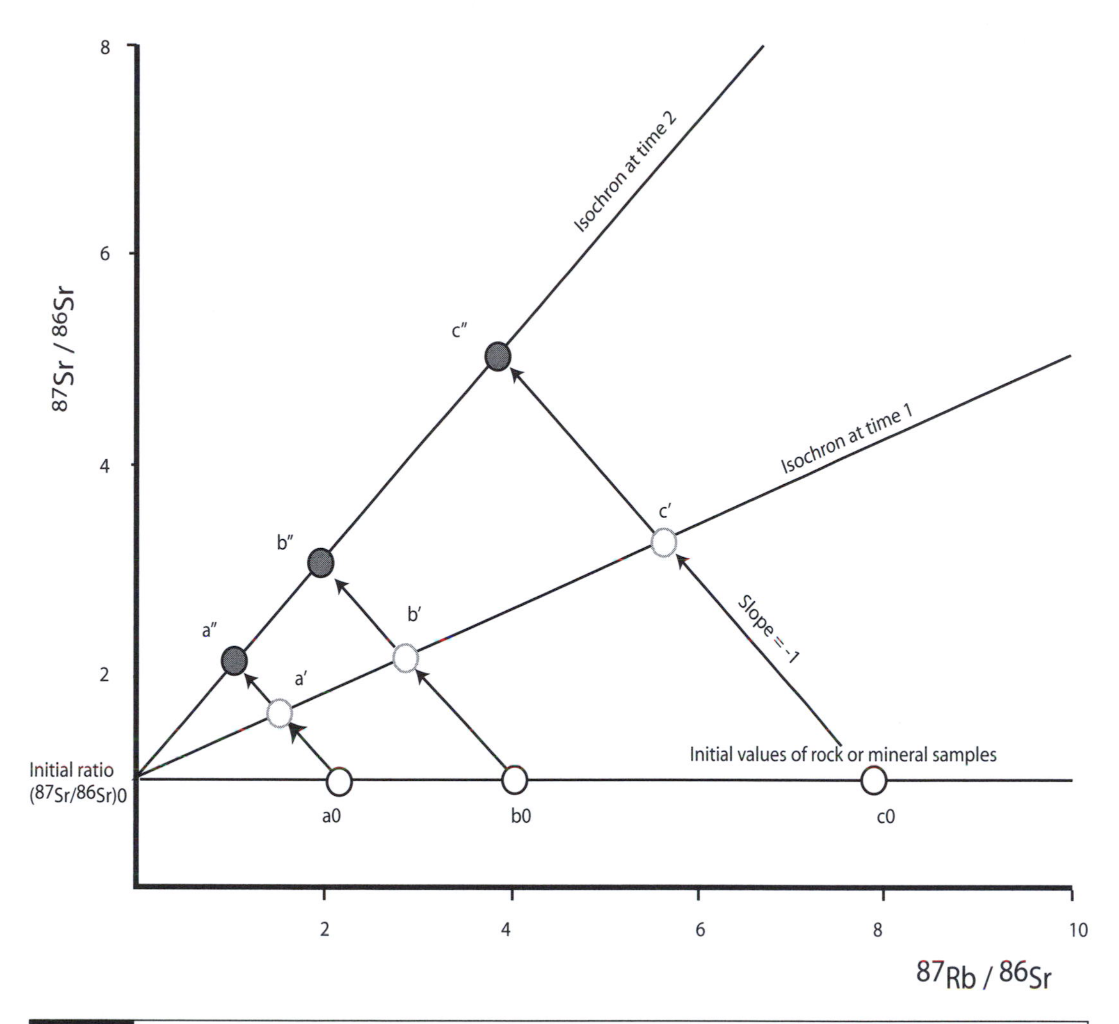

Fig. 3.51 Schematic isochron diagram (from Rollinson, 1993, fig. 6 with permission of Pearson Education Ltd.), showing the isotopic evolution of a suite of rock or mineral samples (a,b,c) over time. At $t = 0$, each rock or mineral has the same initial ratio ($^{87}Sr/^{86}Sr)_0$ but distinct $^{87}Sr/^{86}Sr$ ratios (isotopic dispersion),owing to chemical differences. Between $t = 0$ and time 1, individual rock or mineral samples evolve along a line with a slope of -1, reflecting the decay of individual atoms of ^{87}Rb to ^{87}Sr. The slopes of the isochrons give the age of the rock or mineral series. The intercept on $^{87}Sr/^{86}Sr$ gives the initial ratio.

unconformably deposited on an Archaean basement formed from the Still-well Gneiss; and (2) Archaean Stillwell and Colbeck Gneiss samples experienced a second *c.* 1300 Ma metamorphism that extensively recrystallized the metapelitic gneiss and disturbed the Rb-Sr isotopic system, but had limited effect on the orthogneiss. Selection of one of these alternatives requires more data.

The Rb-Sr isotopic system has provided excellent data in geologically simple terranes. Data provided from whole-rock Rb-Sr isochrons have been less conclusive in high-grade terranes that have experienced polyphase deformation and metamorphism. Collins & Shaw (1995) provided an overview of geological relationships and Rb-Sr age data from the regionally extensive Arunta Block, central Australia, in the context of a large U-Pb zircon age database.

(A)

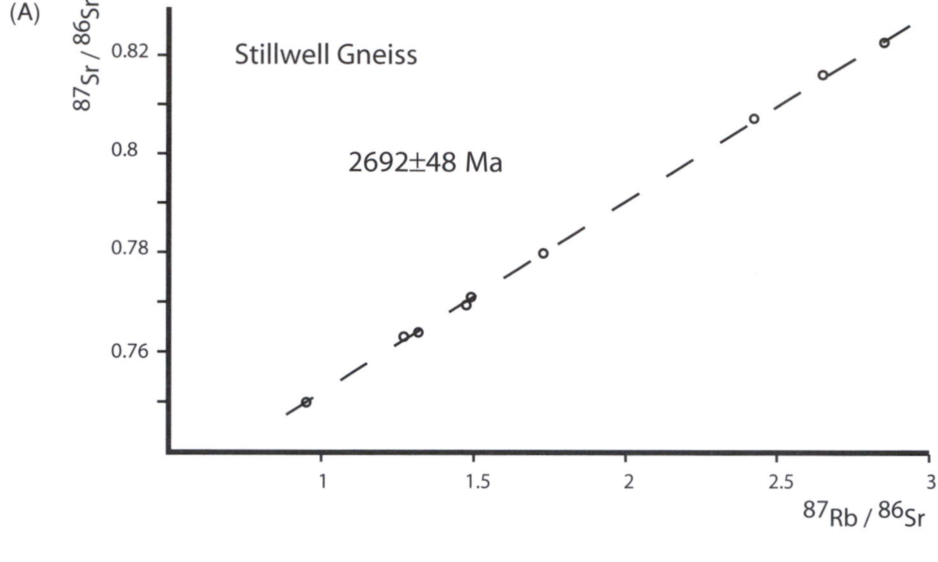

(B)

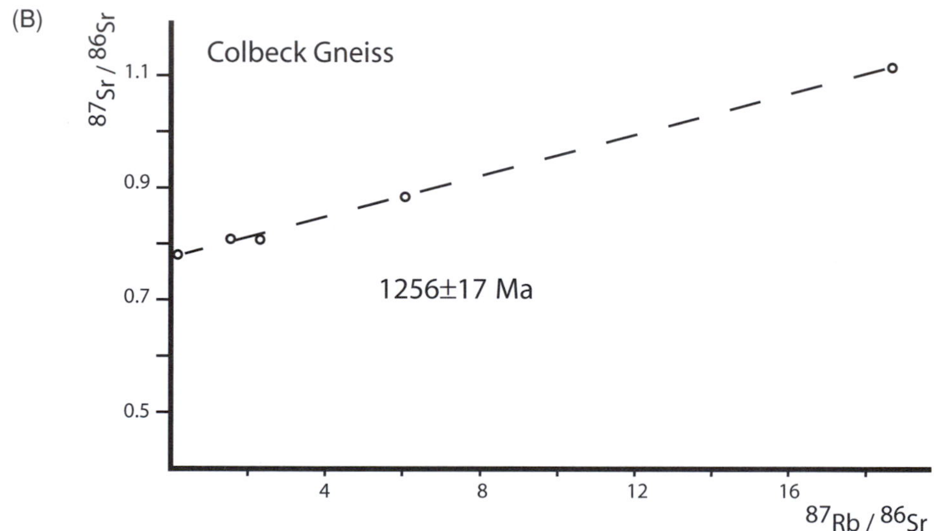

Fig. 3.52 Rb-Sr isochron diagram for two suites of whole-rock samples collected from Kemp Land, East Antarctica (Clarke, 1988). The Stillwell Gneiss (A), a charnockitic orthogneiss, occurs interlayered with the Colbeck Gneiss (B), a metapelitic gneiss rich in garnet, sillimanite, biotite and K-feldspar. Though the Archaean and Proterozoic ages given by the two rock types each have regional context, it is not possible to choose between several geological interpretations without further data.

The Sm-Nd isotopic system has been successfully applied to the dating of mafic metamorphic assemblages (Gebauer, 1990), especially those containing garnet, as this mineral preferentially partitions Sm and Nd (and other rare earth elements), leading to a good spread in $^{143}Nd/^{144}Nd$ values for mineral separates. The comparatively high closure temperature for Sm and Nd diffusion in silicate minerals makes the technique applicable for dating high-grade metamorphic assemblages (e.g., Stowell *et al.*, 2001). However, this can also cause complexity at low-to-intermediate temperature conditions, at which new minerals that pseudomorph pre-existing igneous or metamorphic minerals can partially or completely assume isotopic ratios of the pre-existing mineral, without whole-rock isotopic equilibration (Thoni & Jagoutz, 1992; Jagoutz, 1994).

3.7.4 U-Pb dating of zircon in metamorphic rocks

The U-Pb dating of zircon has become the favoured dating technique (Harley & Kelly, 2007) for several reasons. The zircon crystal lattice is retentive with respect to U, Th, Pb and all of the intermediate daughter elements, and zircon is a common accessory mineral in a wide variety of rocks. The concentrations of U in zircon range from a few hundred to a few thousand parts per million. U^{4+} (ionic radius = 1.05Å) isomorphously substitutes for Zr^{4+} (ionic radius = 0.87Å) in the zircon structure. Pb^{2+} (ionic radius = 1.32Å) is excluded from the zircon structure during mineral growth, because it has a large radius and a comparatively low charge. Therefore, zircon contains very little or no Pb at formation. The rare earth element composition of zircon can also retain fingerprints of sources of magmatic rocks from which it crystallizes (Scherer *et al.*, 2007) or the presence of other key metamorphic minerals, such as garnet, when it formed (Harley *et al.*, 2007; Rubatto & Herman, 2007). Also, the zircon crystal lattice appears to very robust and to have a very high closure temperatures for U and Pb diffusion ($T > 700°C$), dependent on fluid composition and abundance. This feature enables the zircon to preserve evidence for multiple episodes of zircon growth (igneous or metamorphic) through subsequent metamorphic and deformation events. Commonly, new zircon growth on pre-existing zircon grains during metamorphism records evidence of the new metamorphic event. Zircon dissolution may also accompany a given metamorphic event. In addition, the U-Pb isotopic system has several decay schemes, which provide internal checks on data quality.

Uranium has three naturally occurring isotopes: ^{238}U, ^{235}U and ^{234}U, all being radioactive. ^{238}U decays through a *uranium series* in a sequence of relatively short lived 'steps' before reaching a stable end product, ^{206}Pb. ^{235}U decays to ^{207}Pb via a series of decay schemes called the *actinium series*. These are the two main schemes of interest in geochronology. Pb has four naturally occurring isotopes: ^{208}Pb, ^{207}Pb, ^{206}Pb and ^{204}Pb. ^{208}Pb is a decay product of thorium; only ^{204}Pb is not radiogenic, though it is very weakly radioactive with a half-life of 1.4×10^{17} years. ^{204}Pb is used as a stable reference isotope in all ion microprobe U-Pb calculations.

In any U-bearing mineral containing both ^{235}U and ^{238}U, two ages can be calculated from the two independent decay schemes. Another independent decay scheme involving ^{232}Th decaying to ^{208}Pb is not commonly addressed. Ages obtained from the two decay schemes should be consistent, assuming no loss (or gain) of parent or daughter isotopes since crystallization of the mineral. If the ages are not consistent, the mineral host possibly did not remain isotopically closed during its history. Such a mismatch turns out to be comparatively common, and is called 'discordance'; such discordant ages are discussed further below.

A commonly used method for the plotting of U-Pb data and the calculation of geological ages uses the *concordia* or *Wetherill* (Wetherill, 1956) method. This is based on two measured isotopic ratios: $^{206}Pb/^{238}U$ and $^{207}Pb/^{235}U$. Applying the $^{207}Pb/^{235}U$ scheme to equation 3.17 and using ^{204}Pb as the reference stable non-radiogenic isotope gives:

$$\left(\frac{^{206}Pb}{^{204}Pb}\right) \text{now} = \left(\frac{^{206}Pb}{^{204}Pb}\right) 0 + (e^{\lambda t} - 1) \quad (3.19)$$

Table 3.2. Uranium and lead isotopes commonly used in geochronology

Parent	Daughter	Half life	Abundance
^{234}U	^{230}Th	0.02455067 Myr	0.0056%
^{235}U	^{231}Th	703.8193 Myr	0.7205%
^{238}U	^{234}Th	4468.124 Myr	99.274%
^{204}Pb		Stable	1.48%
	^{206}Pb		23.6%
	^{207}Pb		22.6%
	^{208}Pb		52.3%

Rearrangement of equation 3.15 and removal of the ^{204}Pb reference gives:

$$\frac{^{206}\mathrm{Pb}^*}{^{238}\mathrm{U}} = (e^{\lambda t} - 1) \quad \text{where } \lambda^{238}\mathrm{U} = 1.551 \times 10^{-10} a^{-1} \tag{3.20}$$

or the equivalent equation for ^{207}Pb/^{235}U:

$$\frac{^{207}\mathrm{Pb}^*}{^{235}\mathrm{U}} = (\lambda t - 1) \quad \text{where } \lambda^{235}\mathrm{U} = 9.8485 \times 10^{-10} a^{-1} \tag{3.21}$$

These two quantities increase at different rates (Table 3.2) with increasing time (t). The asterisks on ^{207}Pb and ^{206}Pb are added to signify the assumption that all of the relevant isotope has come from the parent. If these two equations (3.20, 3.21) are solved for different values of t and the results are plotted against each other, the increase may be seen as a curve called a *concordia* curve (Fig. 3.53A). For an ideal system in which no Pb or U is lost or gained, calculated ages would lie on the curve and so would be 'concordant'; they are said to lie '*on concordia*'. As long as no Pb or U is lost or gained, a closed isotopic system progressively accumulating daughter Pb evolves along the concordia curve (Fig. 3.53B). The age of the system at any time after closure is indicated by a location on the curve.

Now consider a U-Pb system on concordia at age t_0 that experiences an episode of Pb loss (or gain of U) as a result of a geological event (Figs. 3.53C,D). This results in the point representing the system moving along a straight line to the origin from t_0. If all of the radiogenic Pb accumulated in the system is lost, the point returns to the origin, and the system is reset, preserving none of the earlier history. If the system loses only a fraction of the radiogenic lead, it is represented by points lying somewhere between t_0 and t_1, on a chord. All such points on the chord are discordant dates (as mentioned above), and the chord is called '*discordia*'.

The total loss of all the radiogenic Pb in a system is extreme, and commonly U-bearing minerals such as zircon lose only a fraction of their radiogenic Pb as a consequence of geological disturbance. A number of zircon crystals from a given rock may lose varying amounts of radiogenic Pb, in spite of having been exposed to the same set of conditions, owing to differences in grainsize, U concentration and radiation damage to the crystal lattice. This is fortunate, as it results in the spread of analyses from different zircon crystals along discordia. In practice this means that the position of the discordia can be determined by fitting a straight line to

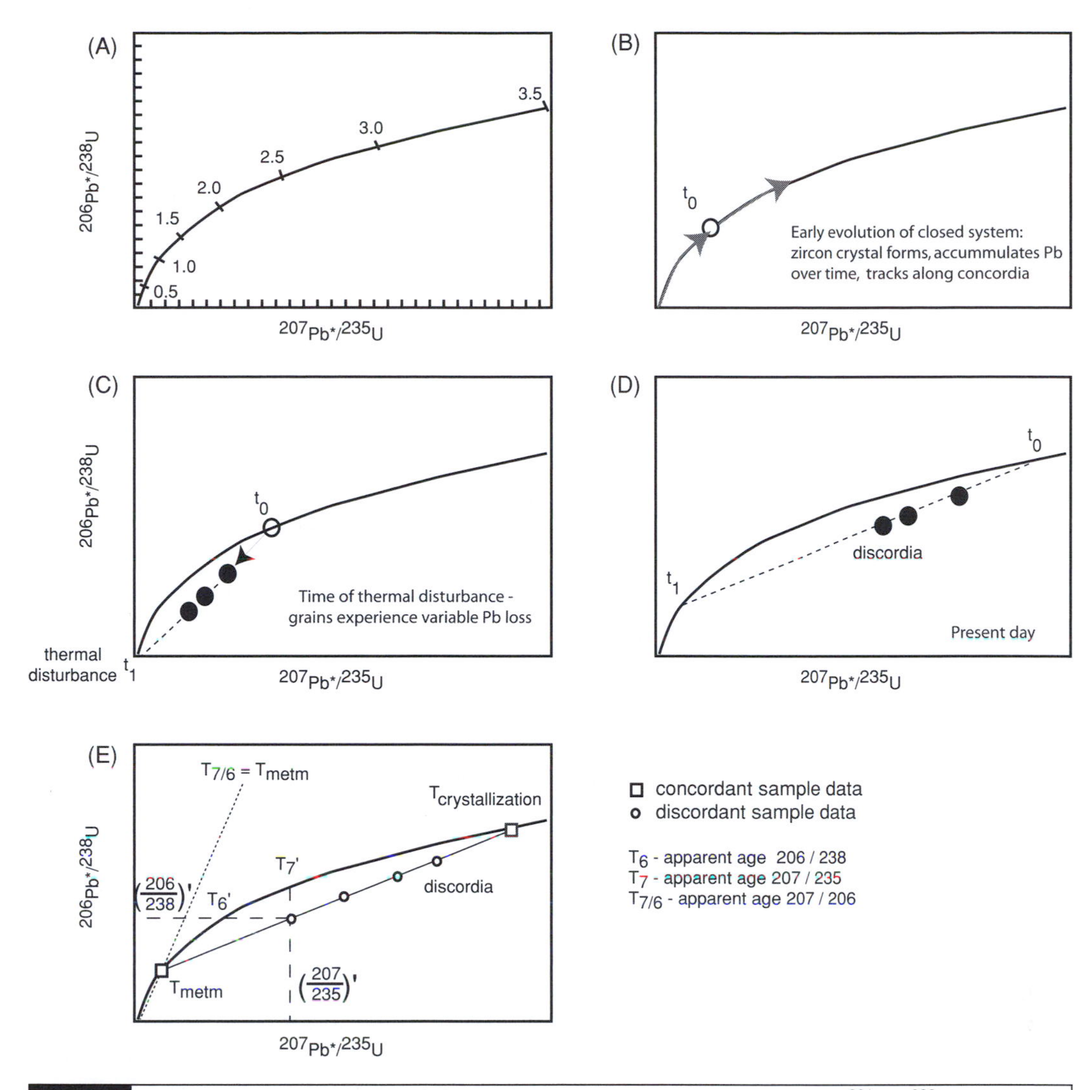

Fig. 3.53 (A) Generalized concordia or Wetherill (1956) diagram showing the evolution of ($^{206}Pb^*/^{238}U$) and ($^{207}Pb^*/^{235}U$) with time, after Faure (1986). Numbers on the concordia curve are in billions of years. (B-C) Sequence illustrating the evolution of a sample with time. The sample crystallized and cooled rapidly at t_0 and evolved along the concordia as a closed system, as shown in (B). At time t_1, a thermal disturbance (due to metamorphism in this example) resulted in the partial and variable loss of lead daughter product, such that different grains in one sample moved down the chord, as shown in (C). After this thermal disturbance, radiogenic lead again accumulated without loss, so that the sample evolved as shown in (D). The final position of the chord defined by the isotopically disturbed samples lies between the concordia points t_0 (upper intercept) and t_1 (lower intercept). (E) Apparent $^{206}Pb^*/^{238}U$ and $^{207}Pb^*/^{235}U$ ages can be obtained from individual discordant data, presenting two episodes of U-Pb isotope evolution.

data points representing zircon crystals that have lost varying proportions of their radiogenic lead. Extrapolating discordia leads to two intersections on concordia (Fig. 3.53E). These intersections are commonly inferred to date the original crystallization of the zircon and the time elapsed since isotopic closure following the episode of lead loss. Recent Pb loss is marked by discordia with a lower intercept at the origin. Because the effects of Pb-loss are comparatively common, otherwise apparently disparate data may

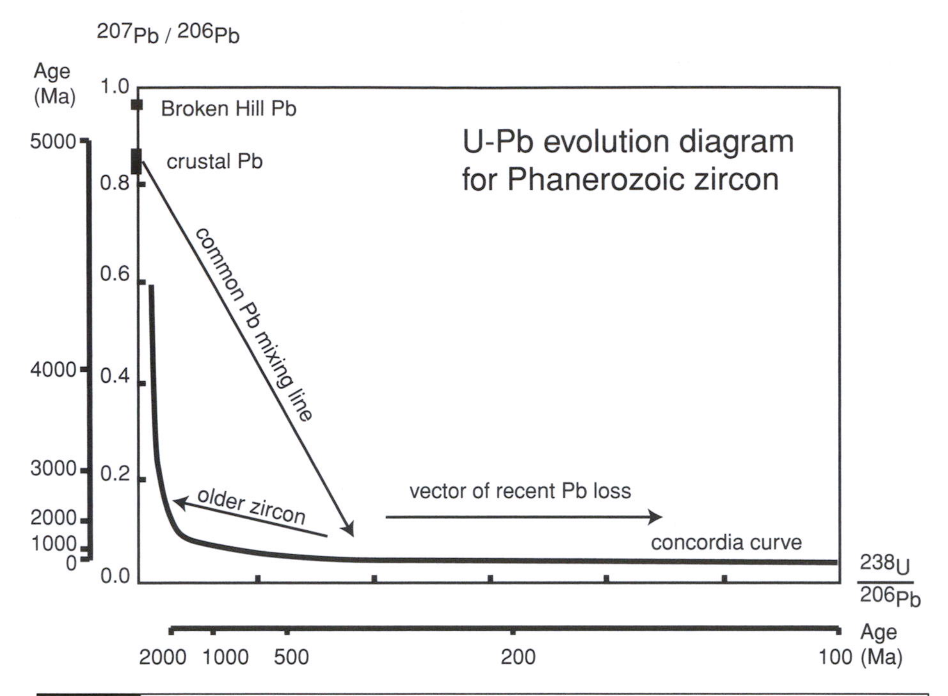

Fig. 3.54 Concordia diagram showing the evolution of $^{207}Pb/^{206}Pb$ and $^{238}U/^{206}Pb$ with time. This plot is more useful for identifying Pb-loss events and ^{204}Pb contamination in Phanerozoic zircon than the Wetherill diagram (Fig. 3.53) because of the low ^{235}U content in these zircon grains. Redrawn from Claoué-Long *et al.* (1995).

share a common $^{207}Pb/^{206}Pb$ ratio and may be combined in Precambrian age calculations.

The Wetherill diagram is well suited to plotting Precambrian U-Pb data, but is less useful for Palaeozoic U-Pb data, owing to the slope of concordia in this era making it more difficult to identify discordia, and the scarcity of ^{235}U when the zircon formed. Instead, $^{238}U/^{206}Pb$ is plotted against $^{207}Pb/^{206}Pb$ on a Terra-Wasserberg diagram (Fig. 3.54). Discordia consequent to recent Pb loss defines a horizontal line on this diagram.

Several methods have been developed for the U-Pb dating of zircon. Conventional methods have involved rock crushing, mineral separation and the dissolution of zircon grains, to yield very precise ages for zircon by thermal ionization mass spectrometry (TIMS). However, as zircon grains can be complexly zoned (see below), this age, though precise, can be a mix of two or more stages of zircon growth. Hence, analytical methods involving the focussing of an ion or laser beam onto polished surfaces of individual mounted zircon grains have been developed; this energy 'sputters' the targeted part of the grain, forming a plasma that includes mixed oxides of U, Th and Pb, which is relayed to a mass spectrometer.

Use of the ion microprobe (Compston *et al.*, 1982, 1984; Claoué-Long *et al.*, 1995) has revolutionized zircon geochronology, because, though the results are less precise than TIMS results, a number of analyses can be made on a single zircon grain. A single zircon grain or a population of zircon grains may resolve the ages of a number of geological events (e.g., Black *et al.*, 1987; Kröner *et al.*, 1987). An alternative method of *in situ* analysis using the electron microprobe, which can identify element, but not

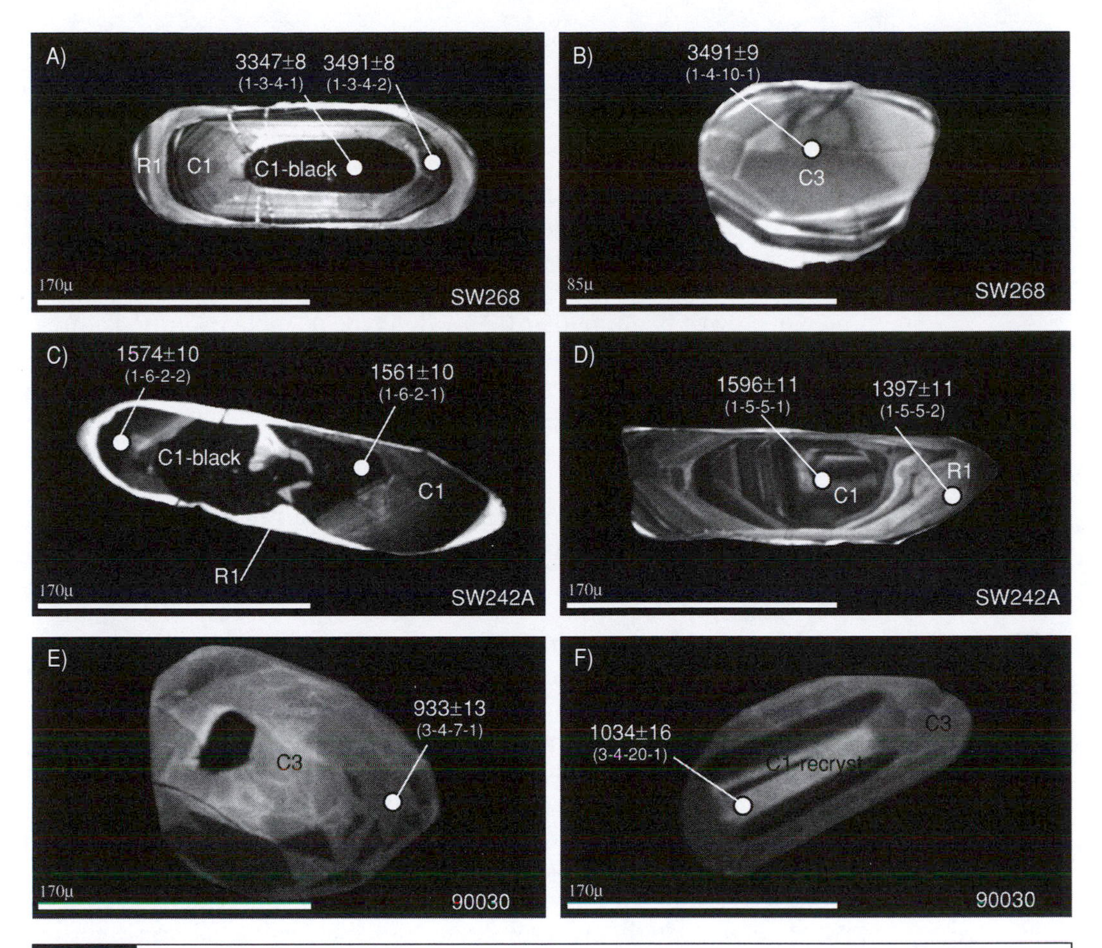

Fig. 3.55 Cathodoluminescence (CL) images of zircon grains from orthogneiss samples SW268, SW242A and 90030 from MacRobertson and Kemp Lands, east Antarctica (Halpin *et al.*, 2007a, fig. 3, with permission of *Contributions to Mineralogy and Petrology*). (A) Black core surrounded by a high-CL oscillatory zoned C1 core and a faintly zoned R1 rim. Ages quoted show the metamict (high-U) black core as being younger than the surrounding oscillatory zoned core (SW268), which is a spurious result. (B) Faintly zoned core with moderately high CL and zonation at rim (SW268). (C) Black inner core surrounded by an outer core and a high CL rim that truncates oscillatory zoning and embays the core (SW242A). (D) Oscillatory zoned core surrounded by a higher CL rim (SW242A). (E) Sector-zoned, 'soccer-ball' type C3 core that encloses a small, possibly inherited core (90030). (F) Partially oscillatory zoned core surrounded by an outer core (90030).

isotopic chemical composition, has been described by Suzuki *et al.* (1995, 1999).

Preparation for the analysis of most geochronology samples involves the separation of zircon grains from crushed rocks using magnetic separation and heavy liquids. If the analysis method uses an ion microprobe or laser ablation, the zircon grains are commonly hand picked and mounted in epoxy resin discs and polished, so as to just reveal their central portions. The mounted grains are then commonly examined using electron back-scatter and cathodoluminescence (CL) imagery in a scanning electron microscope; zircon CL is sensitive to a subtle variations in rare earth element content (Fig. 3.55). Such studies, made prior to mounting in ion microprobe or laser ablation analytical facilities, identify patterns that can be used to interpret early and late growth stages in individual grains and igneous or metamorphic growth stages. The procedure can also identify zircon can grains with high U contents caused by extensive lattice damage, which yield spurious data; these are called *metamict* zircon grains.

Euhedral zircon grains with CL responses revealing oscillatory zoning are commonly interpreted as having grown in the presence of melt (i.e., magmatic). Rounded zircon grains with homogeneous, sector or fir-tree zoned patterns are commonly interpreted as having grown during a metamorphic event. Zircon grains can also show sector zoning, with less common planar growth banding (Fig. 3.55). This type of mixed zoning is interpreted as resulting from fluctuating crystal growth rates, and is commonly observed in zircon that has crystallized from or grown in the presence of melt during high-grade metamorphism (Vavra *et al.*, 1996; Rubatto & Gebaner, 2000; Rubatto *et al.*, 1998; Schaltegger *et al.*, 1999; Vavra *et al.*, 1999; Kelly *et al.*, 2002; Halpin *et al.*, 2007). The ability to precisely resolve stages of zircon growth requires knowledge of field relationships at sampling sites, to give geological significance to ages resolved by the U-Pb zircon data (Fig. 3.56).

As CL imaging may not give a unique solution for zircon showing evidence of mixed igneous/metamorphic growth stages, further detail can be revealed by plotting Th against U for all analyses, or total U/Th against age (Fig. 3.57). In addition, efforts are now being made to fully characterize the REE composition of zircon grains from metamorphic rocks, analysed using laser-ablation ICPMS studies; such information can be used to infer whether zircon grew in, for example, the presence of plagioclase or garnet (Rubatto, 2002; Kelly & Harley, 2005; Harley *et al.*, 2007; Rubatto & Herman, 2007). Procedures used to plot U-Pb zircon data using the computer program ISOPLOT have been described by Ludwig (2003); information can also be obtained at: http://www.geo.cornell.edu/geology/classes/Geo656/Isoplot%20Manual.pdf).

3.7.5 Dating of monazite in metamorphic rocks

Monazite is a lanthanum-cerium light rare earth element phosphate that is common in many igneous and metamorphic rocks. It is useful for geochronology because it contains relatively large amounts of uranium and thorium, which decay radioactively to lead. Because monazite contains Th and U, without much common Pb, it can be used for U-Pb dating. It is highly reactive during metamorphism (Smith & Barreiro, 1990; Kingsbury *et al.*, 1993; Bingen *et al.*, 1996; Franz *et al.*, 1996; Kohn & Malloy, 2004) and partial melting (Bea *et al.*, 1997; Pyle & Spear, 2003), and its presence in a variety of crustal rock-types, such as metaluminous granites, peraluminous granites and metapelitic schists and gneisses (Parrish, 1990), make it extremely versatile and potentially powerful for understanding metamorphic processes (Kelly *et al.*, 2006).

Monazite may be involved in metamorphic reactions with other minor minerals (Smith & Barreiro, 1990; Bingen *et al.*, 1996; Pan, 1997; Finger *et al.*, 1998) and with major minerals (Pyle & Spear, 1999, 2003; Foster *et al.*, 2000; Pyle *et al.*, 2001; Yang & Rivers, 2002; Fraser *et al.*, 2004; Kohn & Malloy, 2004), and may be affected by hydrothermal alteration (Poitrasson *et al.*, 1996). Such interactions typically result in specific trace element zoning signatures in both minor and major minerals (Lanzirotti, 1995; Spear & Kohn, 1996; Bea *et al.*, 1997; Pyle & Spear, 1999, 2003; Foster *et al.*,

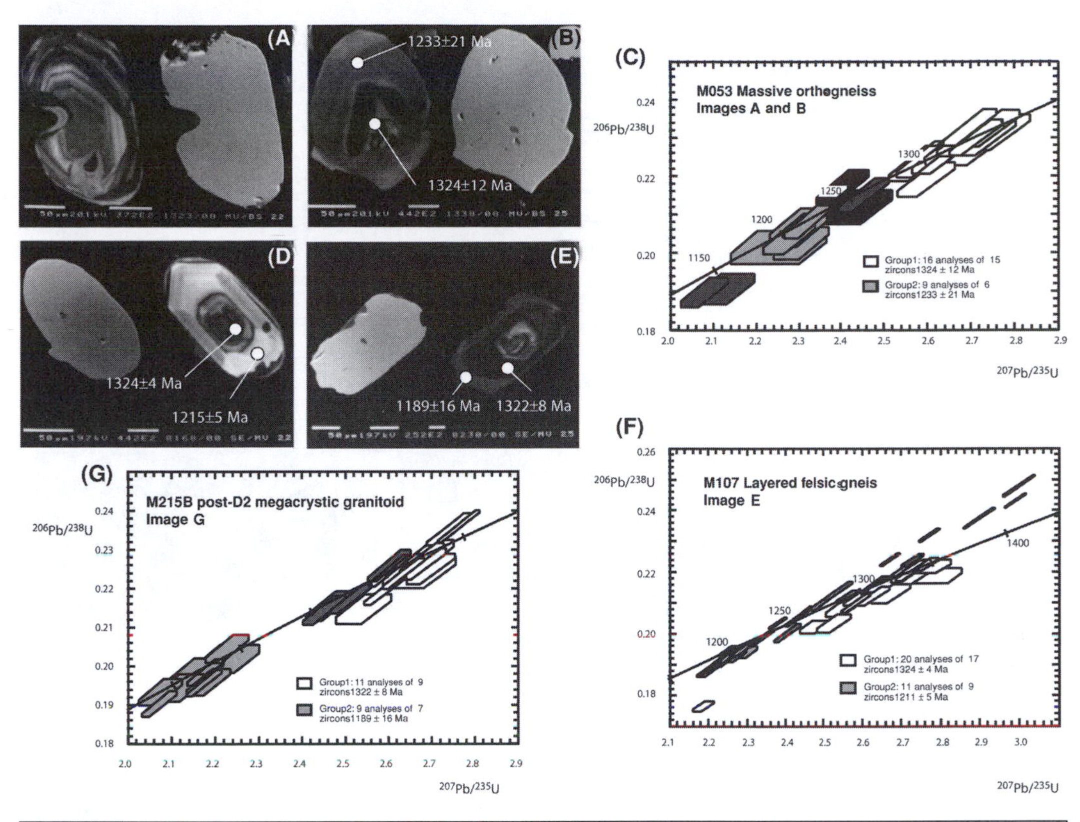

Fig. 3.56 Paired back-scattered electron and cathodoluminescence (CL) images and concordia plots of SHRIMP analyses of a suite of zircon grains from the western Musgrave Block, central Australia. Interpreted dates are $^{207}Pb/^{206}Pb$ ages. (A), (B), (D) and (E) are from White *et al.* (1999, fig. 3), with permission of the *Journal of Metamorphic Geology*. (C), (G) and (F) are redrawn from White *et al.* (1999, figs 4, 7, 9). (A) Concentrically zoned cores from felsic orthogneiss sample M053 that yield a *c.* 1325 Ma age have been interpreted as dating crystallization of the orthogneiss protolith. (B) These cores have 10–50 micron-wide rims with high-CL response and weak internal CL structure, which may be parallel to or truncate concentric zoning in the core. Analyses from these rims form a tight concordant *c.* 1200 Ma cluster shown in (C) that has been interpreted as dating granulite facies metamorphism in the area. Several analyses were excluded from the age calculations because they fall between the clusters or experienced recent Pb loss. (D) Small, moderate- to high-CL cores from a layered orthogneiss inferred to represent *c.* 1325 Ma igneous zircon, successively enclosed by metamorphic layers of euhedral low-CL zircon with weak, fine, concentric layering and a discordant, unzoned, anhedral, moderate to high-CL *c.* 1215 Ma rim [data plotted in (F)]. The reversely discordant analyses (lying above concordia) are inferred to be due to structural damage caused by high uranium content, enabling Pb mobility. (E) Concentric igneous banded zircon cores that yield a *c.* 1325 Ma age were inferred to be inherited from source rocks for a *c.* 1190 Ma granitoid that was emplaced comparatively late in the structural history of the terrane.

2000; Yang & Rivers, 2002), providing important geochemical markers of growth and breakdown during metamorphism.

Monazite can be dated using an ion microprobe or laser ablation facility, in a similar manner to that described above for zircon, and, where early lead contamination can be discredited, it can also be analysed using an electron microprobe (Montel *et al.*, 1996; Cocherie *et al.*, 1998; Williams *et al.*, 1999; Crowley & Ghent, 1999; Williams & Jercinovic, 2002; Jercinovic & Williams, 2005; Pyle *et al.*, 2005). As with zircon, scanning electron microscope (SEM) imaging of monazite is crucial for revealing internal complexity that may result from single or multiple growth events

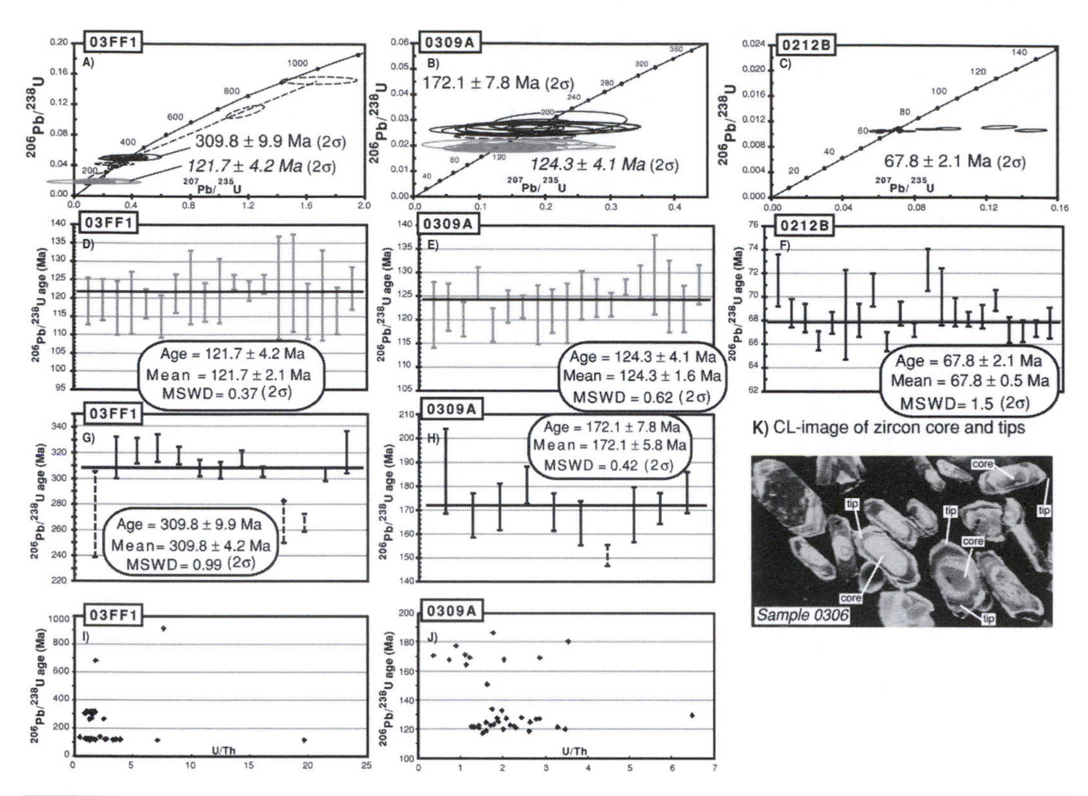

Fig. 3.57 U-Pb isotopic data for zircon collected using laser-ablation ICPMS on Cretaceous rocks from Fiordland, New Zealand (Marcotte *et al.*, 2005, fig. 11; copyright Elsevier; published with permission). The samples were taken from dioritic dykes that cut an early foliation, but are openly folded. The U-Pb zircon data indicate two populations: (1) *c.* 300 and *c.* 170 Ma ages inferred to be the age of crystallization of inherited cores, and (2) *c.* 120 Ma rim ages interpreted as the age of growth of the zircon rims during metamorphism and deformation. Plots (A) (B) and (C) are concordia plots; error ellipses are at the 1σ level, and ages are shown at the 2σ level. Plots (D), (E), (F), (G), and (H) show $^{206}Pb/^{238}U$ ages, which are the important ages for samples as young as this. Two plots are shown for the samples that yielded two age groups. The ages are shown with two uncertainties: those labeled 'age' include all errors, whereas the 'mean' includes only the random (measurement) errors. The larger uncertainty is reported as the age of the grains in the text. Note that these uncertainties are shown at 2σ levels, whereas the error bars for individual analyses are shown at 1σ levels. Plots (I) and (J) of U/Th versus $^{206}Pb/^{238}U$ ages are also shown, to help discriminate between igneous and metamorphic zircon. (K) shows a CL image of zircon grains with zoned cores and overgrowths (tips).

(e.g., Zhu & O'Nions, 1999). Mapping of Y, Th, U and Pb in monazite (Williams & Jercinovic, 2002; Yang & Pattison, 2006) is usually carried out, to detect chemical zonation.

High-resolution compositional mapping and dating of monazite with the electron microprobe is a very promising technique for absolutely timing metamorphic assemblages and foliations (Williams & Jercinovic, 2002), and so providing firmer evidence for *P-T-D-t* paths. Because diffusion of major and minor elements is extremely slow in monazite, it can retain chemical and geochronological information during younger metamorphic events. For example, Williams & Jercinovic (2002) were able to distinguish several stages of metamorphism in both the 1650–1700 Ma and 1400 Ma metamorphic/deformation events in the southwestern United States, and four separate pulses of monazite growth in the 1900–1800 Ma event in northern Saskatchewan, Canada.

3.7.6 K-Ar and Ar-Ar dating

Potassium is a major constituent of many rock-forming minerals, including most micas, feldspars and clay minerals. It has three naturally occurring isotopes (^{39}K-93.26%; ^{40}K-0.02%; ^{41}K-6.73%), including ^{40}K which undergoes branched decay to ^{40}Ca and ^{40}Ar. K-Ar and the related Ar-Ar dating methods are very versatile radiometric dating tools, applicable to finding the age of a wide variety of potassium-bearing rocks or minerals from diverse origins (Faure, 1986). In particular, white micas offer considerable potential for dating metamorphic events because they are stable under a wide range of conditions and they commonly define superimposed foliations (e.g., Dunlap *et al.*, 1991). Care needs to be exercised in K-Ar and Ar-Ar dating because of: (1) the possibility of inherited argon being present in detrital grains at very low-grade conditions, yielding incorrect apparently old ages, and (2) the comparative ease of partial or complete loss of radiogenic argon from crystal lattices during deformation or residence at elevated temperatures, yielding incorrect apparently young ages. Argon begins to diffuse out of white micas at temperatures above approximately 350°C (Dodson & Mclelland-Brown, 1985). Most metamorphic rocks experience *P-T* conditions above the temperature of argon retention, which is distinctive for each mineral. The measured age of a mineral such as white mica commonly records the time since cooling below the closure temperature for argon diffusion, rather than dating a discrete deformation or metamorphic event.

^{40}K concentration can be measured by flame photometry on one half of a sample, and the ^{40}Ar measured by mass spectrometry on the other half of the sample. This has the disadvantage of requiring a comparatively large sample. The growth of radiogenic ^{40}Ca and ^{40}Ar in a K-bearing system closed to K, Ar and Ca can be expressed by:

$$^{40}\mathrm{Ar} + {}^{40}\mathrm{Ca} = {}^{40}\mathrm{K}(e^{\lambda t} - 1) \tag{3.22}$$

where λ is the total decay constant of ^{40}K. There are separate decay constants for the branched scheme, such that

$$\lambda = \lambda_a + \lambda_c = (0.581 + 4.962) \times 10^{-10} = 5.543 \times 10^{-10} a^{-1} \tag{3.23}$$

where λ_a refers to the decay of ^{40}K to ^{40}Ar and λ_c to the decay to ^{40}Ca. This corresponds to a half-life of 1.250×10^9 yr (decay constants after Steiger & Jäger, 1977). As the fraction of ^{40}K that decays to ^{40}Ar is given by λ_a/λ_c, assuming that there is no initial ^{40}Ar, the growth of radiogenic ^{40}Ar atoms in a K-bearing mineral or rocks can be expressed as:

$$^{40}\mathrm{Ar} = \frac{\lambda_a}{\lambda}\,{}^{40}\mathrm{K}(e^{\lambda t} - 1) \tag{3.24}$$

A K-bearing mineral is thus dated by measuring the concentration of ^{40}K and accumulated ^{40}Ar according to:

$$t = \frac{1}{\lambda} \ln\left[\frac{^{40}\mathrm{Ar}}{^{40}\mathrm{K}}\left(\frac{\lambda}{\lambda_a}\right) + 1\right] \tag{3.25}$$

The age calculated from equation 3.25 relies on the analysed mineral or rock having neither gained or lost ^{40}Ar or ^{40}K following formation, and an appropriate correction is made for the presence of atmospheric ^{40}Ar.

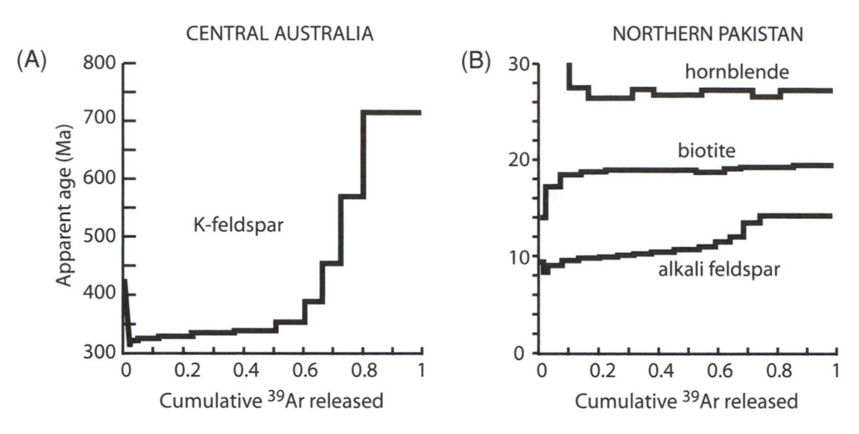

Fig. 3.58 ^{40}Ar/^{39}Ar mineral age spectra for two examples from metamorphic terranes; redrawn from Zeitler (1989). (A) Ar diffusion-loss profile for K-feldspar grains from the Arunta Block, central Australia, inferred to have experienced post-metamorphic cooling at c.1100 Ma, based on ^{40}Ar/^{39}Ar hornblende and muscovite ages, and a second mild thermal event at 350 Ma. The concave-up form of the age spectrum was inferred by Zeitler (1989) to match theoretical Ar loss spectra predicted for samples with multiple grainsizes. (B) ^{40}Ar/^{39}Ar age spectra for slowly cooled, alkali feldspar, biotite and hornblende grains' interpreted as having formed at the same time, in a sample from the eastern Kohistan island-arc terrane, northern Pakistan. The different mineral ages reflect distinct closure temperatures for Ar diffusion, and can be used to assess cooling rates.

The ^{40}Ar/^{39}Ar dating method (Merrihue & Turner, 1966; Dalrymple & Lanphere, 1971, 1974; McDougall, 1974; Dallmeyer, 1979) is based on the same decay system, but only requires measurements of Ar isotopic ratios, which can be achieved using one sample. It has become the most reliable method for obtaining cooling ages in metamorphic terranes. Samples are irradiated in a nuclear reactor, partially transforming the ^{39}K in a mineral to ^{39}Ar. As the natural ^{39}K/^{40}K ratio is known, the reactor-produced ^{40}Ar/^{39}Ar ratio can be measured by mass spectrometry, to determine ^{40}K concentration in the mineral. As both isotopes can be measured on the same sample, less material is needed than for K-Ar dating. The number of ^{39}Ar atoms formed in a sample by neutron irradiation depends on the number of ^{39}K atoms, the period of irradiation and the neutron flux density (Faure, 1986, pp. 93–95). Because of its slow rate of decay, ^{39}Ar can be treated as though it were stable during the short period of time involved in the analyses.

To try and identify samples that have experienced partial to complete Ar loss, step heating is applied during Ar collection and ^{40}Ar/^{39}Ar ages are calculated for each gas aliquot (Fig. 3.58). In numerous instances, samples known to have experienced episodic reheating or slow cooling yield Ar-Ar gas age spectra that reflect modelled diffusion profiles (Harrison, 1981, 1983; Harrison *et al.*, 1986; Zeitler, 1989; Lister & Baldwin, 1996; Fig. 3.58A). Alternatively, in rocks that have experienced multiple metamorphic and deformation events, detailed analysis of different white mica grainsize fractions can indicate an age dependence on grainsize, which is commonly interpreted as reflecting variable effects of dynamic recrystallization on argon diffusion (Reuter & Dallmeyer, 1989; West & Lux, 1993). Plateaus defined by individual mineral data on plots of apparent ^{40}Ar/^{39}Ar age versus cumulative ^{39}Ar released are commonly interpreted as reliable cooling ages (Fig. 3.58), but the data need be assessed for Ar gain (Wijbrans & McDougall, 1986; Zeitler, 1989). Laser Ar-extraction systems are now commonly used to date different generations of a mineral in situ, with the advantage of retaining microstructural context (e.g., York *et al.*, 1981; Sutter & Hunting, 1984).

Chapter 4

Partial Melting during High-Grade Metamorphism

4.1 Introduction

Many silicate rocks begin to melt at temperatures above 650°C in the presence of free water (*H_2O-saturated melting*). At higher temperatures, melting is promoted by combined water in the structures of minerals such as muscovite, biotite and hornblende, even in the absence of free water (*H_2O-undersaturated melting*). In both situations, chemical components of quartz, K-feldspar and plagioclase combine with H_2O to form a felsic melt, and mafic components may also enter the melt, especially at higher temperatures.

Partial melting during high-grade metamorphism is known as anatexis ('*differential anatexis*'). Partly melted rocks (called *anatectic migmatites*) occur in the highest grade areas of regional metamorphic terranes, as well as close to igneous intrusions in some of the hotter contact metamorphic aureoles. Typically migmatites are rocks of complex, deformed appearance (Fig. 4.1), though very variable.

Migmatites form in a variety of rock compositions, ranging from metapelitic (Fig. 4.2), through felsic (Fig. 4.1) to mafic (Fig. 4.3). For example, metapelites contain muscovite and biotite, and mafic rocks contain hornblende – all minerals that can contribute H_2O for melting. However, some rocks are more *fertile* than others, in the sense that they can produce a higher proportion of melt. For example, though the melting of metapelites (Grant, 1983, 1985a,b) can produce strongly aluminous felsic magma, the most probable sources for the abundant, moderately peraluminous granites of south-eastern Australia are rocks of felsic composition (Clemens & Wall, 1981, 1984). These rocks are rich in quartz and both feldspars, as well as containing enough biotite to provide water for the water-undersaturated felsic magmas.

4.2 Melting reactions

Information on partial melting reactions has been obtained from direct experimental melting, evaluation of phase diagrams and thermodynamic modelling. We begin by considering equations in simplified chemical systems, and then discuss more complicated, realistic compositions.

Fig. 4.1 Granulite facies metapsammitic migmatite, Darling Range, north of Broken Hill, Australia, showing isoclinally folded, quartz-rich bed (light coloured), which possibly was chert originally. Knife 9 cm long.

Fig. 4.2 Granulite facies metapelitic migmatite, Round Hill, north of Broken Hill, Australia, showing garnet-rich neosomes that have undergone boudinage during solid-state deformation (i.e., after the melt had crystallized to quartz and feldspar). The coarse grainsize of the garnet reflects a low ratio of nucleation rate to growth rate (Vernon, 2004), with the result that growth of garnet controlled the development of other reaction products in the melting reaction. For example, the melt was forced to form in situ around the growing garnet grains (Powell & Downes, 1990; Hand & Dirks, 1992; Vernon & Paterson, 2001; White *et al.*, 2004). Compare with the structure of metapsammitic migmatites in the same area (Fig. 4.14). Coin diameter 2.8 cm.

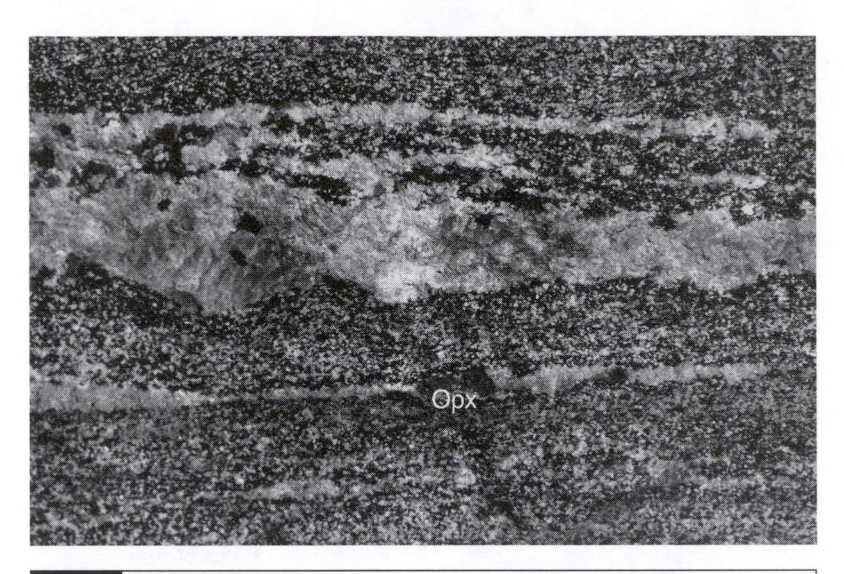

Fig. 4.3 Stromatic amphibolite migmatite, Adirondack Mountains, New York, USA, showing poorly developed, hornblende-rich melanosomes at the margins of the leucosomes. Some large grains of peritectic orthopyroxene (Opx) are also present. Base of photo 20 cm.

4.2.1 Water-saturated melting reactions

The simplest reaction involving H_2O-saturated melting (melting in the presence of free water) of quartzofeldspathic rocks is represented by the equation:

$$Qtz + Kfs + Pl + H_2O = L,$$

which is plotted on a simplified *P-T* diagram in Fig. 4.4A. 'L' represents a hydrous silicate liquid solution, which is generally referred to as 'melt', though it is not the same as one-component melts of metal, ice or an individual mineral, which result from a phase change without reaction with other phases. The above melting reaction is an example of *congruent melting*, because all the reacting components go into solution, without the formation of new solid products, so that the liquid produced has the same chemical composition as the reactant assemblage (including H_2O).

An example of a more complex water-saturated partial melting reaction is provided by migmatites in the contact metamorphic aureole of the Ballachulish Igneous Complex, Scotland, where semipelitic layers were the most susceptible to melting, producing migmatites. The equation for the inferred reaction (Pattison & Harte, 1988) is:

$$Ms + Qtz + Pl + Bt + H_2O = L(melt).$$

Part of the H_2O was derived from dehydration of biotite in adjacent metapelitic layers, though most had an external source. Evidently, the temperature was not high enough for long enough to promote sufficient H_2O-undersaturated melting of biotite, and the pressure was too low to permit H_2O-undersaturated melting of muscovite (Fig. 4.4A), which may be a common situation in contact metamorphism.

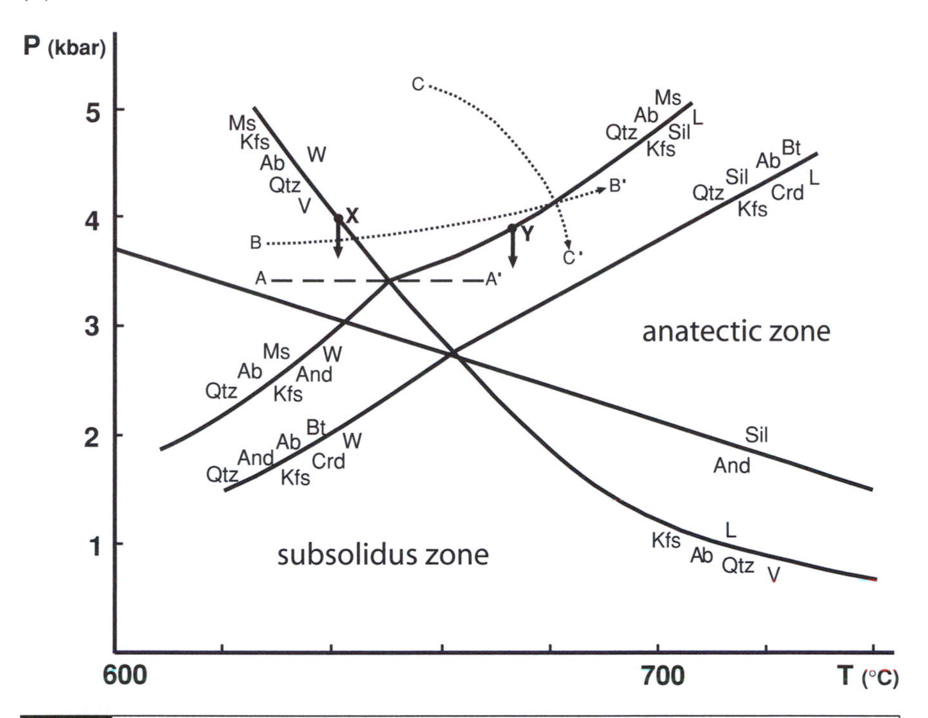

Fig. 4.4 (A) *P-T* 'grid' of some experimentally determined melting reactions – largely after Thompson (2000) – shown as univariant curves for simplicity, though they are at least divariant in natural rocks, owing to solid solution. W = supercritical H_2O fluid in the 'subsolidus (melt-free) field'; L = silicate liquid ('melt') in the 'anatectic (melt-stable) field'. Curves shown represent the melting of Qtz + Ab + Kfs in the presence of free water (water-saturated melting curve or 'wet granite melting curve'), as well as the dehydration breakdown of Ms + Qtz + Ab and the dehydration breakdown of Bt + Als+ Ab + Qtz, both in the melt-free field (left) and melt-stable field (right). The melt-stable field is the *P-T* region above the 'wet granite melting curve', in which melt may be present in many crustal rock compositions (Brown, 2002). Muscovite can melt at pressures above line A-A′, whereas muscovite breaks down (to form solids + H_2O) and disappears before the melt-stable area is reached at pressures below this line. The negative slope of the 'wet granite melting curve' means that a melt on the curve (for example, at X) cannot move upwards in the crust (i.e., undergo *P* decrease) without entering the field of solids and so crystallizing (see arrow), with the result that water-saturated melts typically do not form high-level granite bodies. In contrast, water-undersaturated melt (for example, at Y) may undergo *P* reduction without crystallizing (arrow), and so water-undersaturated melts are the main contributors to high-level granite plutons. During *T* increase (path B-B′) at *P* above line A-A′, muscovite melts before biotite, forming melt that can escape readily, owing to the large ΔV of the melting reaction, whereas biotite-derived melts (formed at higher temperatures) may remain in the source rock (Vernon *et al.*, 2003) or require deformation to assist their escape (Holyoke & Rushmer, 2002). Water-absent melting curves may be approached on *P-T-t* paths involving P increase (path B-B′) or *P* decrease (path C-C′). (B) *P-T* pseudosection in the system KFMASH for a particular pelitic rock composition, showing multivariant fields of assemblages resulting from dehydration reactions at lower temperatures and partial melting at higher temperatures. W = H_2O in the subsolidus field; L = silicate liquid (melt) in the anatectic field. These fields are broadly similar to those shown in (A), but the multivariant assemblage fields are much more realistic for natural rocks. From White *et al.* (2007, fig. 4), with permission of the *Journal of Metamorphic Geology*.

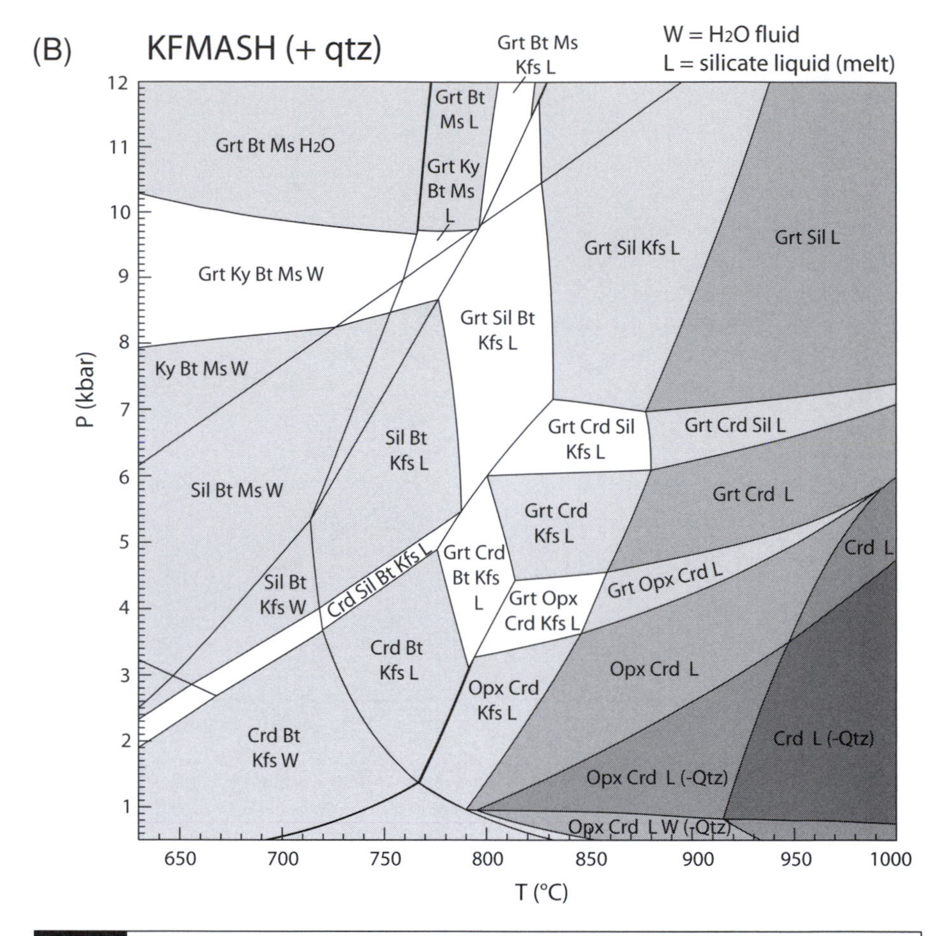

Fig. 4.4 (*cont.*).

The following example of another more complex H_2O-saturated partial melting reaction was inferred for migmatites in the Connemara Schists, Ireland, by Yardley & Barber (1991):

$$Bt + Sil + Kfs + Qtz + H_2O = L(melt).$$

Water-saturated melting reactions are characterized by a negative ΔV, and the experimentally determined negative slope of the *P-T* curve (the '*water-saturated granite melting curve*') implies that a H_2O-saturated melt cannot rise in the crust without crystallizing, as shown in Fig. 4.4A. This is in contrast to the situation for H_2O-undersaturated melts (Section 4.2.2). However, if the water-saturated 'granite' melting curve is crossed during heating without melting (owing to scarcity of free water) and free water is added at some higher temperature (e.g., in shear zones), H_2O-saturated melting may occur (Vernon *et al.*, 2003; Ward *et al.*, 2007). Because the resulting melt is above the solidus (Fig. 4.4A), it can rise in the crust to form an intrusion.

Free water is unlikely to be abundant at depths at which major felsic magmas are believed to be formed (Yardley & Valley, 1997, 2000), except in high-strain zones, which suggests that H_2O-undersaturated melting is a much more realistic way of producing large amounts of felsic magma.

Yardley (1977c) made the point that small amounts of other components, such as CO_2, in the predominantly hydrous fluid may have a large effect on the initial melting temperature, so that temperatures higher than 700°C may be required for H_2O-saturated anatexis in such situations.

4.2.2 Water-undersaturated melting reactions

Field, petrological and experimental evidence indicates that the most common melt-forming reactions in crustal rocks occur at granulite facies *P-T* conditions (800°C–1,000°C) in the absence of a free fluid (e.g., Wyllie *et al.*, 1976; Wyllie, 1977; Thompson, 1982; Conrad *et al.*, 1988; Clemens & Watkins, 2001; Vigneresse, 2004; Clemens, 2005a). Water-undersaturated partial melting reactions are also known as *dehydration-melting reactions* (Thompson, 1982) or *vapour-absent melting reactions* (Grant, 1985a,b). Water released by dehydration of hydrous minerals enters the melt.

Simplified reactions

We begin with a discussion of simplified equations that illustrate the overall style of reactions involved in H_2O-absent melting. The following kinds of generalized fluid-absent melting reactions (where L = H_2O-undersaturated felsic silicate liquid or 'melt') are commonly presented (e.g., Clemens, 2005a), largely as a result of melting experiments and petrographic observations. Though these equations are useful for illustrating general principles, the reactions are mostly multivariant in actual rocks. Therefore, evaluation of partial melting conditions in common rocks requires more sophisticated approaches, such as the use of *P-T* grids in more compositionally comprehensive systems and especially of *P-T* pseudosections, as discussed below.

(1) Ms + Pl + Qtz = Als + Kfs + L (Ms breakdown in medium-grade metapelites);

(2A) Bt + Als + Qtz = Grt/Crd + Kfs + L (Bt breakdown in high-grade metapelites without Pl; Grt at higher pressures, Crd at lower pressures, Crd + Grt at intermediate pressures);

(2B) Bt + Pl + Als + Qtz = Grt/Crd + Kfs + L (Bt breakdown in high-grade metapelites with Pl; Grt at higher pressures, Crd at lower pressures, Crd + Grt at intermediate pressures);

(3) Bt + Pl + Qtz = Opx (+Cpx + Grt) + L (in feldspathic metapsammites and metatonalites, the occurrence of minerals in brackets depending on the rock composition and the *P-T* conditions of the melting reaction); and

(4) Hbl + Qtz = Pl + Opx + Cpx (+Grt) + L (in metabasalts and meta-andesites, the occurrence of minerals in brackets depending on the rock composition and the *P-T* conditions of the melting reaction).

The composition of the liquid depends on the reactants and hence on the bulk composition of the rock. For example, melts produced by reactions (1) and (2A) are rich in Kfs, melts produced by reactions (2B) and (3) contain both Kfs and Pl, and melts produced by reaction (4) are rich in Pl.

The solid products of these reactions result from the fact that these melting reactions are *incongruent*. That is, the rocks do not melt to enter a liquid of the same composition, but instead produce a felsic liquid plus solid crystals (*peritectic minerals*). For example, Waters & Whales (1984) and

Waters (1988) described rocks in south Namaqualand, South Africa, in which H_2O-unsaturated partial melting formed anhydrous granulite facies assemblages, in contrast to adjacent unmelted rocks containing hydrous amphibolite facies assemblages. The reactions involved incongruent melting of mica and amphibole, to form H_2O-undersaturated silicate liquid (melt) plus anhydrous (peritectic) solid minerals, such as garnet, cordierite or orthopyroxene in metapelites and orthopyroxene in mafic rocks.

The felsic components in the liquid come mainly from Qtz and Bt, as well as Pl, if present, the Bt supplying all the K and H_2O. The excess Mg and Fe from the Bt not needed for the felsic liquid combine with the Als components to form the aluminous peritectic minerals, Grt and/or Crd. The amount of H_2O available to enter the liquid is limited by the hydroxyl formerly in the structure of the Bt, with the result that the liquid is undersaturated with respect to H_2O.

The above equations are generalized and may vary, depending on pressure and the bulk composition of the rocks involved. For example, the proportions of reactants in reactions (2A) and (2B) depend on whether Grt (higher *P*) or Crd (lower *P*) is produced. Most metapelites are K-rich, so that the components of the product Kfs in reactions (2A) and (2B) are typically in excess of the amount that can be accommodated in the liquid; therefore, solid Kfs forms. If it is sodic enough, all the Pl may be consumed, because it is generally a minor mineral in metapelites.

The second equation (2B) requires more expansion if the sodic components of a more calcic plagioclase are dissolved in the melt, producing plagioclase that is more calcic than the reactant plagioclase as a solid product, as follows:

$$\mathrm{Bt} + \mathrm{Pl_1} + \mathrm{Als} + \mathrm{Qtz} = \mathrm{Grt/Crd} + \mathrm{Kfs} + \mathrm{Pl_2} + \mathrm{L(melt)},$$

where Pl_1 is more sodic than Pl_2.

For migmatitic mafic rocks of the Kapuskasing Structural Zone, Ontario, Canada, the following variation on reaction (4) can be written with Pl as a reactant, rather than a product, and taking Ti into account by forming titanite:

$$\mathrm{Hbl} + \mathrm{Pl} + \mathrm{Qtz} = \mathrm{Cpx(Di)} + \mathrm{Grt} + \mathrm{Ttn} + \mathrm{L}. \quad \text{(Hartel \& Pattison, 1996)}$$

Moreover, detailed consideration of the proportions and chemical compositions of the minerals involved leads to the formulation of the following more complete equation for this reaction (Hartel & Pattison, 1996):

$$1.0\mathrm{Hbl} + 0.34\mathrm{Pl} + 0.32\mathrm{Qtz} = 0.37\mathrm{Di} + 0.48\mathrm{Grt} + 0.04\mathrm{Ttn} + 0.70\mathrm{L}.$$

Though it represents the chemical balance between the phases involved, the equation does not necessarily describe the progressive stages in the development of this final state. The observed assemblage may represent the integrated effects of both the quartz-consuming dehydration melting reaction and more subtle mineral exchange reactions in the hornblende and plagioclase (Hartel & Pattison, 1996, p. 606). This example emphasizes the important point that a chemical equation does not necessarily indicate an actual reaction, though it may indicate a net change from one mineral assemblage to another (e.g., Carmichael, 1969).

K-feldspar-bearing neosomes/leucosomes (see Section 4.3 for terminology) with garnet and/or cordierite are good indicators that H_2O-undersaturated melting involving biotite breakdown has taken place (Tracy

& Robinson, 1983, p. 171; Waters & Whales, 1984; Waters, 1988; Powell & Downes, 1990; Vernon *et al.*, 1990, 2003; Fitzsimons, 1996; Johnson *et al.*, 2001; White *et al.*, 2004). This observation is very common in high-grade metasedimentary terranes, suggesting that water, where present, occurs in amounts insufficient to promote water-saturated melting at granulite facies conditions. Once H_2O-undersaturated melting begins, the melt tends to act as a fluid 'sink', and so dehydrate the adjacent rocks (Tracy & Robinson, 1983, p. 173), thereby promoting dehydration reactions, which in turn provide water for more melting (e.g., Mogk, 1990). This is a complementary (cyclic or 'feedback') chemical system.

Owing to the H_2O-undersaturated nature of most partial melts and the general absence of free water-rich fluid in granulite facies rocks, the amount of H_2O stays constant unless melt is lost from the rocks (White & Powell, 2002). This situation is in contrast to subsolidus dehydration reactions, for which liberated H_2O continually escapes, allowing reactions to proceed.

Separation of the hydrous felsic liquid from the solid products of reactions (2), (3) and (4) dehydrates the rocks, and leaves dry granulite facies residues (*restites*) that are depleted in the components of the felsic melt (see Section 4.8). 'Depleted' chemical compositions are characteristic of granulite facies terranes from which large amounts of felsic melt have been extracted. Melt-depleted rocks may also be richer in Mg than they were initially, if Fe is preferentially concentrated in the melt (Droop & Bucher-Nurminen, 1984; Johnson *et al.*, 2001).

The above relationships lead to the inference that free water is generally absent from the middle and lower crust (e.g., Yardley & Valley, 1997, 2000). However, water may be introduced locally, especially along shear zones (Vernon & Ransom, 1971; also see Chapter 5), where it may support H_2O-undersaturated melting (Johnson *et al.*, 2001) or cause local H_2O-saturated melting (Vernon *et al.*, 2003).

Experiments and observations of high-grade metapelitic rocks indicate that equations (2A) and (2B) represent the most common type of generalized partial melting reaction for metapelites and felsic compositions at middle to lower crust *P-T* conditions (Holdaway & Lee, 1977; Clemens & Wall, 1981; Thompson, 1982; Tracy & Robinson, 1983; Grant, 1985a,b; Waters, 1988; Vielzeuf & Holloway 1988; Le Breton & Thompson, 1988; Whitney, 1988; Patiño-Douce & Johnston, 1991; Carrington & Harley, 1995; Stevens *et al.*, 1997; Johnson *et al.*, 2001). As noted previously, magmas produced by such reactions can rise through the crust without fully crystallizing (Fig. 4.4A), and are responsible for many high-level peraluminous ('S-type') granites (e.g., Clemens & Wall, 1981). This is because H_2O-undersaturated reactions generally have positive *P-T* curve slopes, and commonly a relatively large interval exists between the temperature of the melt-producing reaction and the 'wet granite' solidus, at which the magma freezes (Fig. 4.4A). The magma can ascend without solidifying, as long as its temperature is in this interval.

These reactions generally involve an increase in volume (positive ΔV), as shown in Fig. 4.4A), which may assist the resulting liquids to segregate from their source rocks (Section 4.8). This applies especially to H_2O-absent melting of muscovite, which involves a large positive ΔV, with the result that the melt rapidly increases the local fluid pressure and fractures the adjacent mineral grains, as shown experimentally (Connolly *et al.*, 1997;

Rushmer, 2001; Holyoke & Rushmer, 2002), allowing the melt to easily escape. In contrast, H_2O-absent melting of biotite involves only a small positive ΔV or even a small negative ΔV, with the result that the melt tends to remain in the source rock (Rushmer, 2001; Holyoke & Rushmer, 2002), unless its escape is assisted by structural anisotropy and especially by tectonic deformation.

As shown in Fig. 4.4A, simplified H_2O-undersaturated *P-T* reaction curves may be crossed during increase or decrease of pressure (Thompson, 1982; Brown, 2002; Grant, 1985a,b; Ashworth, 1985), depending on the tectonothermal history or *P-T-t* path (see Section 3.5). Melting during decompression is possible for tectonically thickened terranes that are slowly heated by conduction and/or radioactive decay and then exhumed (e.g., Teyssier & Whitney, 2002), but not for terranes that remain at normal crustal thickness and which melt during heating at constant or slightly/moderately increasing pressure, requiring mantle heat to cause melting (see Section 4.15).

Several melting reactions may be encountered during prograde metamorphism, each successive step occurring at higher temperature and lower water activity, $a(H_2O)$, as well as producing melts of different chemical composition (Powell, 1983, p. 138). The reduction in $a(H_2O)$ is due to solution in the melt of hydroxyl released from the melting minerals, which effectively extracts water from the rocks, as noted previously. If free water is introduced into the rocks at this time, it is dissolved in the water-undersaturated melt, keeping the rocks effectively dry, unless excess water is present. In other words, the $a(H_2O)$ is controlled ('*internally buffered*') by the melting reactions themselves.

More realistic reactions

Because the minerals involved in melting reactions have complex compositions, the melting reactions in most rocks are more complex than those discussed so far. Melting may occur over a range of temperatures, depending on compositional factors such as the rock Mg/Fe ratio, the Ti and F contents of biotite or amphibole, and the Na content of any plagioclase involved (Le Breton & Thompson, 1988; Vielzeuf & Holloway, 1988; Patiño-Douce & Johnston, 1991; Carrington & Harley, 1995; Stevens *et al.*, 1997). Moreover, less abundant chemical components, such as Ti and Fe^{3+}, need to be considered in some situations, for example, in many metapelitic rocks. In fact, the *P-T* conditions of melting are very sensitive to rock composition, and to properly evaluate this effect, *P-T pseudosections* (Chapters 1 and 2) should be used, provided the most compositionally comprehensive chemical system possible is used, especially for making quantitative *P-T* inferences (White *et al.*, 2007), as discussed below.

Much can be learnt from *P-T* grids ('petrogenetic' grids or *P-T* projections; see Chapter 2) in complex systems (Fig. 4.4A). Stable univariant and invariant equilibria in the system constitute the framework around which more complicated systems can be constructed (White *et al.*, 2007). For example, in the system K_2O-FeO-MgO-Al_2O_3-SiO_2-H_2O (KFMASH), one of the main biotite-breakdown reactions (discussed previously) is: Bt + Sil + Qtz = Grt + Crd + Kfs + L, whereas in the system K_2O-FeO-MgO-Al_2O_3-SiO_2-H_2O–TiO_2-O_2 (KFMASHTO), in which Ti and Fe^{3+} are taken into account, two equivalent reactions appear, namely: Bt + Sil + Mag + Qtz = Grt + Crd + Kfs + Ilm + L and Bt + Sil + Qtz = Grt + Crd + Kfs + Ilm +

Rt + L (White *et al.*, 2007). However, though petrogenetic grids provide useful information on the overall *P-T* limits of mineral assemblages in broad compositional systems, they cannot provide accurate *P-T* estimates of assemblages for specific rock compositions.

In contrast, *P-T* pseudosections (Fig. 4.4B) show relationships between multivariant assemblages over a *P-T* range for a particular whole-rock composition (e.g., White *et al.*, 2007), as discussed in Chapters 1 and 2. They relate partial melting to realistic metamorphic assemblages. As many components as possible should be included in the modelled whole-rock composition, because the inclusion of minor components, such as Ti and Fe^{3+}, in the calculations may have major effects on phase relationships, including assemblages resulting from key melting reactions (White *et al.*, 2007). The pseudosection shown in Fig. 4.4B reveals that, though the broad melting reaction: Bt + Sil + Qtz = Grt + Crd + Kfs + L – equation (2A) mentioned previously – dominates the system before the appearance of orthopyroxene and sillimanite, the mineral assemblages have large variances. Though pseudosections are potentially very useful for inferring the *P-T* evolution of specific metamorphic rocks, care is required, because (1) open-system (metasomatic) processes or melt gain/loss complicate the choice of a suitable whole-rock composition, (2) volumes of equilibration may be relatively local, necessitating an estimation of appropriately local whole-rock composition, and (3) a whole-rock composition as close as possible to the chemical analysis of the rock produces the most reliable results (White *et al.*, 2007).

4.2.3 Volumes of melt produced

Voluminous granites can be produced by fluid-absent melting (Clemens & Vielzeuf, 1987; Sawyer, 1996). For example, fluid-absent biotite dehydration melting can produce 20%–35% melt in pelitic compositions, 10%–40% melt in quartzofeldspathic rocks, and 20%–45% melt in intermediate igneous compositions (Clemens & Vielzeuf, 1987), the amount of melt depending on (1) the volume & composition of the biotite and (2) the solubility of water in the melt, which varies with pressure. Stevens *et al.* (1997) found experimentally that up to 35% melt can be produced over a temperature range of as little as 15°C between 830°C and 875°C in Ti-free pelitic compositions, whereas the main production of melt occurs more gradually between 830°C and >900°C in Ti-bearing compositions. Fluid-absent hornblende dehydration melting can produce 5%–10% melt in quartzofeldspathic rocks, 2%–14% melt in intermediate igneous compositions, and 15%–40% melt in mafic rocks (Clemens & Vielzeuf, 1987).

4.2.4 Compositions of partial melts

The lowest-temperature liquid fractions produced by partial melting of metapelites, felsic rocks and intermediate-mafic rocks are of 'haplogranitic' composition, i.e., rich in normative Qtz, Kfs and Pl (Winkler & von Platen, 1961; Winkler, 1979; Grant, 1983, Wyllie, 1977; Fyfe *et al.*, 1978; Abbott & Clarke, 1979). However, experiments on granitic compositions (Brown & Fyfe, 1972; Winkler *et al.*, 1975; Wyllie, 1977; Winkler, 1979; Clemens & Wall, 1981) indicate that, with increasing temperature, anorthite and mafic components become more soluble in the liquid, and that the more mafic granitic melts form at progressively higher temperatures. High-temperature, fluid-absent melting produces melts with appreciable amounts of dissolved Mg, Fe and Ca, as mafic as granodiorite (Clemens, 1984). For example, realistic granitic liquid compositions, with 66.6% to 73.2% SiO_2, 1.4% to 7.1% FeO + MgO and 3.2% to 3.7% H_2O, have been produced by experimental melting of pelitic compositions at 1 GPa and 875°C to 1050°C (Vielzeuf & Holloway, 1988). Similarly, liquids with 62% to 76% SiO_2, 2.22% to 6.05% FeO + MgO and 1% to 8% H_2O have been produced by melting of felsic compositions at 100 to 320 MPa and 875°C to 1090°C (Rutherford *et al.*, 1985). Melts may vary widely in composition, depending on the source rocks, pressure and H_2O activity (Patiño-Douce, 1996). In the more advanced melting of metapelites, Al_2O_3 may enter the liquid in excess of the amount required by feldspars, so that the melt becomes corundum-normative (Abbott & Clarke, 1979; Grant, 1985b, p. 97).

Questions have been asked about whether equilibrium is approached during partial melting. Chemical equilibrium depends on diffusion and/or surface-controlled processes, both of which are time-dependent at a given temperature (Berger & Rosenberg, 2002). Johannes (1983) found that chemical equilibrium in quartz-feldspar systems is experimentally attainable and hence likely at natural conditions above 850°C, but that it is experimentally unattainable and hence doubtful at natural conditions below 700°C. Berger & Rosenberg (2002) stated that equilibration of major elements between melt and residue is common, especially with regard to feldspar, but not for trace elements (e.g., Bea *et al.*, 1994). Equilibrium during partial melting may not be achieved between melt and residual solids, owing mainly to the removal of part or all of the melt during deformation (Lappin & Hollister, 1980; Kenah & Hollister, 1983; Brown & Earle, 1983; Kriegsman, 2001). Zeng *et al.* (2005) found evidence of Nd isotope disequilibrium during anatexis in the southern Sierra Nevada, California, USA. Sawyer (1991) concluded that for chemical disequilibrium between melt and residual solids to occur, the rate of melt segregation must be greater than the rate of chemical equilibration, and vice versa; more rapid equilibration would be favoured by higher melting temperatures and more rapid segregation would be favoured by deformation. On the other hand, dynamic recrystallization and/or deformation by diffusion creep (see Section 6.2) may assist chemical equilibration by facilitating chemical exchange (Yund & Tullis, 1991), and penetration of melt along grain boundaries is also assisted by deformation (Berger & Rosenberg, 2002). Melt-present granular flow (i.e., dissolution accompanying grain-boundary sliding during partial melting) can also promote chemical equilibrium between melt and solid residue (Berger & Rosenberg, 2002). A complication is that metasomatic (typically retrograde) reactions can be responsible for post-anatectic compositional changes in migmatites, for example, by alteration of feldspar (Ashworth, 1985).

4.2.5 Distinguishing between water-saturated and water-undersaturated melting

Leucosomes/neosomes (see Section 4.3 for terminology) consisting of quartz, K-feldspar, plagioclase and biotite, without anhydrous minerals, are consistent with water-saturated melting, whereas leucosomes/neosomes containing essentially anhydrous peritectic minerals, such as garnet, cordierite or orthopyroxene, are consistent with water-undersaturated melting, though similar leucosomes can be formed by water-saturated melting above the saturated granite solidus (Ward *et al.*, 2007).

Fig. 4.5 Typical stromatic migmatite (metatexite), showing leucosomes with mafic selvedges (melanosomes), Sturt Highway, north of Alice Springs, central Australia. Most of the leucosomes are parallel to the foliation, but one is transgressive, indicating local melt movement.

4.3 Terminology of anatectic migmatites

An unnecessarily complicated terminology for migmatites has developed, and so we propose to use only the most necessary terms. The melted part of an anatectic migmatite is referred to as the *neosome*, which consists of melt plus crystals produced as products of a partial melting reaction ('peritectic crystals'), as discussed in Section 4.2.2. More discussion on the term 'neosome' is presented at the end of this section.

The apparently unmelted part of the migmatite used to be called the 'palaeosome', on the assumption that it represents the unaltered rock (Mehnert, 1968). However, it is generally chemically modified (Johannes, 1985, 1988), as melt is typically extracted from it during the partial melting process (White *et al.*, 2004). These days it is generally referred to as the *mesosome*, and consists of unreacted minerals, together with any solid products of the melting reaction left behind at or near the reaction site; it may also have small pockets of residual melt (e.g., Sawyer, 2001).

Together, the unmelted and peritectic minerals constitute *restite*, or the solid minerals resulting from a melt-producing reaction, as defined by Chappell & White (1991). Unreactive rocks that escape melting (in the same area in which anatexis occurs in other rocks) are referred to as *resister* or *resistate*.

If the melt segregates from restite and then crystallizes, it forms *leucosome* (which is light coloured, owing to an abundance of quartz and feldspar), the segregated solid material being referred to *melanosome* (which is dark coloured owing to a relative abundance of mafic peritectic minerals, such as biotite, hornblende, garnet or cordierite). Commonly the melanosome forms selvedges on layers (*stromata*) of leucosome, as shown in Figs. 4.1 and 4.5, but in some migmatites, it segregates in patches or discontinuous lenses (Figs. 4.2, 4.6, 4.7). Evidence of arrested progressive segregation of restite from melt is preserved in some migmatites (Figs. 4.6, 4.7, 4.8). Though melanosomes are generally interpreted as resulting from segregation of restite from neosome, some may result from accumulation of mafic minerals precipitated from the neosome (although generally these are present in amounts too small to account for the observed melanosome

Fig. 4.6 Local segregation of former felsic melt from more mafic neosome in migmatite, Cadney Creek, Arunta Inlier, central Australia. Knife 9 cm long.

thickness), and others may result from hydration reactions between restite in the adjacent mesosome and residual melt in the neosome; some mesosomes may result from all three processes (Kriegsman, 2001).

Leucosomes consist largely of melt or crystals precipitated from the melt (Sawyer, 1987; Barbey *et al.*, 1996; Sawyer *et al.*, 1999), with or without some peritectic minerals, such as K-feldspar and garnet or cordierite resulting from reactions such as: Bt + Qtz + Sil = Crd/Grt + Kfs + melt (equation 2A). Leucosomes with peritectic minerals imply either local bulk flow of magma (e.g., Brown *et al.*, 1999) or growth of leucosome around the peritectic grains (Section 4.7.2); leucosomes without peritectic minerals imply that such bulk flow of magma has not occurred (Sawyer *et al.*, 1999).

Fig. 4.7 Garnet-bearing neosome, from which felsic leucosome has partly segregated, Cadney Creek, Arunta Inlier, central Australia. Knife 9 cm long.

Many, if not most, observed leucosomes have lost melt, as indicated by their mineral assemblages and chemical compositions (e.g., Sawyer *et al.*, 1999; White & Powell, 2002). For example, if leucosomes formed by partial melting of metasedimentary rocks are rich in plagioclase, but poor in K-feldspar and quartz, a reasonable explanation is that residual melt remaining after crystallization of some of the early plagioclase was lost from the leucosome. Therefore, leucosomes should not be assumed to represent original melt compositions in the absence of confirmatory mineralogical/chemical evidence.

Though its conceptual definition is clear enough, the term 'neosome' is mostly not very useful in practice, because it is too general, covering any new material produced in the migmatite-forming process. For example, in some granulite facies migmatites, practically the whole rock consists of melt plus peritectic minerals (White *et al.*, 2004), and so could be called 'neosome'. Originally 'neosome' was meant to be the complement of 'palaeosome', but, as noted previously, this term is unsatisfactory, as it implies the original rock in an unaltered state, which rarely applies. The common (though not universal) abandonment of 'palaeosome' devalues the usefulness of 'neosome'. However, it is still a useful term when referring to mobile material prior to segregation of melt and restite, where evidence of this process can be observed (e.g., Vernon *et al.*, 2003; see also Section 4.7.1 and Figs. 4.6, 4.7 and 4.8), which is an unusual situation for most migmatites. Modern usage favours 'leucosome' as a general, descriptive term for the lighter-coloured component of migmatites and 'mesosome' for the darker-coloured component. No genetic implication would be made in this usage, apart from accepting that leucosome represents some degree of segregation of mobile material. It commonly represents a melt or part thereof, but this is not an essential part of this definition, and so the term can be used for both anatectic and non-anatectic migmatites.

(A)

(B)

Fig. 4.8 (A) Stromatic amphibolite migmatite, south of Black Hill, near Alice Springs, central Australia, showing hornblende-bearing neosome stromata (running 'north-south', as viewed in the photo), and transgressive leucosome with evidence of segregation of former melt from hornblende restite. (B) Magnified view of centre part of (A).

4.4 Non-anatectic migmatites

Some light-coloured layers (leucosomes) in migmatitic rocks may not be due to anatexis, but may represent subsolidus *segregation layering* (differentiated layering, hydrothermal mass-transfer) formed by migration of components, probably mainly in a fluid. This is a form of *metamorphic differentiation* (e.g., Vernon, 2004, pp. 359, 393, 485), as discussed in Section 5.14. As explained by Yardley (1977c, p. 293), metamorphic compositional segregation may occur in the presence of a small amount of hydrous fluid, in response to chemical potential gradients between grains of mesosome minerals and lower-energy grains of the same minerals growing at preferred sites (e.g., fractures). Chemical components dissolved in the mesosome are transported by *diffusion* (transfer of chemical components down chemical potential gradients) and/or *infiltration* (fluid flow down pressure gradients), and precipitated on the new grains, forming non-anatectic leucosomes. The compositions of non-anatectic leucosomes reflect the local bulk-rock composition and the chemical potential differences, but not the composition of the material dissolved in the fluid. In other words, the resulting leucosome composition depends only on the minerals stable in that environment, which determine the chemical components that can be fixed in the leucosome, the other components remaining in solution.

As an example, Kretz (1966b) inferred that quartz-feldspar veins in an amphibolite were formed by metamorphic segregation, because of the same composition of the plagioclase in the leucosome and mesosome. Yardley (1977c) used a similar argument to infer a non-anatectic origin for trondhjemitic leucosomes in the Huntly-Portsoy area of north-east Scotland.

In the New England, USA, vein-like, foliation-parallel leucosomes composed of quartz and plagioclase without K-feldspar were formed in the staurolite zone at temperatures too low for melting (Tracy, 1978, 1985, p. 214). Sawyer & Robin (1986) and Sawyer & Barnes (1988) observed a continuous suite of light-coloured concordant veins with similar morphology and structure throughout the Quertico metasedimentary belt, Canada, in rocks formed at greenschist to upper amphibolite facies conditions. At lower grades, the veins are quartz-rich, but in the highest grade migmatitic terranes, the vein (leucosome) compositions are quartzofeldspathic. Regardless of the metamorphic grade, mafic selvedges on the veins contain only quartz, plagioclase and biotite, and do not represent residua left after partial melting. Because of these features, Sawyer & Robin (1986) suggested that the leucosome veins were formed by stress-induced mass transfer of mobile chemical components at subsolidus conditions, which they termed 'tectonic segregation.' These layers are in contrast to discordant leucosomes in the same rocks, which are of granitic composition and appear to be products of partial melting.

4.5 Distinguishing anatectic from non-anatectic migmatites

Evidence that a leucosome is due to anatexis includes: (1) *temperatures appropriate for partial melting*, as indicated by mesosome mineral assemblages, though other components (such as CO_2) diluting the hydrous fluid may force initial water-saturated melting temperatures to exceed the minimum for pure water (Yardley, 1977c), (2) *cotectic igneous compositions*, though if the melt or part of the melt is removed during deformation, the remaining leucosome may not be of magmatic composition, (3) *boudinage of mesosome in leucosome*, owing to the greater strength of the mesosome (e.g., Mehnert, 1968, pp. 21–23), which is consistent with a magmatic condition of the leucosome, (4) *igneous microstructures* (Vernon & Collins, 1988; Vernon *et al.*, 1990; Vernon, 1999, 2004; Sawyer *et al.*, 1999; Vernon & Johnson, 2000; Brown, 2001; Gibson, 2002, p. 68; Marchildon & Brown, 2003), as shown in Fig. 4.9, (5) *aligned euhedral feldspar crystals*, reflecting magmatic flow (Fig. 4.9D), with or without imbrication (Sawyer *et al.*, 1999,

(A)

(B)

Fig. 4.9 (A) Igneous microstructure in leucosome, characterized by crystal faces of plagioclase (Pl) against quartz (Qtz), Snowy Mountains, south-east Australia. Crossed polars; base of photo 1.75 mm. (B) Igneous microstructure in leucosome, characterized by crystal faces of K-feldspar (now microcline-microperthite, owing to exsolution and mild deformation) and cordierite against quartz, Mount Stafford, central Australia. Crossed polars; base of photo 4.4 mm. From Vernon & Collins (1988, fig. 7), with permission of *Geology*. (C) Igneous microstructure in leucosome, characterized by crystal faces of K-feldspar (Kfs) against quartz (Qtz), Cooma Complex, south-east Australia. The feldspar crystal faces have been preserved despite deformation and recrystallization of the quartz. Crossed polars; base of photo 1.5 mm. From Vernon & Johnson (2000, fig. 20) with permission of *Journal of the Virtual Explorer*. (D) Aligned feldspar crystals in granite sheet (neosome), reflecting magmatic flow, Wuluma Hills, central Australia. Knife 9 cm long. From Collins *et al.* (1989, fig. 9), copyright Elsevier; published with permission.

Fig. 4.9 *(cont.).*

p. 227), and (6) *presence of garnet, cordierite or orthopyroxene in the leucosome* of metasediment-derived migmatites (indicating water-undersaturated melting involving breakdown of biotite), or pyroxene in the leucosome of amphibolite-derived migmatites (indicating water-undersaturated melting involving breakdown of hornblende), as shown in Figs. 4.2 and 4.3, respectively. Yardley (1978) suggested that anatectic leucosomes in metasedimentary rocks should have K-feldspar and more sodic plagioclase than in the parent rocks.

4.6 Stages of melting and melt migration

During anatexis: (1) the metamorphic temperature becomes high enough for melting, (2) melt segregates from crystalline residuum, (3) magma (melt and entrained crystals, if any) may remain as leucosomes in the anatectic migmatites or be transported to sites where (4) magma batches are collected. These magma batches may or may not migrate further and amalgamate to form plutons (Sawyer *et al.*, 1999). Steps 1 and 2 occur in the anatectic zone, whereas, though steps 3 and 4 begin in the anatectic zone, they may end anywhere between there and Earth's surface. In the following sections, we consider these processes in terms of two broad 'stages', namely the *melting (anatectic) stage*, and the *transport and local accumulation stage*.

4.7 Melting (anatectic) stage: Accumulation and movement of melt in source migmatites

The detailed processes involved in the formation of migmatites are complicated. Not only can several melting reactions occur during progressive metamorphism in the same terrane, but melts produced at deeper levels may also ascend and be trapped in the terrane being observed (Powell, 1983. p. 138; Vernon *et al.*, 2003; Slagstad *et al.*, 2005). In addition, though partial melting is the main process, other processes, such as regional or local metasomatism, diffusional segregation and magma injection may also occur, and several of these processes may interact (Johannes, 1983; Olsen, 1983; Grant, 1985a,b). However, Kriegsman (2001) has cast some doubt on the mass-balance techniques commonly used to infer metasomatism during partial melting.

4.7.1 Melt retained in weakly deformed source rocks: Formation of patch and bedded migmatites

Melt initiates at reaction sites, which are multimineral grain boundaries, along which reactant minerals are in contact (Mehnert, 1968; Mehnert & Büsch, 1982; Sawyer, 2001; Brown, 2005b). Microstructural evidence of such reaction sites has been presented by Sawyer (2001, fig. 1), and consists mainly of films along grain boundaries and cuspate patches of former melt that crystallized to quartz, K-feldspar or plagioclase, commonly with corroded relics of these minerals, as well as elongate patches of former melt pseudomorphing partly melted biotite (Fig. 4.10). Small dihedral angles (<60°) are typical of felsic melt in contact with solid grains, as observed in rocks (e.g., Clemens & Holness, 2000; Sawyer, 2001) and determined experimentally (Jurewicz & Watson, 1984, 1985; Laporte, 1994).

Melt can stay in its source rock if space is available, so that it is not forced out of the rock during deformation. It may fill or force open cavities along grain boundaries or along microfractures produced by concurrent deformation. The resulting neosome may form patches or discontinuous veinlets (with or without local segregation into leucosome & melanosome) either *in situ* or after local movement, forming *patch migmatites* (McLellan, 1988), *bedded migmatites* (Vernon *et al.*, 1990, 2003; Greenfield *et al.*, 1996; Vernon & Johnson, 2000), and some *stromatic (layered) migmatites*. In bedded migmatites, the neosome is retained in the source beds, (typically metapelites), thereby preserving sedimentary bedding. Examples are bedded migmatites at Mount Stafford, central Australia (Vernon *et al.*, 1990; Greenfield *et al.*, 1996; White *et al.*, 2003) and at Cooma, south-eastern Australia (Vernon & Johnson, 2000; Vernon *et al.*, 2003), as shown in Figs. 4.11–4.13.

Because these neosomes/leucosomes are relatively undeformed, they commonly retain igneous microstructures, especially (1) euhedral K-feldspar, plagioclase and cordierite against quartz (Vernon & Collins, 1998;

Fig. 4.10 Polygonal aggregate of quartz and feldspar, with scattered grains of orthopyroxene (high relief) and biotite (elongate crystals) in a granulite-facies migmatite from the Ashuanipi Subprovince, Canada. Veinlets and elongate, cuspate pockets of former melt occur along some of the grain boundaries (especially centre and top-left of centre), and spectacular evidence of partial melting of biotite (preserving the elongate grain shape) is also evident (centre). Photo by Ed Sawyer. Crossed polars, with one-wave quartz plate; base of photo 3 mm. After Vernon (2004, fig. 4.82).

Fig. 4.11 Bedded migmatite, Cooma Complex, south-east Australia, showing leucosome patches and lenses aligned in a foliation oblique to bedding, but confined to metapelite beds. Space for the leucosome appears to have been made by boudinage of the metapelitic beds, which were stronger than the metaspammite beds. From Vernon *et al.* (2003, fig. 7), with permission of Springer.

Fig. 4.12 Bedded migmatite, Cooma Complex, south-east Australia, showing leucosome patches and lenses aligned in a foliation oblique to bedding, but confined to metapelite beds. Many of the quartzofeldspathic leucosome lenses are poorly defined, owing to incomplete separation of melt from peritectic cordierite grains (dark coloured). The metapelite beds have resisted deformation, owing to abundance of coarse-grained K-feldspar, cordierite and andalusite, whereas the metapsammite beds have been thinned and contorted. The greater strength of the metapelitic beds has resulted in their fracturing and boudinage, providing space for the leucosome. From Vernon *et al.* (2003, fig. 10), with permission of Springer.

Fig. 4.13 Bedded migmatite, showing extensively melted metapelitic beds (light coloured, owing to abundant quartz and K-feldspar, with dark coloured porphyroblasts of cordierite and sillimanite pseudomorphs of andalusite) and darker coloured, unmelted metapsammite beds, Mount Stafford, Arunta Inlier, central Australia.

Fig. 4.14 Granulite facies metapsammitic migmatite, Darling Range, north of Broken Hill, Australia, showing abundant leucosome that has moved locally though fractured and disrupted mesosome. This structure is in marked contrast to that shown by metapelitic migmatites in the same area, in which the leucosome has segregated into layers (Fig. 4.2). Knife 9 cm long.

Vernon & Johnson, 2000; Vernon *et al.*, 2003; Marchildon & Brown, 2003, fig. 5), as shown in Fig. 4.9, and (2) euhedral inclusion-free overgrowths precipitated from the neosome melt on inclusion-rich grains of the same minerals in the adjacent mesosome, such as cordierite and K-feldspar in bedded migmatites at Cooma, Australia (Vernon & Johnson, 2000; Vernon *et al.*, 2003) and cordierite and K-feldspar in stromatic migmatites at Broken Hill, Australia (White *et al.*, 2004, fig. 6a).

Even if it doesn't escape from the rock, melt may be locally redistributed, and can move into low-pressure sites, such as local fracture openings, fold hinges and interboudin necks (Figs. 4.11, 4.13). For example, melt may occupy interboudin necks formed where a strong layer or bed fractures and undergoes boudinage while the adjacent layers or beds flow in a ductile manner (Pattison & Harte, 1988; Williams *et al.*, 1995; Brown & Rushmer, 1997; Vernon & Johnson, 2000; Vernon *et al.*, 2003).

Migmatites that fracture extensively may develop melt-filled fracture networks. For example, metapsammitic migmatites at Broken Hill, Australia, are heavily veined with leucosome (Fig. 4.14), space having been provided by dilational openings formed during brittle deformation. Melt-rich material moved on the outcrop scale, but was largely or completely retained in the host rock. This occurred because the metapsammites were stronger than the interbedded metapelites, which were anisotropic in structure and more ductile; they became stromatic migmatites, in which the neosome/leucosome developed and/or segregated in layers (Fig. 4.2), and was partly squeezed out during deformation of the mica- and sillimanite-rich, foliated mesosomes, leaving residual leucosome lenses (White *et al.*, 2004), as discussed in the next section.

In some migmatites, leucosome that develops preferentially along fractures may have gradational boundaries with the mesosome (Fig. 4.15). Pattison (1991) inferred that similar light-coloured, diffuse veins in

Fig. 4.15 Gradational leucosomes that appear to have formed along fractures, Helsinki, Finland. Knife 9 cm long.

Fig. 4.16 Gradational contacts between *in situ* leucosome and partly melted metasedimentary breccia, Mount Stafford, Arunta Inlier, central Australia. Knife 9 cm long. From Vernon *et al.* (1990, fig. 11.21), with permission of Unwin Hyman Ltd.

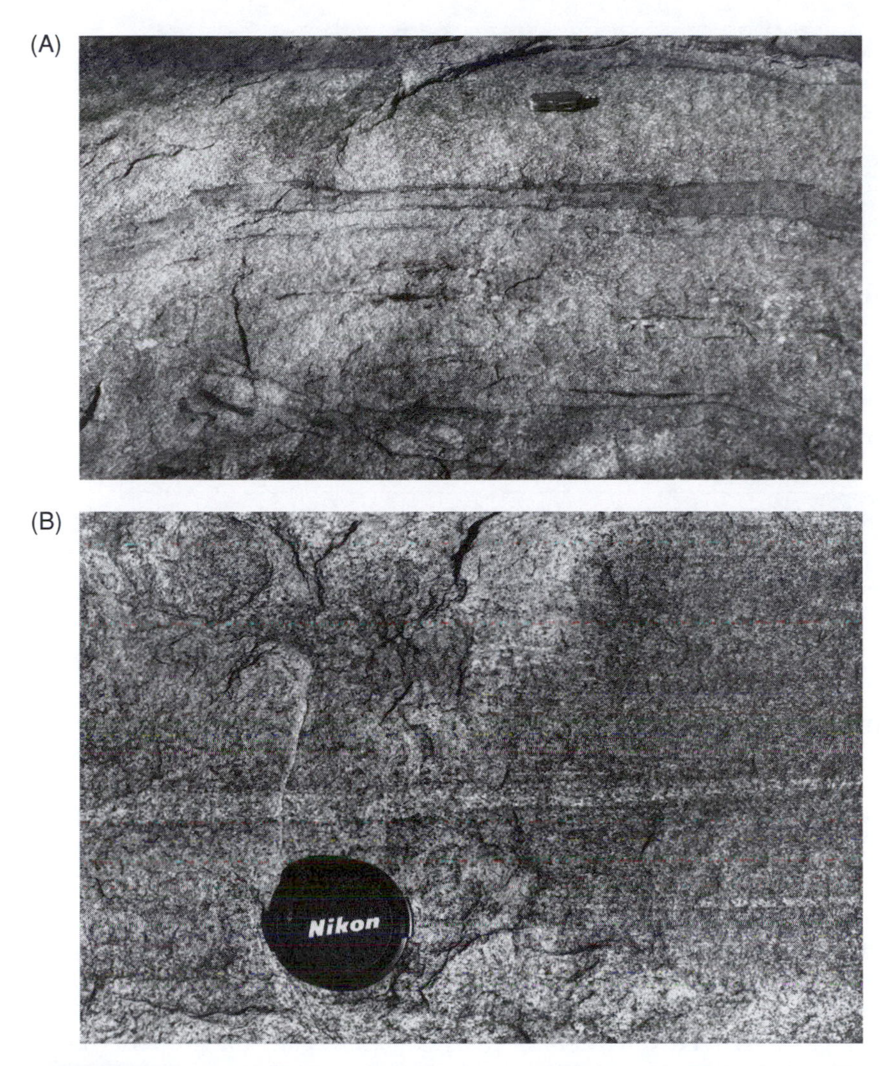

Fig. 4.17 Granite formed by in situ melting, Wuluma Granite, Arunta Inlier, central Australia (Collins *et al.*, 1989). (A) Arrested dispersal of biotite-rich bedding relics, in their original orientation, into the granite, which shows alignment of feldspar crystals, owing to local magmatic flow. (B) Diffuse contact with a gneissic enclave (right), a remnant felsic layer continuing across the boundary.

metagabbro were formed by fluids penetrating cracks, metasomatizing the metagabbros, and marking a transition to infiltration-driven, localized anatexis, which produced tonalitic leucosomes with sharp boundaries. In other migmatites, extensive development of leucosome by in situ melting may leave partly melted remnants grading into the neosome (Figs. 4.16 to 4.18).

If the melt accumulation rate exceeds the loss rate, the increased proportion of melt may cause the original foliated or bedded structure to weaken and disintegrate into fragments that passively rotate in flowing magma, forming a 'schlieren migmatite' or 'primary diatexite' (Fig. 4.19). Melt accumulation is promoted by rapid temperature increase and/or water influx (Sawyer, 1998), as well as by weak or no deformation.

Fig. 4.18 Evidence of in situ melting of metasedimentary rocks, May Lake, Sierra Nevada, California, USA, showing lateral gradation from foliated schist and metatexite to similarly foliated leucogranite (pale). Knife 9 cm long.

4.7.2 Melt retained in more strongly deformed source rocks: Formation of stromatic migmatites

In migmatites undergoing stronger deformation, which is the most common situation, the melt-rich material moves into lower-pressure sites in developing folia, where it forms light-coloured (quartz-feldspar-rich) layers ('stromata') of leucosome (Robin, 1979; McLellan, 1988; Sawyer, 1994, 1996, 1999, 2001; Brown, 1994; Brown *et al.*, 1995a,b; Brown & Rushmer, 1997; Guernina & Sawyer, 2003). The resulting rock is called a *metatexite* (Brown, 1973) or *stromatic migmatite*, which is often thought of as the 'typical' migmatite. The leucosomes form discrete veins or lenses mainly concordant with the foliation in the mesosome (Figs. 4.1, 4.3, 4.5, 4.20). Generally the solid peritectic crystals either remain at the melting site or segregate as relatively mafic selvedges (*melanosome*) on the leucosome (Fig. 4.5), though some may be carried along in mobile melt (Brown *et al.*, 1999), at least for short distances. Stromata may be referred to as 'neosome' if it can be shown that both former melt and restite are present, but more commonly they are referred to as 'leucosome', because at least local segregation of melt from restite is usual, and some melt loss from the source rock is also common (Section 4.8).

Stromatic migmatites typically show structural evidence of deformation, such as folding, development of several foliations and boudinage (Figs. 4.1, 4.20). The development of stromata is commonly connected with the deformation, which also assists melt to leave the system on the outcrop scale. However, the formation of stromata is complicated, and the following processes have been proposed, some of which do not necessarily involve deformation, and not all of which involve anatexis. Criteria by which the various types can be distinguished have been discussed by Hyndman (1972, p. 291) and Yardley (1978), though several of the proposed processes may operate simultaneously.

(1) Physical segregation of dispersed melt from mesosome: Tectonic deformation may initiate and open local small fractures ('microfractures') along

(A)

(B)

Fig. 4.19 (A) Schlieren migmatite (diatexite), Mount Stafford, Arunta Inlier, central Australia. From Vernon *et al.* (1990, fig. 11.17, with permission of Unwin Hyman Ltd). Knife 9 cm long. (B) Schlieren migmatite (diatexite), contact aureole of the Kameruka Granodiorite, Bega Batholith, south-east Australia. Knife 9 cm long. (C) Diatexite (bottom half of photo) in contact with and interfingering with metatexite, at 'Diatexite Hill', Southern Cross Mine area, north-west of Broken Hill, Australia. This relationship has been interpreted as a transition from metatexite to diatexite (White *et al.*, 2005). Knife 9 cm long.

(C)

Fig. 4.19 *(cont.)*.

grain boundaries (e.g., where the minerals concerned respond differently to local stresses) and along potentially weak surfaces, such as bedding planes and tectonic folia, facilitating local movement of melt (Brown, 1994; Brown *et al.*, 1995a; Sawyer, 1999, 2001). Local openings can also form in the mesosome where the rate of local extraction of melt is less than the rate of local accumulation of melt, causing a local increase in melt pressure. This process is assisted by a positive ΔV of the melting reaction, which is typical of H_2O-undersaturated melting reactions (Clemens & Mawer, 1992; Brown *et al.*, 1995b; Brown & Rushmer, 1997). In addition, shearing along irregular fracture surfaces can produce dilatant 'jogs.' Melt flows into (is

Fig. 4.20 Folded metatexite, Adirondack Mountains, New York, USA. The large leucosome layer at the bottom was probably injected into its present position, and underwent boudinage during later solid-state deformation.

Fig. 4.21 Leucosomes that nucleated on and concentrated around peritectic garnet porphyroblasts in felsic gneiss, Waterworks Hill, Broken Hill, New South Wales, Australia. Knife 9 cm long.

sucked into) all these openings, owing to the local lowering of pressure. Migration of the melt-filled openings or of the melt itself (which is possible if an interconnected porosity is developed) to bedding-parallel or foliation-parallel fractures may form stromata.

(2) 'Segregation' by in situ melting: In H_2O-undersaturated melting, the melt may be localized by the reaction itself, rather than by deformation, though both processes can occur together. This process of *focussed melt production* (White *et al.*, 2004) may produce patchy or lenticular leucosomes, but commonly produces stromata (Figs. 4.2, 4.21). The following two situations may apply.

(a) The concentration of melt may be controlled by a *product* (peritectic) mineral (Waters, 1988; Stüwe & Powell, 1989; Powell & Downes, 1990; Hand & Dirks, 1992; White *et al.*, 2004). For example, garnet in high-grade migmatites at Broken Hill, Australia, was formed by the H_2O-undersaturated melting reaction: Bt + Sil + Qtz = Grt + Kfs + melt. Because the garnet nucleated with difficulty, it grew as isolated grains, with the result that continued reaction required diffusion of chemical components from reaction sites to the growing garnet, producing aggregates of melt ± peritectic K-feldspar localized around the garnet (Powell & Downes, 1990; White *et al.*, 2004), as shown in Figs. 4.2 and 4.21.

(b) The concentration of melt may be controlled by a *reactant* mineral (Vernon *et al.*, 1990; White *et al.*, 2003; Johnson *et al.*, 2001). For example, andalusite in migmatites at Mount Stafford, central Australia, was partly dissolved by the H_2O-absent melting reaction: And + Bt + Qtz = Crd + Kfs + melt. Because the andalusite occurs as scattered porphyroblasts, the other reactant components had to migrate to the dissolving porphyroblasts, forming leucosome rims on the andalusite (Vernon *et al.*, 1990; White *et al.*, 2003).

(3) Injection of magma: 'Lit-part-lit injection' parallel to bedding and/or foliation (Sederholm, 1967) can produce 'injection migmatites' (Fig. 4.22),

(A)

(B)

Fig. 4.22 (A) Metatexite consisting of thick layers of leucosome (light) and mesosome (darker), Helsinki, Finland. Many of the leucosomes, especially the thicker ones, were probably injected into their present position. Later deformation has produced folding, as well as boudinage of the felsic layers, which, when crystallized, were stronger than the mesosome. (B) Injection migmatite, Swakane area, Cascade Range, Washington, USA, some of the leucosomes being connected to a transgressive pegmatitic granite.

the magma either being locally derived or introduced from an intrusive magmatic source.

(4) Diffusional exchange in melt without bulk compositional change: Diffusional segregation of chemical components in the presence of melt can develop and/or intensify felsic layers, according to Maaløe (1992), who described stromatic migmatites consisting of foliation-parallel leucosomes with melanosome borders south-east of Oslo, Norway. The neosome (leucosome plus melanosome) has a similar biotite content to that of the mesosome, suggesting that melt did not infiltrate from the mesosome to form the neosome. Maaløe (1992) suggested that melt initially formed at grain corners, where it remained isolated, except where a crack formed parallel to the foliation in response to deformation. This enabled local segregation of the trapped melt into the opening and possibly promoted some melting

Fig. 4.23 (A) Early stage of the formation of a stromatic migmatite by folding and transposition of a bedded migmatite, Cooma Complex, south-east Australia. The deformation has produced sub-parallel, discontinuous layers, lenses, boudins and rootless thickened fold hinges of leucosome. From Vernon *et al.* (2003, fig. 12), with permission of Springer.
(B) Stromatic migmatite with residual quartzite beds, Snake Creek, Cooma Complex, south-east Australia. Though the leucosomes have been elongated into stromata by strong deformation, they do not cross into the quartzite beds, indicating that the rock was deformed when the leucosomes were solid and that the rock originally was a bedded migmatite. Knife 9 cm long. From Vernon *et al.* (2003, fig. 11), with permission of Springer.

by addition of small amounts of water. The melt film enabled diffusion of chemical components between the developing leucosome and the melt in the immediately adjacent mesosome. Diffusion of felsic components to the leucosome enabled it to widen by promoting melting of grains at the contact, concentrating biotite passively at the boundary between the leucosome and mesosome.

(4) Metasomatism: Some leucosomes appear to have been formed by complex processes, involving intensification of existing layering by

(A)

Fig. 4.24 (A) Leucosome stromata parallel to fold axial surfaces, transecting bedding, Helsinki, Finland. Knife 9 cm long. (B) Leucosome stromata parallel to fold axial surfaces, transecting bedding and stromatic migmatite, May Lake, Sierra Nevada, California, USA. Knife 9 cm long. (C) Fine-grained neosome layers parallel to axial surfaces of folds in pre-existing metatexite; paving slab of unknown origin Coin 2.8 cm diameter.

introduction of aqueous fluid or melt, typically with interaction between the introduced fluids and mesosome (Mehnert & Büsch, 1982; Johannes, 1988; Olsen, 1982, 1983, 1985).

(5) Solid-state deformation: Some stromatic migmatites appear to result from solid-state deformation of bedded migmatites (Vernon *et al.*, 2000, 2003). For example, anatectic, metapelite-derived, bedded migmatites in the Cooma Complex, south-east Australia, show progressive stages of deformation by folding and transposition to produce stromatic migmatites (Fig. 4.23). The leucosomes do not transect bedding surfaces, but remain confined to metapelitic beds, indicating that the leucosomes were solid during the deformation (Vernon *et al.*, 2003).

4.7.3 Axial-surface leucosomes

Though most leucosomes are parallel to the foliation in stromatic migmatites, they may also occur oblique to bedding or the dominant

Fig. 4.24 *(cont.)*.

foliation, especially parallel to the axial surfaces of folds (e.g., Hopgood & Bowes, 1978; Vernon *et al.*, 1989; Allibone & Norris, 1992; Anhaeusser, 1992, fig. 6D; Hand & Dirks, 1992, fig. 4; Brown, 1994; Johnson *et al.*, 1994; Williams *et al.*, 1995, fig. 5e; Brown & Rushmer, 1997, fig. 8a; Vernon & Paterson, 2001), as shown in Fig. 4.24. This is not intuitively expected, because the local principal compressive stress (σ_1) is generally inferred to

(A)

(B)

Fig. 4.25 (A) Neosome injected into low-pressure site in boudinaged stromatic migmatite, Mount Hay, Arunta Block, central Australia. Base of photo approximately 1 m. (B) Leucosome occupying dilatant sites formed by fracturing (boudinage), Olary area, eastern South Australia. Knife (towards left of photo) 9 cm long.

be approximately perpendicular to the fold axial surfaces and the leucosome veins (though stresses cannot be confidently inferred from strains). Concentration of leucosome in antiformal fold hinges (Brown & Rushmer, 1997; Sawyer *et al.*, 1999) can be formed by various processes, as outlined by Vernon & Paterson (2001), including local *magma injection* and *in situ leucosome formation*.

The most probable *magma injection* mechanisms (Vernon & Paterson, 2001) are: (1) fluid (magma) pressure overcoming σ_1 plus the tensile strength of the rock parallel to σ_3, which can occur during folding if the plane perpendicular to σ_1 has a prominent foliation (axial-surface foliation) and σ_1–σ_3 is small (Lucas & St-Onge, 1995; Vernon & Paterson, 2001), which may well apply in the lower to middle crust (Etheridge, 1983) where anatexis is most common; (2) magma filling dilatant sites or being forced into axial-surface folia during temporary relaxation, which may involve reversals of the local stress; (3) magma filling transient, dilatant jogs opened during relative movement along the axial surfaces of strongly asymmetrical folds or local shear zones, without stress reversals on the fold scale; (4) injection of magma into extensional shear zones in fold limbs of an oblique extensional crenulation-foliation; and (5) boudinage of stronger folded layers, allowing leucosome to form vein-like bodies parallel to the fold axial surface (Figs. 4.12, 4.25), as discussed by Vernon *et al.* (1990), Allibone & Norris (1992), Anhaeusser (1992), Williams *et al.*, (1995), Greenfield *et al.* (1996), Collins & Sawyer (1996), Brown & Rushmer (1997), Pattison & Harte (1988), Vernon & Johnson (2000) and Vernon *et al.* (2003).

The most probable mechanisms for *in situ leucosome formation* in fold axial surfaces (Vernon & Paterson, 2001) are: (1) local fluids flowing or diffusing into low-pressure sites or forcing openings parallel to axial-surface foliations during temporary relaxation in a folding event, promoting melting at those sites; and (2) water-absent melting reactions occurring preferentially along axial-surface folia, controlled by a low nucleation rate of peritectic minerals (e.g., garnet), which necessitates coalescence of melt around large grains of these minerals (Powell & Downes, 1990; Hand & Dirks, 1992; White *et al.*, 2004), as shown in Fig. 4.21; elongation of the melt parallel to the foliation may be promoted by the deflection of the folia around the large peritectic grains during deformation, forming elongate low-pressure sites ('pressure shadows'), as discussed by Vernon & Paterson (2002, pp. 190–1).

4.7.4 Multiple leucosomes

Leucosomes occupying more than one S-surface (bedding and/or foliation surface) are common in migmatites (e.g., Barr, 1985). For example, leucosomes may occupy the axial surfaces of the folds that deform apparently earlier leucosome stromata (Figs. 4.1, 4.24). Various transgressive leucosome relationships may occur (Figs. 4.1, 4.15, 4.22, 4.26, 4.27), and the leucosomes may vary in microstructure and chemical composition (e.g., Sawyer *et al.*, 1999, p. 230).

Though these structures are complex and hard to interpret, partly because evidence tends to be obscured by repeated deformation (Ashworth, 1985), they are nevertheless potential time markers in a sequence of deformation/melting events, provided they can be separated and radiometrically dated accurately enough (Hopgood & Bowes, 1978; Hopgood *et al.*, 1983). Earlier leucosomes may show a strong solid-state foliation and evidence of transposition into a dominant foliation, whereas later leucosomes may be less foliated and transgressive to a dominant foliation (Ashworth, 1979).

Complex leucosome relationships could be taken to indicate: (1) the presence of melt in successive deformation events (i.e., a single melting event spanning more than one deformation 'event'), (2) progressive melting phases in a single deformation event, or (3) separate melting events, which could represent either events markedly separate in time or separate batches formed by melting of different rocks in the same general area during a single melting 'event' (though separated slightly in time).

If both sets of leucosomes have the same composition (Fig. 4.1), the simplest and probably the most likely interpretation is the first one. For example, in migmatites of southern Brittany, France, petrographic continuity between leucosome stromata and transgressive veins or dykes of leucogranite suggests that all this material underwent final crystallization simultaneously (Marchildon & Brown, 2003). McLellan (1984) suggested

(A)

(B)

Fig. 4.26 (A) Mobile diatexite, with rectangular K-feldspar crystals (at least some of which precipitated from the melt), flow-segregation layering and many mesosome relics, cutting metatexite, Wuluma Area, Arunta Inlier, central Australia. Some of the metatexite leucosomes have biotite-rich melanosome borders. (B) Mobile fine-grained diatexite that has transected and disrupted metatexite; paving slab of unknown origin. Coin 2.8 cm diameter.

(A)

Fig. 4.27 (A) Transgressive veins of mobile, metapsammite-derived neosome in feldspathic metapsammite, Spring Creek, Cooma Complex, south-east Australia. This represents the earliest stage in the development of water-saturated neosome that accumulated (and probably mixed with added deeper magma) to form diatexite (Cooma Granodiorite). From Vernon *et al.* (2003, fig. 16), with permission of Springer. Knife 9 cm long. (B) Irregular vein of mobile, metapsammite-derived neosome transgressing, interfingering with and beginning to dismember previously formed, solidified metapelite-derived migmatite, Agnew's Dam locality, Cooma Complex, south-east Australia (Vernon *et al.*, 2003), with permission of Springer. Knife 9 cm long.

that apparent re-folding structures can be produced by a single deformation event involving partly melted rocks.

However, where the second set of leucosomes has a different composition, the second interpretation would be reasonable, as for the Cooma Complex, south-east Australia (Vernon *et al.*, 2003), where neosomes mobilized from metapsammite-derived migmatites transect earlier-formed, solidified leucosomes in metapelite-derived migmatites (Fig. 4.27B), and in the Wuluma Hills, central Australia (Sawyer *et al.*, 1999, p. 230), where transecting leucosomes have differences in composition and microstructure.

In some deep-seated migmatite terranes that remain hot for long periods, numerous transecting leucosomes may reflect a protracted structural history in the presence of melt. For example, in the classical migmatites of southern Finland (Sederholm, 1967), leucosomes were formed during at least eleven, apparently separate deformation events (Hopgood *et al.*, 1983). Detailed isotopic analysis, based on relative chronology of leucosomes determined from field structural analysis, indicates that the main phase of neosome formation occurred over 5–20 Ma, and that leucosomes may have continued to form for at least 100 Ma (Hopgood *et al.*, 1983). Similarly, in the Wuluma Hills migmatites, Arunta Block, central Australia, SHRIMP U-Pb dating of zircon in complex, transecting leucosomes suggests

(B)

Fig. 4.27 *(cont.)*.

that melt was present and mobile for at least 4 Ma, and possibly for 50 Ma (Sawyer *et al.*, 1999, p. 234).

Some multiple-pulse migmatite complexes may act as conduits for melts derived from melting zones at deeper structural levels, as well as generating local melts (Vernon *et al.*, 2003; Slagstad *et al.*, 2005).

4.8 Transport and local accumulation stage

4.8.1 Melt segregation

Segregation of melt is controlled by the proportion of melt and the stress field (Vigneresse, in press). Segregation processes depend on: (1) the *degree of melting*, which increases with increasing temperature, and may also increase if external free water is added (in amounts too small to change the water-undersaturated nature of the melting), (2) *deformation*, which assists melt segregation (Brown & Rushmer, 1997; Rutter, 1997; Sawyer 2000, 2001; Brown *et al.*, 1995a,b; Sawyer *et al.*, 1999; Bons *et al.*, 2004), and (3) the *interplay between rates of melting and melt segregation* (Sawyer, 1994). Segregation of melt can be rapid enough to prevent chemical equilibration between melt and solid residue, as discussed previously; for example, felsic melts undersaturated in Zr are formed by rapid segregation in migmatites in Québec (Sawyer, 1991).

4.8.2 Melt segregation pathways in migmatites

Connected leucosome networks of stromata, transgressive veins and dykes, interboudin partitions and shear zones are generally regarded as evidence of melt frozen while being transported, the melt having flowed down local pressure gradients during deformation (McLellan, 1988; Sawyer, 1991, 1994; Anhaeusser, 1992; Brown, 1994; Brown *et al.*, 1995a,b; Williams *et al.*, 1995; Brown & Solar, 1999; Marchildon & Brown, 2003). However, though these networks probably represent melt transfer paths, they are generally not melt accumulation sites, because the leucosomes commonly consist of early crystallized minerals (i.e., they may be thought of as 'felsic cumulates'), residual melt having been lost from the rocks (Section 4.3).

Evidence of concentration of melt in fold hinges (generally considered to be lower pressure sites than fold limbs, where melt tends to be squeezed out of leucosomes), interboudin zones and shear zones is common (Allibone & Norris, 1992; Ashworth, 1976; Barr, 1985; McLellan, 1988, Mogk, 1990; Sawyer *et al.*, 1999; Marchildon & Brown, 2003). Several generations of melt, from different local sources, may accumulate in low-pressure sites, such as major fold hinges (especially antiforms), and may show evidence of *in situ* crystallization, producing more evolved magmatic compositions (richer in K, with higher Rb/Sr) than those of the leucosomes in the escape paths (Sawyer *et al.*, 1999, pp. 234–5).

Detailed observations by Sawyer (2001) on the Ashuanipi migmatite, Québec, Canada, have shown that initial segregation of melt (formed by the inferred reaction: $Bt + Qtz + Pl = Opx + oxides + melt$) occurs at openings (caused by externally imposed stress or local magma pressure) along grain boundaries. The evidence consists of elongate lenses and cuspate patches of former melt (now quartz, K-feldspar or plagioclase) at biotite-quartz-plagioclase boundaries, and former melt pseudomorphs of biotite (Sawyer, 2001), as shown in Fig. 4.10. The lenses do not occupy obvious planar fractures, but probably resulted from opening of microfractures following grain boundaries. Melt then segregated to leucosome stromata through vein networks and more obvious fractures. This process led to the effective draining of the mesosome, except for local melt pockets that remained behind. Most of the restite stayed in the mesosome, as the leucosome contains less than 5% $Bt + Opx$. Larger fractures and shear zones helped remove melt from the rock, and 640 000 km^3 of leucogranitic melt were extracted in this way (Guernina & Sawyer, 2003).

Evidence of removal of melt from migmatites includes: (1) *leucosomes that have lost melt*, on the basis of mineralogical or chemical evidence (Section 4.3); (2) *lack of retrograde metamorphism of mafic peritectic minerals*, such as garnet and cordierite, which is generally taken to imply that H_2O-bearing residual melt was removed from the leucosome (Ellis & Obata, 1992; White & Powell, 2002), (3) *layers (interpreted as former leucosomes) very rich in peritectic minerals*, such as anomalously aluminous gneiss layers in migmatites of central Massachusetts, USA (Tracy & Robinson, 1983), layers of coarse-grained peritectic orthopyroxene in migmatites in the Wuluma Hills, central Australia (Sawyer *et al.*, 1999, p. 227), and layers rich in coarse-grained

(A)

(B)

Fig. 4.28 (A) Leucogranite with unmelted and migmatitic remnants, formed by local melt accumulation and intrusion in a migmatite terrane, Helsinki, Finland. Knife 9 cm long. (B) Leucogranite with minor amounts of biotite in flow schlieren (left and bottom-left of centre) in migmatite complex, Wuluma Hills, central Australia. Knife 9 cm long.

peritectic garnet in migmatites at Broken Hill, Australia (White *et al.*, 2004), and (4) *depleted bulk rock compositions*, for example abundance of peritectic minerals (garnet, cordierite, orthopyroxene) in migmatites with amounts of leucosome that are too small to account for these abundances. All these pieces of evidence support the inference that removal of melt from restite is a very efficient process. The melt fraction is kept small, because it is either either continually drained (Sawyer *et al.*, 1999, p. 234) or is drained episodically but frequently, in small batches.

Thus, melt or melt-rich material may migrate over distances larger than on the outcrop scale. The melt-rich material forms bodies, including small transgressive leucogranite veins and 'dykes', extension-gashes on fold limbs, 'saddle-reef' leucogranites occupying openings at fold hinges, thick intrusive leucogranite granitic sills, sheets or dykes (Fig. 4.22), and more irregular leucogranite bodies, with or without mesosome remnants (Fig. 4.28). Some of these rocks may be referred to as 'injection migmatites' (Fig. 4.22B).

4.8.3 Effects of deformation on melt segregation

Movement of melt within and out of partly melted source rocks is generally considered to require the assistance of deformation (Hollister & Crawford, 1986; Davidson *et al.*, 1994; Brown & Rushmer, 1997; Rutter, 1997; Kisters *et al.*, 1998; Sawyer *et al.*, 1999; Sawyer, 2001; Petford *et al.*, 2000; Rushmer, 2001; Marchildon & Brown, 2001, 2002, 2003; Brown, 2005a,b).

Vigneresse *et al.* (1996) postulated a 'liquid percolation threshold' (LPT), and a 'melt escape threshold' (MET), between which is a complex transition stage that depends on strain rate. Melt remains in the parent rock until its local abundance exceeds the 'melt-escape threshold' (Vigneresse *et al.*, 1996), which depends on several factors (especially melt viscosity and density), but which can be lowered by deformation forming and/or opening fractures. Numerical modelling indicates that melt segregation is discontinuous, and that melt can segregate only if the melt proportion falls within a restricted range, namely 5–10 vol.%, according to Vigneresse (in press). Below this range, melt cannot segregate because melt patches are not sufficiently connected, and above this range, the rock becomes too weak to permit segregation.

Rosenberg & Handy (2005) reviewed experimental investigations, and inferred that (1) at c. 7 vol.% melt, the proportion of melt-bearing grain boundaries increases rapidly, marking a 'melt connectivity transition' (MCT), and (2) at 40–60 vol.% melt, a solid-supported aggregate changes to a melt-supported aggregate (= suspension or magmatic flow), with consequent loss of strength, marking a 'solid to liquid' transition (SLT).

As noted in Chapters 1 and 5, liquid produced in a chemical reaction may build up pressure and force its way out of a confined solid aggregate by forming local fractures that may preferentially form along planes of potential weakness, such as grain boundaries, bedding planes and foliation surfaces. The process is known as *hydraulic fracturing* ('hydrofracturing') if dilute fluids are involved, and may be referred to as *melt-enhanced embrittlement* (Davidson *et al.*, 1994) or *magma fracturing* (Clemens & Mawer, 1992) where silicate liquids are involved. For example, owing to the positive ΔV of most water-undersaturated melting reactions, accumulating melt may build up local pressure and force its way into potential openings by 'magma fracturing' (Clemens & Mawer, 1992; Rushmer, 1995; Watt *et al.*, 2002; Bons *et al.*, 2004; Brown, 2005b). For this to occur, the melt pressure must exceed the minimum normal lithostatic stress plus the minimum tensile strength of the rock.

If the rate of melt production exceeds the rate that melt can escape, the melt pressure can increase rapidly enough to promote melt-enhanced embrittlement (Davidson *et al.*, 1994). The process may be assisted by external deformation, creating low-pressure sites (for example, the formation of extensional fractures, as well as openings where local stress heterogeneities develop because of variable responses of minerals with different strengths and orientations), into which liquid is sucked. Melt migrates to lower pressure (lower mean stress) sites, such as boudin necks (Allibone & Norris, 1992; Vernon & Paterson, 2001; Williams *et al.*, 1995, fig. 6c; Collins & Sawyer, 1996; Vernon & Johnson, 2000; Vernon *et al.*, 2003), 'pressure-shadows' or 'strain-shadows' adjacent to porphyroblasts (Vernon & Paterson, 2001; Williams *et al.*, 1995, figs. 4–6), and other local dilation sites. Melt may also migrate into intergranular fractures (openings along grain boundaries formed during deformation) sub-parallel to the shortening direction, inducing sliding of grains relative to each other and thereby weakening the rock (Rosenberg, 2001, p. 65).

If peritectic grains are large enough, they may become jammed in melt paths, assisting separation of melt from restite. However, garnet grains completely enclosed in some leucosomes (Brown *et al.*, 1999) indicate that peritectic grains may be transported in some situations. Peritectic crystals that nucleate and grow on existing grains of the same minerals in the adjacent mesosome should tend to remain when melt moves, and so would assist separation of melt and restite. Evidence of such overgrowths consists of euhedral inclusion-free rims (precipitated from the melt) on inclusion-rich grains of the same minerals in the adjacent mesosome, such as cordierite and K-feldspar in bedded migmatites at Cooma, south-eastern Australia (Vernon & Johnson, 2000; Vernon *et al.*, 2003) and garnet and K-feldspar in stromatic migmatites at Broken Hill, Australia (White *et al.*, 2004, fig. 6a).

4.8.4 Experimental melt segregation

Observations of rapidly quenched, experimentally melted felsic aggregates have suggested that melt migrates away from grain boundaries perpendicular to the maximum compressive stress direction (σ_1) in response to small (as low as 100 MPa) applied deviatoric (non-hydrostatic) stress, resulting in melt pools elongate parallel to σ_1 (Gleason *et al.*, 1999). This occurs without grossly fracturing the rock, though transient, grain-scale microfractures may form during the melt segregation.

Important information has been provided by 'see-through' experiments (synkinematic microscopy) on the deformation of partly melted mineral analogues (Rosenberg & Handy 2000). These compounds have melting temperatures low enough for processes to be observed in the microscope during deformation. The compounds employed in these experiments were norcamphor (melting point, MP $\sim$ 98°C) and benzamide (MP 132°C), which have a eutectic temperature of ~42°C. The strain rate was 10^{-5} s^{-1}. Melt-bearing shear bands developed parallel to a shear zone boundary, and became high-strain zones, grainsize reduction occurring by intergranular fracturing. Strain also localized the melt by opening up transient spaces, owing to differential strengths of adjacent grains in relation to the locally imposed stresses. The low-pressure sites sucked in available melt, and so concentrated melt in the high-strain rocks (Rosenberg & Handy, 2000). The main conclusions inferred from these experiments are: (1) *deformation controls melt* flow, shear bands acting as melt channels, which facilitate segregation of melt out of the sample, and (2) *melting controls deformation*, onset of melting inducing nearly instantaneous localization of deformation.

The experiments of Rosenberg & Handy (2000) probably provide reliable analogues for segregation of felsic melt in shear zones, producing relatively leucocratic granites (owing to efficient extraction of melt from the restite). However, grain-scale segregation may not be a viable process for shear zones feeding major upper crustal plutons rapidly; it may be most applicable to more local, irregular shear zones in migmatite complexes ('ductile fractures').

4.9 Diatexites

Although the most common situation is for the melt to move, leaving behind all or most of the restite, the entire magma (melt plus crystals) may move as a diatexite if enough melt is locally accumulated (e.g., Brown *et al.*, 1999), as noted previously. Diatexites ('dirty' granites) are typically very

(A)

(B)

Fig. 4.29 (A) Diatexite with remnants of metapelite-derived metatexite (bottom half of photo), Cooma Complex, south-east Australia. The diatexite contains small, dispersed relics of leucosome (light) and mesosome (dark) from earlier-formed, solidified, metapelite-derived migmatite (like that in the large remnants), which was intruded and dismembered by the mobile diatexite. (Vernon *et al.*, 2003). (B) Advanced development of a diatexite (the Cooma Granodiorite) with small and large relics of metapelite-derived migmatite, Snowy Mountains Highway near Cooma, south-east Australia. From Vernon *et al.* (2003, fig. 21), with permission of Springer.

heterogeneous in composition, and many show transitions from, or enclose patches of, stromatic migmatites (metatexites), as shown in Figs. 4.19 and 4.29. Diatexites form over a wide range of H_2O activity, from fluid-present to fluid-absent, and the processes involved may be complicated (Sawyer, 1998).

4.9.1 Primary diatexites

If the rate of accumulation of melt exceeds the rate of extraction (for example, owing to relatively rapid melting &/or absence of applied stress), enough melt may accumulate for the solid rock to lose cohesion, forming *primary diatexites* (Sawyer, 1994), as shown in Fig. 4.19. Local melt accumulation, coupled with deformation, may produce *schlieren migmatites* (*nebulitic migmatites, nebulites*), in which the layered structure of metatexites or bedded migmatites is disrupted by flow en masse (Fig. 4.19). These rocks show contorted and diffuse structures, in which leucosomes and mesosomes cannot be resolved.

If enough melt accumulates, the whole mass may move as a restite-rich magma or primary diatexite (Greenfield *et al.*, 1966; Brown, 1973; Sawyer, 1998; Vernon *et al.*, 1990, 2003; Percival, 1991; Milord *et al.*, 2001; Milord & Sawyer, 2003; White *et al.*, 2005). A mass of diatexite is gravitationally unstable, and so, if it is held long enough at depth in the crust (that is, before cooling causes the mass to solidify), it can be convectively overturned with overlying rocks and may ascend as a diapiric pluton (Wickham, 1987b). However, conventional diapiric, en masse movement is limited by the viscous-brittle transition (Vigneresse *et al.*, 1996), and many studies suggest that diatexite bodies move only relatively small distances, for example, at Cooma, south-east Australia (Vernon *et al.*, 2003). However, arguments in favour of diapiric granite emplacement have been presented by Weinberg & Podlachikov (1994), Weinberg (1996, 1999) and Miller & Paterson (1999). In particular, diapiric ascent driven by buoyancy plus regional stress, involving multiple pulses and sheet-like magma bodies, with heterogeneous host rock deformation by brittle and ductile processes, may be more realistic than simple 'hot-Stokes' diapiric models (Miller & Paterson, 1999). Rapid, dyke-like ascent of magma conceivably could result in transport of more restite than slower ascent mechanisms, but flow generally leads to segregation of melt and solids (Section 4.9).

4.9.2 Secondary diatexites

Melt released from source rocks may build up local pressure, disrupting and collecting solid material, to form *secondary diatexite* (Percival, 1991; Sawyer, 1998, 2001). For example, extensive diatexites in the Superior Province of Québec and Labrador, Canada, studied in detail by Percival (1991), Sawyer (1998, 2001) and Guernina & Sawyer (2003), started as leucogranitic melts released from source rocks into accumulated melt zones, where they dislodged solid grains from the wall-rocks as they moved, forming large areas of secondary diatexite (30,000 sq km). These are Opx-Bt diatexite and Grt-Bt diatexite, for both of which solid restite (Opx + Bt or Grt + Bt, respectively) was entrained by mobile melt. These secondary diatexites are suitable parent magmas for leucogranites observed at higher crustal levels (Sawyer, 1998; Guernina & Sawyer, 2003). Evidently, the moving secondary diatexites left most of the restite in the source region.

Similarly, in the Wuluma Hills, central Australia, leucogranitic melts, which left restite in or near the source rocks during segregation and transport in shear zones, accumulated in low strain zones in fold hinges. During transport, the melts eroded wall rocks and developed xenocrystic biotite-rich schlieren, to form secondary diatexites (Sawyer *et al.*, 1999). Later batches of accumulating melt remained leucocratic, as they did not contact biotite-rich wall rocks.

4.9.3 Segregation of solids from flowing melt in diatexites

Liquid-solid segregation in flowing melt occurs because of velocity gradients. Some migmatites show evidence of the separation of crystals from flowing melt in the form of arrested progressive development of leucocratic and melanocratic patches in leucosomes (Figs. 4.6, 4.7), melanosome selvedges to leucosome stromata (Fig. 4.8), and layers (*schlieren*) of accumulated mafic crystals (Sawyer, 1998; Milord & Sawyer, 2003). The 'classical' melanosome borders on leucosome stromata (Fig. 4.5) are generally taken to imply local segregation.

As a result of such segregation, some leucosomes represent the former melt fraction of the neosome. However, as noted previously, the chemical compositions of some leucosomes and migmatites indicate that they have lost some melt during syn-melting deformation, with the result that leucosome compositions cannot be assumed to represent melt compositions, in the absence of supporting chemical evidence.

Processes by which crystals separate from flowing liquid (e.g., Tobisch *et al.*, 1997) include: (1) concentration of solids in zones of restricted flow or reduced flow velocity, (2) rotation, collision & imbrication of solids, forming schlieren (e.g., Sawyer, 1988; Milord & Sawyer, 2003), (3) dispersive pressure on grains (the 'Bagnold effect'), especially those with high aspect ratios, forming schlieren (Barrière, 1981; Sawyer, 1998), and (4) collection of solids at wall irregularities. The Bagnold effect results from dispersive pressure between grains in a fluid undergoing shear strain, leading to a velocity gradient. The gradient is at a maximum at the walls of a channel, zero at the centre, and the crystals migrate to the zone of lower shear strain at the centre (Fig. 4.30). Wall effects are important in the early stages, when the velocity gradient is restricted to the near-wall zone, producing grain layers close to the walls if the liquid has partly crystallized at this stage.

A long-standing problem is the extent to which solid material (restite and/or resister) is removed from melt in source areas, and whether much restite can be carried along with melt during magma transport. Many studies of partial melting and melt removal in migmatite terranes indicate that very felsic (leucogranite) magma is released for transport, especially where melt removal is assisted by deformation, which is generally inferred to be very efficient (e.g., Sawyer, 1994, 1996, 1998, 2001; Brown & Rushmer, 1997; Rosenberg & Handy, 2000; Brown, 2001, 2007). For example, studies of equilibration rates between monazite and melt have indicated that Himalayan tourmaline leucogranites were extracted very rapidly – in less than 50 kyr and possibly less than 7 kyr (Ayres *et al.*, 1997).

Segregation of solids from melt appears to be efficient, even in flowing diatexites (e.g., Sawyer, 1998). For example, in the St Malo area, France, the change from stromatic migmatite, through diatexite to leucogranite occurs over about 5 km, reflecting effective segregation of biotite from melt in flowing diatexite (Milord *et al.*, 2001; Milord & Sawyer, 2003).

Dykes might be a good way of moving restite-rich magma rapidly though the crust without segregation from the melt, allowing restite to constitute part of high-level plutons. However, the crystal content of the parent diatexite would need to be low enough to allow sufficient fluidity of the magma. Moreover, calculations show that flow differentiation into crystal-rich and crystal-poor zones is very likely (Petford & Koenders, 1998). In addition, Clemens *et al.* (1997) and Annen *et al.* (2006) suggested that crystals melt during ascent of felsic magma, owing to the shallower slopes of P-T mineral stability curves, compared with the steep slopes of magma ascent paths. Increasing evidence suggests that many magmas that reach

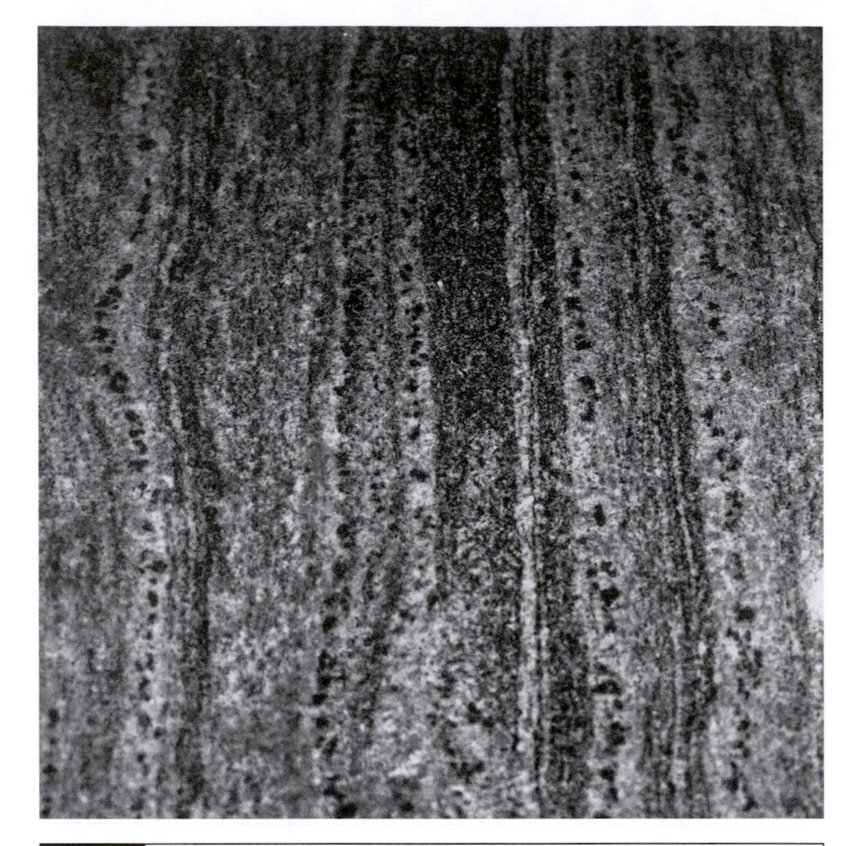

Fig. 4.30 Concentrates of garnet grains in the centres of leucosomes, consistent with segregation by the 'Bagnold effect' during melt flow. Polished slab; original locality unknown. Base of photo 30 cm.

high crustal levels to form intrusive granite plutons are crystal-poor and relatively fluid. For example, inferred feeder dykes for the Kameruka Granodiorite of the Bega Batholith, south-east Australia are phenocryst-poor (Collins *et al.*, 2000). Evidently, most restite remains in or near the source region (Sawyer, 1996).

4.10 Near-source reaction processes in migmatites

Reaction processes occurring at or near the source may be complex, making it difficult to use high-level magma compositions to infer source compositions. For example, different magma batches may mix, felsic magma may mix with more mafic magma introduced into the source area (e.g., from the mantle-derived mafic magma source commonly inferred to be necessary for many crustal anatectic events, as discussed in Section 4.15), and crystals may be lost and/or gained by flowing magma.

A good example is provided by migmatites and granites in the Glenelg River area, western Victoria, Australia (Kemp & Gray, 1999; Kemp, 2001, 2004) where leucosomes can be traced, compositionally and structurally, into local granitic plutons. Segregation of melts was very efficient, with little entrainment of restite, except for diatexites, which show effects of restite entrainment and unmixing, but close to the source (Kemp, 2001). Some restite-free melt accumulated, incorporated and dissolved restite, as well as mixing with other felsic magma batches – processes that increased diversity (Kemp, 2001). Also, various local mafic magmas mixed with leucosome-derived magmas to produce the observed compositional range of granitic magmas (Kemp, 2001).

Another example of complexity is afforded by migmatitic Delamerian granites of south-east South Australia (Foden *et al.*, 2002), which have been intruded by and mixed with I-S granites formed by basalt-sediment interaction at greater depth. The I- & S-type granites show a compositional continuum. The S-type granites were formed by in situ partial melting of metasediments to form diatexites, which mingled and mixed with I-type granitic magmas and mafic magmas that were the local heat source. The I-types were formed by mixing of mantle-derived mafic magma with metasedimentary partial melts.

The following reaction processes involving melt can operate at or near the source of melting, all potentially increasing magma compositional diversity: (1) segregation of restite from melt at or near the source (Sawyer *et al.*, 1999), (2) mixing of local anatectic magma batches (as at Mount Stafford, central Australia (Greenfield *et al.*, 1996, 1998) and the Glenelg River complex, western Victoria, Australia (Kemp & Gray 1999; Kemp, 2001, 2004), (3) mixing of local more mafic magmas with the felsic magmas, as in the Glenelg River complex, western Victoria, Australia (Kemp & Gray 1999; Kemp, 2001, 2004), the Delamerian of South Australia (Foden *et al.*, 2002), the Bindal Batholith, Norway (C. G. Barnes *et al.*, 2004) and the Murrumbidgee Batholith south-east Australia (Collins & Richards, 2001; Richards & Collins, 2002), (4) contamination with xenocrysts during flow, as in the Cooma Complex, south-east Australia (Vernon & Johnson, 2000; Vernon *et al.*, 2003), the Wuluma Hills, central Australia (Sawyer *et al.*, 1999), and Québec, Canada (Percival, 1991; Sawyer, 1999), and (5) fractional crystallization, producing compositional zoning in leucosomes (Sawyer, 1987). Regional metasomatism accompanying anatexis is a potential complication, though generally it is considered to have a relatively minor effect (Ashworth, 1985, p. 7).

4.11 Migmatites and granite plutons

Migmatites may grade into local intrusions ('microplutons') 10–50 m across, as in the classic Finnish Precambrian migmatite terrane (Sederholm, 1967; Hopgood & Bowes, 1978), the Black Forest, Germany (Mehnert, 1968), north-east Brittany (Brown, 1979, 2007), the Gwenoro Dam area, Zimbabwe (Condie & Allen, 1980), the Taylor Valley, Antarctica (Allibone & Norris, 1992), and the Pyrenees (Wickham, 1987a). Irregular peraluminous granite bodies, some with cordierite and/or garnet, occur throughout the section of migmatitic metasedimentary gneisses around Inzie Head, north-east Scotland, suggesting that these granites were locally derived (Johnson *et al.*, 2001, p. 100). However, clear connections between specific major high-level granite plutons and migmatite complexes are commonly not recognized.

Deformation-assisted melt segregation from source rocks is such an efficient process (Sawyer, 1991; Davidson *et al.*, 1994; Rutter & Neumann, 1995;

Petford *et al.*, 2000) that (1) large granitic magma chambers are unlikely to form and are generally not observed in migmatite terranes (Petford *et al.*, 2000; Brown, 2004, 2005b), though small granite plutons may be present in local accumulation sites (e.g., Allibone & Norris, 1992; Sawyer *et al.*, 1999; Johnson *et al.*, 2001), and (2) restite is likely to be left behind by the mobile melt, as discussed previously.

Migmatite complexes have often been considered to be the most likely sources of high-level granites, and detailed studies on the processes of melt segregation have been carried out, as reviewed by Brown (2007). However, this interpretation is questionable for many large high-level plutons (e.g., Vernon, 2005, 2007), for the following reasons. (1) Doubts exist about the accumulation of enough melt to feed the dyke-like transport zones that are widely inferred to be important contributors to felsic plutons. (2) Many granites show evidence of some mixing with more mafic magma (as indicated, for example, by microgranitoid enclaves), which is generally unavailable or minor in migmatite terranes. (3) Melt that segregates from migmatites is typically poor in mafic components, which tend to remain behind, whereas many high-level granites contain higher proportions of mafic minerals. The common occurrence of weakly hydrated to anhydrous residual granulites, commonly showing evidence of depletion in melt components, provide general support for the idea that melt lost from the lower crust is emplaced as leucogranite in the upper crust, leaving most or all of the granulite restite behind (Guerina & Sawyer, 2003; Brown, 2001, 2007), as discussed in the next section.

4.12 Migmatite complexes as sources for high-level leucogranites

Escape of melt, leaving restite in the source migmatites, typically leads to the formation of *leucogranites*, which are very common products of crustal partial melting. For example, melt extracted from anatectic rocks accumulated as leucogranite plutons higher in the crust in the Variscan orogen of France (Gapais *et al.*, 1993; Marchildon & Brown, 2003; Brown, 2007). In the Shuswap complex, British Columbia, Canada, most of the leucosomes in migmatites are cumulates, and chemically related leucogranites represent evolved or residual melts (Hinchey & Carr, 2006). Other examples of high-level leucogranites include the Himalayas (Le Fort, 1981; Le Fort *et al.*, 1987; Weinberg & Searle, 1998; Zeitler & Chamberlain, 1991; Harris *et al.*, 1995; Ayres *et al.*, 1997; Butler *et al.*, 1997), the Pyrenees (Wickham, 1987a,b); Québec (Percival 1991; Sawyer, 1998, 2001; Guernina & Sawyer, 2003); the Taylor Valley, Antarctica (Allibone & Norris, 1992); Maine, USA (Tomascak *et al.*, 1996; Pressley & Brown, 1999; Solar & Brown, 2001); the Black Hills, South Dakota, USA (Redden *et al.*, 1990; Nabelek *et al.*, 1992; Nabelek & Bartlett, 2000); southern Brittany, France (Strong & Hanmer, 1981; Milord *et al.*, 2001; Marchildon & Brown, 2003; Brown, 2005b, 2007); southern India (Braun *et al.*, 1996); the Adirondacks, USA (Selleck *et al.*, 2005); the Zambezi Belt, Zimbabwe (Carney *et al.*, 1991a, p. 468; Gilotti & McClelland, 2005); the Higo metamorphic terrane, Japan (Obata *et al.*, 1994); Broken Hill, Australia (White *et al.*, 2004), as shown in Fig. 4.1; Mount Hay, central Australia (Collins & Sawyer, 1996); the Palmer area,

South Australia (White, 1966; Fleming & White, 1984); the Wuluma Hills, central Australia (Sawyer *et al.*, 1999), as shown in Fig. 4.28B; the Limpopo Belt, South Africa (Anhaeusser, 1992); the Shuswap complex, south-eastern British Columbia, Canada (Vanderhaeghe, 1999; Hinchey & Carr, 2006); and the classic migmatite terrane of southern Finland (Sederholm, 1967; Ehlers *et al.*, 1993; Väisänen & Hölttä, 1999), as shown in Figs. 4.22A and 4.28A.

Many leucogranites occur in collisional orogenic belts, in which basaltic magma is not a major contributor to the formation of granitic magmas (e.g., Nabelek & Liu, 2004). These peraluminous muscovite leucogranites typically have ≥70% SiO_2, and can be produced by pure crustal melting of metasediments, as shown by experimental data (Patiño-Douce, 1999). Many may be formed by H_2O-absent melting during decompression caused by the tectonic collapse of thickened intracontinental crust (Harris & Massey, 1994; Patiño-Douce, 1999). However, some leucogranites occur in terranes in which synchronous basaltic magma has been active. Most leucogranite occurrences presumably reflect the efficient segregation of melt from restite in the source rocks.

4.13 Migmatite complexes as sources for typical S-type granites

4.13.1 Sources of mafic components

In contrast to leucogranites, S-type granites, such as those typical of the Lachlan Fold Belt, south-east Australia, are richer in Ca and mafic components, and may be poorer in SiO_2 (typical range: 67%–76%) than typical collisional peraluminous leucogranites (Patiño-Douce, 1999). Though garnet may be present, S-type granites are characterized by LPHT mafic minerals, such as cordierite, aluminous orthopyroxene and spinel, suggesting that they formed at LPHT metamorphic conditions. Processes capable of producing less silicic, more mafic magmas must be involved, such as high-temperature melting, entrainment and remelting of restite, interaction with mantle-derived magmas, or all of these processes (Clemens & Wall, 1981; Patiño-Douce, 1995, 1999; McCarthy & Patiño-Douce, 1997; Stevens, 2007).

High-temperature melting can produce more mafic granites (Section 4.2.4). For example, melting experiments indicate that peraluminous ('S-type') high-level granites and equivalent volcanic rocks in the Lachlan Fold Belt, south-east Australia were formed by high-temperature melting (850°C–950°C) of felsic rocks (Clemens & Wall, 1981, 1984), these temperatures being much higher than those indicated by migmatite complexes in the region.

Entrainment and melting of restite, for example, peritectic garnet, can explain the mafic composition of some S-type granites (Stevens, 2007).

Interaction with mantle-derived magmas is indicated by chemical, mineralogical, microstructural and isotopic studies of felsic magmas, crystals in granites, and microgranitoid enclaves (Harker, 1909; Wilcox, 1944; King, 1964; Wager *et al.*, 1965; Blake *et al.*, 1965; Gamble, 1979; Grove *et al.*, 1982; Reid *et al.*, 1983; Vernon, 1983, 1984, 1990a, 1991a; Myers *et al.*, 1984; Gray, 1990; Davidson *et al.*, 1988, 1990, 2001; Hildreth & Moorbath, 1988;

Clarke *et al.*, 1988; Elburg & Nicholls, 1995; DePaolo *et al.*, 1992; Grunder, 1995; Elburg, 1996a,b,c; Rossiter & Gray, 1996; Collins, 1996, 1998, 2002; Keay *et al.*, 1997; Maas *et al.*, 1997; Castro *et al.*, 1999; Kemp & Gray, 1999; Collins *et al.*, 2000; Waight *et al.*, 2000a,b, 2001; Kemp, 2001, 2004; Collins & Richards, 2001; Gray & Kemp, 2001; Foden *et al.*, 2002; Gagnevin *et al.*, 2005a,b). Microgranitoid enclaves ('mafic enclaves'), which are very common in high-level granites, show typical microstructural features of mixed magmas, such as mantled feldspar and quartz xenocrysts (Vernon, 1983, 1990a, 1991a, 2007).

Experiments indicate that magma mixing at high pressure produces melts of the wrong composition (Patiño-Douce, 1999). However, mixing of mafic magma and crustal metasedimentary material at low pressure (probably no more than 6 kbar, or around 20 km depth) produces magmas of both observed and modelled compositions that are peraluminous for up to 50% basalt involvement (Patiño-Douce, 1999).

4.13.2 Are diatexites suitable sources of high-level S-type granites?

Both primary and secondary diatexites may lose restite as they flow, becoming more leucocratic (e.g., Sawyer, 1998). For example, in the Wuluma Hills, central Australia, flowing leucocratic material, which was extracted from the enclosing source rocks (granulite facies felsic gneiss), eroded the host rocks and incorporated xenocrystic material to form secondary diatexites, which themselves show good evidence of restite removal to form local leucogranite (Sawyer *et al.*, 1999). In addition, as noted previously, in the Superior Province of Québec and Labrador, Canada (Percival, 1991; Sawyer, 1998, 2001; Guernina & Sawyer, 2003), extensive primary diatexites began as leucogranitic melts extracted from source rocks, and then collected solid wall-rock material as they moved, becoming secondary diatexites, which were the immediate source for abundant high-level leucogranites. These restite-segregating diatexites fed dykes that produced highly fractionated, restite-free plutons 20 km higher in the crust (Sawyer, 1998).

Diapiric ascent of diatexites could produce the more mafic high-level granite compositions (e.g., Chappell & White, 1974, 1991; White & Chappell, 1977), but conventional diapirs are unlikely above the brittle-ductile transition (Vigneresse *et al.*, 1996). Moreover, diatexites appear to be less common than leucogranites in and close to migmatite complexes, and tend to lose restite relatively close to their source (Sawyer, 1996, 1998), becoming progressively more leucocratic. In addition, as mentioned previously, restite crystals partly or completely dissolve during ascent, because ambient pressure reduction takes crystals into the liquid stability field (Clemens *et al.*, 1997; Annen *et al.*, 2006); anhydrous peritectic minerals (e.g., garnet or pyroxene) react with the melt and hydrous minerals (e.g., biotite or hornblende) eventually precipitate from the liquid (Beard *et al.*, 2005), with cooling and/or gas loss (Annen *et al.*, 2006).

4.13.3 Deeper source?

Though local mixed magmas close to the source (e.g., Kemp & Gray, 1999; Kemp, 2001, 2004; Foden *et al.*, 2002) can produce more mafic granites, mafic magma is not sufficiently abundant in typical migmatite complexes to account for the observed volume of chemically appropriate high-level granite. Evidently, a more extensive, deeper source of mafic magma is necessary.

Some migmatite terranes show evidence of deeper sources of abundant magma, as well as the formation of magma in the migmatite terrane itself, at the exposed level. For example, in migmatites in the Mount Hay area, Arunta Block, central Australia, pervasive, vein-like transfer of local migmatite leucosomes produced relatively small higher-level plutons of leucogranite and leucotonalite. In contrast, large, high-level intrusions of charnockite were fed by steep sheets and dykes of hot magma transecting the stromatic migmatites and fed by magma from an unidentified deeper source (Collins & Sawyer, 1996). Similarly, in the Cooma Complex, south-eastern Australia, the small Cooma Granodiorite was derived partly from local metapsammite-derived neosome contaminated with solid metapelite-derived migmatite fragments, but required addition of similar deeper magma (Vernon *et al.*, 2003). Moreover, the contiguous, closely related Murrumbidgee Batholith required major transfer and mixing of magmas from deeper sources than the Cooma Complex (Richards & Collins, 2002; Healy *et al.*, 2004). In the Ashuanipi complex, Québec-Labrador, Canada, older migmatites are cut by sheets and batholiths of peraluminous diatexite, which consists of enclaves of migmatite and tonalitic gneiss in a coarse- to medium-grained groundmass of Opx-Bt $\pm$ Grt granodiorite; the melt was extracted from an unknown deeper source and entrained solid material from both the source area and intruded rocks (Percival, 1991). In the Trois Seigneurs Massif, Pyrenees, France, locally derived migmatites are intruded by biotite granite with a deeper source, though from within the same sequence, and with addition of magma of unknown provenance (Wickham, 1987a).

In view of these considerations, conventional migmatite complexes may not be appropriate sources for large, high-level granite plutons, especially the abundant more mafic varieties (Vernon, 2005, 2007). For example, peraluminous ('S-type') high-level granites in the Lachlan Fold Belt, south-eastern Australia, are restite-poor, show isotopic evidence of magma mixing, contain microgranitoid enclaves formed by mixing of peraluminous felsic magma with more mafic magma at depth (Vernon, 1983, 1990a, 1991a, 2007), and crystallized from high-temperature (850°C–950°C) magmas (Clemens & Wall, 1981, 1984). Deep crustal, low-pressure melting fostered by mafic magma, with some magma mixing, may be appropriate for these high-level granite magmas (Vernon, 2005, 2007).

What would a deeper source look like, with respect to typical migmatite complexes? It might contain a higher proportion of mafic rock, and might show evidence of a higher proportion of melting, hotter magmas, more melt depletion, repeated magma mixing, repeated mafic magma injection and partial remelting of crystallizing intrusions, as well as overall structural and petrological complexity.

The upper part of a possible source region for S-type granite magmas (and some I-type granite magmas) may be represented by the Hidaka Metamorphic Belt, Hokkaido, northern Japan, as discussed by Vernon (2007). High-temperature, S-type, garnet-orthopyroxene tonalite sheets have intruded and been deformed in granulite facies rocks (orthogneisses and paragneisses) in island-arc crust at the base of a tilted crustal section (Shimura *et al.*, 1992, 2004). The deepest exposed rocks are garnet-cordierite-biotite gneiss, garnet-orthopyroxene gneiss, orthopyroxene-cordierite gneiss, orthopyroxene-plagioclase gneiss and orthopyroxene amphibolite, all cut by anatectic leucosomes with euhedral orthopyroxene.

As well as garnet-orthopyroxene tonalites, abundant gabbros and diorites also intrude the metamorphic rocks in the Hidaka Metamorphic Belt, and complex field relationships indicate mixing and mingling of mafic and felsic magmas (Owada *et al.*, 2003), as well as mixing between I- and S-type tonalites. This kind of terrane could produce high-temperature S-type felsic magmas, as well as metasedimentary xenoliths and hybrid magmas suitable for the microgranitoid enclaves that are typical of high-level S-type granites. Such a terrane might be a potentially suitable site for much of the magma mixing that is a common feature of many or most felsic magmas, as evidenced by abundant chemical, isotopic, mineralogical and microstructural data, as discussed previously, though additional magma mixing and mingling may occur at higher crustal levels, both in conduits and replenished plutons (Reid *et al.*, 1983; Vernon *et al.*, 1988; Vernon, 1990a, 1991a; Wiebe, 1994, 1996), and possibly also in intermediate-level magma chambers (Vigneresse, in press).

4.14 Magma transport and formation of large granite bodies

In the source area, melt is transferred through local melt networks to storage networks that feed major drainage channels (Brown & Solar, 1999; Brown, 2004), such as dykes, drainage networks or major shear zones. However, detailed consideration of the processes involved in the transfer of large amounts magma from partly melted rocks to form high-level intrusive bodies is outside the scope of this book.

Processes suggested are: (1) compaction and filter-pressing, which calculations have shown to be unlikely to produce granite plutons in a geologically reasonable time, though it may be capable of producing outcrop-scale segregations (Richter & McKenzie, 1984), (2) rapid transfer of magma through narrow, brittle, dyke-like openings (Clemens & Mawer, 1992; Petford *et al.*, 1993; Clemens *et al.*, 1997), (3) slower transport of melt through less regular fracture networks (Weinberg, 1999), (4) mobile 'hydrofractures' transporting batches of magma that merge and accumulate to form larger bodies (Bons *et al.*, 2001, 2004), (5) complex diapiric ascent (Miller & Paterson, 1999), which may be an effective transport mechanism in many instances, and (6) transfer of magma in shear zones (e.g., Brown & Rushmer, 1997).

Brown & Rushmer (1997) suggested that increasing amounts of melt promote growth of shear zones, and that deformation assists accumulation of melt in shear zones (i.e., a 'feedback mechanism'). This could occur by the opening of minute spaces between grains undergoing different types and degrees of deformation, melt being sucked into these lower-pressure sites. Activation of mylonitic zones existing prior to the melting could also help to transfer magma. Evidence of melt transfer into shear zones includes petrographic continuity between felsic shear zones and stromata in the adjacent migmatite (Brown, 1994, 2005a,b).

All these mechanisms allow the ascent of separate magma batches that accumulate to form larger intrusive bodies (plutons), many of which show evidence of accumulation of several pulses (Stephens, 1992; Vernon &

Paterson, 1993; Paterson & Vernon, 1995; Wareham *et al.*, 1997; Pressley & Brown, 1999; Bons *et al.*, 2004). Tectonic deformation and structural heterogeneities would assist all these processes.

4.15 Heat source for crustal melting

The main mechanisms of crustal melting (Thompson, 1999) are: (1) decompression of thickened crust, which is mainly applicable to the formation of collision-type leucogranites, (2) upwelling of asthenosphere beneath the crust in extensional orogens, and (3) massive underplating and intrusion of lower crust by mantle-derived mafic magma. Calculations suggest that though upper amphibolite-facies migmatite terranes may result from crustal melting in response to typical regional metamorphic geotherms in thickened crust, the formation of substantial bodies of granitic magma at granulite-facies conditions in crust of average thickness requires an additional external heat source, such as radioactive crustal layers or mantle heat (e.g., Thompson, 1999; Collins, 2002). This is commonly postulated to be in the form of mafic magma intrusions and/or underplating of the lower crust by mafic magma (Patchett, 1980; Phillips *et al.*, 1981; Wyllie, 1983; Clemens, 1988, 2003, 2005a; Huppert & Sparks, 1988; Bergantz, 1989; Vielzeuf *et al.*, 1990; Clarke *et al.*, 1990; Vernon *et al.*, 1990, 1993a; Collins & Vernon, 1991, 1992; Powell *et al.*, 1991; Percival, 1991; Sandiford & Powell, 1991; Bergantz & Dawes, 1994; Finger & Clemens, 1995; Grunder, 1995; Keay *et al.*, 1997; Patiño-Douce, 1999; Thompson, 1999; Bodorkos *et al.*, 2002; Collins, 2002; Vigneresse, 2004; Bonin, 2004).

Mantle-derived heat is especially likely for partial melting in LPHT metamorphic terranes, such as the Lachlan Fold Belt (south-eastern Australia), the Broken Hill area (western New South Wales, Australia), the Mount Isa Block (Queensland, Australia) and Archean greenstone belts. Though crustal extension may explain mantle-derived heat in some terranes, metamorphism commonly is accompanied by deformation and crustal thickening. In such situations, the heat could be due to asthenospheric upwelling induced by (1) convective thinning of mantle lithosphere during crustal thickening or (2) detachment and foundering of dense mantle lithosphere (Loosveld & Etheridge, 1990; Sandiford & Powell, 1991; Collins, 2002; Bonin, 2004). Thompson (1999) calculated that granites involving the highest degrees of crustal melting (up to 60 volume percent) require large volumes of pre-heated crust and widespread access of external mantle-derived heat, such as would be caused by delamination of thickened orogenic roots and slab break-off.

Lithospheric weakening and/or delamination tend to counter the effect of crustal thickening and limit the amount of uplift (Powell *et al.*, 1991). For example, in the Lachlan Fold Belt, evidence of limited thickening and uplift is provided by exposures of volcanic equivalents of Siluro-Devonian granites generated during the metamorphism; uplift would have resulted in their erosion. Weakening of the lower crust may have been assisted by H_2O-absent melting of biotite-bearing assemblages, which involves only a small or no volume increase (Section 4.2), with the result that melt may remain in the source region, unless expelled by deformation.

4.16 Retrograde reactions in cooling migmatites

If melts begin to crystallize (forming mainly anhydrous quartz and feldspar) in their source rocks, they release as free water part of the hydroxyl in the melt, which originated from the dehydration-melting of muscovite, biotite or hornblende. This water can induce hydration of the peritectic minerals in the leucosome (e.g., cordierite, garnet or orthopyroxene), as well as anhydrous minerals in the adjacent mesosome (Ashworth & McLellan, 1985; Stevens & Clemens, 1993; Stevens, 1997). This process is often referred to as 'back-reaction' (Kriegsman & Hensen, 1998; Kriegsman, 2001; Waters, 2001), though a preferable term is 'retrograde reaction' (R. W. White, personal communication), because the prograde reaction is rarely reversed, as explained below.

The clearest microstructural evidence of retrograde reactions in migmatites is a hydrous mineral, or an assemblage containing a hydrous mineral, occurring as a rim separating leucosome from a peritectic mineral produced during earlier partial melting (Kriegsman, 2001), such as biotite rims on peritectic garnet or cordierite, and amphibole rims on peritectic pyroxene. For example, in the Cooma Complex, southeastern Australia (Vernon & Johnson, 2000; Vernon *et al.*, 2000, 2003) and the Mount Stafford area, central Australia (Vernon *et al.*, 1990), water-undersaturated partial melting of metapelites involved biotite breakdown, forming neosomes with cordierite, K-feldspar and quartz in 'bedded migmatites'. Much of the cordierite and K-feldspar in the mesosomes was extensively replaced by biotite-andalusite-quartz symplectite (Vernon, 1978b), probably in response to release of water from the crystallization of melt in these *in situ* neosomes, as represented by the equation: $\text{Crd} + \text{Kfs} + H_2O = \text{Bt} + \text{And} + \text{Qtz}$. The replacing aggregate is an amphibolite facies assemblage, suggesting that the water was released at relatively high temperatures, which is consistent with water release during cooling of the neosome magma, rather than introduction of external water.

This example suggests that some retrograde reactions resulting from melt crystallization in LPHT migmatites reverse the H_2O-undersaturated partial melting reactions that produced the melt. However, this situation is generally unlikely to occur, for the following reasons. (1) If some of the melt is selectively removed, hydration reactions that involve the remaining melt may not reverse the partial melting reactions (Kriegsman, 2001; Brown, 2002). (2) In terranes experiencing decompression on clockwise P-T-t curves (Section 3.4.4), partly melted rocks may cross different retrograde reaction curves from those crossed during prograde metamorphism; the greatest likelihood of retrograde reactions reversing prograde reactions would occur during approximately isobaric cooling (Brown, 2002), though even for that situation the next point needs to be considered. (3) Because mineral compositions and proportions change during multivariant prograde melting, a cooling path would have to cross a reaction boundary at exactly the same P and T as when the reaction occurred during prograde metamorphism; otherwise the mineral compositions and proportions would be different from those of the prograde reactants. (4) Because most assemblages are mainly in high-variance P-T fields and only rarely experience univariant conditions, as shown by P-T pseudosections (Figs. 1.10, 4.4B), the same minerals can reflect too large a range in P-T

conditions to permit the inference that a retrograde *P-T* path has duplicated the prograde path. Effects similar to the results of back-reaction may be caused by infiltration of H_2O-rich fluid during cooling of migmatites (Fitzsimons, 1996). However, widespread retrograde metamorphism in regional metamorphic terranes, producing lower amphibolite and especially greenschist facies assemblages (Vernon & Ransom, 1971; Corbett & Phillips, 1981; Johnson *et al.*, 2001), is likely to be due to influx of relatively large volumes of pervasive, lower-temperature H_2O-rich fluid, after leucosome magmas crystallized (Johnson *et al.*, 2001).

The mafic minerals in many neosomes show little or no retrograde alteration, and many granulite facies terranes show preservation of abundant high-grade minerals. The usual interpretation is that the hydrous melt escaped from the source rocks, leaving behind the solid reaction products (Powell & Downes, 1990; Ellis & Obata, 1992; White & Powell, 2002; Brown, 2002). Grant (1985b) and Waters (1988) discussed alternative interpretations to the melt loss hypothesis, namely: (1) shielding of the anhydrous minerals by crystallization, and (2) nucleation problems and/or sluggish reaction rates during falling temperature. With regard to interpretation (1), if crystallization of the melt initiates by heterogeneous nucleation on the peritectic minerals, the precipitated quartz and feldspar could armour the peritectic minerals against reaction with any water released from the crystallizing melt, especially before the melt becomes saturated in water.

4.17 Deformation of migmatites

If leucosomes contain melt during deformation, feldspar crystals may show no internal deformation (Fig. 4.9), owing to passive rotation (Vernon, 2000a). During stronger deformation, mesosomes tend to be preserved as resistant pods or boudins, the weaker, melt-containing leucosome flowing around and between them. However, once leucosomes have solidified, they typically become stronger than the nesosome, owing to the strength of quartz and feldspar aggregates, especially where they are coarser-grained, as in pegmatitic leucosomes (e.g., Mehnert, 1968, pp. 21–23). As a result, the leucosomes may form residual pods, lenses and boudins in more strongly deformed former mesosome (Figs. 4.31–4.33).

Strong solid-state deformation of migmatites (Figs. 4.31–4.33) commonly produces augen gneisses and compositionally laminated mylonites (see Sections 6.6, 6.11) anastomosing around residual pods of less deformed material (e.g., Paterson *et al.*, 1989; Vernon, 1999, 2000a, 2004; Vernon & Paterson, 2002). Mafic and felsic pods, boudins and lenticular layers of variable thickness may occur in migmatites that are strongly deformed in the solid-state (mylonitized), depending on local minerals and grainsize.

For example, Carney *et al.* (1991) described interlayered felsic-mafic migmatites that were converted to spectacular laminated mylonites at 820°C and 12 GPa in the Zambezi Belt, northeastern Zimbabwe. The rocks show residual non-deformed and weakly deformed pods of mafic garnet granulite, around which anastomose strongly laminated mafic-felsic mylonites with intense isoclinal refolding, along with thicker pegmatite layers (former leucogranitic material segregated from the migmatites). Carney *et al.* (1991) suggested that syndeformation injection of felsic melt into the mylonite zones may have assisted cataclastic deformation by magma

(A)

(B)

Fig. 4.31 (A) Augen gneiss and mylonite formed by strong deformation of a felsic migmatite, originally formed by partial melting of a megacrystic granite, Stephens Creek area, Broken Hill district, New South Wales, Australia. The leucosomes have been converted to thin, elongate lenses by the solid-state deformation. Knife 9 cm long. (B) Mylonite formed by intense solid-state deformation of a felsic migmatite, preserving a few lenticular relics of coarse-grained K-feldspar in a formed leucocratic, pegmatitic leucosome layer, Adirondack Mountains, New York, USA.

(hydraulic) fracturing. However, recrystallization/neocrystallization, leading to extreme grainsize reduction, has obliterated all former microstructures in the mylonitic rocks.

Many migmatites show 'ptygmatic folding' (Fig. 4.34), described originally by Sederholm (1907), which consists of tortuous, disharmonically folded leucocratic veins that show no obvious relationship to major regional structures (Agostino, 1971), as described by Mehnert (1968, pp. 26–35). Some have suggested that ptygmatic folding is a characteristic feature of migmatites, and that it develops during and as a result of melt

(A)

(B)

Fig. 4.32 (A) Myonlite with isoclinal folds, formed by intense solid-state deformation of mafic migmatite, Dingo Rock Hole, Arunta Inlier, central Australia. Former leucosomes have been deformed into elongate lenses. (B) Same locality, showing more intense mylonitic deformation, with some lenticular leucosome relics. (C) Mylonite resulting from the most intense deformation at this locality. (D) Same locality, showing boudins of leucosome formed by dismembering during tight isoclinal folding.

injection (e.g., Wilson, 1952). However, identical folding can be observed in rocks that have not melted (e.g., Ashworth, 1985, fig. 1.4). Moreover, the experiments of Agostino (1971) showed that ptygmatic folds can be produced by compressional deformation of originally planar veins, provided the viscosities of the vein and host rock are sufficiently different. The intensity of deformation of the veins is a function of the original inclination of the veins relative to the major compressive stress. If this

(C)

(D)

Fig. 4.32 (*cont.*).

inclination is ≥45°, no deformation occurs, even during extensive deformation of adjacent veins originally inclined at smaller angles. Later cross-cutting veins may be folded (if oriented appropriately), even though older straight veins remain undeformed (Agostino, 1971).

Fig. 4.33 Myonlite with isoclinal folds, formed by intense solid-state deformation of felsic migmatite. Paving slab, original locality unknown. Coin 2.8 cm diameter.

Fig. 4.34 'Ptygmatic' folding of leucosome, Purnamoota Road, Broken Hill area, New South Wales, Australia. Knife 9 cm long.

Fig. 4.35 Veins of pseudotachylite (dark) in felsic granulite, locally passing into pseudotachylite breccia (top), eastern Musgrave Ranges, central Australia. Knife 9 cm long.

4.18 Friction and impact melting

Where a large amount of local heat energy is generated very rapidly, as in fault movements in cool rocks or in meterorite/comet impacts, the affected rocks may partly melt. Rapid cooling commonly produces dark coloured glassy rocks called *pseudotachylites* (*pseudotachylytes*), which typically occur in veins with sharp contacts with the original rock (Fig. 4.35).

The term 'pseudotachylite' (Shand, 1916) implies that though the rock looks like tachylite (basalt glass), it is chemically different, many bulk-rock analyses indicating a felsic composition. However, careful microprobe analyses of the melted portion (free of unmelted fragments) or recalculation of bulk chemical analyses by removing clast compositions, commonly reveals a relatively mafic composition (Sibson, 1975; Magloughlin, 1989; Maddock, 1992; Camacho *et al.*, 1995). The reason may be because the internal structures of pyroxene and calcic plagioclase are more disturbed and compositionally altered by shock deformation than those of quartz or K-feldspar, with the result that the pyroxene and plagioclase melt preferentially (Camacho *et al.*, 1995).

Pseudotachylite forms by extremely rapid fragmentation, frictional heating, partial melting and cooling, the cooling being fast enough to preserve melt. The original minerals undergo 'shock' (extremely rapid) deformation, and some partly melt. Pseudotachylites commonly show

(A)

(B)

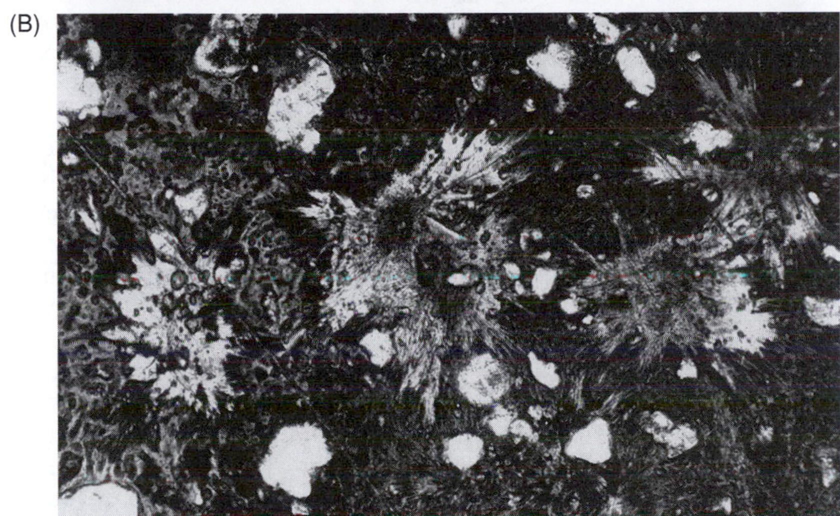

Fig. 4.36 (A) Spiky outgrowths on partly melted fragments of plagioclase (colourless) and radiating, fern-like, dendritic aggregates of metastable high-temperature clinopyroxene (pigeonite) in pseudotachylite, eastern Musgrave Ranges, central Australia. Plane-polarized light; base of photo 0.7 mm. From Vernon (2004, fig. 3.43). (B) Spherulitic aggregates that grew in glass with abundant gas bubbles (vesicles) and some unmelted residual fragments of quartz and cloudy feldspar in pseudotachylite, eastern Musgrave Ranges, central Australia. Plane-polarized light; base of photo 4.4 mm. From Vernon (2004, fig. 3.44).

structures reflecting flow of the liquid, and injection of frictionally produced melt into shear zones may produce intricate flow structures resembling 'sheath folds' in mylonites (Berlenbach & Roering, 1992). Chilling of the melt produces glass and dendritic crystal forms, reflecting strong supersaturation (Maddock, 1992; Camacho *et al.*, 1995), as shown in Fig. 4.36A. Vesicular, amygdaloidal and spherulitic microstructures may also be formed (Maddock *et al.*, 1987; Camacho *et al.*, 1995), as shown in Fig. 4.36B. High-temperature metastable minerals, such as pigeonite, may be preserved, owing to the rapid cooling (Camacho *et al.*, 1995). High-pressure SiO_2 polymorphs (coesite and stishovite) may occur in pseudotachylite formed in quartz-rich rocks (Martini, 1978).

4.18.1 Friction melting

Rapid brittle sliding between large rock volumes, especially in the absence of water, can build up frictional heat at a fast enough rate to cause partial melting in faults (Philpotts, 1964; Sibson, 1975; Grocott, 1981; Masch *et al.*, 1985; Maddock, 1983, 1986, 1992; Magloughlin, 1989; Camacho *et al.*, 1995; Spray, 1995; Curewitz & Karson, 1999), as well as at the base of thrust blocks (Scott & Drever, 1953) and some large landslides (Masch *et al.*, 1985; Legros *et al.*, 2000). Pseudotachylite has been produced in frictional sliding experiments (Spray, 1987, 1995).

Friction melts in faults reflect very rapid heat buildup and dissipation on slip surfaces during seismic shearing (McKenzie & Brune, 1972; Sibson, 1975; Grocott, 1981). In contrast, the non-melted, foliated-lineated fabrics of mylonites are due to slower, aseismic shearing (Sibson, 1990).

Spectacular pseudotachylites (Fig. 4.35) occur in the eastern Musgrave Ranges, central Australia (Camacho *et al.*, 1995), where felsic granulites (consisting mainly of quartz, plagioclase, K-feldspar and orthopyroxene) were involved in a thick mylonite zone and later faulted. The dry rocks accumulated heat rapidly, and melted to form huge amounts of pseudotachylite (4% of a zone 1 km thick), which consists of rock and mineral fragments enclosed in glass with stretched gas bubbles (Fig. 4.36). Chilling formed spectacular dendritic overgrowths on plagioclase on fragments of original plagioclase, tendril-like branches of pigeonite on fragments of orthopyroxene, feathery dendrites of pigeonite throughout the former melt, and spherulitic aggregates in altered glass (Fig. 4.36).

4.18.2 Impact melting

Major meteorite and comet impacts can damage and fragment Earth rocks, producing distinctive structures and ejecta deposits that are formed in short time periods. For example, the largest known preserved impact structure, the deeply eroded, 2023 Ma Vredefort Dome, south-west of Johannesburg, South Africa, had an initial diameter of 200–300 km. It shows the following evidence of 'impact metamorphism': (1) shatter cones, (2) high-pressure SiO_2 polymorphs, coesite and stishovite, (3) quartz and zircon with planar deformation features, which are characteristic of shock deformation, as shown in Fig. 4.37, (4) dykes of impact melt breccia, and (5) abundant veins of pseudotachylite (Martini, 1978; Killick & Reimold, 1990; Reimold & Gibson, 1996, 2006; Koeberl, 2006).

Such meteorite impacts may cause practically instantaneous heating by release of energy from decay of the shock wave (Shand, 1916; Spray & Thompson, 1994). Extremely high temperatures (*c.* 1350°C) may induce shock melting of individual mineral grains in situ, preserving the original structure of the rock, or may induce melting of several minerals, which may be followed by local mixing of the resultant melts (Gibson, 2002, p. 68). Somewhat lower impact temperatures may induce partial melting of metamorphic assemblages, analogous to anatexis during high-grade regional or contact metamorphism. For example, Gibson (2002) described melting of Archaean metapelitic granulites in the Vredefort Dome, where temperatures of >900°C were produced within a radius of a few kilometres of the

Fig. 4.37 Several sets of planar deformation features (PDFs) in shocked quartz from an impact site, Gosse's Bluff, central Australia. Specimen by courtesy of Tom Bradley. From Vernon (2004, fig. 4.10B). Crossed polars; base of photo 0.5 mm.

centre of the impact structure. The peritectic melting product assemblage was: aluminous Kfs + Crd + Spl ± Crn ± Sil, and small, segregated bodies of peraluminous biotite leucogranite were produced. Rapid isobaric cooling following shock heating and exhumation gave rise to local equilibration volumes, indicated by the relatively small amounts of melt produced, as well as by complex, fine-grained coronas around residual garnet and biotite, formed by incomplete, local reactions between unmelted minerals (Gibson, 2002).

The fallout units (*impactites*) resulting from meteorite and comet impacts (e.g., Glikson, 2004) include fragmental ejecta, such as nanodiamonds (Carlisle & Bradman, 1991), grains of shocked quartz (Bohor *et al.*, 1984), and grains of high-pressure SiO_2 polymorphs, such as stishovite (Bohor *et al.*, 1984) and coesite (Glass & Wu, 1993), as well as '*spherules*', which are silicate melt droplets up to a few millimetres across (many modified after deposition) expelled from impact sites and scattered widely. 'Spherule' is a general term embodying 'tektite' (>1 mm across), 'microtektite' (≤ 1 mm across) and 'microkrystite'. *Tektites* (O'Keefe, 1976; McCall, 2001) are spherical, ellipsoidal, disk-like, dumbbell-shaped, and irregularly shaped blobs of silicate glass formed by bolide (meteorite or comet) impact, typically showing sculptured external forms that reflect spinning, and commonly with some modification by ablation during transport in the atmosphere (Baker, 1963). Many show contorted flow layering, and some are vesicular (Glass, 1967; McCall, 2001). Tektites occur as uncrystallized glass objects in four widespread 'strewnfields' on Earth's surface, resulting from impact events as old as 34 Ma (O'Keefe, 1976; McCall, 2001). Their isotopic characteristics, especially with regard to ^{26}Al (Glass & Heezen, 1967) and ^{10}Be, are more consistent with an origin by impacted Earth rocks, rather than impacted Moon rocks, though some controversy exists on this point. *Microtektites* have been observed in rocks as old as the Cretaceous-Tertiary (K-T) boundary and even the Permian-Triassic boundary

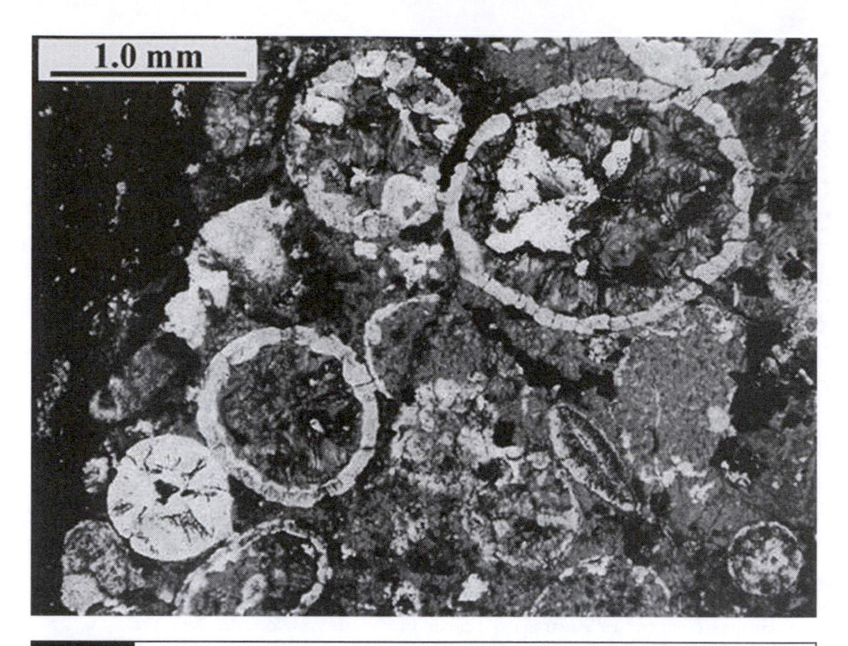

Fig. 4.38 Aggregate of microkrystite spherules showing K-feldspar rims and stilpnomelane-rich cores, both minerals resulting from devitrification, Brockman Iron Formation, Hamersley Basin, Western Australia. Plane-polarized light. From Glikson & Allen (2004, fig. 3A), copyright Elsevier, published with permission.

(Miura *et al.*, 2001). They vary from smooth to pitted, may be vesicular, and are very variable in shape, from spherical to ellipsoidal, teardrop-shaped, dumbbell-shaped, pear-shaped, cylindrical and irregular (Glass, 1967; Glass & Zwart, 1979). *Microkrystites* (Figs. 4.38, 4.39) are spherules that have undergone partial or complete crystallization (Glass & Burns, 1988). In older rocks, especially Precambrian rocks, secondary minerals have replaced the original material, with the result that no glass remains. The general term 'spherule', if used alone without qualification, generally refers to microkrystites in older rocks. Several microkrystite-rich accumulations (Fig. 4.38) in Precambrian rocks have been described (Lowe & Byerly, 1986; Simonson, 1992; Kyte *et al.*, 2003; Lowe *et al.*, 2003; Glikson & Allen, 2004; Simonson & Glass, 2004; Rasmussen & Koeberl, 2004; Addison *et al.*, 2005; Hassler *et al.*, 2005; Scally & Simonson, 2005; Glikson, 2005a,b, 2006, 2007; Glikson & Vickers, 2006; Koeberl, 2006).

Many spherules show internal microstructures reflecting strongly supersaturated (quench) crystallization, owing to rapid cooling in the atmosphere, for example, intersecting acicular crystallites resembling 'microspinifex structure' (Fig. 4.39), skeletal to dendritic crystals, and aggregates of fibrous crystallites radiating inwards from spherule margins. Similar microstructures have been observed in partly devitrified glassy tektites from the K-T boundary layer (Simonson, 1992), as well as in the chondrules of chondritic meteorites, which were formed as quenched melt droplets in the protoplanetary nebula of the solar system (King, 1983; Lowe & Byerly, 1986; Hewins *et al.*, 1996). They match microstructures produced by experimental quench crystallization of silicate melt droplets and glass fragments (Lofgren, 1971, figs. 2, 3; Hewins, 1983; Simonson, 1992). Such structures may be preserved as pseudomorphs in the secondary

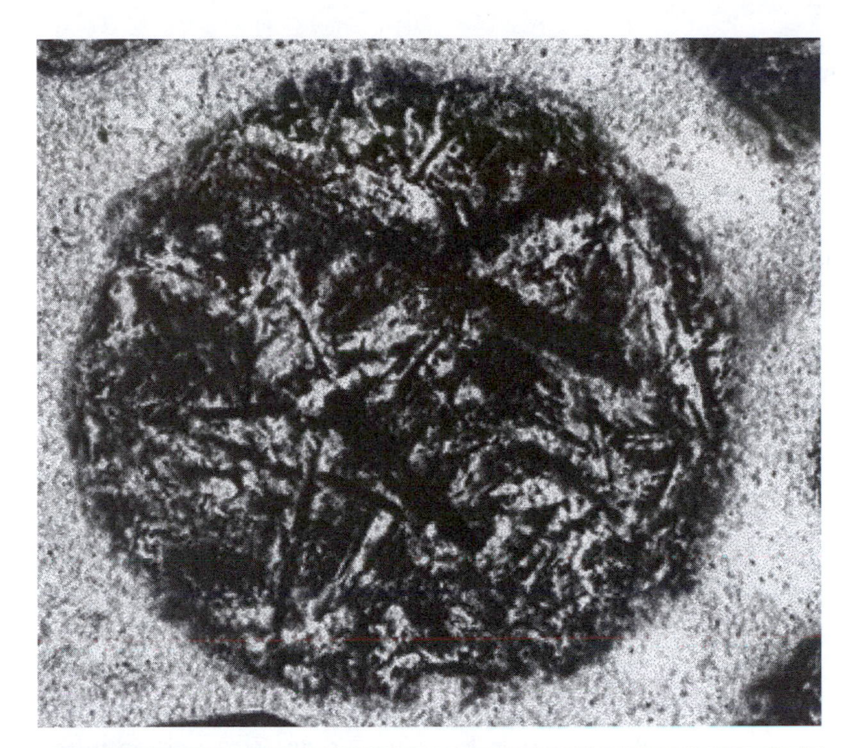

Fig. 4.39 Microkrystite with pseudomorphs of probable primary quench minerals, such as olivine and/or pyroxene, Pilbara Craton, Western Australia. Plane-polarized light; diameter of spherule 0.6 mm. From Glikson *et al.*, (2004, fig. 4D), copyright Elsevier, published with permission.

(authigenic) minerals of microkrystites, discussed below (Lowe & Byerly, 1986; Simonson, 1992, 2003; Glikson & Allen, 2004; Glikson, 2005a,b; Hassler *et al.*, 2005; Scally & Simonson, 2005), as shown in Fig. 4.39. In these microkrystites, the shapes of the original microlites are outlined in minute tan-coloured granules that may consist of Ti oxide (Scally & Simonson, 2005). That these are pseudomorphs of primary minerals and not the result of secondary growth processes is indicated by the fact that microkrystite fragments do not show aggregates radiating inwards from broken surfaces, but only from primary spherule surfaces (Scally & Simonson, 2005). These random and inwards-radiating fibrous microstructures constitute evidence of the impact origin of spherules, in contrast to an origin as volcanic spherulites or varioles, which show evidence of crystals radiating outwards, not inwards; see Vernon (2004) for a review of these microstructures. Some microkrystites also show apparent vesicular microstructures (e.g., Simonson, 2003).

Owing to the inherent instability of glass during weathering in Earth's atmosphere and at near-surface conditions, older (e.g., Cretaceous-Tertiary) microtektites may be altered to fine-grained clay minerals. Minerals constituting Precambrian microkrystites (K-feldspar, stilpnomelane, chlorite, 'sericite' silica minerals, carbonate minerals, sulphide minerals and clay minerals) are all or nearly all of secondary origin (Figs. 4.38, 4.39). For example, the fibrous and radiating aggregates of secondary K-feldspar common in many Precambrian spherules (Fig. 4.39) appear to be pseudomorphs of primary quench minerals (Glikson & Allen, 2004; Glikson, 2005a,b), the K possibly being derived from seawater (Glikson & Allen, 2004).

The secondary minerals (such as K-feldspar or stilpnomelane) may enclose minute particles (crystallites) of inferred primary minerals, such as Ni-chromite with skeletal and dendritic habit (Glikson, 2005a,b, fig. 6.9; Glikson, 2007, fig. 3), reflecting rapid growth at high supersaturation (e.g., Vernon, 2004), and sub-micromillimetre equant particles ('nanonuggets'), observable in the scanning electron microscope (e.g., Glikson, 2005a,b, fig. 9), of Ni metal, Ni oxide, Ni sulphide, Ni arsenide, Ni-Mg sulphide, NiCoAsS and Ir (Glikson & Allen, 2004; Glikson, 2005a,b), as well as Ni-bearing magnetite, Si-bearing magnetite or Ni-Si-magnetite (Glikson, 2005a, fig. 10), which have been reported in inferred meteoritic fall-out units at the Cretaceous-Tertiary (K-T) and Permian-Triassic boundaries (Miura *et al.*, 2001). The Ni chromite may contain inclusions of PGE 'nanonuggets' (Glikson, 2005a). All these minute crystallites appear to have precipitated directly from the original material of the spherule, later to be incorporated in the secondary minerals. The Ni chromite has high Co, V and Zn abundances unknown in terrestrial chromite (Byerly & Lowe, 1994).

Most spherules are spherical, though some are elongate-ellipsoidal, flattened (Lowe & Byerly, 1986), dumbbell-shaped (Lowe & Byerly, 1986; Glikson *et al.*, 2004) or, less commonly, irregular in shape (Hassler *et al.*, 2005). Some elongate and dumbbell-shaped spherules may represent original aerodynamically induced shapes, involving viscous deformation in the atmosphere (Glikson & Allen, 2004; Glikson *et al.*, 2004). However, other dumbbell-shaped spherules may be due to welding of hot spherules in contact (Lowe & Byerly, 1986) or to flattening on impact while viscous (Scally & Simonson, 2005).

Local concentrations of spherules (Fig. 4.38) suggest rapid accumulations in single-event fall deposits (Lowe & Byerly, 1986), but commonly spherules are exhumed and redeposited by currents, so that they occur dispersed among terrestrial detritus (Glass & Heezen, 1967; Simonson, 1992; Lowe & Byerly, 1986; Lowe *et al.*, 2003).

Because shocked quartz is uncommon in Precambrian impactites, owing to the mafic nature of most of the impacted rocks, spherules are important for identification of fallout units. A connection between spherules and shocked quartz was established by Rasmussen & Koeberl (2004), who found a fragment of shocked quartz in a spherule in the Jeerinah Fomation of the Pilbara Craton, Western Australia. Other evidence that the spherules are of impact origin includes: (1) the presence of non-terrestrial minerals, as discussed previously, (2) high abundance and non-terrestrial ratios of platinum group elements (PGE), and (3) meteoritic isotope ratios, where preserved (Glikson, 2005a,b).

As discussed by Scally & Simonson (2005), spherules deposited in the K-T boundary layer within *c.* 4000 km of the Chicxulub (México) impact consist of glass without crystals, showing flow banding and vesicles; non-spherical splash forms (such as teardrop and dumbbell shapes) are common, similar to those of younger tektites and microtektites. In contrast, spherules deposited more than 7000 km from the Chicxulub impact site are smaller, entirely spherical, and mostly partly crystallized (i.e., they are technically microkrystites), most commonly with dendritic clinopyroxene.

Many have suggested an origin as ballistic droplets of silicate melt for tektites, and an origin as melt droplets condensed from of clouds of silicate vapour ('melt mist') for microtektites and microkrystites (Lowe *et al.*, 2003; Scally & Simonson, 2005; Glikson, 2005a,b; Koeberl, 2006, p. 215). Byerly & Lowe (1994) inferred that impact spherules condense from a vapour formed from a mixture of target crust and meteorite, with subsequent modification during atmospheric re-entry. The anomalous Cr isotopes and large Ir anomalies of Early Archaean spherules in the Barberton region of South Africa are consistent with an origin involving impacts of projectiles of carbonaceous chondrule type (Kyte *et al.*, 2003; Lowe *et al.* 2003). Similarly, Simonson *et al.* (2000) described a spherule layer enriched in Ir and platinum group elements (PGEs) in the Transvaal Supergroup, South Africa, and Rasmussen & Koeberl (2004) described a spherule layer enriched in Ir in the Pilbara craton, Western Australia. The relatively high Ni, Cr and Co contents of most Precambrian microkrystites (preserved despite secondary alteration) reflect mafic/ultramafic target crust, which, together with the estimated diameter of inferred impact craters, implies the existence of lunar maria-scale impact basins in ocean-style crust in the Archaean and Palaeoproterozoic (Glikson, 2005b).

4.19 Melting of sulphide-rich rocks

Sulphide-rich rocks appear to have undergone partial melting and mobilization in some high-grade metamorphic terranes, for example, Broken Hill, Australia and Hemlo, Canada (Lawrence, 1967; Mavrogenes *et al.* 2001; Frost *et al.*, 2002, 2005; Tomkins & Mavrogenes, 2001, 2002; Tomkins *et al.*, 2004, 2007; Sparks & Mavrogenes, 2005). Partial melting is suggested by rounded multi-mineral sulphide inclusions in high-grade silicate minerals, and mobilization of sulphide melt is suggested by the occurrence of minerals inferred to have crystallized from a polymetallic sulphide melt (especially sulphosalts, native metals and tellurides) as dyke-like bodies, fracture-fillings and aggregates in extensional sites, such as boudin necks (Lawrence, 1967; Tomkins *et al.*, 2004). Experiments on the system sphalerite-galena-chalcopyrite-S have shown that melting occurs at 700°C–730°C, which confirms that sulphide partial melting can occur at upper amphibolite facies metamorphic conditions, and that S-rich fluid promotes melting (Stevens *et al.*, 2005). This interpretation is supported by experiments on inclusions in garnet and quartz in the Broken Hill Lode, consisting of galena, sphalerite, arsenopyrite, chalcopyrite, tertrahedrite-tennantite, argentite, bornite, dyscrasite (Ag_3Sb) and gudmundite (FeSbS), which show homogenization at 720°C $\pm$ 10°C – well below the peak metamorphic temperature. Tomkins *et al.* (2007) suggested that rocks with sulphosalts or tellurides can begin to melt at lower greenschist to amphibolite facies conditions, whereas rocks without these compounds can begin to melt at upper amphibolite facies, or well into granulite facies conditions if galena is absent. Masive Pb-Zn-(Cu) deposits may start to melt at low to middle amphibolite facies conditions if pyrite and arsenopyrite coexist and at upper amphibolite facies conditions if they do not. Such melting can produce >0.5 vol.% melt, which may form mobile sulphide magma dykes (Tomkins *et al.*, 2007).

Chapter 5

Fluids and Metasomatism

5.1 Fluids in Earth's crust

Fluids are intimately involved with most metamorphism and deformation, because prograde metamorphic reactions produce free fluid, which must be removed from the reaction sites for reactions to proceed (e.g., Norris & Henley, 1976), and retrograde reactions require addition of fluid (Chapter 1). Progressive metamorphism of originally water-rich metapelites causes preliminary *dehydration* by breakdown of hydrous clay minerals to form mica and chlorite (*dehydration reactions*). This is followed by further dehydration of these hydrous minerals to form anhydrous minerals (garnet, andalusite, sillimanite, K-feldspar), through the amphibolite or hornblende hornfels facies, to the much less hydrous or even anhydrous rocks of the granulite or pyroxene hornfels facies, liberating water-rich fluid (Chapter 1). Similarly, metamorphism of hydrous mafic and ultramafic rocks (at zeolite and greenschist facies conditions) causes them to undergo dehydration by breakdown of hydrous minerals, through the same facies sequence as for metapelitic rocks (Chapter 1). Carbonate rocks undergo progressive *decarbonation* (with accompanying dehydration if hydrous minerals are also present), liberating CO_2-rich fluid. Dehydration and decarbonation reactions are collectively referred to as *devolatilization* reactions.

The free fluid produced as a result of prograde devolatilization reactions, locally and rapidly increases the fluid pressure, occupying available pores ($P_{fluid} = P_{solids}$). Calculations suggest that the proportion of free water liberated from metapelites at 500 °C and 5 kbar is about 12 volume percent. If the fluid pressure exceeds the local minimum confining stress (σ_3) plus the local tensile strength of the rock, *hydraulic fracturing* ('hydrofracturing') may occur (e.g., Beach, 1977; Cox & Etheridge, 1989; Ferry, 1994a), creating new pore spaces. The tensile strength may be reduced by relatively weak surfaces, such as bedding planes or tectonic folia (e.g., Etheridge, 1983) and is likely to be very small at metamorphic conditions (Walther & Orville, 1982). Though the volume percentage of fluid in the rock remains small, mobile cavities are opened by the locally increased fluid pressure. This process is generally assisted by the formation of local dilatant cracks caused by variations of the geometrical and mechanical properties of adjacent mineral grains during tectonic deformation (Etheridge *et al.*, 1983, 1984). Because all these openings are sites of lower pressure, they suck in fluid, which can be transported through rocks if the openings migrate.

Movement of fluid requires permeability, rather than isolated pores. *Porosity* is the percentage of pore space in the total volume of a rock, regardless of whether or not the voids are connected. *Permeability* refers to the ease with which fluids pass through a porous medium. A rock is said to be 'permeable' if it permits an appreciable amount of fluid to pass through in a given time, and 'impermeable' if the rate of flow is negligible. Porosity does not give a measure of permeability (though non-porous rocks are obviously impermeable), because permeability also depends on the size, shape and connection of the pores. Connectivity may be increased by interlinked microfractures (see below), and also by the tendency of fluid (hydrous fluid or silicate melt) to spread along grain edges to form small dihedral angles (θ fluid versus grain/grain) in rocks undergoing grain-boundary movement (Chapter 3). As explained by Smith (1953), if the dihedral angle is $60°$ or less, fluid spreads along all grain edges, forming a fluid network, although the rock remains solid. Note that the fluid does not cover all the grain boundaries, which would lower the effective normal compressive stress across the boundaries and so force grains apart, markedly weakening the rock. If fluid pores are static long enough to establish dihedral angles with solid grains, a continuous grain-edge (not grain-surface) film of fluid could theoretically form, which could help transport chemical components by diffusion. This situation evidently applies to some partly melted rocks (e.g., Clemens & Holness, 2000; Sawyer, 2001), as discussed in Section 4.7.2. However, in deforming rocks, fluid and melt are likely to be mobile and governed by tectonically produced openings.

Transient hydraulic and tectonic microfratcures along grain boundaries are commonly inferred to provide the main pathways for removal of fluid generated by devolatilization reactions (e.g., Etheridge *et al.*, 1983), typically leaving no microstructural record. Some rocks do show relatively clear evidence of fluid flow along grain boundaries, for example, late fibrous sillimanite in the Cooma Complex (Vernon, 1979). However, even in the absence of such evidence, a reasonable inference is that fluids move through transient opening along grain boundaries during both prograde and retrograde metamorphism.

Prograde sequences of dehydration reactions reduce the H_2O content of progressively higher-grade metamorphic rocks, so much so that highest grade rocks are typically 'dry' or H_2O-poor, though the transient fluid pressure (melt pressure) may be locally high (Etheridge *et al.*, 1983). Furthermore, partial melting involving breakdown of hydrous minerals at granulite facies conditions produces melt undersaturated in H_2O, with the result that any free water is dissolved in the melt. This melt may move out of the rocks, leaving the remaining granulites essentially anhydrous (Chapter 4). If the melt crystallizes in the source area, hydration of the previously dehydrated rocks may occur (e.g., Vernon, 1978b; Vernon *et al.*, 1990; White & Powell, 2002), but commonly anhydrous minerals remain in migmatitic granulite facies rocks, suggesting melt loss (White & Powell, 2002), as discussed in Section 4.16.

The most effective fluid flow occurs in *channels* (Ferry, 1987; Brantley *et al.*, 1990; Cartwright & Valley, 1990; Oliver *et al.*, 1990; Tobsich *et al.*, 1991; Cartwright *et al.*, 1995; Sibson, 1996; Oliver, 2001; J. D. Barnes *et al.*, 2004), such as fracture networks, mica- and sillimanite-rich cleavage folia, and shear (mylonite) zones, in which deformation increases permeability by continually opening local cracks and cavities (sites of lower pressure), owing to differences in the mechanical properties of adjacent mineral grains and aggregates. Such *channel (channelled, focussed) flow* also occurs during collection of melt in migmatites, which segregates from grain boundaries into progressively more throughgoing veins and fractures (see Chapter 4).

The distribution of fluid in the crust depends on hydraulic gradient, existing anisotropy of permeability caused by bedding and foliations, and stress-developed permeability (Sibson, 1996). Field evidence indicates that *mesh structures*, consisting of faults interlinked with extensional fractures, may be generated by infiltration of pressured fluids into stressed heterogeneous rocks with varying mechanical properties. Such fracture networks may allow passage of large amounts of fluid, some of which may contribute to hydrothermal ore deposits (e.g., Newhouse, 1942; Phillips & Groves, 1983; Groves & Phillips, 1987; Cox *et al.*, 1991; Sibson, 1996; Ague, 1997; Read & Cartwright, 2000).

Examples of channelled fluid flow are provided by three Proterozoic terranes (Mount Isa and the Reynolds Range in Australia, and the Grenville Province in Canada), in which time-integrated fluid flow varied by an order of magnitude on the millimetre to metre scales, and fluids commonly were channelled across strike of the metasedimentary rocks. These features indicate that permeabilities during metamorphism varied considerably over small distances, reflecting variable concentrations of microfractures, through which the fluid flowed (Cartwright *et al.*, 1995). Similarly, in the Harlech Dome, north Wales, considerable variation in mineral and oxygen isotope composition between units implies little or no pervasive movement of fluid at the metamorphic peak, though focussed (channelled) fluid flow occurred afterwards, in response to uplift and unloading, depositing gold in quartz veins where the fluids intersected graphitic shales (Bottrell *et al.*, 1990). Channelled flow on the kilometre scale also occurs in many terranes (e.g., Ferry, 1987).

Another example of channelled fluid flow is evidence presented by Wickham *et al.* (1994) for alteration due to CO_2-rich fluid flow along major Proterozoic shear zones, producing ankerite, $CaFe(CO_3)_2$, and other carbonate minerals replacing silicate minerals at a temperature of at least 500°C; calcite replaced ankerite during infiltration of H_2O-rich fluid at lower temperatures. Channelled flow is also responsible for quartz-carbonate veins in Archaean greenstones and higher-grade rocks in Western Australia (Barnicoat *et al.*, 1991), and for wollastonite-bearing layers formed by layer-parallel infiltration of water-rich ($X_{CO_2} = 0.1$–0.3) fluid from local pegmatites at *c.* 700°C into Proterozoic granulite facies marbles in the Reynolds Range, central Australia (Cartwright & Buick, 1995).

Channelled flow is especially common in retrograde metamorphism of high-grade metamorphic rocks with very low to negligible porosity. Fracture networks and/or shear zones allow infiltration of hydrous fluids (e.g., Vernon & Ransom, 1971; Vernon, 1976, 2004; Corbett & Phillips, 1981; Hobbs *et al.*, 1984; Yardley *et al.*, 2000; Buick *et al.*, 2000; Doyle & Cartwright, 2000; Read & Cartwright, 2000).

Fluid travelling out of rocks undergoing prograde metamorphism enters more continuous channel networks, such as fractures and other deformation zones, some of which may be marked by quartz veins (Walther & Orville, 1982). Such fluids may cause hydration in drier rocks at some other (commonly higher level) location, for example, in previously dehydrated rocks.

Metamorphosed igneous intrusions may be relatively brittle, and so may undergo extensive fracturing, in contrast to adjacent metasedimentary rocks undergoing ductile folding, with the result that fluid flows into lower pressure openings in and around the intrusion, but not through the metasedimentary rocks (Oliver *et al.*, 1990).

In the upper crust (above about 10–15 km), temperatures and confining pressures are relatively low, so that rocks are strong enough to maintain pores, which commonly contain fluid ($P_{fluid} < P_{solids\ (total)}$). As the degree of compaction and intensity of metamorphism (involving deformation and recrystallization/neocrystallization; see Section 6.3.2) increase with depth, the porosity decreases, and free water tends to be forced out of the rock (for example, as tectonic cleavages form; see Sections 6.6 to 6.12), so that porosity is minimized, and eventually $P_{fluid} = P_{solids}$.

The inferred low porosity (perhaps no more than 1 volume percent) of most non-deformed regional metamorphic rocks implies that the amount of free fluid is also small. However, as noted previously, deformation can open transient pores and microfractures, which promote fluid transfer of small amounts of fluid, and more extensive fracture networks can transport larger amounts of fluid. Even where the porosity is small, large volumes of fluid can pass through deforming rocks with time (Etheridge *et al.*, 1983, 1984; Sibson, 1996). For example, because of the low solubility of gold transported as complexes in hydrothermal fluids (*c.* 10 parts per billion), a flow of *c.* 1 km^3 of aqueous fluid is needed to precipitate 10 tonnes of gold in a medium-sized ore deposit (Sibson, 1996).

5.2 Evidence of fluid flow in rocks undergoing metamorphism

Many rocks show evidence that H_2O-rich fluid commonly moves through metamorphic terranes during deformation and metamorphism, as does CO_2-rich fluid in terranes with calcareous rocks. The evidence (detailed below) consists of: (1) primary fluid inclusions, (2) veins, (3) replacement of anhydrous by hydrous minerals (hydration reactions), (4) evidence of isotopic exchange between metamorphic minerals and fluids, (5) hot spring fluids, (6) partial solution, solution seams and the development of slaty and crenulation cleavages, and (7) major changes in mineral assemblages in carbonate rocks.

(1) Primary fluid inclusions typically are inferred to represent trapped interstitial fluid at the time of metamorphism, but they are uncommon in metamorphic rocks because prograde metamorphism and especially retrograde metamorphism tend to destroy them. Therefore, most fluid inclusions in metamorphic rocks are secondary, and mainly reflect conditions during the uplift history of the rock (e.g., Hollister, 1969; Barker, 1990). Inclusions in metamorphic minerals commonly are small (less than 0.02 mm across) and their fluids are rich in H_2O, with varying amounts of anhydrous gases such as CO_2 and CH_4, cations such as Si^4, Al^{3+}, Fe^{2+}, Mg^{2+}, K^+, Ca^{2+} and Na^+, and anions such as Cl^-, Br^- and $(SO_4)^{2-}$. Water-poor, high-grade metamorphic rocks may have either no fluid inclusions or contain inclusions rich in CO_2, N_2 or NaCl.

(2) Veins provide good evidence of fluid movement in metamorphic rocks, and are especially common in deformed rocks, where they typically mark the former presence of fluid-transporting fractures. In metasedimentary sequences, they mostly consist of quartz or carbonate minerals, though some are complex, depending on the source rocks, wall rocks and the effects of P and T on the solubility of chemical components. The rocks adjacent to veins may show evidence of reaction with the transported fluids (e.g., Hewitt, 1973; Ague, 1991, 2002; Ague & Rye, 1999). Studies of veins can provide information on relationships between metamorphism, fluid flow and deformation (Bebout & Barton, 1989, 1993; Ague, 1991, 1994, 1997; Selverstone *et al.*, 1991, 1992; Dipple & Ferry, 1992a; Oliver, 1996; Agard *et al.*, 2000; Masters & Ague, 2005). Veins are parts of complicated chemical systems that depend on P-T conditions, fluid composition, fluid volume and the scale of chemical transport (Miller & Cartwright, 2006). The following are the two main mechanisms of vein formation.

(1) Veins may result from broad-scale fluid flow systems (Wood & Walther, 1986; Ferry, 1991, 1994a,b; Ferry & Dipple, 1991; Bebout & Barton, 1993; Oliver, 1996; Masters & Ague, 2005). In such an *open system*, the fluid may come from outside the rocks being considered, possibly from great distances, transporting dissolved chemical components. Precipitation and dissolution reactions result from chemical disequilibrium between this external fluid and the rocks penetrated by the fluid. The fluid infiltration can be either *single pass* or *multipass*, which involves recycling by convection or pumping (Sibson, 1975; Fyfe *et al.*, 1978; Etheridge *et al.*, 1983, 1984).

(2) Alternatively, veins may result from local-scale fluid transport in a *closed system* (Heinrich, 1986; Philippot & Selverstone, 1991; Marquer & Burkhard, 1992; Selverstone *et al.*, 1992; Yardley & Bottrell, 1992; Cartwright *et al.*, 1994; Slater *et al.*, 1994; Henry *et al.*, 1996; Becker *et al.*, 1999; Cartwright & Barnicoat, 1999; Widmer & Thompson, 2001; Miller & Cartwright, 2006). This may occur in the following two ways. (a) Locally formed fluid may *flow (advect)*. As temperature gradients would be small at the outcrop scale, chemical interaction results mainly from pressure gradients (Widmer & Thompson, 2001), which may be enhanced by deformation (Ague, 1994). (b) Veins ('accretionary veins') may be filled by *diffusion* through a stagnant fluid, possibly from nearby sites at which pressure-solution is taking place (Durney & Ramsay, 1973). Deposition is inferred to occur on grain boundaries subjected to relatively low normal stress in a differentially stressed aggregate, producing a vein by dilation ('force of crystallization'). The diffusion is driven by chemical potential gradients between the vein and enclosing rock (Widmer & Thompson, 2001). An example of accretionary veining is provided by kyanite-rich veins in metabasalts at the blueschist-eclogite facies transition in the Zermatt-Sass Zone, Switzerland (Widmer & Thompson, 2001); Al and Si depletion haloes adjacent to the veins reflect local compositional segregation.

An example of locally sourced fluids is provided by albite veins in metabasic rocks that underwent retrograde metamorphism in alpine Corsica (Miller & Cartwright, 2006). The veins have alteration haloes of actinolite + epidote + chlorite + albite, and were formed during decompression and exhumation of a high-pressure metamorphosed ophiolite containing hydrous minerals (such as lawsonite, phengite, paragonite, chloritoid, carpholite and glaucophane), which provided much of the fluid. Stable isotope data indicate that the fluids responsible for the albite veins

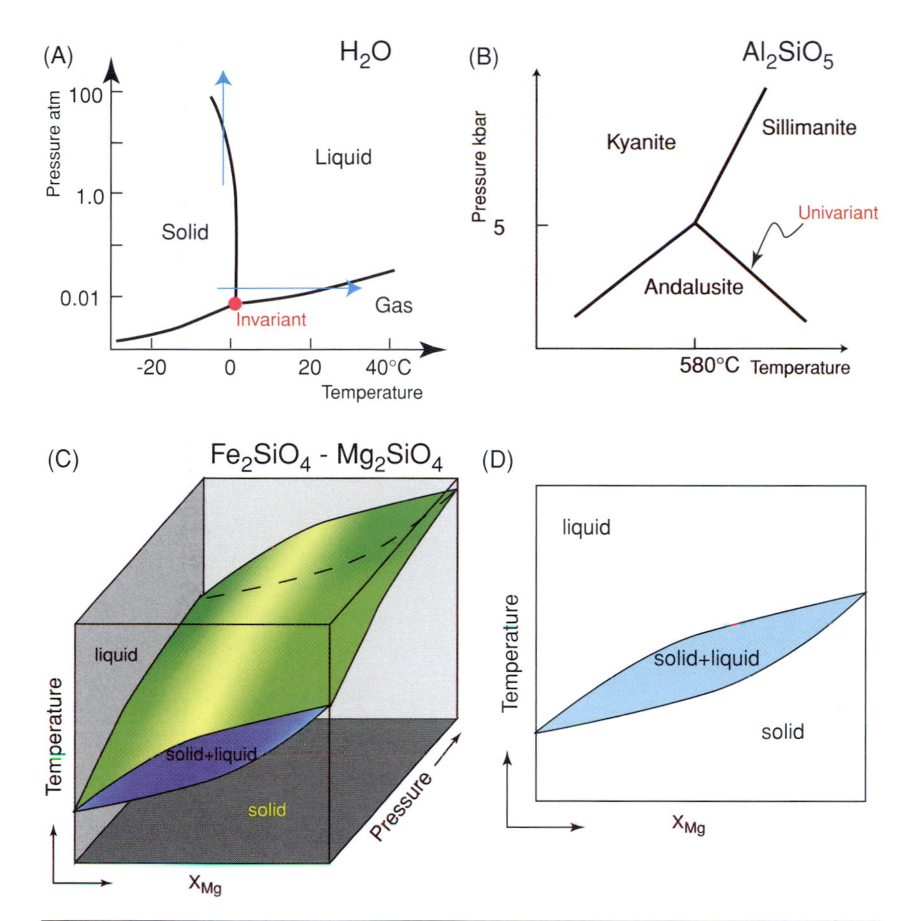

Fig. 1.3 (A) *P-T* diagram for the one-component system, H_2O. The vertical blue arrow indicates a phase transition from solid to liquid with increasing *P*, such as occurs at the base of large glaciers. For pure H_2O, the point at which ice, water, and vapour all coexist is fixed or invariant. (B) *P-T* diagram for the one-component system Al_2SiO_5, showing the relative stability ranges of the three minerals andalusite, sillimanite, and kyanite. The *P-T* conditions at which two of the minerals can coexist are univariant reaction lines, and the *P-T* condition at which all three phases coexist is an invariant point. (C) Diagrammatic three-dimensional illustration showing the three parameters (*P*, *T*, *X*) needed to define solid–liquid relationships in the two component system Fe_2SiO_4-Mg_2SiO_4 (fayalite-forsterite), where X = mole fraction of one of the end-member components. (D) Two-dimensional slice, at constant *P*, through Figure, 1.3C.

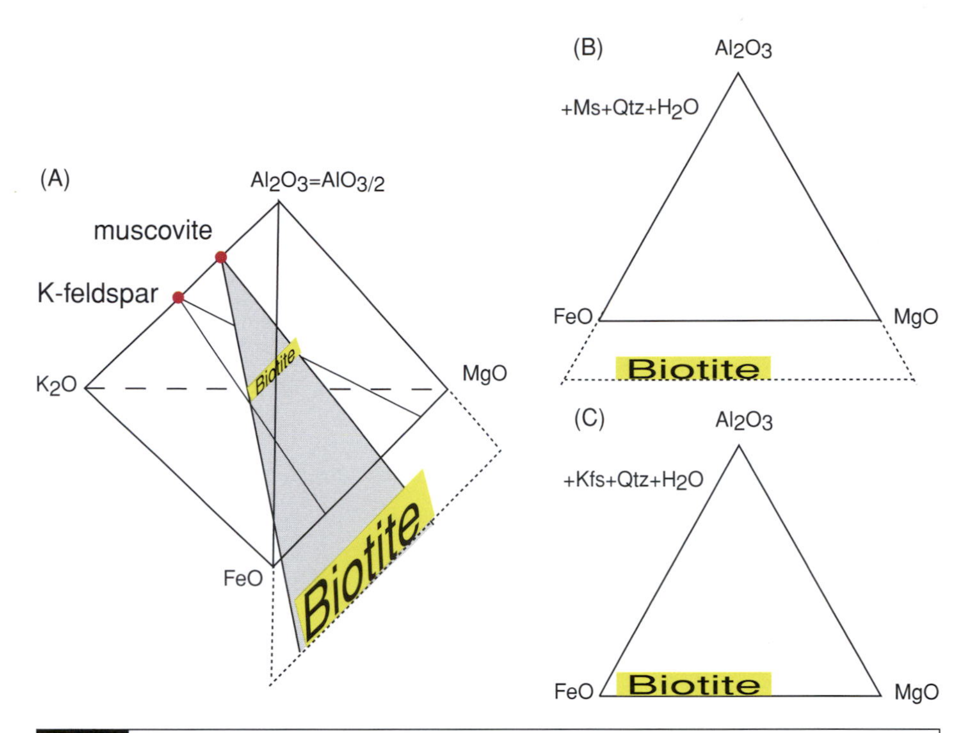

Fig. 1.4 (A) K_2O-Al_2O_3-FeO-MgO tetrahedron illustrating how projecting the position of biotite, which lies in the tetrahedron body, from muscovite onto the Al_2O_3-FeO-MgO face of the diagram causes biotite to plot at negative Al_2O_3. (B) Two-dimensional Al_2O_3-FeO-MgO (AFM) diagram showing the position of biotite projected from muscovite. (C) Two-dimensional Al_2O_3-FeO-MgO (AFM) diagram showing the position of biotite projected from K-felspar.

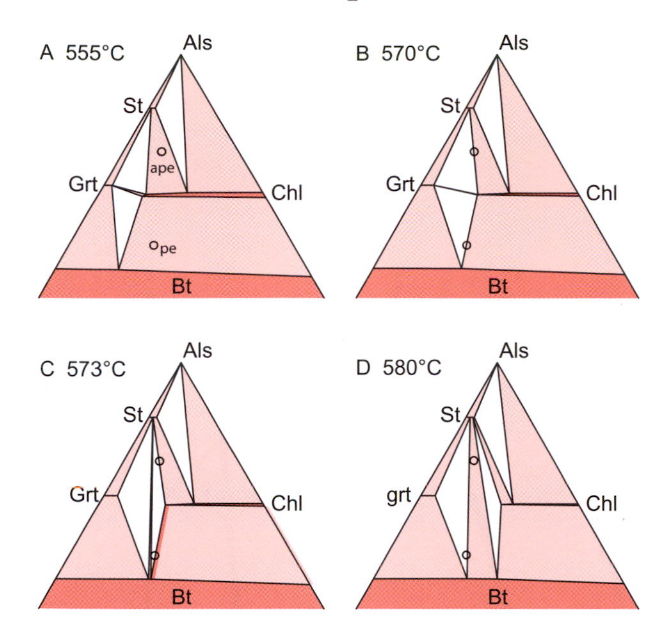

Fig. 1.5 AFM diagrams illustrating the compositional and mineral relationships predicted by phase diagram modelling of metapelitic equilibria in KFMASH using THERMOCALC, for $P = 6$ kbar and $T = 555\,^{\circ}\mathrm{C}$–$580\,^{\circ}\mathrm{C}$, assuming muscovite, water and quartz are in excess; modified from figures in Worley & Powell (1998a). The projected position of a common 'greywacke' (feldspathic sandstone) is shown in all four diagrams as 'pe'; Figure 1.5A additionally shows the projected position of an aluminous metapelite, 'ape'. The white triangles represent divariant assemblages, the light red shading represents trivariant assemblages, and the dark red shading represents quadrivariant assemblages.

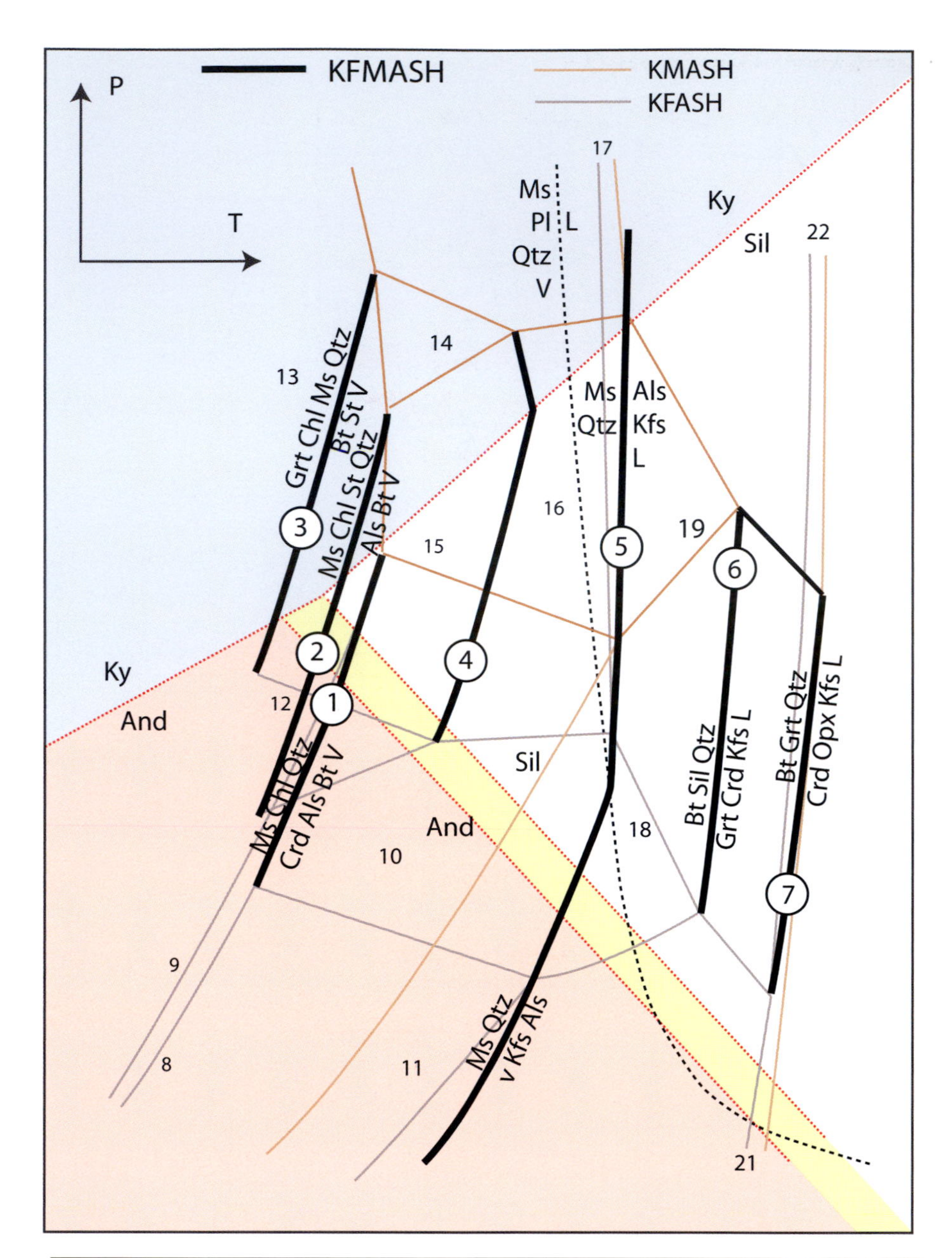

Fig. 1.8 Schematic petrogenetic grid for a portion of the system KFMASH involving the reactions listed in Table 1.2; redrawn from fig. 33 of Pattison & Tracy (1991). Dashed lines represent aluminosilicate relationships, the And = Sil reaction occurring over an interval, which reflects the common interpretation that andalusite metastably persists into the sillimanite field. V = supercritical H_2O fluid; L = silicate liquid (melt).

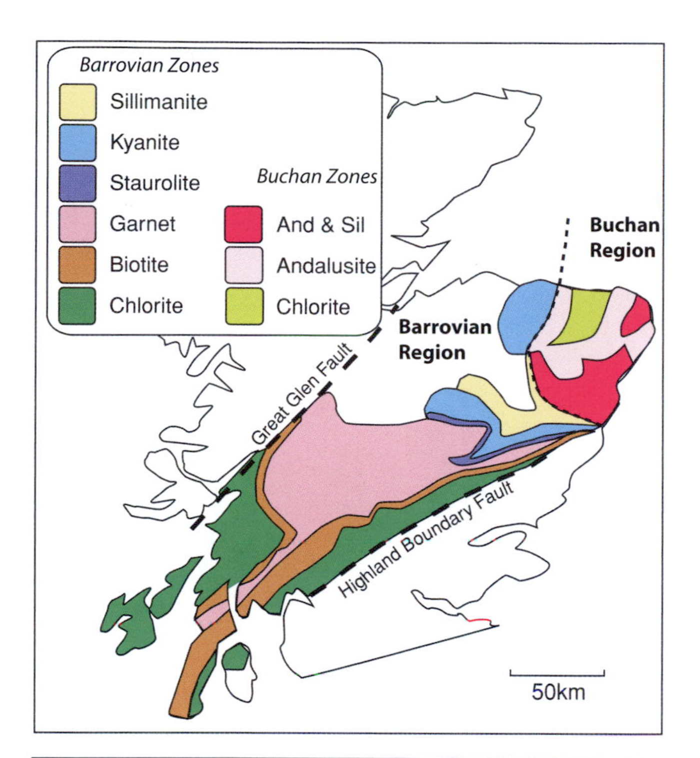

Fig. 1.11 Map of the Grampian Highlands in Scotland, showing the distribution of metamorphic zones, identified on the basis of mineral occurrence, as well as the broad subdivision into medium-*P* metamorphic rocks, commonly referred to as the Barrovian region, and low-*P* metamorphic rocks, commonly referred to as the Buchan region.

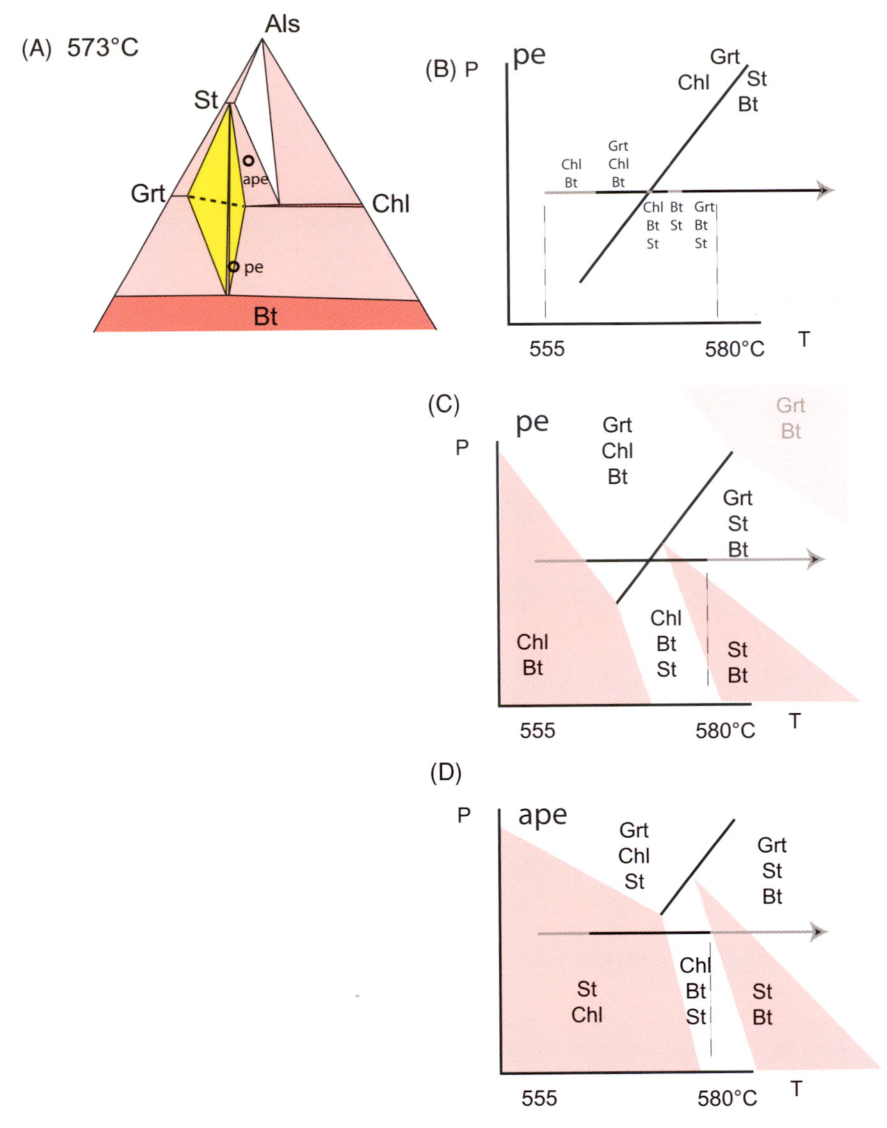

Fig. 1.12 Construction of a qualitative KFMASH pseudosection using the example of isobaric heating of a typical metapelite, 'pe', from $T = 555\,^{\circ}\text{C}$–$580\,^{\circ}\text{C}$, across the staurolite isograd, as discussed in the text. (A) AFM diagram showing mineral relationships at $P = 6$ kbar and $T = 573\,^{\circ}\text{C}$, with muscovite quartz and water in excess. The yellow diamond, defined by the positions of staurolite, garnet, biotite and chlorite, outlines the range of projected rock compositions that would be involved in the univariant reaction. (B) Summary of the mineral assemblages that the 'greywacke' rock composition 'pe' would produce during isobaric heating at $P = 6$ kbar. Divariant equilibria indicated by thick black horizontal line; trivariant equilibria indicated by thin horizontal line. (C) Qualitative *P-T* pseudosection constructed from Fig. 1.5B by extrapolating relationships of the various divariant and trivariant mineral equilibria. The relative geometry of the phase equilibrium boundaries can be calculated by Schreinemaker's analysis (Zen, 1966), using univariant equilibria around an invariant KFASH or KMASH point involving all the minerals present in the KFMASH reaction. (D) Qualitative *P-T* pseudosection constructed for the same heating path as for the aluminous metapelite composition 'ape' in Fig. 1.5A. Though both rock compositions underwent the univariant reaction at the same *T*, they have different starting assemblages and different divariant reaction sequences.

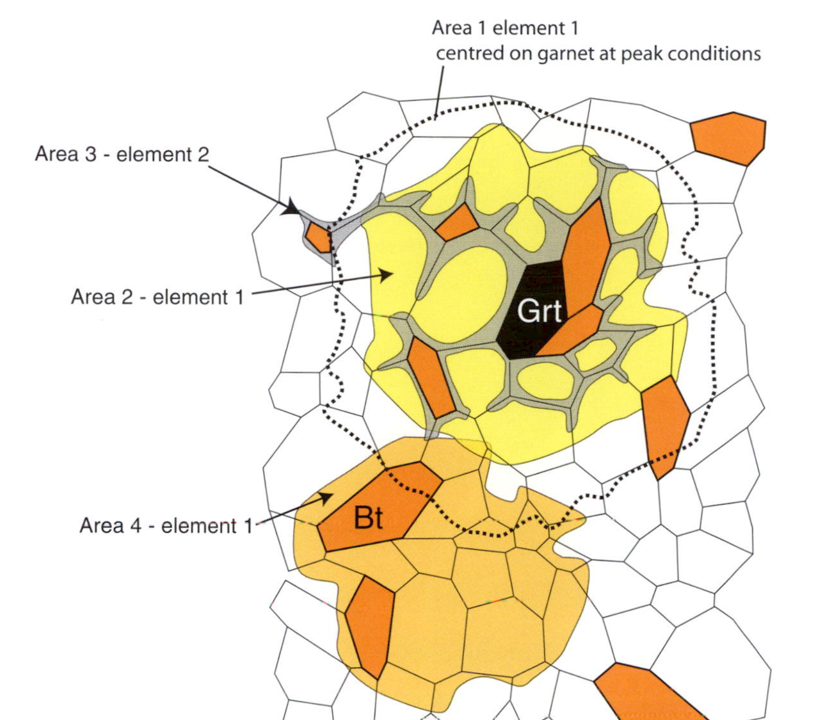

Fig. 1.14 Schematic thin section, modified from Stüwe (1997, fig. 2), consisting of four different minerals, namely garnet (black, Grt), biotite (orange, Bt) and a quartz-feldspar matrix (white). Superimposed are idealized equilibrium areas, (and volumes by extrapolation) for a given time at a given temperature (coloured lobate regions). Areas 1 to 3 overlap, and are centred on a garnet grain; they indicate that the equilibrium volume contracts with cooling and may be quite distinct for different elements. Area 1 indicates the volume for an element for which the ratio of grain boundary diffusion to intracrystalline (volume) diffusion is large at peak conditions; area 2 indicates similar relationships at lower T. Area 3 indicates the volume of another element for which the ratio of intracrystalline to grain boundary diffusion is large. Thus, area 1 extends along grain boundaries and includes only biotite grains that can easily be reached along the grain boundaries. Area 4 is a region, approximately the same size as area 2, centred on quartz and feldspar grains. Each of the four areas would have different chemical compositions.

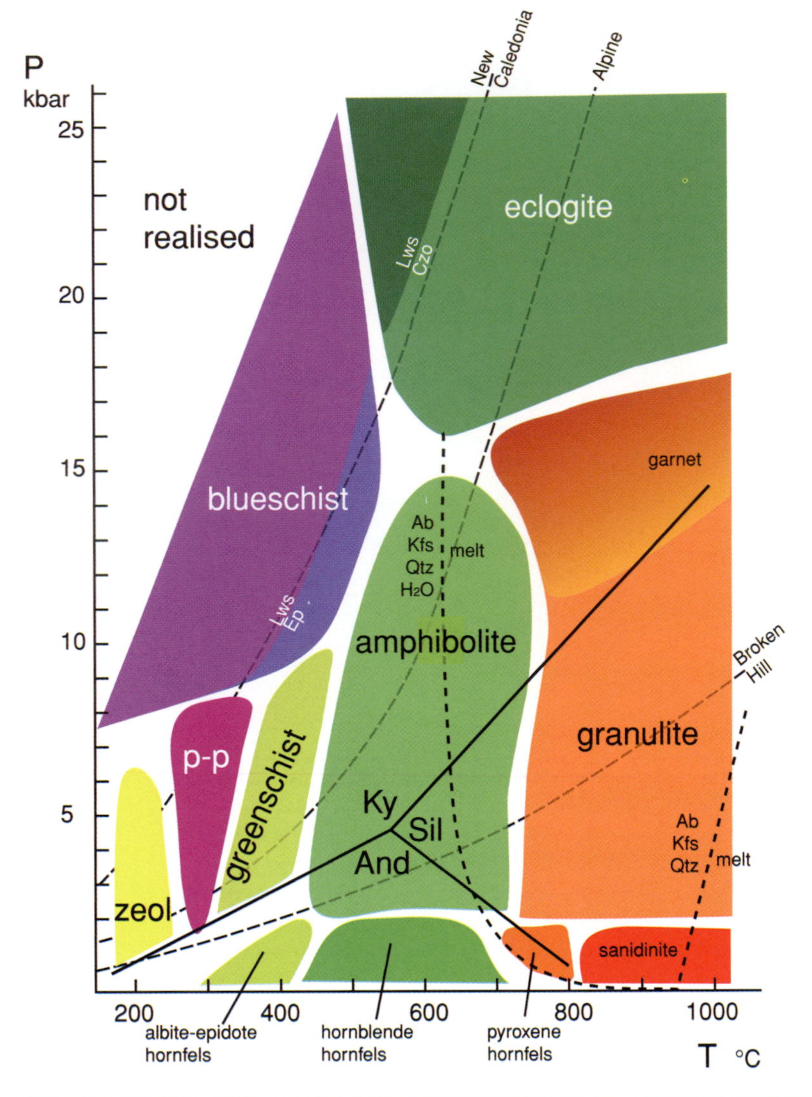

Fig. 1.15 *P-T* diagram showing the generalized distribution of the main metamorphic facies; derived from Eskola (1915, 1921) and Turner (1981). Andalusite is common and almandine garnet is rare in the low-*P* facies, though considerable overlap exists between the low-*P* (typically contact metamorphic) facies and the equivalent intermediate-*P* (regional metamorphic) facies (shown in the same colours). The blueschist and eclogite facies may be subdivided on the basis of the breakdown reaction of lawsonite to form epidote. The low-*P* limit of garnet in the granulite field, delineating the 'garnet granulite subfacies of Green & Ringwood (1967), is sensitive to whole-rock composition, and so is shown as a shaded zone. Representative geotherms are shown for: subduction metamorphism from Eocene rocks in northern New Caledonia; convergent continental metamorphism, such as is responsible for forming the European Alpine metamorphism; and exceptionally high heat flow in low-pressure regional metamorphism, evident at Broken Hill, Australia. Also shown are phase relationships for the pure Al_2SiO_5 system and the water-saturated melting curve for a simplified 'granite' composition. Abbreviations: p-p = prehnite-pumpellyite; zeol = zeolite.

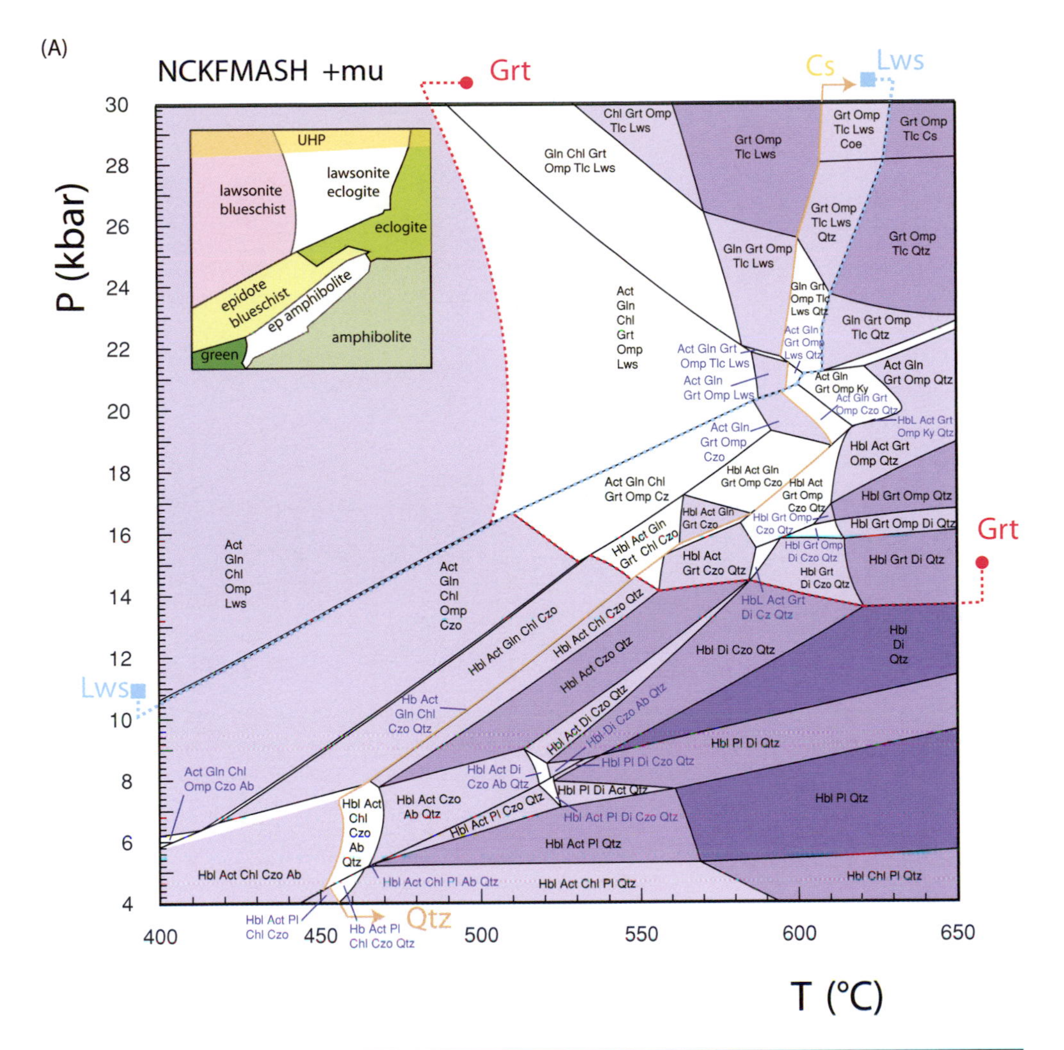

Fig. 1.16 (A) *P-T* pseudosection in NCKFMASH, calculated using THERMOCALC (Section 2.3) with H_2O and muscovite in excess, for a metabasite from New Caledonia inferred to be a subtly altered MOR basalt; modified, with additional material, from Clarke *et al.* (2006, fig. 2a). Univariant reactions experienced by this rock composition are shown as continuous lines. Divariant fields are unshaded six phase fields, trivariant fields are shaded five phase fields and so on. Quartz is present only at high-*T* conditions, and garnet is present at high-*P*/*T* conditions. A narrow band of three amphibole assemblages occurs in the upper epidote blueschist and lower eclogite subfacies conditions, and a glaucophane-hornblende solvus is at intermediate eclogite facies conditions. High-*T* amphibolite facies equilibria are metastable with respect to orthoamphibole-bearing equilibria; so the diagram is less reliable for those conditions. Inset shows the inferred extent of common metamorphic subfacies based on the calculated positions of the mineral equilibria. The lower limit of the ultra high-pressure (UHP) field is delineated by the reaction quartz = coesite. (B, C) *P-T* pseudosections calculated using THERMOCALC for interlayered metapelitic and metapsammitic portions of turbidites at Mount Stafford, central Australia; modified from White *et al.* (2003, figs. 5 and 7). In calculating the mineral equilibria, there is a need to fix water contents in the modelled whole-rock compositions at the solidi, and the water-absent sub-solidus fields are inappropriate for interpreting the pre-melting prograde evolution. The pseudosections also show the calculated mode isopleths in mole precent for several minerals and silicate melt. Insets show details of the mode changes that occur near the solidi, at conditions within the areas covered by the *P-T* boxes.

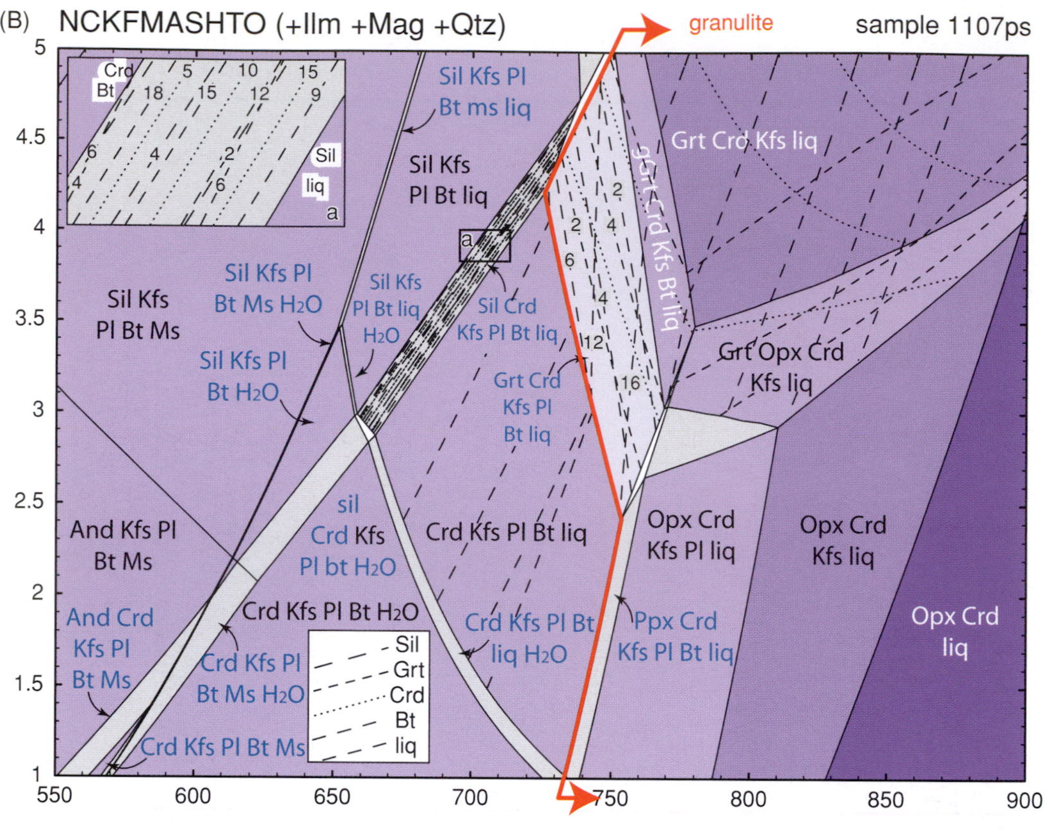

(B)
NCKFMASHTO (+Ilm +Mag +Qtz)
granulite
sample 1107ps
Sil Kfs Pl Bt ms liq
Sil Kfs Pl Bt liq
Grt Crd Kfs liq
gGrt Crd Kfs Bt liq
Sil Kfs Pl Bt Ms H2O
Sil Kfs Pl Bt Ms
Sil Kfs Pl Bt liq H2O
Sil Crd Kfs Pl Bt liq
Sil Kfs Pl Bt H2O
Grt Crd Kfs Pl Bt liq
Grt Opx Crd Kfs liq
And Kfs Pl Bt Ms
sil Crd Kfs Pl bt H2O
Crd Kfs Pl Bt liq
Opx Crd Kfs Pl liq
Opx Crd Kfs liq
Crd Kfs Pl Bt H2O
Crd Kfs Pl Bt liq H2O
Ppx Crd Kfs Pl Bt liq
Opx Crd liq
And Crd Kfs Pl Bt Ms
Crd Kfs Pl Bt Ms H2O
Crd Kfs Pl Bt Ms
Sil
Grt
Crd
Bt
liq

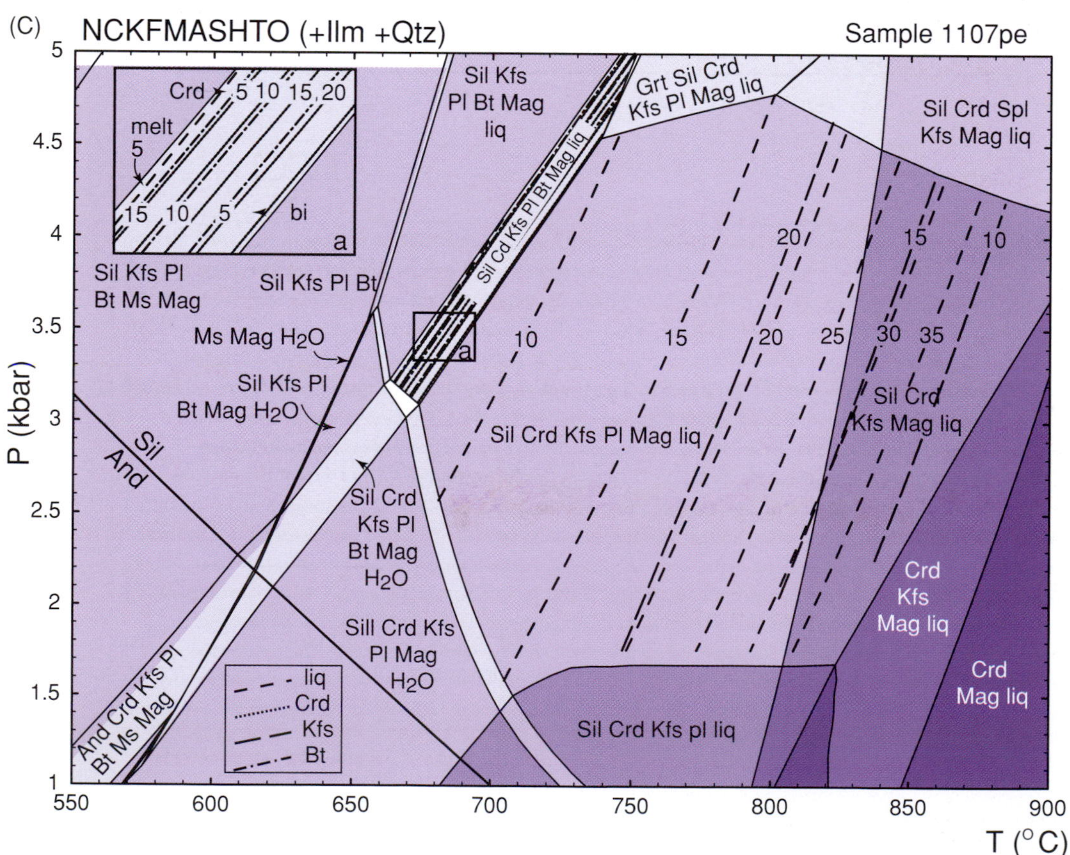

(C)
NCKFMASHTO (+Ilm +Qtz)
Sample 1107pe
Sil Kfs Pl Bt Mag liq
Grt Sil Crd Kfs Pl Mag liq
Sil Crd Spl Kfs Mag liq
Sil Cd Kfs Pl Bt Mag liq
Sil Kfs Pl Bt Ms Mag
Sil Kfs Pl Bt
Ms Mag H2O
Sil Kfs Pl Bt Mag H2O
Sil Crd Kfs Pl Mag liq
Sil Crd Kfs Mag liq
Sil And
Sil Crd Kfs Pl Bt Mag H2O
Sill Crd Kfs Pl Mag H2O
Crd Kfs Mag liq
Crd Mag liq
Sil Crd Kfs pl liq
And Crd Kfs Pl Bt Ms Mag
liq
Crd
Kfs
Bt
P (kbar)
T (°C)

Fig. 1.16 (*cont.*).

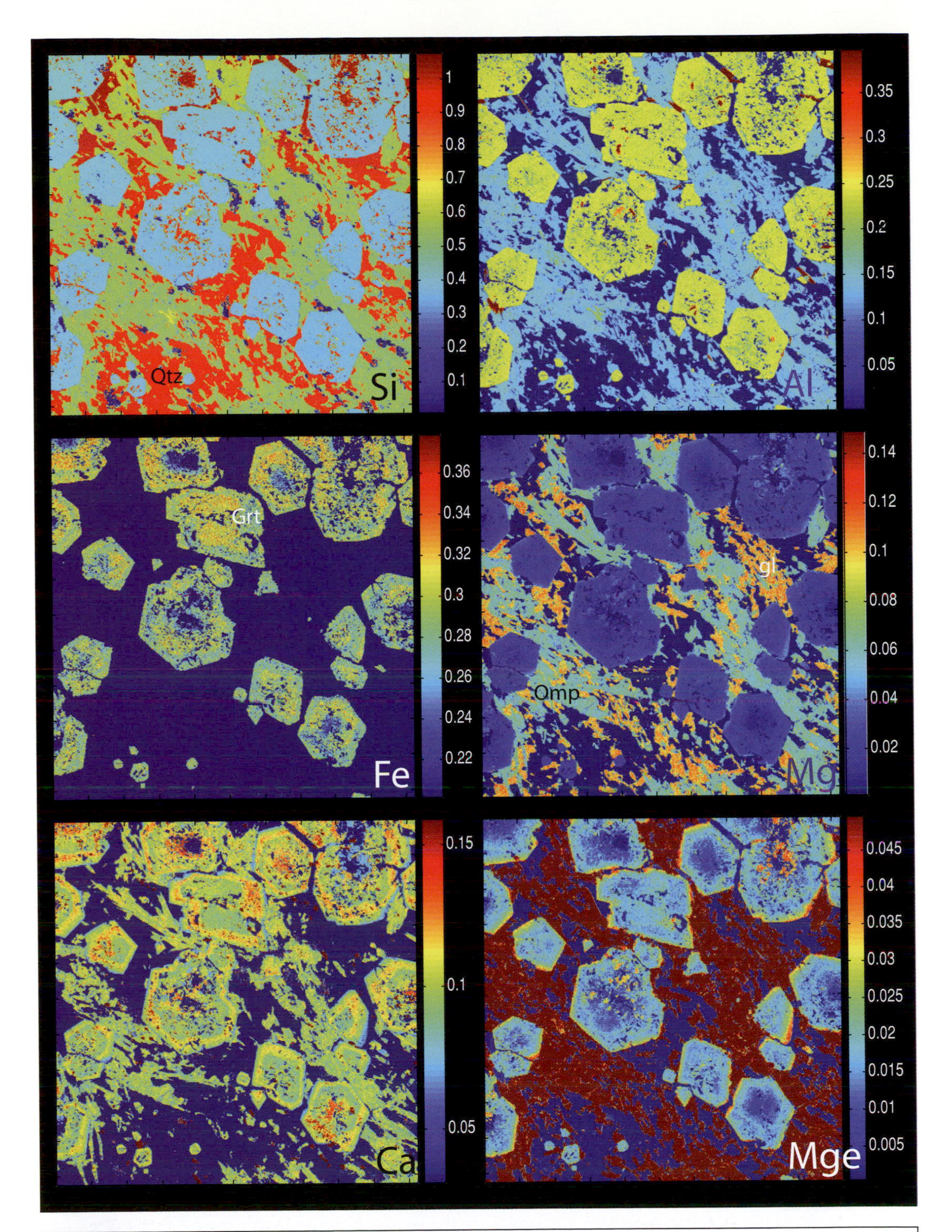

Fig. 2.1 Maps of element concentrations (weight precent; for example, 0.15 = 15%), illustrating complex chemical zoning in garnet from a basaltic eclogite in New Caledonia (Clarke *et al.*, 1997). Garnet growth has been inferred to be related to a change in prograde conditions, from approximately 475 °C to 590 °C and from 15 to 19 kbar (Marmo *et al.*, 2002), with Fe-rich, Mg-poor cores grading to Mg-rich rims (see Mge map with restricted Mg range). Most of the matrix of the rock is composed of random omphacite, glaucophane and quartz grains (see Mg map), omphacite having formed late in the prograde history (Carson *et al.*, 1999). The Ca map shows an unusual spike in garnet Ca content in the outer part of the grains. The imaged area is approximately 1.5 cm across. Maps of X-ray intensity were converted to weight per cent following the method of Clarke *et al.* (2001). A subtle effect from a small focussing error is evident in decreased concentration of iron from the top to the bottom of each image.

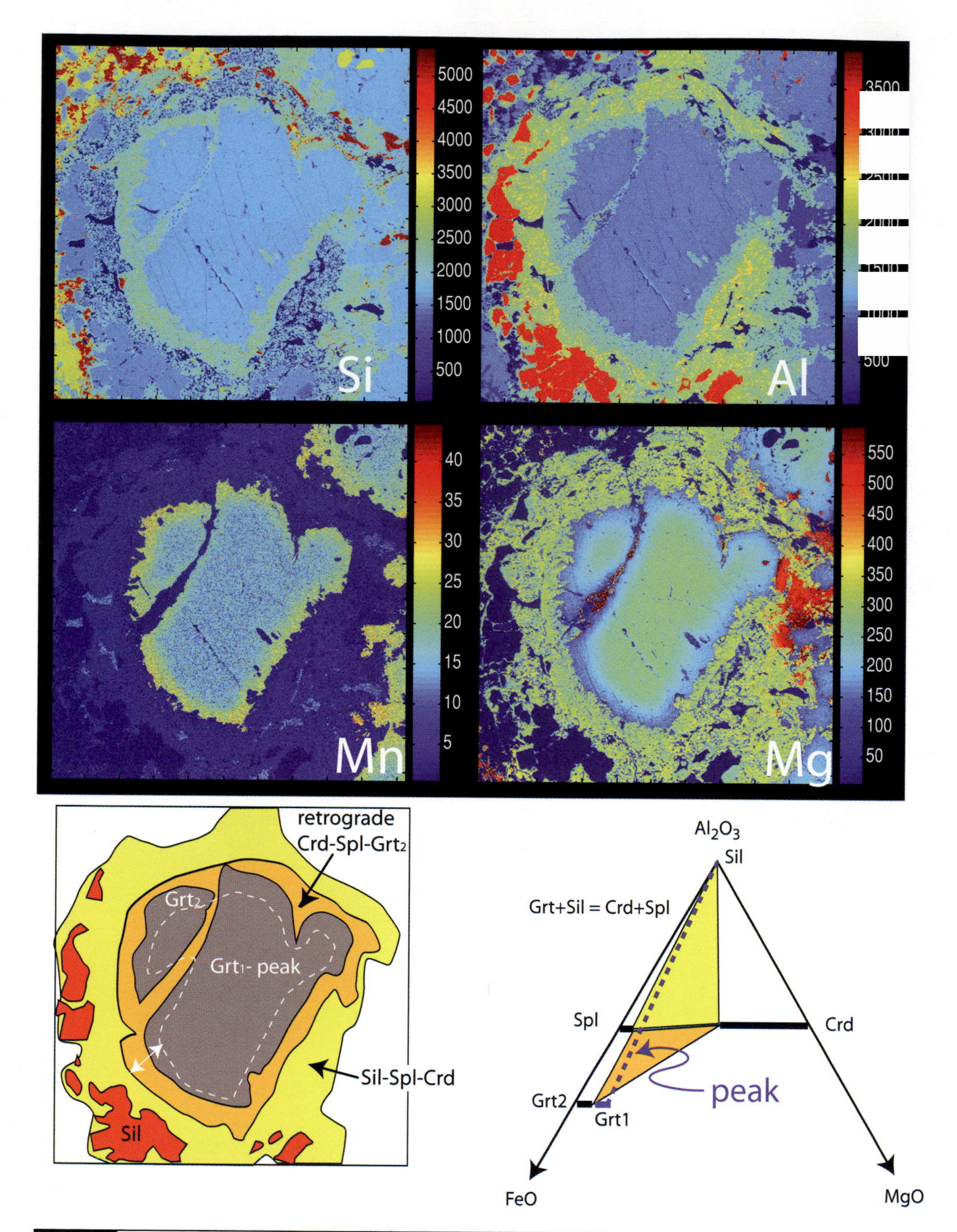

Fig. 2.2 Maps of X-ray counts for Si, Al, Mn and Mg in a granulite facies reaction corona involving spinel and cordierite that separates garnet from sillimanite; Cohn Hill, western Musgrave Block, central Australia. The reaction has been inferred to be the result of decompression from peak metamorphic conditions (Clarke & Powell, 1991). Image approximately 2.5 cm across.

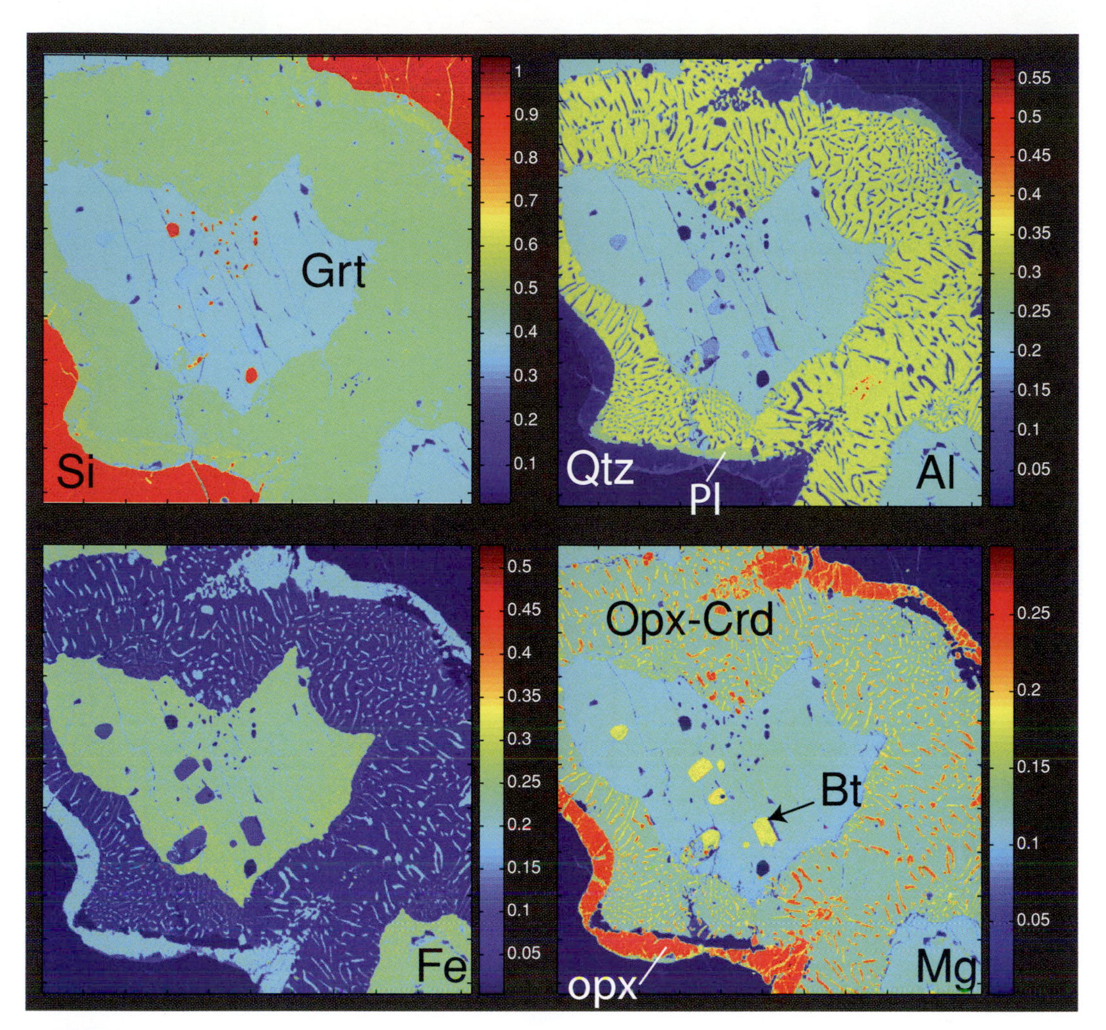

Fig. 2.3 Maps of element concentrations (weight precent), showing a granulite facies reaction corona involving orthopyroxene and cordierite that separates garnet from quartz, the Rauer Group of islands, east Antarctica. The reaction has been inferred to be related to decompression with cooling from peak metamorphic conditions (Kelsey *et al.*, 2003). The biotite inclusions in garnet were inferred to be part of the peak assemblage. Image area approximately 2 cm across. Maps of X-ray intensity were converted to weight per cent following the method of Clarke *et al.* (2001).

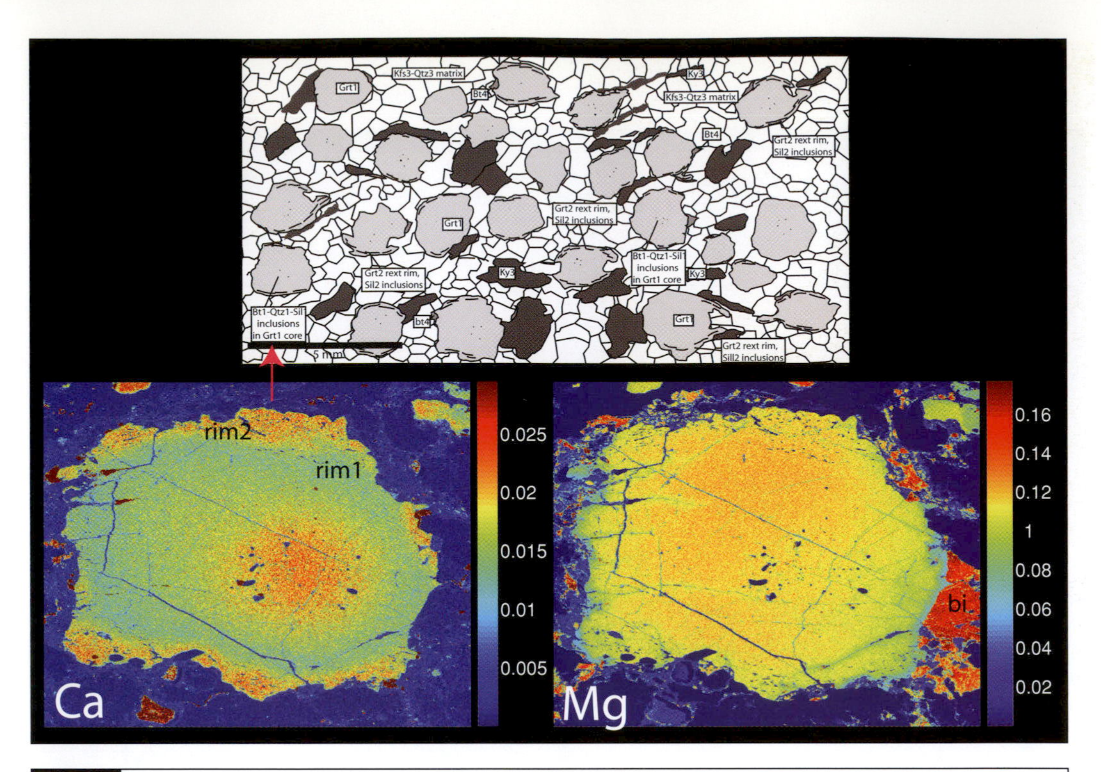

Fig. 2.4 Ca and Mg maps showing patterns of complex chemical zoning in garnet that have been inferred to reflect differences in diffusion rates of these elements at high-grade conditions. The sample is a UHT metapelitic gneiss from South Harris, Scotland (Baba, 1999); thin section sketch modified, with additional material, from Hollis *et al.* (2006, fig. 3).

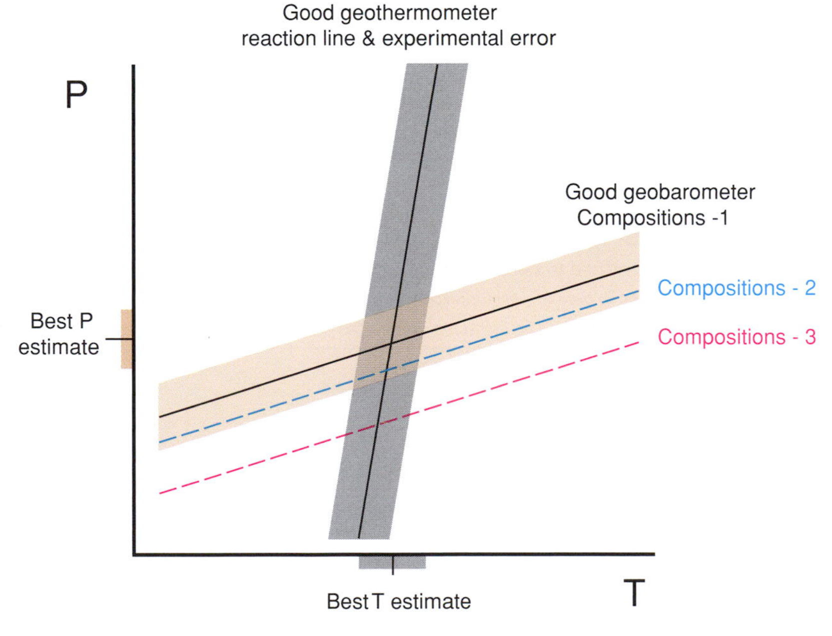

Fig. 2.5 *P-T* diagram indicating the position of two theoretical exchange reactions, the slope of the reaction being strongly influenced by $\Delta V_{reaction}$. The slope determines whether it is more suitable as a geothermometer or geobarometer. By using minerals with exchange reactions suitable as thermometers and barometers, a more robust estimate of *P* and *T* conditions can be obtained from the intersection of two (or more) reactions. The effect of mineral composition differences, in terms of displacing an exchange reaction to different *P-T* conditions, is shown for the geobarometer; subtle differences may result in displacements that lie within experimental error, as shown for compositions 1 and 2.

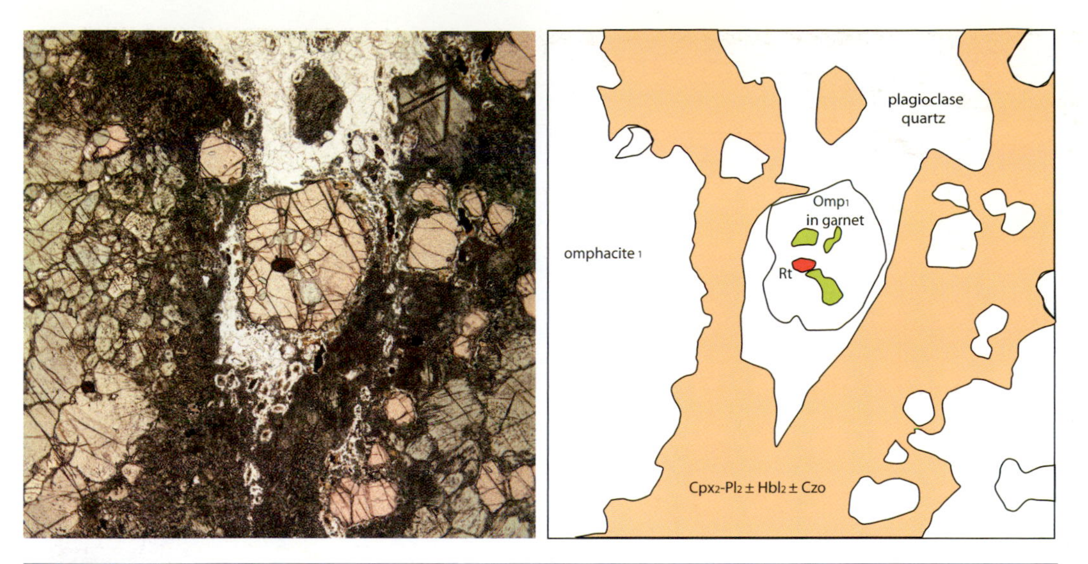

Fig. 2.6 Plane-polarized light photomicrograph and sketch of part of sample 0669, a garnet granulite from Fiordland, New Zealand. An early paragenesis is inferred to comprise the large garnet, plagioclase, omphacite and quartz grains, with a secondary paragenesis formed from comparatively fine-grained symplectites of sodic diopside, plagioclase 2, and quartz. Rims of garnet grains surrounded by symplectite have compositions distinct from those of grain cores, and are inferred to be part of a secondary paragenesis. Field of view 6 mm across.

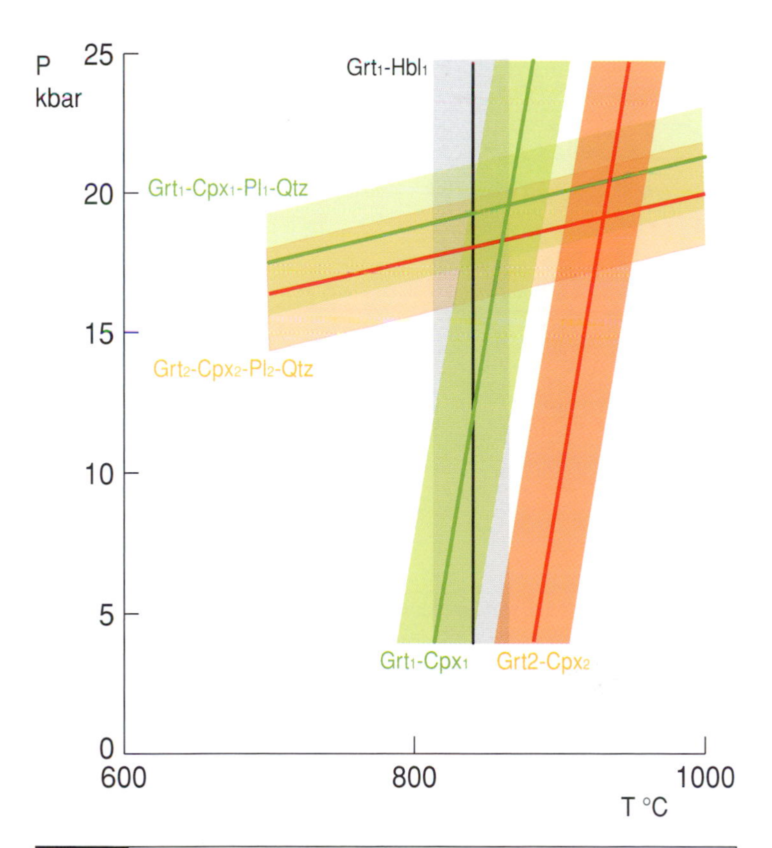

Fig. 2.7 *P-T* diagram showing the position of various exchange equilibria for the two parageneses discussed in the text for sample 0669, a garnet granulite from Fiordland, New Zealand (see Fig. 2.6). Nominal error ranges of 50 °C and 1.9 kbar are shown for the positions of the exchange equilibria used as thermometers and barometers, respectively. The *P-T* evolution of the reaction microstructure can be related to minor decompression with heating, but the *P* estimates lie within the error ranges.

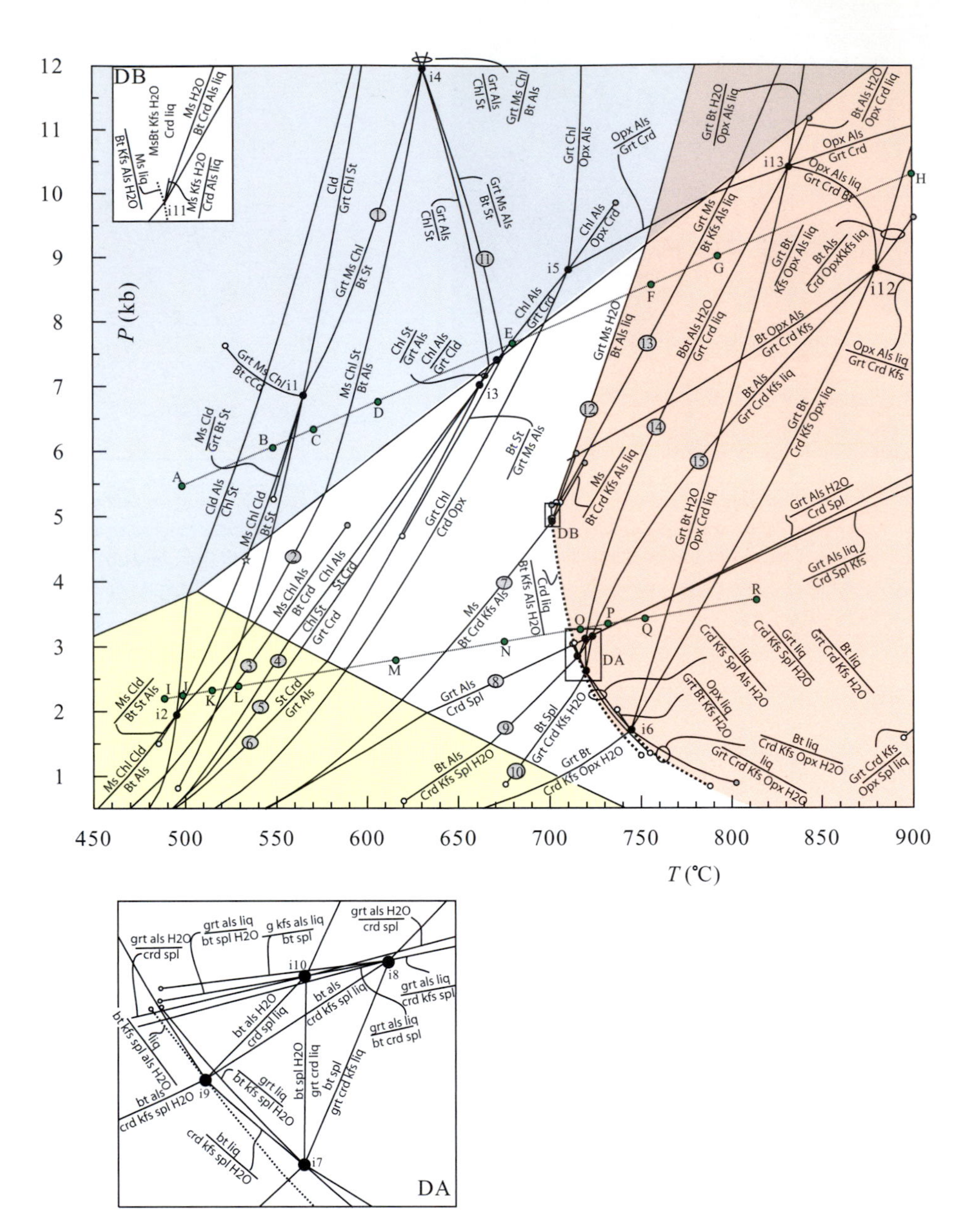

Fig. 2.8 *P-T* projection for the system KFMASH, with quartz and H_2O in excess for subsolidus conditions, and for just quartz in excess for supersolidus conditions (pale red); modified from Wei *et al.* (2004, fig. 2.1). Solid dots with labels i1 to i13 refer to invariant points, and grey-shaded circles identify reactions discussed by Wei *et al.* (2004). Unfilled stars indicate the location of singularities, where one or more phases swap sides of the reaction. Smaller open dots refer to points invariant in KFASH, and smaller grey dots to points invariant in KMASH. Heavy dotted lines represent the H_2O-saturated solidus in KFMASH. Larger grey dots with letters A-R are *P-T* locations of the compatibility diagrams shown in Fig. 2.9. Inset DA shows detail around invariant points i7 to i10, and inset DB shows detail around invariant point i11. For the thermodynamic data used to construct the diagram, the andalusite field (yellow) does not intersect the solidus, but the andalusite-sillimanite transition is now generally considered to plot at higher pressure, such that partial melting of metapelitic rocks can occur in the andalusite field (Vernon *et al.*, 1990).

were derived from within the metabasaltic parts of the ophiolite, although fluid flow occurred on greater than the outcrop scale (Miller & Cartwright, 2006).

The source of vein minerals in regional metamorphic terranes is commonly obscure. As discussed in Section 5.3, fluid may be released from many sources, such as cooling magma bodies and metamorphic devolatilization (e.g., dehydration) reactions. Some veins are composed of components inferred to be dissolved during the development of foliations in deforming rocks, and others are formed by interaction of metamorphic minerals with fluids (Ague, 1991, 1994, 1997).

(3) Hydration reactions: Introduced H_2O-rich ($\pm$ CO_2-rich) fluids react with anhydrous mafic and ultramafic igneous rocks (basalts, gabbros and peridotites), especially in seafloor settings (e.g., Hemley *et al.*, 1980; Seyfried *et al.*, 1988), forming hydrous lower-grade metamorphic rocks of the zeolite, blueschist, greenschist and albite-epidote hornfels facies (Section 1.8.1).

(4) Isotopic exchange between metamorphic minerals and fluids is indicated by ratios of stable isotopes, especially O, H and C (Section 5.3). Infiltration of a fluid that is out of isotopic equilibrium with metamorphic rocks shifts (*resets*) the isotopic ratio of an element (Phillips & Groves, 1983; Baumgartner & Rumble, 1988; Read & Cartwright, 2000; Buick *et al.*, 2000; Doyle & Cartwright, 2000), though kinetic retardation of the equilibration process can obscure the resetting (Lassey & Blattner, 1988; Nabelek, 2002). For example, vein quartz may be out of oxygen isotopic equilibrium with wall-rock quartz in regional metamorphic metapelites (Ague, 1991). Carbonate rocks are especially useful, because (1) they react with infiltrating fluid, assisting determination of fluid compositions (e.g., Ague, 2003) and (2), fluid infiltration usually causes marked resetting of their isotope ratios because their initial oxygen isotope ratios are different from those of other crustal rocks (Baumgartner & Ferry, 1991; Nabelek, 1991; Valley *et al.*, 1990; Dipple & Ferry, 1992a; Cartwright & Buick, 1995; Cartwright *et al.*, 1995; Cartwright & Barnicoat, 2003).

(6) Hot spring fluids may have compositions reflecting equilibration at low-grade metamorphic P and T, for example, metal-bearing saline brines in the Salton Sea, California, which are pumped from greenschist facies metamorphic rocks (e.g., Muffler & White, 1969; Fyfe *et al.*, 1978), and 'black smokers' at mid-ocean ridges.

(7) Partial solution, solution seams and the development of slaty and crenulation cleavages: Much mesoscopic and microscopic evidence exists for solution, transfer and redeposition of chemical components during the formation of foliations in metamorphic rocks (Plessman, 1964; Williams, 1972; McClay, 1977; Gray, 1977, 1981; Beach, 1977; Beutner, 1978; Mosher, 1980; Gregg, 1985; Vernon, 1987b, 2004). For example, the following primary structures may be partly removed by solution, the remnants being transected by dark seams of residual material: fossils (Durney, 1972, p. 316; Siddans, 1972, fig. 17), ooliths (Ramsay & Huber, 1987), worm burrows (Wright & Henderson, 1992), pebbles (Mosher, 1980), sand volcanoes (Wright & Henderson, 1992), dewatering pipes (Wright & Henderson, 1992), quartz and other detrital grains (Williams, 1972; Bell, 1978; Gray, 1978), and early quartz veins (Wright & Henderson, 1992).

Anastomosing *solution seams* (typically fine-grained concentrations of residual material that is insoluble and/or stable in high-strain folia) are common in slates (Section 6.8), constituting an essential part of slaty cleavage (e.g., Williams, 1972; Hobbs *et al.*, 1976; Vernon, 2004, pp. 390–393), as well as in schists (e.g., Talbot & Hobbs, 1968; Williams, 1990; Vernon, 2004), and may also occur in high-grade metamorphic rocks, for example, sillimanite seams and folia truncating zoned plagioclase (Vernon *et al.*, 1987b) and mica folia truncating zoned garnet (Vernon, 1978a). Sillimanite folia may result from concentration of Al and Si by base-cation leaching in acidic fluids during deformation (Vernon, 1979, 1987b, 1998, 2004; Vernon *et al.*, 1987), as discussed in Section 5.12.6. Another example of solution transfer is the formation of concentrations of mica and/or graphite at high-stress and/or high-strain sites adjacent to porphyroblasts of minerals such as garnet or staurolite, owing to solution and removal of other minerals, especially quartz and feldspar (Vernon, 2004, fig. 5.33).

Slaty and crenulation cleavages show evidence of transfer of chemical components between P (mica-rich, graphite-rich or sillimanite-rich) domains and Q (quartz ± feldspar-rich) domains (see Sections 6.8 and 6.9). Mass transfer in fluid is a major mechanism for the development of foliations, especially slaty cleavage and discrete crenulation cleavage in lower-grade metamorphic rocks (Durney, 1972; Siddans, 1972; Alvarez *et al.*, 1976; Rutter, 1983; Groshong, 1988; Wright & Henderson, 1992). It is generally referred to as 'pressure-solution' (Sorby, 1908) or 'solution-transfer' (Durney, 1972), though preferable terms are 'stress-induced solution transfer' or 'dissolution-precipitation creep', as discussed by Vernon (2004). Chemical components are inferred to be dissolved from sites of high normal compressive stress and deposited at sites of low stress (e.g., Durney, 1972). Differential strain conceivably may also contribute to the process. Solid diffusion is too slow a process to promote this activity, implying that diffusion through static fluid and/or flow (advection) of fluid are required.

Material inferred to be dissolved by pressure- or strain-solution is redeposited at lower-pressure or low-strain sites, respectively, such as veins and Q domains. Relatively clear features indicating deposition from solution in Q domains include: 'beard structures' and fibre growths (Gray & Wright, 1984; Gray & Willman, 1991; Waldron & Sandiford, 1988), quartz overgrowths on detrital quartz grains (Powell, 1969; Williams, 1972; Gregg, 1985; Sutton, 1991), and quartz or carbonate fringes on pyrite crystals. 'Pressure shadow' is a common name for these structures. Evidence of precipitation in some P domains has also been observed (Waldron & Sandiford, 1988; Sutton, 1991). Overgrowths are more clearly seen in the Q domains of slaty and penetrative domainal cleavages, which commonly partly preserve original detrital microstructures (e.g., Williams, 1972). Evidence of precipitation is less clear in most crenulation cleavages, owing to typically more extensive recrystallization/neocrystallization.

(8) Major changes in mineral assemblages in carbonate rocks: Though relatively unimportant compared with fluid processes involved in the hydration of basalts and serpentinites or the dehydration of sedimentary and mafic igneous sequences in subduction zone environments, calcsilicate rocks readily undergo reaction with fluids, producing new mineral assemblages. Therefore, they are sensitive reflectors of the effects of fluids. For example, because metacarbonate rocks are at least partly out of equilibrium with adjacent rocks (metapelites, metapasmmites and meta-igneous rocks), infiltrating hydrous fluids tend to promote reactions, especially at contacts between the carbonate rocks and adjacent rocks (Vidale, 1969; Hewitt, 1973; Vidale & Hewitt, 1973; Joesten, 1974a,b; Thompson, 1975;

Brady, 1977; Yardley *et al.*, 1991a; Ague, 2002, 2003). In addition, certain minerals and assemblages, such as vesuvianite, wollastonite, grossular, and zoisite + quartz, generally occur in calcsilicate rocks where infiltration of H_2O-rich fluid has occurred (e.g., Baumgartner & Ferry, 1991; Cartwright & Oliver, 1992), and so can be used to map fluid pathways (e.g., Cartwright *et al.*, 1995). Another major example is the formation of 'skarns' (Fe-rich calcsilicate rocks), resulting from interaction of calcareous rocks and chemically complex fluids, as discussed in Section 5.12.8.

5.3 Sources of fluids in metamorphic rocks

5.3.1 General comments

Fluids infiltrating deforming regional metamorphic rocks may be derived from devolatilization of hydrous and/or carbonate minerals during prograde metamorphism at greater depth (e.g., Rumble & Hoering, 1986; Selverstone *et al.*, 1991), which is known loosely as 'metamorphic fluid.' For example, dehydration of serpentinite by regional metamorphism in the Alps released fluids responsible for hydration of largely dry rocks (McCaig *et al.*, 1990, 1995; Selverstone *et al.*, 1991; J. D. Barnes *et al.*, 2004). However, magmatic fluids released during crystallization of deep intrusions and/or meteoric fluid may also be present.

Each of these three sources has relatively distinct oxygen and hydrogen isotope characteristics (Sheppard, 1986), which can be revealed by stable isotopic investigations of rocks, hydrous minerals and fluid inclusions. The $^{18}O/^{16}O$ and D/H ($^2H/^1H$) ratios provide two independent signatures of the origin of water (Sheppard, 1986). Ratios are used instead of absolute values, because relative differences are more precise (Rumble, 1982b). For oxygen isotopes, normal seawater oxygen is taken as the standard. The δ notation is used, where:

$$\delta = 1000[(R_{sample}/R_{standard}) - 1],$$

R being the ratio of the less abundant to the more abundant isotope, for example, $^{18}O/^{16}O$ and D/H. For oxygen isotopes, the standard is taken to be 'standard mean ocean water' (SMOW). Thus, the $\delta^{18}O$ value reflects the per mil (per thousand) difference between the $^{18}O/^{16}O$ of the sample and that of SMOW.

Most meteoric waters have $\delta^{18}O = 0$ to $-10‰$ (‰ is the symbol for 'per mil' or 'parts per thousand') and $\delta D = +10$ to $-70‰$ (Sheppard, 1986, p. 168). In contrast, most I-type granites have $\delta^{18}O = +5.5‰$ to $+10$ and $\delta D = -50‰$ to $-90‰$; waters in equilibrium with such magmas are estimated to have $\delta^{18}O = +5.5‰$ to $+9.5‰$ and $\delta D = -40‰$ to $-80‰$ (Sheppard, 1986, p. 174). Metasedimentary rocks typically have $\delta^{18}O = +8$ to $+26$ and $\delta D = -40‰$ to $-100‰$, and metamorphosed oceanic mafic rocks typically have $\delta^{18}O = +3‰$ to $+14‰$ and $\delta D = -35‰$ to $-70‰$, which, after applying H and O isotope fractionation factors between minerals and fluids, suggests a calculated isotopic range for metamorphic waters of $\delta^{18}O = +3‰$ to $+20‰$ and $\delta D = 0$ to $-70‰$ (Sheppard, 1986, p. 173). For example, the $\delta^{18}O$ of vein quartz in high-grade metasedimentary rocks in New Hampshire, USA, is $+13.5‰$ to $+14.5‰$, which reflects fluid compositions that are typical of regional metamorphism (Chamberlain & Rumble, 1988). However, the isotopic composition of metamorphic fluids varies from one area to another, depending on rock types and their history of

Table 5.1. | Stable isotope ranges

	$\delta^{18}O$	δD
meteoric waters	0 to −10	+10 to −70
igneous waters	+5.5 to +9.5	−40 to −80
metamorphic waters	+3 to +20	0 to −70

rock-fluid interaction. The foregoing generalizations are summarized in Table 5.1.

The simplified normal ranges of $\delta^{18}O$ for some common rock types are shown in Fig. 5.1. Igneous rocks have the lowest and narrowest ranges of $\delta^{18}O$, and sedimentary rocks have the highest values. Metamorphic rocks have $\delta^{18}O$ values that overlap those of igneous and sedimentary rocks. Sedimentary rocks exchange ^{18}O for ^{16}O during metamorphism (Shieh, 1974).

In view of some overlaps in δD and $\delta^{18}O$ ranges, specific fluid sources in a particular region may be difficult or impossible to determine. For example, meteoric and metamorphic waters may have similar O and especially H isotopic compositions to those of magmatic waters, though a magmatic source can be ruled out for terranes without igneous intrusions (Jenkin *et al.*, 1994; Clark *et al.*, 2006). Moreover, the isotopic composition of a meteoric fluid may be modified by reaction with wall rocks or mixing with metamorphic fluid (Jenkin *et al.*, 1994). Therefore, though it may be possible to infer a largely external source for the fluid in a particular terrane, it may not be possible to identify the specific source (Sheppard, 1986, p. 178).

The $\delta^{18}O$ of rocks and minerals decreases and the isotopic composition becomes more homogeneous with increasing metamorphic grade, approaching igneous values in metasedimentary rocks at the highest grades (Garlick & Epstein, 1967; Shieh & Taylor, 1969; Dontsova, 1970; Shieh, 1974; Rye *et al.*, 1976; Fleck & Criss, 1985; Wickham & Taylor, 1985, 1987; Chamberlain & Rumble, 1988; Buick *et al.*, 2000). For example, in New Hampshire, USA, whole-rock $\delta^{18}O$ is 7‰ in chlorite zone rocks and 3‰ in sillimanite zone rocks (Chamberlain & Rumble, 1988). This is due to a tendency towards oxygen isotope exchange equilibrium between different

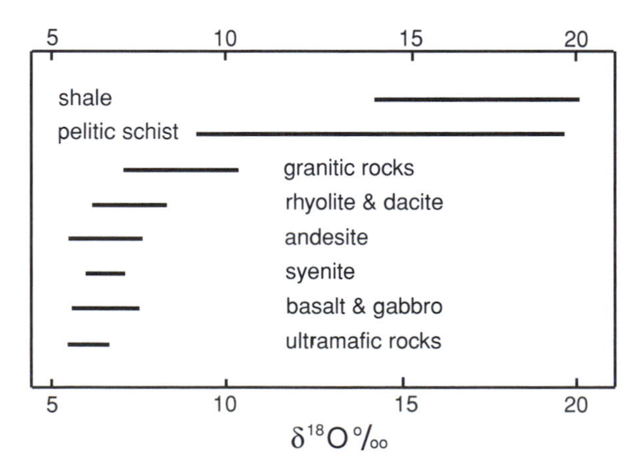

Fig. 5.1 Normal $\delta^{18}O$ ranges for some common rock types, after Shieh (1974, fig. 1).

rock-types, involving reaction with a pervasive, oxygen-rich fluid. However, steep gradients in $\delta^{18}O$ (as much as 10 per mil over a few centimetres) at the margins of metacarbonate layers may occur in dry, high-grade (amphibolite-granulite facies) terranes with very small amounts of pore fluid (Wickham & Taylor, 1987). In contact metamorphic aureoles, the oxygen isotope ratios are typically constant, except for samples from a very narrow exchanged zone adjacent to the actual contact (Sheih, 1974) or to fracture zones transporting magmatic fluid (Buick & Cartwright, 2000).

5.3.2 Meteoric fluid

Various studies have presented isotopic evidence (for example, markedly low $\delta^{18}O$ values) of surface waters penetrating upper crustal and mid-crustal, dehydrated rocks, especially along shear zones formed well after peak prograde metamorphism (Sheppard & Taylor, 1974; Taylor, 1977; Wickham & Taylor, 1985, 1987; McCaig, 1988; McCaig *et al.*, 1990, 1995; Koons & Craw, 1991; Fricke *et al.*, 1992; Jenkin *et al.*, 1992, 1994; Morrison, 1994; Wickham *et al.*, 1994; Miller & Cartwright, 1997; Cartwright & Buick, 1999; Doyle & Cartwright, 2000; Read & Cartwright, 2000; Yardley *et al.*, 2000; Clark *et al.*, 2006).

However, fluids percolating down to mid-crustal depths must overcome low permeability, fluid buoyancy and gradients of increasing pressure (from hydrostatic pressure in the shallow crust to lithostatic pressure in the middle crust). The fluid would also be required to travel in the direction of increasing temperature, which has been proposed for some areas (Lobato *et al.*, 1983). Mechanisms proposed for deep penetration of meteoric fluid involve transport of fluid in active fault or fracture zones, especially shear zones. Such high-strain deformation zones permit much larger flow rates than in the adjacent rocks, owing to (a) localized fluid channelling and (b) high permeability, owing to opening of transient pores and microfractures caused by local strain heterogeneities. The proposed mechanisms include: (1) fluid flow driven by topography, the fluid entering active thrusts at depth in rocks experiencing elevated temperatures caused by rapid uplift along shear-zone thrusts (Bottrell *et al.*, 1990; Koons & Craw, 1991; Jenkin *et al.*, 1992, 1994; Barker *et al.*, 2000), (2) magmatically driven convection systems (McCaig *et al.*, 1990; Cartwright & Buick, 1999), (3) seismic pumping, by which dilatant openings in large shear zones lower the fluid pressure sufficiently to suck down upper crustal fluids (Sibson, 1981; McCaig, 1988; McCaig *et al.*, 1990; Cartwright & Buick, 1999; Yardley *et al.*, 2000), (4) thrusting of metamorphosed over non-metamorphosed rocks, providing meteoric fluid by dehydration of hydrous minerals in the underlying rocks (Lobato *et al.*, 1983), (5) burial of blocks of rock with former fault zones containing hydrous minerals produced by the previous action of meteoric fluid, providing a source of isotopically light water at depth (Clark *et al.*, 2006), and (6) penetration of fluid down fractures formed by extension in the upper plate of a metamorphic core complex, the increase in permeability and porosity consequent on extensive faulting being capable of transferring large amounts of fluid (Taylor, 1988; Morrison, 1994). Yardley *et al.* (2000) proposed that retrograde alteration of many basement rocks is due to pumping down of fluid along fractures from overlying sedimentary basins, the driving force being the large gradient in the chemical potential of water and other fluid components between water-saturated rocks and largely anhydrous rocks at depth that are unstable in the presence of hydrous fluid.

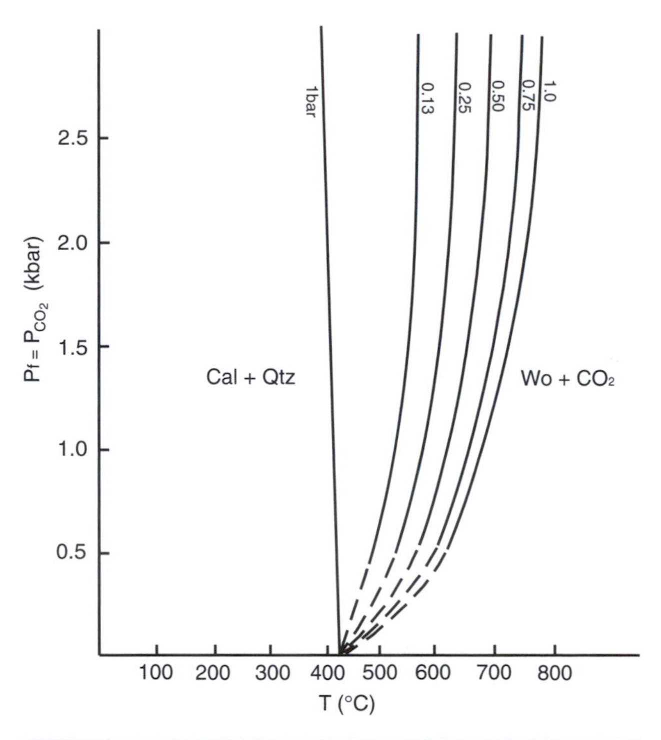

Fig. 5.2 Experimentally determined univariant equilibrium curves for the reaction: Cal + Qtz = Wo + CO_2 plotted on a P_{fluid} ($= P_{CO_2}$) − T diagram, showing variations in the positions of the curves with varying composition of the fluid, expressed as mole fraction (X) of CO_2. After Winkler (1967).

5.3.3 Interaction with seawater

Many metamorphic rocks show chemical or isotopic evidence of interaction with seawater (e.g., Humphris & Thompson, 1978; Ito & Anderson, 1983; Mottl, 1983; Wickham & Taylor, 1985; Schiffman & Smith, 1988). Some areas preserve isotopic evidence of such interaction overprinted by later metamorphism. For example, in high-grade regional metamorphic rocks in the Broken Hill area, Australia, $\delta^{18}O$ values of up to 16‰ regularly decrease to as low as 7‰ in rocks close to the main orebody (Cartwright, 1999). Because these low values are similar to those in volcanogenic massive sulphide deposits, they probably represent hydrothermal fluid circulation through rocks at or near the seafloor, well before the main metamorphism (Cartwright, 1999).

5.4 Effect of fluid pressure on reaction curves

As noted previously, the fluid produced by devapourization reactions must be removed from the reaction site for the reaction to proceed (e.g., Norris & Henley, 1976). If the diffusion rate of the volatile component (e.g., H_2O or CO_2) is low, the reaction is constrained to occur on the univariant curve in a *P-T* grid with $P_{fluid} = P_{solids}$ (e.g., Fig. 5.2), whereas if the fluid escapes readily from the reaction site, the reaction proceeds at the lowest temperature, in an attempt to produce more fluid.

5.5 Effect of fluid composition on mixed volatile reactions

Because the compositions of minerals in carbonate rocks are relatively simple, *P-T* grids might be expected to realistically portray phase relationships, in contrast to the necessity for *P-T* pseudosections for metapelitic and mafic compositions involving minerals with extensive solid solution and consequent continuous reactions (Chapters 1 and 2). However, metamorphism of carbonate rocks commonly involves a fluid containing the two principal components, H_2O and CO_2, which introduces a complicating factor.

The first attempt to represent mineral equilibria in siliceous limestone and dolomite compositions was the *P-T* ('petrogenetic') grid of Bowen (1940). However, anomalies in the order of appearance of minerals predicted by the grid soon became apparent, as outlined by Ferry (1994a). For example, though field-based investigations commonly indicate that the lowest-grade reaction in this system is between dolomite and quartz to produce tremolite and calcite, some areas show evidence that talc and calcite are the products (Tilley, 1948), and other areas indicate that diopside is the product (Engel & Engel, 1960). In addition, periclase forms at lower temperatures than wollastonite in some areas, but not in others. The dilemma was resolved by the recognition of the importance of fluid composition, giving rise to the *P-T-X* 'petrogenetic model' and the routine use of *T-X* sections, where *X* refers to the composition of the fluid in terms of the mole fraction of H_2O or CO_2 (Greenwood, 1961, 1962, 1967). Metz & Trommsdorff (1968) produced a *T-X* grid for metamorphosed siliceous dolomites, showing that if the metamorphic fluid is rich in H_2O, dolomite and quartz react to form talc and calcite; if the metamorphic fluid is very rich in CO_2, dolomite and quartz react to form diopside; and if the metamorphic fluid is compositionally intermediate (i.e., has a 'mixed volatile' composition), dolomite and quartz react to form tremolite and calcite. Similarly, periclase forms at lower temperatures than wollastonite in the presence of fluid very rich in H_2O, and the reverse is true if the fluid is rich in CO_2.

If the fluid contains both H_2O and CO_2 (i.e., has a 'mixed volatile' composition), a change in the proportion of one of these components (for example, by addition of the other component) either increases or decreases the reaction rate, in an attempt to restore the equilibrium fluid composition. For example, the rate of a biotite-forming reaction producing CO_2 in layered rocks with varying carbonate-mica proportions in New England, USA, was increased locally by infiltration of H_2O-rich fluids (Hewitt, 1973; Baumgartner & Ferry, 1991). The H_2O altered the fluid composition (that is, it raised X_{H_2O} and lowered X_{CO_2}), with the result that more biotite was produced, in order to make more CO_2. Many detailed studies have produced evidence of layer-parallel fluid flow in these rocks, coupled with across-layer infiltration and/or diffusion, which is responsible for different degrees of reaction progress in different layers (Leger & Ferry, 1993; Evans & Bickle, 1999, 2005; Ague, 2002, 2003). Dilution of H_2O by CO_2 causes *P-T* curves for dehydration reactions to move to lower *T* (at fixed *P*) even if CO_2 takes no part in the reaction (Fig. 5.3). The same effect is caused by the addition of inert components (e.g., argon) to water, which changes

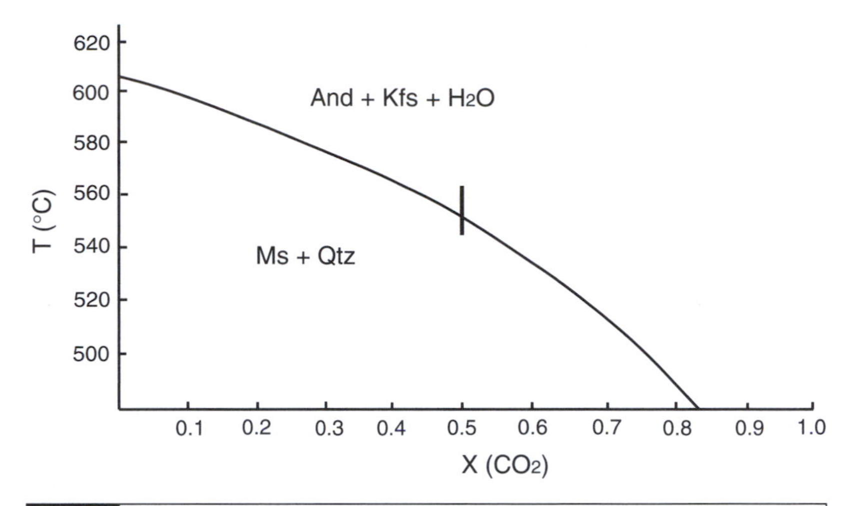

Fig. 5.3 T-X_{CO_2} plot of a calculated curve of the reaction: Ms + Qtz = Kfs + And + H_2O at P_{fluid} = 2 kbar, with an experimentally determined equilibrium bracket. Though CO_2 does not take part in the reaction, if it dilutes the fluid phase the reaction temperature decreases. After Kerrick (1972).

the position of the analcite dehydration curve, as shown in Fig. 5.4 (Greenwood, 1961).

The contrasting effects of mobile and immobile product fluid are shown for the reaction Cal + Qtz = Wo + CO_2 in Fig. 5.5. If the product CO_2 is mobile, its ready removal stimulates the reaction to proceed, in order to produce more fluid and attempt to restore local equilibrium, so that the reaction proceeds at constant X_{CO_2} with increasing T. On the other hand, if the CO_2 is immobile, local increase in P_{CO_2} forces the reaction to proceed at progressively higher X_{CO_2} until one of the reactants is used up.

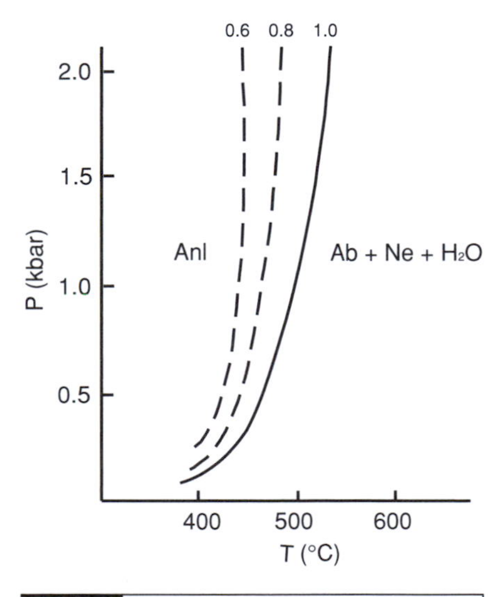

Fig. 5.4 Variable positions of the P-T curve for the reaction: Anl = Ab + Ne + H_2O for different mole fractions of H_2O in a fluid composed of water and argon. After Greenwood (1961).

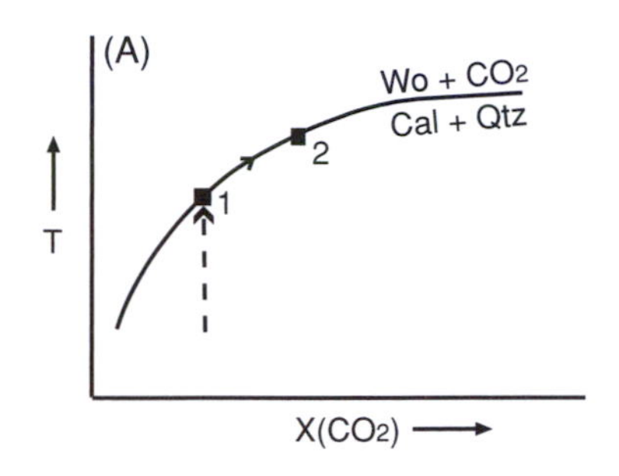

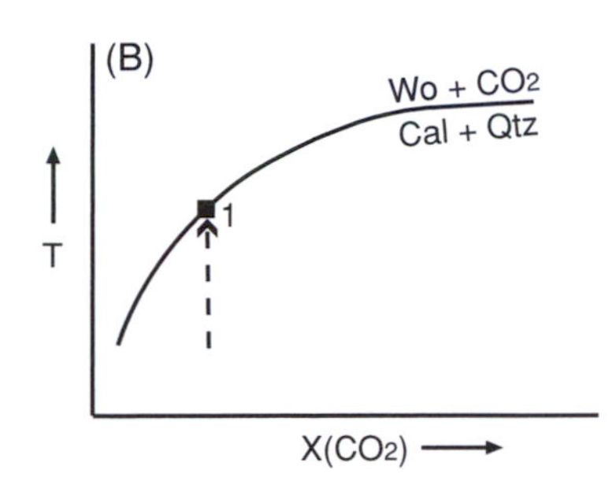

Fig. 5.5 Diagrammatic T-X_{CO_2} plot of the reaction: Cal + Qtz = Wo + CO_2, showing two extreme situations, namely (A) *CO_2 immobile*, so that as calcite and quartz begin to react at point 1, CO_2 is released and builds up local pressure, with the result that the reaction is forced to proceed with increasing T and X_{CO_2}, until one of the reactants is exhausted at point 2; and (B) *CO_2 mobile*, so that the reaction proceeds at a fixed T (point 1), which is lower than for (A). Compare with Fig. 5.12.

The effect of dilution of H_2O by CO_2 and vice versa can also be shown in a P-T-X_{CO_2} diagram (Fig. 5.6). The P-T curves for $P_{H_2O} = P_{total}$ and $P_{CO_2} = P_{total}$ for two simple devolatilization reactions are shown on the top and bottom faces of the three-dimensional block, each curve sloping down to lower temperatures with progressive dilution of the fluid by the other component. Examples of T-X_{CO_2} equilibrium reaction plots are shown in Fig. 5.7;

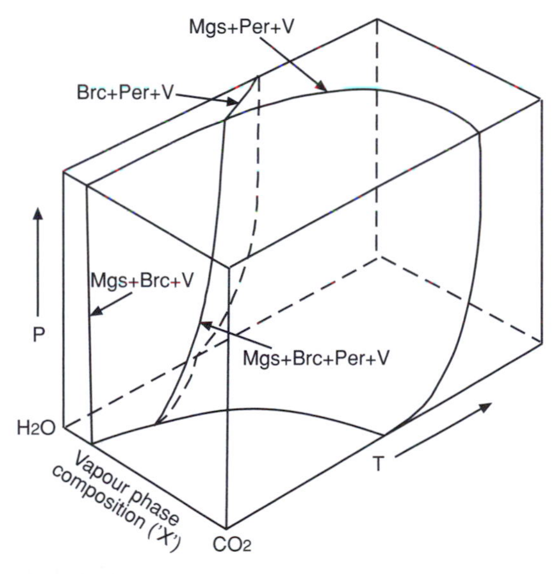

Fig. 5.6 Three-dimensional P-T-X diagram, after Wyllie (1962), of the system MgO-CO_2-H_2O, showing plots of three divariant (3-phase) surfaces meeting in a univariant (4 phase) reaction line, and dividing the diagram into three trivariant (2-phase) volumes. The divariant surfaces are: MgO + V (top-right), $MgCO_3$ + V (front) and $Mg(OH)_2$ + V (rear). The P-T curve shown on the front face represents the reaction: $MgCO_3 = MgO + CO_2$ at $P_{CO_2} = P_{total}$ and the P-T curve shown on the back face represents the reaction: $Mg(OH)_2 = MgO + H_2O$ at $P_{H_2O} = P_{total}$.

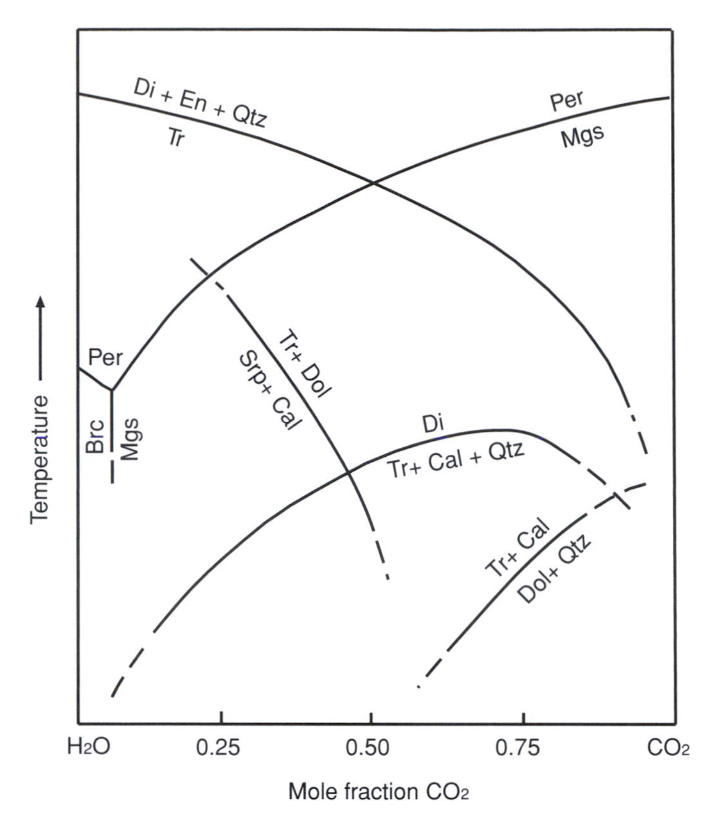

Fig. 5.7 Diagrammatic T-X_{CO_2} plot of curves of the following reactions, after Greenwood (1962).

$Tr = 2Di + 3En + Qtz + H_2O$

$Mgs = Per + CO_2$

$Tr + 3Cal + 2Qtz = 5Di + 3CO_2 + H_2O$ (maximum T at $X_{CO_2} = 0.75$, owing to the 3:1 ratio of CO_2 to H_2O)

$Brc + CO_2 = Mgs + H_2O$

$5Dol + 8Qtz + H_2O = Tr + 3Cal + 7CO_2$

$4Srp + 9\ Cal + 5CO_2 = Tr + 7Dol + 7H_2O$

some of the curves involve the production of single-component fluids, others involve the production of two-component fluids, and still others have different fluids on each side of the equation.

T-X_{CO_2} grids may be expanded, by thermodynamic modelling, into T-X_{CO_2} *pseudosections* for more complex systems, (e.g., mafic rocks with carbonate minerals), which illustrate the effect of changing the activity of a fluid component on mineral assemblages for a particular whole-rock composition (Elmer *et al.*, 2006), as shown in Fig. 5.8.

In view of the foregoing discussion, isograds based on thermal devolatilization equilibria alone in terranes with calcareous rocks should be treated with some caution, unless the composition of the fluid is known. A consequence of variations in fluid composition in a metamorphic terrane is illustrated by the Whetstone Lake area, Ontario, Canada (Carmichael, 1970). This area has a normal sequence of isograds based on prograde dehydration reactions, intersected by another isograd involving mixed volatiles (Fig. 5.9). The position of this intersecting isograd depends on the proportion of CO_2 and H_2O in the fluid, and is displaced to lower

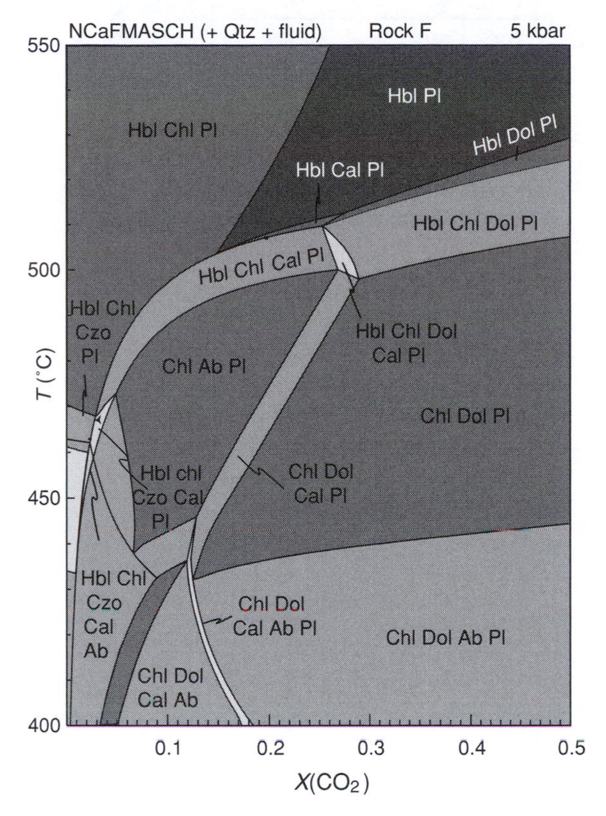

Fig. 5.8 T-X_{CO_2} pseudosection for a particular carbonate-bearing mafic whole-rock composition in the system NCaFMASCH (+ quartz and fluid), from Elmer *et al.* (2006, fig. 8), with permission of the *Journal of Metamorphic Geology*.

temperatures close to a granite intrusion, which may have supplied water. For this reason, the sequence of isograds intersected by traverses across the map varies (Figs. 5.9, 5.10). The CO_2/H_2O ratio of metamorphic fluids can vary in space and time in a metamorphic terrane, and field studies have shown that the variation commonly is on scales as local as individual beds or layers (Butler, 1969; Hewitt, 1973; Rumble, 1978).

The fluid in graphite-bearing metapelitic rocks may contain CH_4, CO_2, H_2O, H_2 and N_2, so that the concentrations of all these components may need to be considered for a full understanding of reactions involving fluids of this kind.

5.6 Effect of fluid composition on 'ionic' reactions

Though many inferred metamorphic reactions are concerned with the *thermal stability* of minerals, ionic components dissolved in fluid may contribute to the *ionic stability* of minerals (Carmichael 1969; Eugster, 1970). For example, the thermal stability of muscovite plus quartz (producing the gas, steam) is expressed by the following equation.

$$\underset{\text{Ms}}{KAl_3Si_3O_{10}(OH)_2} + \underset{\text{Qtz}}{SiO_2} = \underset{\text{Kfs}}{KAlSi_3O_8} + \underset{\text{Sil}}{Al_2SiO_5} + H_2O \qquad (5.1)$$

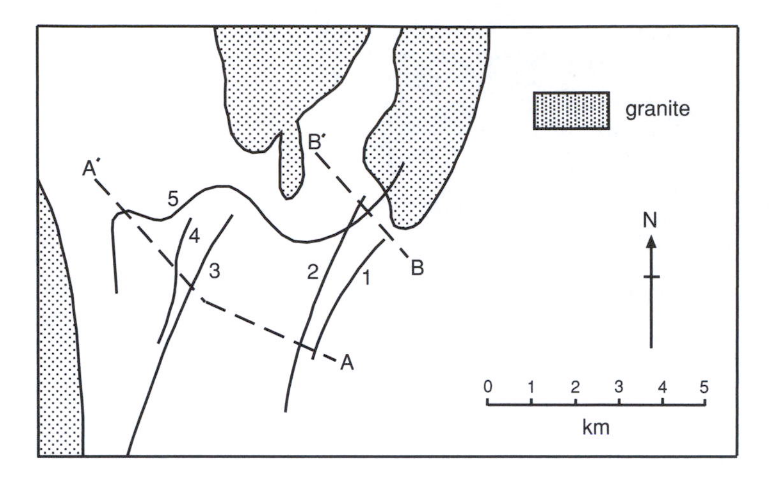

Fig. 5.9 Map of isograds in the Whetstone Lake area, Ontario, Canada (Carmichael, 1970, fig. 22, with permission of Oxford University Press), reflecting the inferred reactions:

Chl + Ms + Grt = St + Bt + Qtz + H_2O
Chl + Ms + St + Qtz = Ky + Bt + H_2O
Ky = Sil
St + Ms + Qtz = Sil + Grt + Bt + H_2O
Bt + Cal + Qtz = calcic Am + Kfs + CO_2 + H_2O

The ionic stability of muscovite plus quartz is expressed by the following equation.

$$KAl_3Si_3O_{10}(OH)_2 + 6SiO_2 + 2K^+ = 3KAlSi_3O_8 + Al_2SiO_5 + 2H^+ \qquad (5.2)$$

Combining the equilibrium constants for both these reactions, Eugster (1970) derived equilibrium constants for the following two reactions

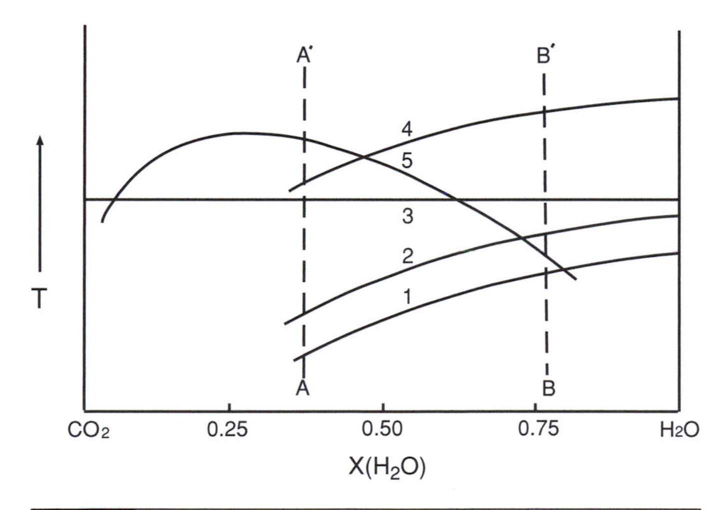

Fig. 5.10 T-X_{CO_2} plot at a constant P_{total} ($= P_{H_2O} + P_{CO_2}$), showing schematic curves for reactions (1) to (5) of Fig. 5.9. Traverses AA′ and BB′ correspond to those shown in Fig. 5.9. Because no fluid is involved in reaction 3, the curve is unaffected by the fluid composition. After Carmichael (1970, fig. 22), with permission of Oxford University Press.

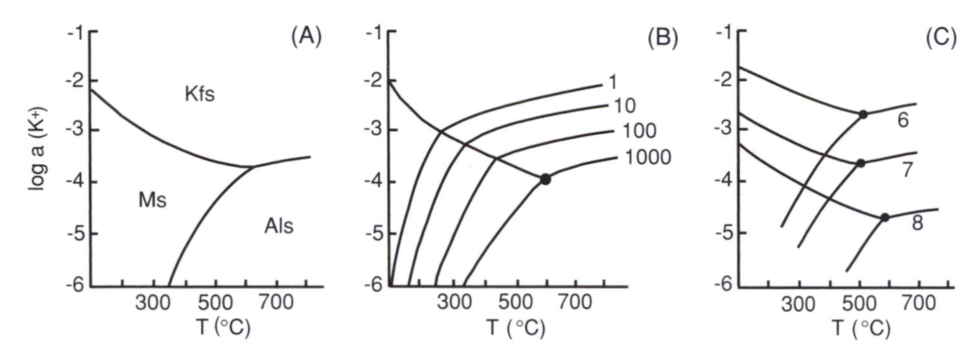

Fig. 5.11 (A) T-a_{K^+} diagram at constant pH and f_{H_2O} = 1 kbar, with quartz and fluid present in excess (Eugster, 1970). Reactions (2), (3) and (4) (see text) intersect at a 5-phase point. (B) Same diagram, showing changes in the positions of curves (3) and (4) with changing f_{H_2O}. (C) Same diagram, showing how the positions of the curves change markedly with changing pH.

involving both 'stable gas' (steam) and ionic components.

$$2KAl_3Si_3O_{10}(OH)_2 + 2H^+ = 3Al_2SiO_5 + 3SiO_2 + 2K^+ + 3H_2O \quad (5.3)$$
$$2KAlSi_3O_8 + 2H^+ = Al_2SiO_5 + 5SiO_2 + 2K^+ + H_2O \quad (5.4)$$

The equilibrium constants (log K) of both these reactions are given by: log $K = (a^2_{K^+} \times f_{H_2O})/(a^2_{H^+})$, where a = chemical activity of ionic components (closely related to concentration) and f = fugacity (closely related to partial pressure) of steam. This provides enough information to evaluate phase relationships between the solid minerals (muscovite, K-feldspar and sillimanite) and an H_2O fluid plus the ions K^+ and H^+. A diagram illustrating the variation of T and a_{K^+} (Fig. 5.11) can be constructed by fixing P, pH and f_{H_2O}. The diagram shows the positions of equations (5.2), (5.3) and (5.4). K-feldspar is stable at high a_{K^+} for all temperatures, sillimanite is stable a low a_{K^+} and high T, and muscovite is stable at low a_{K^+} and low T. If a_{K^+} in solution is reduced (e.g., by metasomatic introduction of K-poor solutions or by crystallization of another nearby K-rich mineral), K-feldspar breaks down to form muscovite by reaction (2). An identical result is achieved by increasing pH (e.g., by increasing H^+ or decreasing $(OH)^-$ of the fluid, perhaps by crystallization of a nearby hydrous silicate mineral), which expands the muscovite stability field at the expense of K-feldspar. The competing processes may occur simultaneously. Similarly, replacement of an Al_2SiO_5 mineral by muscovite, expressed by equation (5.4), is favoured by decreasing T and increasing f_{H_2O}, pH or a_{K^+}. Therefore, it cannot be assumed that replacement of K-feldspar or Al_2SiO_5 (sillimanite, andalusite or kyanite) by muscovite in quartz-bearing rocks represents a simple reversal of prograde reaction (1).

5.7 Infiltration of fluid or not?

5.7.1 General comments

Important questions are: do metamorphic mineral assemblages control the composition of coexisting fluid (*internal buffering*), do external reservoirs of fluid control the mineral assemblages with which they are in communication, or can both these mechanisms occur to varying degrees?

Internal control (buffering) of fluid composition: Variations in buffer mineral assemblages between different layers or beds can control the oxygen activity ('fugacity') of coexisting fluid during metamorphism (Eugster, 1959). For example, Chinner (1960) found that more oxidized beds (characterized by hematite and magnetite) are enriched in muscovite and depleted in biotite and garnet in the Barrovian metapelitic schists of Glen Clova, Angus, Scotland. The differences in oxidation state of adjacent beds were inherited from a pre-metamorphic condition, and so the oxygen fugacity of the very small amount of fluid present was determined (buffered) by the minerals.

Similarly, different reactions involving hydrous silicate minerals and carbonate minerals can buffer the composition of coexisting H_2O-CO_2 fluids at different local values during metamorphism (Greenwood, 1967). The porosity in calcsilicate rocks is generally so small (in the absence of deformation) that even if introduced H_2O is present, devolatilization reactions produce CO_2 so fast that it displaces the H_2O, and the reaction is effectively constrained to occur at $X_{CO_2} = 1$ (Greenwood, 1975). This may explain why many regional and especially contact metamorphic areas of calcsilicate rocks are regularly zoned over large areas, under the dominant control of P and T (e.g., Kennedy, 1949; Trommsdorff, 1966; Moore & Kerrick, 1976; Rice, 1977; Rice & Ferry, 1982; Ferry, 1991). An example is the Notch Peak contact aureole, Utah, USA, in which unchanged oxygen and carbon isotope ratios in calcite and silicate minerals at all metamorphic grades, except above the wollastonite isograd (which was infiltrated by magmatic fluids), indicate a lack of infiltration of disequilibrium fluids (Nabelek, 2002). Prograde metamorphic reactions may buffer the X_{CO_2} of metamorphic fluid in carbonate rocks, even where infiltration of hydrous external hydrous fluid occurs (Ferry, 1983a).

External control of fluid composition: Nonetheless, decarbonation reactions promoted by hydrous fluid do occur (e.g., Hewitt, 1973; Ferry, 1994a,b), showing that such rocks are infiltrated by aqueous fluid during metamorphism (Rice & Ferry, 1982). Thus, a rock can record evidence of equilibrium with H_2O-rich fluid while simultaneously undergoing a decarbonation reaction if it is continually infiltrated by H_2O-rich fluid, which dilutes the CO_2 produced by the reaction, and transports the CO_2 away from the reaction site, thereby allowing the reaction to proceed. A complete spectrum exists between the extremes of internal buffering of fluid composition without infiltration, through a combination of buffering and infiltration, to infiltration without buffering (Rice & Ferry, 1982). However, internal buffering is by far the most common situation.

Some calcsilicate minerals and assemblages (e.g., vesuvianite, wollastonite, grossular and zoisite + quartz) are consistent with addition of H_2O (e.g., Ferry, 1983a; Baumgartner & Ferry, 1991; Cartwright & Oliver, 1992). However, this situation is relatively uncommon, because if all the CO_2 were flushed out by incoming H_2O, minerals such as wollastonite (Fig. 5.2) and diopside would form consistently at lower temperatures than usually inferred (Winkler, 1967).

The effect of introduced water dominates especially in zones of high permeability, such as deformation zones. For example, the growth of wollastonite may be promoted locally in lower pressure regional metamorphic rocks infiltrated with abundant H_2O during deformation (Buick & Cartwright, 1996).

5.7.2 Fluids in regional metamorphism

Some regional metamorphic terranes show evidence of extensive fluid flow in zones of high permeability, such as shear zones and fracture/breccia zones (e.g., Oliver *et al.*, 1990; Selverstone *et al.*, 1991; Cartwright & Buick, 1999). However, where vein systems are not present, evidence of fluid infiltration in many regional metamorphic areas is more difficult to detect than for contact metamorphism. This has led to the following opposing driving mechanisms for regional metamorphism (Barnett & Chamberlain, 1991): (1) thermal perturbations, which are supported by regular broad-scale metamorphic grade changes and clear relationships between *P-T-t* paths and regional deformation events, and (2) infiltration of chemically reactive fluids that drive devolatilization reactions, which is supported by some petrological and isotopic studies that indicate fluid out of chemical equilibrium with the mineral assemblage. For example, several studies indicate large amounts of fluid flow during prograde metamorphism (Rye *et al.*, 1976; Rumble *et al.*, 1982; Ferry, 1984, 1987; Fleck & Criss, 1985; Wickham & Taylor, 1985; Hoisch, 1987; Oliver & Wall, 1987; Chamberlain & Rumble, 1988; Oliver *et al.*, 1990; Selverstone *et al.*, 1991).

In an attempt to test these two mechanisms, Barnett & Chamberlain (1991) found that the amount of hydrous fluid necessary for fluid-present metamorphism in the calcsilicate Waits River Formation, New England, USA, could be derived from dehydration of the immediately adjacent amphibolites and metapelitic schists, and consequently that differences in metamorphic grade result from thermal perturbations on a scale of tens of square kilometers. However, they also found that grade differences within individual folds are due to differences in the amount of fluid infiltration on a scale of tens or hundreds of metres.

Some prograde regional metamorphic areas show little or no evidence of fluid-rock interaction during metamorphism from low (greenschist facies) to high (granulite facies) grades. For example, metapelites and some marbles of the Reynolds Range, central Australia show little evidence of fluid-rock interaction during prograde metamorphism (M2), though they do show variable oxygen isotope ratios, owing to channel flow of magmatic and/or meteoric fluids prior to M2 (Buick & Cartwright, 1996). However, other marbles in this area have metre-scale wollastonite-bearing layers formed by layer-parallel, channelled infiltration of water-rich ($X_{CO_2} = 0.1$–0.3) fluid from local pegmatites at *c.* 700°C; adjacent wollastonite-free marls with quartz and calcite escaped fluid infiltration (Cartwright & Buick, 1995).

On the other hand, the southern margin of the Pilbara Craton, Western Australia, underwent Palaeoproterozoic regional heating, folding, thrusting and extensive fluid flow, indicated by widespread occurrence of the phosphate minerals, monazite and xenotime. The low-grade metamorphic front migrated over 350 km at about 5 mm/year, as indicated by in situ U-Pb geochronology on authigenic monazite and xenotime (Rasmussen *et al.*, 2003). This study gives an idea of the rate and duration of metamorphism and deformation, with concomitant fluid flow, in some orogenic settings (Rasmussen *et al.*, 2003).

Evidence for fluid infiltration may be ambiguous in some regional metamorphic terranes. For example, oxygen isotopic evidence indicates fluid infiltration for the Waterville limestone, Maine, USA, but whether this occurred before metamorphism (i.e., during diagenesis) or during metamorphism is difficult to determine (Rumble *et al.*, 1991). The presence of four generations of veins in this area confirms that some fluid flow occurred, and indicates that fracture fluid flow dominated over grain boundary fluid flow (Rumble *et al.*, 1991).

The foregoing discussion suggests that, though thermal perturbations probably control most prograde regional metamorphism and that fluid-rock interaction may be relatively minor in many areas, major fluid effects may occur, both regionally and on a relatively local scale, especially in and near fracture and shear zones (Oliver *et al.*, 1990; Cartwright & Buick, 1999). On the other hand, large volumes of veins may be produced by fracturing and fluid migration in individual rock units, without involving large quantities of externally derived fluids (Cartwright *et al.*, 1994).

Many studies of high-grade migmatite terranes suggest that they contain effectively no hydrous fluid, because most partial melting produces H_2O-undersaturated melts (Chapters 2 and 4), with the result that any introduced free water is dissolved in the melt; that is, the melt buffers the water content of the rocks and keeps them effectively dry (White & Powell, 2002).

5.7.3 Fluids in contact metamorphism

Though contact metamorphic aureoles are typically metamorphically zoned, suggesting a dominant control by thermal perturbation, evidence of fluid activity is common around high-level granitic plutons. Isotopic studies are consistent with hydrothermal convection (Criss & Taylor, 1986; Ferry, 1991), and contact metamorphic hydrothermal and 'skarn' (metasomatic calcsilicate) ore deposits (Section 5.12.8) provide evidence of intense local reaction between contact aureole rocks and magmatically derived and/or enhanced fluids.

Calcsilicate rocks tend to be more permeable than carbonate-rich marbles (Ferry, 1991, p. 388). Most devolatilization reactions, especially decarbonation reactions (such as the reaction of quartz and calcite to form wollastonite), involve decreases in the volumes of the solid minerals, forming a potential *metamorphic porosity*, which can contribute to the permeability of these rocks (Rumble *et al.*, 1982; Etheridge *et al.*, 1983, p. 213).

Evidence of hydrous fluid flow is most clearly shown by calcsilicate rocks, which undergo decarbonation reactions that can be affected by addition of H_2O. In contrast, metapelitic and mafic rocks undergo dehydration reactions, the effects of which obscure any effects of introduced H_2O. Though some calcsilicate rocks show evidence of external fluid dominating the effect of temperature (causing local variations in mineral assemblages), most contact metamorphic calcsilicate rocks show regular variation in minerals assemblages, enabling mapping of metamorphic zones that are regularly distributed with respect to the pluton contact. This suggests that the fluid composition is controlled by the mineral assemblages (i.e., is internally buffered), rather than being controlled by an external fluid reservoir, and that the mineral assemblages are controlled by P and T (Yardley, 1989). Though pore fluids react with minerals in contact metamorphic calcsilicate rocks (e.g., Ferry, 1983a,b), local diffusional exchange between compositional layers, rather than fluid infiltration, may apply, especially to thinly layered rocks (Yardley, 1989).

As discussed by Ferry (1991), sequences of mineral assemblages can enable distinction between the controls of externally buffered infiltrating

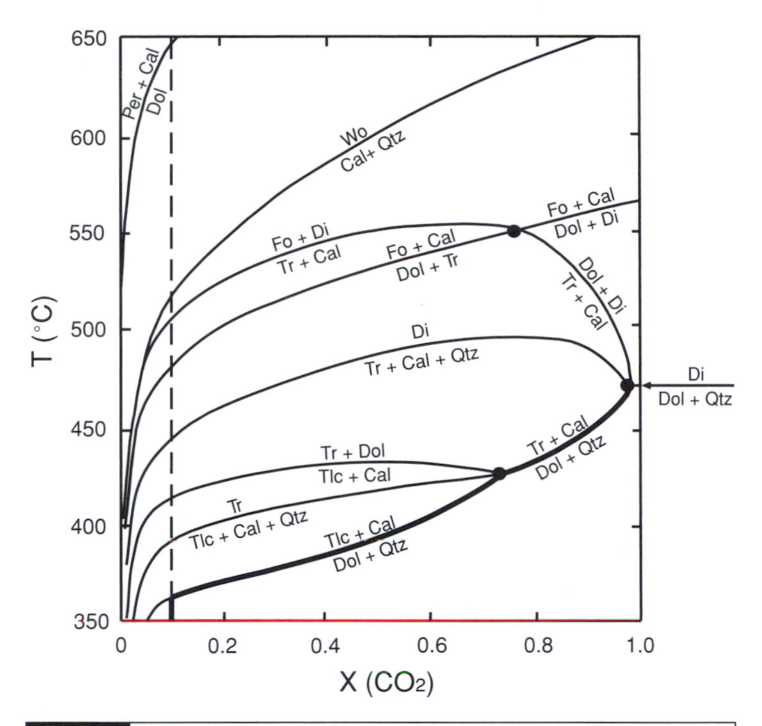

Fig. 5.12 Isobaric (constant-P) T-X_{CO_2} plot (Ferry, 1991, fig. 18, with permission of the Mineralogical Society of America) of some calculated equilibria at 1 kbar in the system CaO-MgO-SiO_2-CO_2-H_2O, representing a model siliceous dolomite composition. Invariant points are shown by large dots. The thick line shows the T-X_{CO_2} evolution of a siliceous dolomite with near-zero porosity and an H_2O-rich fluid of composition $X_{CO_2} = 0.1$ as it is heated from 350°C to 650°C without infiltration of an external fluid (i.e., with fluid composition buffered internally). The vertical dashed line shows the T-X_{CO_2} evolution of the same rock heated from 350°C to 650°C while being infiltrated by a fluid of composition $X_{CO_2} = 0.1$, which enables the reaction to proceed at a constant X_{CO_2} with increasing temperature. Infiltration of an external fluid with low, constant X_{CO_2} has the same effect as the 'CO_2 mobile' situation of Fig. 5.5, because the infiltrating fluid continually flushes out the CO_2 and so forces the reaction to attempt to make more CO_2. Similarly, reaction without fluid infiltration has the same effect as the 'CO_2 immobile' situation of Fig. 5.5, because local buildup of CO_2 forces the reaction to proceed at progressively higher X_{CO_2}.

fluid and internally buffered fluid. As shown in Fig. 5.12, at constant P, the evolution of a siliceous dolomite with near-zero porosity and a fluid of composition $X_{CO_2} = 0.1$ follows an internally buffered path along univariant lines to invariant points, whereas infiltration by a CO_2-H_2O fluid of the same composition ($X_{CO_2} = 0.1$) follows a quite different path across a sequence of alternating univariant assemblages (stable at a single T) and divariant assemblages (stable over a T interval). Therefore, a sequence of univariant assemblages alternating with invariant assemblages is consistent with zero fluid flux, and a sequence of univariant assemblages alternating with divariant assemblages without any invariant assemblages, is consistent with fluid infiltration (Fig. 5.12). However, though divariant calc-silicate assemblages are generally taken to suggest fluid flow, some contact aureoles also have invariant assemblages, suggesting fluid-rock interaction with limited amounts of fluid (Ferry, 1991, p. 379). Most contact metamorphic assemblages in siliceous dolomites are unlike assemblages predicted for rocks heated without fluid infiltration, implying that contact metamorphism of impure carbonate rocks involves both increasing *T* and fluid-rock interaction (Ferry, 1991, p. 387).

5.8 Fluids as heat transporting agents in regional metamorphism

Metamorphic heat may be transported by (1) *conduction* or (2) *advection (flow) of fluids*. For example, highest-grade, granulite facies rocks in New Hampshire, USA, were localized in 'hot spots' by fluids concentrated in zones of high fracture permeability (Chamberlain & Rumble, 1988), and high-grade mylonitic rocks in the Bear Mountains fault zone, California, USA, appear to have been localized by flow of hot fluids in a major shear zone (Vernon *et al.*, 1989). Fluid flow may be a major mechanism of heat transport, but only if large amounts of hot fluid are focussed into narrow channels, through which fluid travels rapidly (Brady, 1988; Chamberlain & Rumble, 1988). Pervasive infiltration of water released from dehydration reactions at depth would not be a likely heat transferring agent, because the release of water is too slow and the amount of water released is too small (Brady, 1988; Chamberlain & Rumble, 1988).

5.9 Metasomatism

Reactions between minerals and compositionally complex fluids during metamorphism may produce new rocks with markedly different chemical compositions, the process being known as *metasomatism*. Lindgren (1925) defined metasomatism as 'an essentially simultaneous, molecular process of solution and deposition by which, in the presence of a fluid phase, one mineral is changed to another of differing chemical composition.' He regarded 'replacement' as a synonym, and thought of most metamorphic changes as metasomatic at the grain scale. However, though many metamorphic reactions obviously involve such grain-scale compositional changes, overall metamorphic reactions commonly do not (neglecting gain or loss of H_2O or CO_2). Therefore, a more useful definition (Goldschmidt, 1922) is: 'metasomatism is a process of alteration which involves enrichment of the rock by new substances brought in from outside. Such enrichment takes place by definite chemical reactions between the original minerals and the enriching substances.' This definition covers the most important aspect of metasomatism, namely that it involves a change in whole-rock chemical composition (e.g., Thompson, 1959). Lindgren (1925) regarded metasomatism as a constant-volume process, but this does not always apply, as discussed in Section 5.15.

Chemical components may move differentially by flow mechanisms or by diffusion down their own chemical potential gradients, established especially by gradients in chemical composition (Vidale, 1969). Components may move in a static fluid (*diffusive mass transfer*) or be transported in a moving fluid (*advective mass transfer*). Solid diffusion is too slow to achieve much mass transfer in a reasonable time (Fyfe & Kerrich, 1985).

Diffusion metasomatism ('bimetasomatism') generally occurs on a local scale between rocks of different composition (e.g., Tilley & Alderman, 1934; Orville, 1969; Vidale, 1969; Vidale & Hewitt, 1973; Schrijver, 1973; Carswell *et al.*, 1974; Joesten, 1974a; Thompson, 1975; Fukuyama *et al.*, 2006), is fostered by high temperature and long duration, and is driven by chemical potential differences (Frantz & Mao, 1976, 1977; Brady, 1977; Joesten, 1977, 1991; Joesten & Fisher, 1988). The situation is analogous to grain-scale zones developed in *coronas* between incompatible minerals (e.g., Griffin, 1971, 1972; Griffin & Heier, 1973; Whitney & McLelland, 1973; Grant, 1988; Johnson & Carlson, 1990; Indares, 1993; White & Clarke, 1977; Keller *et al.*, 2004), discussed in Section 2.2.2.

Infiltration metasomatism, involving advection of fluid, is more common, both in contact and regional metamorphic settings, especially around igneous plutons. In this situation, the distribution and abundance of chemical and mineralogical alteration is controlled by permeability; flow of fluid from one chemical environment to another occurs along active percolation networks (Norton, 1988).

Modal changes (changes in the volume proportions of minerals) and changes in whole-rock chemical composition constitute the most obvious evidence of metasomatism, provided the original rocks can be identified. This is particularly clear where normal metasedimentary or meta-igneous rocks are altered metasomatically along fractures. For example, aluminous selvedges rich in staurolite and/or kyanite may form along quartz veins in metapelites that are free of these minerals away from the veins (Ague, 1991). In the absence of adjacent parent rocks, unusual whole-rock compositions consisting of high-variance mineral assemblages suggest a metasomatic origin.

Field evidence commonly suggests that metasomatism conserves volume (e.g., Lindgren, 1925) or a constant number of anions, effectively preserving the oxygen mineral framework (Barth, 1948). However, constant volume does not apply to many metasomatic changes, as discussed in the next section. In the absence of clear field evidence, other approaches can be used to infer mass transfer (Gresens, 1967a; Brimhall & Dietrich, 1987; Grant, 1986). For example, variations in the concentrations of components can be related to the concentration of a component chosen as being 'immobile' (and hence conserved) at the inferred metamorphic conditions. Typical assumed immobile components are Al and Ti. The assumption appears to be basically true, on the basis of geological and experimental evidence; however, these components are mobile in some environments, e.g., veins of titanite in some gabbroic mylonites (Lafrance & Vernon, 1993) and veins of aluminosilicate in some felsic metamorphic rocks (e.g., Vernon *et al.*, 1987b; McLelland *et al.*, 2002a,b).

5.10 Local equilibrium in metasomatic processes

Lattice (volume) diffusion is not effective for large chemical transport distances, which require advection of, and diffusion through, fluid in fractures and pore spaces, as noted previously. However, solid diffusion, which may be facilitated by deformation, is necessary for the establishment of equilibrium within grains, and hence for the production of constant compositions of mineral solid solutions in a rock. This results in an orderly distribution of elements between coexisting minerals, which is a criterion of chemical equilibrium (Mueller, 1967).

Free mobility of all elements would result in density-stratified bands with few minerals or only one (Mueller, 1967). Instead, compositionally domainal structures, such as sedimentary bedding and tectonic layering, are commonly preserved, indicating that equilibrium is established only on a local scale (Section 1.7). Concentration or even restriction of aluminous metamorphic minerals, such as andalusite, staurolite, cordierite and garnet, to more metapelitic beds or folia is very common, indicating that the least mobile components under the prevailing conditions control the siting of these aluminous minerals. However, this should not be taken to imply that all components of these minerals were immobile during their growth and consequently that pre-metamorphic minerals constituting the layering were converted isochemically into metamorphic equivalents. Though some components may have been relatively immobile, others may have been free to move between or along layers, thereby enabling the complex chemical reactions that typify metamorphic rocks to occur (see Sections 1.7, 3.5.1 and 5.6).

'Mobile' components need to be mobile enough to have their chemical potentials controlled externally, but commonly they move less than some of the 'inert' components (Vidale & Hewitt, 1973). The situation is complex, as pointed out by Thompson (1975, p. 340), who noted that, though in single-phase systems, chemical concentrations usually give a direct measure of relative chemical potentials of the components, in multicomponent systems, relations between component concentrations and chemical potentials are not necessarily obvious or predictable. The presence of a component at a particular concentration in a rock does not necessarily mean that the component was immobile during metamorphism.

Zones of minerals and mineral assemblages can form by diffusion metasomatism between bodies of rock with compositions that are chemically incompatible at the prevailing metamorphic conditions. Examples are sequences of simple silicate mineral assemblages that commonly develop at marble-schist contacts during metamorphism (Thompson, 1959), the simple assemblages suggesting that some chemical components are 'mobile' (Korzhinskii, 1950, 1959, 1966, 1967, 1970), as discussed in following sections. Other examples are mineral zones formed around chert nodules in limestone or dolomite (Tilley & Alderman, 1934; Tilley, 1948, 1951; Reverdatto, 1970; Joesten, 1974a; Moore & Kerrick, 1976; Suzuki, 1977; Hoersch, 1981; Joesten & Fisher, 1988). Reaction between carbonate-rich and carbonate-free metapelitic rocks in open-system conditions can produce thinly layered amphibolite (Orville, 1969).

Differential transfer of the 'non-volatile' components K, Ca, Mg and probably Al has been inferred for experiments that produced metasomatic zones between carbonate-bearing and mica-bearing assemblages (Vidale, 1969); the transfer was initiated by a pre-existing compositional gradient, which, by differential transfer of elements, became a sequence of mineral assemblages, each of which equilibrated with and buffered the pore fluid composition.

In a bimetasomatic zoned sequence, local equilibrium occurs at each point in the diffusion column with respect to diffusing components, which

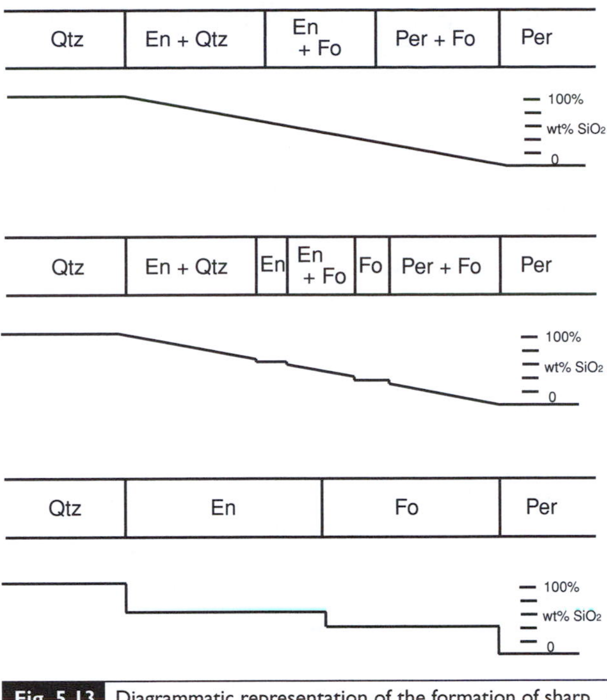

Fig. 5.13 Diagrammatic representation of the formation of sharp compositional zones during bimetasomatism, based on Thompson (1959).

implies that metamorphic reaction rates are much faster than diffusion rates (Brady, 1977, p. 114). Thompson (1959) showed that the establishment of local chemical equilibrium along a chemical potential gradient eliminates incompatibilities between minerals in contact, and tends to produce segregation of components of minerals into compositional layers, each of which consists of mineral assemblages that are compatible with those of adjacent layers. He considered a layer of periclase separated from a layer of quartz by three bimineralic zones, through which the mineral proportions vary continuously, so that the whole-rock chemical composition also varies continuously, at fixed P and T (Fig. 5.13). The mineral assemblages in each zone and the two end-member layers are in local equilibrium, so that no reaction occurs within the zones. However, at the boundary between the Fo + En zone and the adjacent Per + Fo zone, grains of Pe and En in contact react to form the more stable mineral Fo, and at the boundary between the Fo + En zone and the adjacent En + Qtz zone, grains of Fo and Qtz in contact react to form the more stable mineral En. The result is two new monomineralic zones that may grow and eventually consume the original three zones, provided diffusion of Mg and Si is effective enough for long enough. In this way, a zoned sequence of monomineralic or simple assemblages may be constructed during 'bimetasomatism.'

The thermodynamic basis of the concept of local equilibrium in the formation of zones of minerals and mineral assemblages in metasomatic situations has been discussed by many people (Thompson, 1959, 1970; Korzhinskii, 1950, 1959, 1966, 1967, 1970; Brady, 1977; Joesten, 1977, 1991; Weill & Fyfe, 1964, 1967; Rumble, 1982a; Joesten & Fisher, 1988). Thompson (1959) emphasized that (1) reaction between chemically incompatible assemblages involves diffusion-controlled growth of minerals in a sequence of sharp compositional layers, each in local equilibrium with adjacent layers (Fig. 5.13), and (2) metasomatic processes produce high-variance assemblages with few minerals. Korzhinskii (1950, 1959, 1966, 1967, 1970) developed the concept of 'perfectly mobile' components (the chemical potentials of which are defined externally to the reactive system) and 'inert' components (the chemical potentials of which are defined by their concentration in the reactive system itself). Both 'volatile' components (such as H_2O and CO_2) and 'non-volatile' components may be regarded as being mobile, in terms of thermodynamic considerations (Vidale & Hewitt, 1973).

If a system is open to a chemical component (i), the establishment of equilibrium requires that the chemical potential of that component (μ_i) is equalized in both the reactive system and its immediate surroundings (external reservoir of the component concerned). Therefore, μ_i in the reactive system is imposed on the system from outside. This means that, in addition to P and T, an extra variable (μ_i) controls the reactive assemblage, which adds a degree of freedom, which, according to the phase rule (Chapter 1), implies one fewer phase for each mobile component. For example, in some layered calcsilicate rocks inferred to have been formed by metasomatic processes, the number of minerals decreases by one for each component inferred to be mobile in the layer concerned (Vidale & Hewitt, 1973). Therefore, as each component becomes mobile, the phase rule (Section 1.3) is modified from $F = C + 2 - P$ to $F = C + 2 - P - m$, where F = variance C = number of independent chemical components, P = number of coexisting phases, and m = number of components with externally controlled chemical potentials (Korzhinskii, 1950, 1959; Vidale & Hewitt, 1973). Accordingly, the end product of a compositionally zoned rock formed by progressive metasomatism involving increasing solubility (hence mobility) of components might be expected to be a rock with few minerals. This has been used as an argument to explain the common observation of relatively few minerals in metasomatic rocks. For example, metasomatic calscilicate rocks (skarns) typically have high-variance assemblages, as discussed below. An example is the hedenbergite-ilvaite skarn shown in Fig. 5.14.

However, thermodynamic considerations do not take into account kinetic effects, namely the different rates of diffusion of different chemical components, which also control sequences of mineral assemblages, and which can result in multimineral layers in metasomatic situations (Joesten, 1991). In other words, the distinction between 'perfectly mobile' and 'inert' components, suggested by Korzhinskii, (1950, 1959, 1966, 1967, 1970) does not take into account 'mass transport (chemical component) mobility' (Thompson, 1970; Rumble, 1982a; Joesten, 1991).

At the contacts of chemically incompatible compositional layers, the chemical potentials of components are buffered by local reactions (which produce a layer of a new mineral or minerals) and establish gradients that drive exchange of elements to attempt to establish local equilibrium. Local equilibrium requires that the chemical potentials of all components are the same as those in equilibrium with the assemblages of both adjacent layers. In metasomatic conditions, material is continually supplied to and removed from the layer contact by diffusion, with the result that a

Fig. 5.14 Skarn consisting of coarse-grained radiating hedenbergite and black ilvaite, island of Elba, Italy.

buffering reaction occurs, which consumes or produces diffusing components in proportions that maintain local component concentrations (and therefore local chemical potentials) at values in equilibrium with the adjacent assemblage. Thus, the growth of the mineral assemblage of one layer is at the expense of the adjacent layer and therefore movement of the layer boundary. The actual reaction at the layer contacts is controlled by the relative diffusion rates of components diffusing into and out of the reaction site (Joesten, 1991). Theoretical considerations have shown that the number of perfectly mobile components in a mineral assemblage layer is only coincidentally related to the total number of components diffusing through the layer (Thompson, 1970; Joesten, 1974a,b, 1991; Brady, 1977), so that the number of diffusing components cannot be predicted from the variance of the layer assemblage (Joesten, 1991). Another complication is that low-variance multimineral assemblages may represent incomplete elimination of more complex zones during a 'prograde' diffusion cycle, or may represent 'retrograde' reaction between intrazonal assemblages during cooling and uplift (Thompson, 1975).

5.11 Nature of metasomatic replacement reactions

Metasomatic reactions include *exchange reactions* and *net transfer reactions* (Thompson, 1982), though components of both types may occur together (Barton *et al.*, 1991). Simple exchange reactions (for example, K-Na exchange in feldspar and mica) conserve the number of moles and the minerals, but change the chemical composition and the mass, volume changes generally being small. Complex exchange reactions (such as: $3Kfs + 2H^+ = Ms + 6Qtz + 2K^+$) may change the mode, and may or may not conserve the number of moles (Carmichael, 1969). Net transfer reactions (for example, serpentinization of olivine) change the number of moles and generally involve large volume changes.

The traditional view of replacement reactions is that they occur by solid-state diffusion of chemical components at the grain scale, though diffusion in and transport of fluid is generally postulated to apply at broader scales, as mentioned previously. The solid diffusion process is inferred to involve movement of cations through a fixed anionic structural framework, which can explain the preservation of detailed microstructural features and crystallographic relationships. This process can also account for the formation of metasomatic reaction fronts, for example, by reduction of temperature below a critical value for continued diffusion.

However, though solid-state diffusion rates in sulphides may be geologically fast enough, diffusion rates in silicates and oxides are very slow. Diffusion in fluids is much faster, suggesting that fluids may be involved in most replacement reactions. Beus (1983) suggested that replacement is accomplished in a thin film of fluid (a 'zone of interaction'), which changes composition as the reaction proceeds. Recent detailed investigations indicate that the reactant mineral is dissolved in the fluid, and the replacement mineral is precipitated from the fluid, in a *coupled dissolution-reprecipitation replacement* process in which strong anionic bonds are broken by solution and re-formed by precipitation in a fluid film at the reaction interface (Putnis, 2002; Putnis *et al.*, 2005). A sharp compositional gradient at the migrating interface (reaction front) reflects limited volume diffusion, and the process takes place at relatively rapid rates, compared with solid diffusion. Moreover, epitaxial (epitactic) nucleation (Section 1.7) of the precipitating replacement mineral on the reactant mineral surface could promote observed crystallographic relationships between reactant and replacement minerals, though the exact mechanism is not yet well understood. The fluid composition not only governs the proportion of 'nutrient' components available (hence promoting metasomatism), but also enables rapid transport of 'nutrient' and 'waste' components to and from the reaction front. The replacement can be achieved by transport of fluid parallel to the metasomatic 'front', rather then across it (see comments on skarns in Section 5.12.8).

5.12 Types of metasomatism

5.12.1 Introduction

Metasomatism occurs in various styles, many of which can be broadly categorized according to the types of reaction involved, provided a transition from unaltered to altered rock can be observed, and provided reasonable reactions can be inferred. This situation is more common in metasomatism accompanying contact metamorphism than in metasomatism accompanying regional metamorphism, though it may apply to some regional metamorphic settings. However, some regional metamorphic rocks of unusual composition have been inferred to be of metasomatic origin, even though unaltered equivalents cannot be observed. In these situations, reactions cannot be inferred with any degree of reliability, and even a metasomatic origin may be arguable. Styles of metasomatism that can be categorized are discussed in the next section. Some are dominated by exchange reactions (e.g., the simpler examples of Mg metasomatism, Na metasomatism, Na-Ca metasomatism and K metasomatism), whereas others are dominated by net transfer reactions (e.g., calcsilicate skarns, discussed in Section 5.12.8).

5.12.2 Mg metasomatism

Mg metasomatism occurs when added Mg is exchanged for Fe and Ca, accompanied by hydration (Barton *et al.*, 1991). It is most common in the chloritization of shallow marine mafic rocks (Bloch & Hofmann, 1978), and is characterized by the addition of Mg without the leaching of other cations that is typical of hydrolysis reactions (see Section 5.12.6). Oxidation, hydration, Si loss and K gain may also occur.

5.12.3 Na metasomatism and Na-Ca metasomatism

Na metasomatism and Na-Ca metasomatism are characterized by exchange of added Na for Ca (actually NaSi for CaAl) or Na for K, with or without exchange of Ca for Fe and Ca for Mg. Such alteration affects mafic to intermediate igneous rocks in seafloor hydrothermal systems and also dioritic to monzonitic arc intrusions involved in magmatic or external hydrothermal systems (Hollister, 1975; Carten, 1986; Barton *et al.*, 1991). Igneous plagioclase typically is replaced by albite, and mafic igneous minerals typically are replaced by chlorite and actinolite; commonly epidote is also present. Unless metasomatism can be demonstrated, these minerals would be indistinguishable from greenschist facies assemblages formed by isochemical metamorphism. They are also similar to assemblages inferred to be formed by hydrolytic alteration in mafic compositions (see below).

5.12.4 Ca metasomatism

Ca metasomatism (calcic alteration) produces epidote-rich rocks ('epidosites') by exchange of Ca for Mg, K and Na in seafloor basaltic rocks (Schiffmann & Smith, 1988). Plagioclase is replaced by epidote, alkali feldspar is replaced by zeolites, and ilmenite is replaced by titanite and magnetite. Mafic rocks associated with serpentinite bodies commonly undergo Ca metasomatism to form 'rodingite', in which former plagioclase is replaced by grossular ± vesuvianite and other calcsilicate minerals (Section 5.12.7). Evidence of Ca metasomatism has also been observed in former granite converted to epidote-rich mylonite in some high-strain deformation zones in the Wyangala Batholith, central-western New South Wales, Australia (Fig. 5.15). Takagi *et al.* (2007) described quartz-intermediate plagioclase assemblages formed by hydrothermal alteration along fractures in granite.

Evidence of Ca metasomatism has also been observed in contact metamorhic aureoles, some related to Ca-Zn, W-Mo and porphyry copper ore deposits (Kuniyoshi & Liou, 1974; Carten 1986; Dilles & Einaudi, 1992; Seal *et al.*, 1987; Perring *et al.*, 2000).

5.12.5 K metasomatism

K metasomatism (potassic alteration) is characterized by exchange of K for Na and Ca, causing replacement of plagioclase by K-feldspar ± muscovite and replacement of mafic minerals by biotite (Barton *et al.*, 1991). Addition of Si and metals, as well as hydration (producing mica) may also occur. Examples include Tertiary volcanic rocks in the western USA that were

(A)

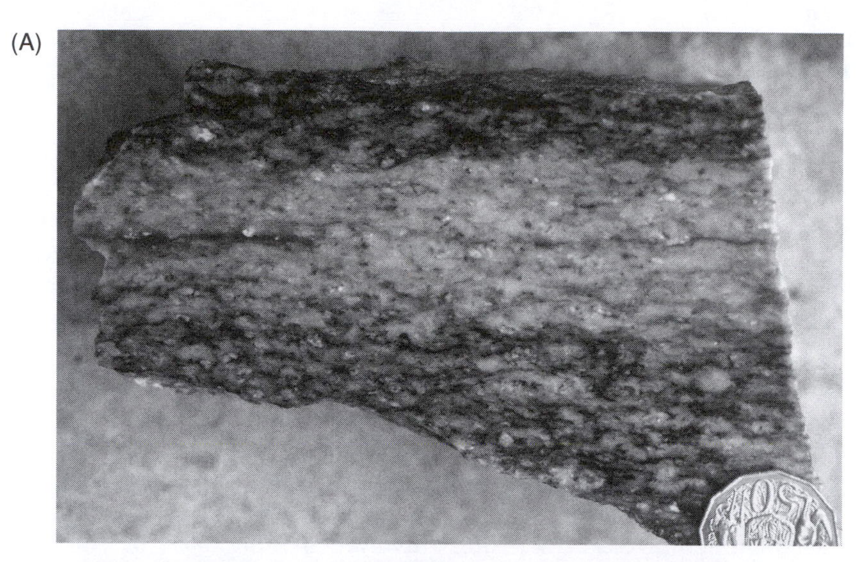

(B)

Fig. 5.15 (A) Metasomatically altered zone (light coloured) rich in quartz, epidote and albite, formed during mylonitic deformation of granite, Wyangala Batholith, Wyangala Dam, near Cowra, central-western New South Wales, Australia. Coin diameter 31 mm.
(B) Metasomatically altered mylonite, same locality, consisting of fine-grained quartz, albite and epidote (high relief). Crossed polars; base of photo 0.7 mm.

metasomatized by low-temperature (<150 °C) fluids at low pressure (<1–2 km) producing secondary K-feldspar, quartz and calcite (Brooks, 1986; Glazner, 1988). The most strongly altered rocks have K-feldspar – Fe-oxide assemblages. The rocks show excellent microstructural evidence of replacement of plagioclase by adularia (a low-temperature form of K-feldspar), marginally and along fractures. The rocks are strongly enriched in K (up to 13.3. wt.% K_2O) and depleted in Na and Mg; addition of Zn, Ba and Mn also occurred. The alteration zones follow faults or breccia zones, suggesting that the metasomatizing solutions were transported along faults. The potassic solutions were either percolating closed-basin brines or resulted from hydrogen metasomatism (see Section 5.12.6) of deeper rocks (Glazner, 1988). Other examples of low-temperature potassic alteration are Proterozoic rhyolite in Texas, USA, in which quartz and albite have been replaced by K-feldspar (Barker & Burmester, 1970), and Triassic zeolite-facies tuffs rich in adularia in southern New Zealand (Coombs, 1954).

5.12.6 Hydrolysis reactions (hydrogen-ion metasomatism; 'hydrolytic' alteration)

In hydrolysis, hydrogen ions are exchanged for other cations ('base cations'), such as K, Na, Ca, Mg and Fe. H_2O, CO_2 and S may also be added. The alteration ranges from minor exchange to complete base-cation leaching. The most common varieties of alteration are known as 'sericitic', 'chloritic' and 'argillic,' which are common at high crustal levels, especially around 'porphyry copper' deposits. Examples of hydrolytic alteration accompanying the formation of metallic ore deposits are discussed in Section 5.17.

The products of hydrolysis reactions are hydrous minerals or H_2O and leached cations (in solution) displaced by H^+ ions. The following are examples of hydrolysis reactions (Hemley & Jones, 1964; Meyer & Hemley, 1967; Vernon, 1969; Barton *et al.*, 1991):

$$3\text{Kfs} + 2\text{H}^+ = 2\text{K}^+ + \text{Ms} + 6\text{Qtz}$$
$$\text{Kfs} + \text{An} + 2\text{H}^+ = \text{Ca}^+ + \text{Ms} + 2\text{Qtz}$$
$$2\text{Ab} + 2\text{H}^+ = 2\text{Na}^+ + \text{Als} + \text{H}_2\text{O} + 5\text{SiO}_2$$
$$2\text{Kfs} + 2\text{H}^+ = 2\text{K}^+ + \text{Als} + \text{H}_2\text{O} + 5\text{SiO}_2$$
$$\text{An} + 2\text{H}^+ = \text{Ca}^{2+} + \text{Als} + \text{H}_2\text{O} + \text{SiO}_2$$
$$2\text{Bt} + 14\text{H}^+ = 2\text{K}^+ + 6(\text{Mg,Fe})^{2+} + \text{Als} + 9\text{H}_2\text{O} + 5\text{SiO}_2$$
$$2\text{Ms} + 2\text{H}^+ = 2\text{K}^+ + 3\text{Als} + 3\text{H}_2\text{O} + 3\text{SiO}_2$$
$$\text{Crd} + 4\text{H}^+ = 2(\text{Mg,Fe})^{2+} + 2\text{Als} + 2\text{H}_2\text{O} + 3\text{SiO}_2$$
$$0.75\ \text{andesine} + \text{K}^+ + 2\text{H}^+ = 1.5\text{Na}^+ + 0.75\text{Ca}^{2+} + \text{Ms} + 3\text{SiO}_2$$

In more complex systems, the residual and displaced ions depend on pH, oxidation state and chemical composition of the fluid (including the activity of CO_2, if present), leading to much reaction complexity, as illustrated by the detailed schemes of cyclical chemical reactions proposed for the low-temperature alteration of glassy mafic rocks by Vallance (1969, fig. 8). The following are some examples of metasomatic hydrolytic alteration.

'Spilitic' alteration is a term for complex hydrolytic chemical changes accompanying Na-Ca metasomatism of mafic rocks, forming patchy concentration of Na-rich (e.g., albite) and Ca-rich (e.g., epidote) minerals in local metasomatic systems, which constitute part of an overall hydrolytic alteration system that commonly also involves chloritic alteration. For example, Smith (1968) described patchy alteration in metabasalts in the Orange area, New South Wales, Australia, involving Na metasomatism (albitization) of plagioclase (forming the rock known as 'spilite', once thought to be of purely igneous origin), with complementary local concentration of Ca to form assemblages with pumpellyite or epidote. The Ca-rich alteration zones are localized around former cavities, intersecting

joints and rupture zones cutting primary structures, as well as in primary vesicular zones and around the margins of pillows. The rearrangement of chemical components is a form of local metasomatism instigated by fluid movement through cracks and cavities at relatively low temperatures during burial metamorphism.

Reed & Morgan (1971) found that integrating the chemical compositions of sodic metabasalt and epidosite (epidote-quartz rock) in their observed 3:1 proportions produced a typical unaltered basalt composition for the region concerned (Shenandoah National Park, Virginia, USA), from which they inferred that both metamorphic rock-types were produced by the hydrothermal metasomatism of basalt through the action of fluids with high oxidation potential (accounting for Fe^{3+} in the epidote) gaining access through fractures. On the basis of chemical analyses across pillows in Mid-Atlantic Ridge basalts hydrothermally altered to produce greenschist facies (chlorite-albite-actinolite-epidote) assemblages, Humphris & Thompson (1978) used mass balance calculations to show that Si and Ca were leached from the basalt, and that Mg and H_2O were added to the basalt.

Vallance (1960, 1965, 1967, 1969, 1974) reviewed and described examples of such alteration, noting that Na metasomatism in the cores of basalt pillows (represented by albitized plagioclase, chlorite and common residual basaltic clinopyroxene, with or without some carbonate or hydrous calcsilicate minerals) are complemented by Ca, Mg and/or Fe enrichment of the pillow margins (represented by various combinations of chlorite, epidote, pumpellyite, prehnite, hydrogarnet, carbonate minerals, tremolite-actinolite, titanite and hematite). Calcsilicate and chlorite-rich concentrations in vesicles and inter-pillow vein-like aggregates testify to the local solution and mobility of chemical components in the alteration process.

The common presence of former glass in these rocks suggests that its relative instability in the presence of percolating hot hydrous fluids instigates the alteration process, which is complex. For example, replacement of basalt glass (i.e., in pillow margins, as well as in interstices between plagioclase and pyroxene in pillow cores) to form chlorite is not a simple hydration reaction, although introduction of water is involved, because Si, Ca, Na, and Ti must be removed (Vallance, 1969, p. 25). These more soluble cations are taken into solution by hydrolysis (hydrogen metasomatism), and are exchanged for a chemically equivalent amount of H^+ from the fluid (Hemley & Jones, 1964; Vallance, 1969, 1974). In broad outline, the Na and Si released from the chloritization of glass leads to exchange of Ca for Na and Si for Al in the albitization of plagioclase. The Ca released from both cores and margins of pillow, migrates to form aggregates of calcsilicate and/or carbonate minerals locally replacing pillow margins and other parts of the flow, especially adjacent to fractures, as well as crystallizing in inter-pillow cavities and vesicles (Smith, 1968; Vallance, 1969).

A number of elements may become mobile during low-temperature hydrous alteration. For example, Wood *et al.* (1976) found that Si, Mg, K, Rb, Sr and light rare earth elements were mobile, whereas Ti, P, Zr, Y, Nb, Hf and some rare earth elements were relatively unaffected as a result of hydrothermal alteration of basalt at zeolite facies conditions.

'Spilitic' alteration develops and may be preserved in low-temperature (local hydrothermal or regional burial metamorphic) environments (e.g.,

Vallance, 1960, 1969; Smith, 1968, 1974; Reed & Morgan, 1971; Jolly & Smith, 1972; Coombs, 1974). Later higher-grade, isochemical, contact or regional metamorphism of these previously metasomatized mafic rocks produces Mg-Al-rich, cordierite-anthophyllite rocks, commonly with closely associated calcsilicate assemblages (e.g., with diopside, hornblende, garnet or wollastonite), as described by Vallance (1967, 1969), Irving & Ashley (1976), James *et al.* (1978) and Reinhardt & Rubenach (1989). These rocks may or may not show structural evidence of their igneous origins, despite anomalously high magnesium contents. Because these assemblages represent far from typical metabasalt compositions, they were once mistakenly thought to result from Ca or Mg-Al metasomatism during the high-grade metamorphism (Vallance, 1967, 1969). Layered alteration zones in metamorphosed pillow basalts are illustrated in Fig. 7.19A.

Argillic alteration results in aluminous assemblages formed by cation leaching in acid solutions that penetrate rocks along grain boundaries or in narrow deformation zones, commonly, though not exclusively, in contact metamorphic settings. In advanced stages, almost all base cations are removed (e.g., Meyer & Hemley, 1967), though some Fe commonly remains as pyrite if S is present in the fluids. Typical minerals produced are clay minerals, pyrophyllite and alunite at low temperatures, and andalusite at higher temperatures (Lowder & Dow, 1978). A regional metamorphic example is the development of late fibrous sillimanite replacing all minerals, from their grain boundaries inwards, in metapelitic rocks at Cooma, south-east Australia (Vernon, 1979).

Advanced argillic alteration is common in volcanic or high-level intrusive rocks in the 'porphyry copper' environment (Espenshade & Potter, 1960; Hemley & Jones, 1964; Gustafson & Hunt, 1975; Lowder & Dow, 1978, Valiant *et al.*, 1983). The fluids responsible for the alteration are H_2O-rich (commonly also rich in NaCl), generally inferred to be of magmatic origin, and many hydrothermal metallic rocks are concentrated from such fluids (Hedenquist & Lowenstern, 1994; Burnham, 1997; Heinrich *et al.*, 1999, 2004). Watanabe & Hasegawa (1986) inferred a hydrothermal origin for veinlets of andalusite and sillimanite in altered rhyolite, which suggests a continuum between hydrothermal alteration and regional metamorphism (e.g., Vernon, 1976, pp. 17, 18). Local base-cation leaching in zones of high-strain and high fluid flow is a possible explanation for at least some concentrations of sillimanite in anastomosing folia that are common in high-grade regional metapelitic schists and gneisses (e.g., Harker, 1932, fig. 106; Vernon, 1987a,b).

5.12.7 Hydration-dominant reactions

Addition of H_2O is typical of retrograde metamorphism, deuteric alteration of igneous rocks, and alteration in porphyry copper systems. Other metasomatic changes are commonly involved, as in the following examples.

'Sericitic' alteration of felsic rocks is a form of hydration, characterized at low temperatures by the replacement of feldspars by fine-grained 'sericite' ('white mica', commonly muscovite) $\pm$ clay minerals (Meyer & Hemley, 1967) and at higher temperatures by coarse-grained white mica ('greisenization'). In the presence of F in the fluid, F-mica, topaz and fluorite may also replace feldspar. With more chemically complex fluids, mafic minerals are altered to Li-F mica, chlorite, oxide minerals and sulphide minerals.

Mass changes include loss of Na and Ca, gain or loss of K and Si, and gain of H_2O, F and S. Chloritic alteration occurs in altered mafic and intermediate igneous compositions, where chlorite takes the place of or accompanies 'sericite'.

Chloritic alteration occurs in intermediate 'porphyhry copper' systems and in marine volcanic systems. Common mass changes are addition of H_2O, loss of Na and Ca, and either gain or loss of Mg and Fe (Barton *et al.*, 1991).

'Propylitic alteration' (propylitization) is common in the hydration of andesitic and other rocks, especially in porphyry copper environments. It involves replacement of plagioclase by albite, epidote and/or carbonate, pyroxene by tremolite/actinolite, chlorite and carbonate, and hornblende and biotite by chlorite, calcite, titanite and Fe oxide minerals. Pyrite, chalcopyrite, piemontite, apatite, hematite, anhydrite or ankerite are present in some examples. K released by the chloritization of biotite may form 'sericite'.

Serpentinization is a common form of hydration in ultramafic rocks, by which olivine and orthopyroxene are replaced by aggregates of serpentine minerals, the Fe component entering fine-grained magnetite or awaruite. The replacement typically is pseudomorphous, and the much lower relative density of serpentine, compared with olivine, means that the excess mass leaves the reaction site, forming chrysotile veins though the ultramafic body.

Rodingite is a light-coloured, tough, commonly flinty, hydrous calcsilicate rock typically replacing gabbro associated with serpentinite, commonly preserving gabbroic microstructures (e.g., Bloxham, 1954; Baker, 1959; Coleman, 1963; O'Hanley, 1996; Li *et al.*, 2007). Granite and metasedimentary rocks (Coleman, 1963; O'Hanley, 1996), as well as eclogite (Li *et al.*, 2007), can also be converted to rodingite. Rodingites commonly occur as xenoliths and boudins in serpentinite, or as alteration zones at serpentinite contacts (Coleman, 1963; O'Hanley, 1996). The formation of rodingite involves hydration and Ca ($\pm CO_2$) metasomatism, forming hydrogrossular, commonly with vesuvianite, prehnite, epidote, clinozoisite, zoisite or diopside, less commonly wollastonite, pectolite or calcite. These minerals mainly replace original minerals, but locally also occur as veinlets. The original plagioclase is progressively replaced by aggregates of hydrogrossular, with or without the other calcsilicate minerals, and the original pyroxene or hornblende may be replaced by chloritic aggregates with some clinozoisite and grossular (Baker, 1959). Original gabbroic microstructures are preserved in the earlier to moderate stages of alteration, but may be obliterated in completely metasomatized rocks (Baker, 1959). The felsic minerals lose Na, K and Si during the rodingite-forming process (Coleman, 1963; O'Hanley, 1996). A common interpretation is that Ca released from the alteration of clinopyroxene during the serpentinization of ultramafic rocks accumulates in fluids until Ca supersaturation results in reaction with mafic rocks (Coleman, 1963; O'Hanley, 1996).

5.12.8 Skarns

Skarns (tactites) are coarse-grained, metasomatically zoned silicate rocks that develop between carbonate-rich rocks (metamorphosed limestones, dolomites and Mn-rich carbonate rocks) and Al- and Si-rich rocks (granitic or other intrusive rocks, shales or schists), especially in contact metamorphic environments, but also in some regional metamorphic situations (Tilley, 1951; Edwards *et al.*, 1956; Zharikov, 1970; Burt, 1974; Kerrick, 1977; Kwak, 1978, 1987; Einaudi *et al.*, 1981; Einaudi & Burt, 1982; Barton *et al.*, 1991; Buick & Cartwright, 2000; Baker & Lang, 2003).

Most skarns are formed by magmatic-hydrothermal activity associated with granitic to dioritic intrusions in orogenic belts, and are of Precambrian to Tertiary age (Einaudi & Burt, 1982). Skarns typically are coarse-grained, generally Fe-rich aggregates of Ca-Mg-Fe-Al silicates formed by metasomatic processes at relatively high temperature, and commonly show metasomatic zoning.

Reaction skarns are small silicate reaction zones developed by local diffusion between incompatible rock types (Burt, 1974), for example, successive rims of tephroite (Mn_2SiO_4) and rhodonite ($MnSiO_3$) formed between layers of rhodochrosite ($MnCO_3$) and chert (SiO_2) during contact metamorphism (Watanabe *et al.*, 1970). *Replacement skarns* ('ore skarns') are extensive silicate replacements of carbonate and adjacent rocks caused by large-volume infiltration of fluids that are out of equilibrium with the host rocks (Burt, 1974). Though they grade into reaction skarns, replacement skarns are commonly relatively large, and many are mined for Fe, Cu, Zn, W and other metals.

Endoskarns are aluminous skarns, rich in epidote or grossular garnet, generally limited in volume, that replace non-carbonate rocks, such as intrusive granite at a contact with metasomatically altered carbonate rocks. For example, enodskarn consisting of plagioclase + clinopyroxene (salite) ± garnet has locally replaced monzodiorite adjacent to metasomatized wall-rocks at the Ann-Mason porphyry copper deposit, Nevada (Dilles & Einaudi, 1992). Another example is endoskarn consisting of diopside, chlorine-rich scapolite and albite formed in a granite sill (the Revenue Granite in the Mount Isa Inlier, northwest Queensland, Australia) shortly after emplacement, through the action of high-temperature, magmatic, Na-, Ca-, Cl-bearing brines; the metasomatism involved addition of Na and Ca, and loss of Fe, Mg and K from the granite (Aslund *et al.*, 1995).

Exoskarns replace carbonate rocks. Exoskarns replacing metadolomite tend to be rich in Mg, whereas exoskarns replacing metalimestone ('Ca-Fe-Si skarns') tend to be rich in Ca, Fe and locally Mn (Burt, 1974). The most common minerals in Ca-Fe-Si exoskarns are andradite, $Ca_3Fe^{3+}{}_2Si_3O_{12}$, and hedenbergite, $CaFe^{2+}{}_2Si_2O_6$, with ilvaite, $CaFe^{2+}{}_2Fe^{3+}Si_2O_7O(OH)$, in lower-temperature examples (Fig. 5.14). Concentrations of sulphide minerals may be found in exoskarns, though the main ore-forming hydrothermal stage may post-date the calcsilicate minerals (e.g., Baker & Lang, 2003).

Metasomatism in Ca-Fe-Si exoskarns results in segregation into zones with few minerals, commonly monomineralic (see Section 5.10), around inferred fluid transfer channels, such as igneous intrusive contacts, sedimentary contacts or fractures. The zones range from a few centimetres to several hundreds of metres thick, reflecting local variations in porosity and permeability of the rocks. The same zoning sequence tends to occur throughout a given skarn, though the zone thicknesses may vary considerably (Burt, 1974). Common zoning sequences are listed in Table 5.2. They

Table 5.2. Zoning sequences in Ca-Fe-Si exoskarns (Burt, 1974)

Zoning sequence
vein/intrusion/endoskarn – andradite – calcite
vein/intrusion/endoskarn – andradite – wollastonite – calcite
vein/intrusion/endoskarn – hedenbergite – calcite
vein/intrusion/endoskarn – ilvaite – hedenbergite – calcite
vein/intrusion/endoskarn – hedenbergite – wollastonite – calcite
vein/intrusion/endoskarn – andradite – hedenbergite – calcite
vein/intrusion/endoskarn – andradite – hedenbergite – wollastonite – calcite

suggest an overall progressive decease in the less soluble components, such as Al, towards the carbonate.

The main chemical additions in skarn formation are Si, Fe and Al, and the main losses are CO_2 and H_2O. Ca and Mg may be added or lost, and alkalies and rare elements, as well as metallic sulphides, are added in some examples. High-*T* skarns, formed mainly by infiltration of magmatic fluids, contain largely anhydrous calcsilicate minerals, such as Ca-Fe-Al garnet, hedenbergite, pyroxenoids and vesuvianite in Ca-rich rocks (metasomatized limestones) and forsterite, diopside, monticellite, magnetite, tremolite and chlorite in Mg-rich rocks (metasomatized dolomites). Oxidized skarns (reflecting oxidized magmatic sources of the fluids) are characterized by Fe^{3+}-rich minerals, such as andradite, ilvaite and Fe oxide minerals, with Fe-poor clinopyroxene, whereas reduced skarns are characterized by grossularite, spessartine and almandine, with Fe-rich clinopyroxene. Hydrous skarns are characterized by hydrous calcsilicate minerals, as well as abundant minor minerals, including sulphide and oxide minerals; the hydrous minerals may replace the minerals of previously formed anhydrous skarns. Later hydrothermal replacement of one calcsilicate mineral by another may occur in skarns (e.g., Joesten, 1974b).

Andradite in some skarns shows complex, euhedral, oscillatory growth zoning (e.g., Yardley *et al.*, 1991b; Yardley & Lloyd, 1995; Jamtveit *et al.*, 1993), reflecting a history of garnet growth in a fluid. This implies a transient porosity of up to a few volume percent during metasomatism, caused by volume changes in net transfer metasomatic reactions. The porosity makes these calcsilicate layers conduits for fluid, which flows parallel to a layer or vein, not perpendicular to it (Yardley & Lloyd, 1995). This suggests that a 'metasomatic front' represents the limits of lateral leakage of fluid into adjacent rocks, and that the assumption of flow across the compositional boundary is not justified. This is supported by the observation that metasomatic rocks may be surrounded by unaltered precursors with no isotopic indication of fluid infiltration (Gerdes & Valley, 1994; Buick & Cartwright, 2000).

5.12.9 Fenitization

Fenitization is the formation of metasomatic rocks around alkaline syenitic, carbonatite and kimberlite intrusions, and involves addition of Na, K and CO_2 and removal of Si from felsic, metapelitic and mafic rocks, with redistribution of Ca, Fe, Mg and Al (Currie & Ferguson, 1971; Rock, 1976; Heinrich, 1980; Kresten & Morogan, 1986; Barton *et al.*, 1991). Quartz is typically removed and carbonates may be added. The silicate minerals produced include sodic pyroxenes, sodic amphiboles, phlogopite, alkali feldspar and feldspathoids.

5.12.10 'Blackwall' alteration

Bimetasomatic reactions occur between ultramafic and silicic and/or aluminous rocks during regional metamorphism (Sanford, 1982; Matthes & Olesch, 1986; Grundmann & Morteani, 1989), as well as between intrusions of felsic magma and ultramafic wall rocks (Arai, 1975; Barton *et al.*, 1991). Diffusion of Si from the silicic rock and Mg $\pm$ Fe from the ultramafic rock forms zones of chlorite $\pm$ biotite between the two rock-types. Further into the ultramafic rock, hydration and silication reactions may produce zones of amphiboles, talc and serpentine minerals, and consequent desilication of the margins can produce corundum (Rossoviskiy *et al.*, 1981). Emerald has been produced by metasomatic exchange at the contact of Be-bearing pegmatite with biotite-talc schist and actinolite schist, and also at the contact of Be-bearing garnet-mica schist or biotite-plagioclase gneiss with serpentinite and talc schist (Grundmann & Morteani, 1989).

5.13 How isochemical is prograde metamorphism?

As mentioned in Chapter 1, metamorphism generally is inferred to be essentially isochemical, apart from loss of volatile components, such as H_2O in the metamorphism of pelites and mafic igneous rocks and CO_2 in the metamorphism of carbonate rocks (Harker, 1893; Barton *et al.*, 1991; Pattison & Tracy, 1991, p. 118). As noted by Harker (1893), metamorphism acts on whatever whole-rock composition is available, regardless of whether or not pre-metamorphic compositional changes (such as weathering or hydrothermal alteration in various geological settings) have occurred.

However, isochemical metamorphism is difficult to infer with confidence for prograde metasedimentary rocks, because of sampling problems across a metamorphic terrane. For example, sampling the same sedimentary bed or unit is generally uncertain, and even if this can be done, it cannot be assumed that the original whole-rock chemical composition remains constant along the entire length of the unit. Arguments for isochemical prograde metamorphism have been presented by Shaw (1956), Butler (1965), Atherton (1977), Yardley (1977d), Moss *et al.* (1995, 1996), Cardenas *et al.* (1996) and Roser & Nathan (1977), and arguments for bulk chemical changes during prograde metamorphism of metasedimentary rocks have been presented by Leake & Skirrow (1958), Ague (1991, 1994, 1997), Ferry (1983b) and Masters & Ague (2005). The problem has been reviewed by Vernon (1998).

Changes in Si and alkalies during metamorphism may occur in high-grade rocks close to some igneous contacts. For example, loss of melt from migmatites adjacent to magmatic heat sources may deplete high-grade rocks in Si and alkalies (Gribble, 1968), possibly accounting for some Si-depleted, quartz-absent assemblages (Tilley, 1924; Grant & Frost, 1990; Pattison & Tracy, 1991, p. 118). In addition, some metamorphic rocks show evidence of cation leaching (Section 5.12.6) caused by acidic solutions

(Vernon, 1979), which may be of magmatic origin (Compton, 1960), though the source is not always evident.

5.14 Metamorphic differentiation

The formation of a new compositional layering by chemical transfer processes during metamorphism is known as *metamorphic differentiation* (Stillwell, 1918; Eskola, 1932). Typically the new layering is related to the effects of deformation, for example folding (Ayrton, 1969). For example, the compositional layering characteristic of schists is related to the stress and/or strain distribution associated with the development of crenulation cleavage. Compositional layering parallel to slaty or crenulation cleavages in schists is common (Turner, 1941; Plessman, 1964; Talbot, 1964a,b; Nicholson, 1966; Williams, 1972, 1990; Cosgrove, 1976; Hobbs *et al.*, 1976, p. 223; Gray, 1977a,b; Marlow & Etheridge, 1977; Gray & Durney, 1979a,b; Hanmer, 1979; Bell & Rubenach, 1983; Passchier & Trouw, 1996; Worley *et al.*, 1997; Vernon, 1998; Williams *et al.*, 2001). It appears to be the result of local redistribution of chemical components related to stress and/or strain variation between crenulation hinges (Q domains) and limbs (P or M domains), as discussed in Section 6.9.

As pointed out by Vidale (1974), the local diffusion involved in metamorphic differentiation is driven by gradients in the activities (or chemical potentials) of some chemical components. These gradients are generated by heterogeneities in the rock. For example, chemical gradients inherited from original compositional layering can drive local diffusion processes and produce segregation layering (Orville, 1969; Vidale, 1969; Vidale & Hewitt, 1973). So can diffusion transport in small local pressure gradients generated in slightly layered rocks by deformation (Vidale, 1974).

In some instances, no relationship between compositional layering and deformation is evident. For example, at Rosetta Head, South Australia, a marked stripy compositional layering oblique to bedding occurs in greenschist to amphibolite facies dolomitic schists and phyllites, preserving delicate sedimentary structures (Talbot & Hobbs, 1968). The layers are rich in quartz or andalusite, and do not show a foliation. They do not appear to be dilational or to result from transposition of an earlier layering. They occur in domains of no obvious differential strain, and probably are the result of metamorphic differentiation in situ (Talbot & Hobbs, 1968, p. 585), though the actual process has not been specified.

Compositional layering in deformed, metamorphosed igneous rocks commonly cannot be related to observed folds. In many mylonites, the layering is due to extreme elongation of different mineral grains or rock bodies, not to metamorphic differentiation (Vernon, 1974; Myers, 1978; Vernon *et al.*, 1983).

5.15 Extent of mass transfer and volume loss in deforming rocks

As discussed previously, the formation of cleavages commonly involves exchange of components between P and Q domains, and, in many instances, transfer of material from P domains out of the rock system on the specimen or outcrop scale (Ague, 1991, 1994, 1997). The extent of volume loss caused by solution during the formation of slaty cleavage at low metamorphic grade and differentiated crenulation-foliations at higher metamorphic grade is a controversial topic (Vernon, 1998). The problem has not yet been resolved, but a recent review by Gray (1997) indicates that volume losses in slates from the Martinsburg Formation, eastern USA, and the Lachlan Fold Belt, south-eastern Australia, vary from 0 to >50%, depending on the different processes responsible for the cleavage. Volume losses tend to be broadly complemented by veins in deformed rocks (Beach, 1974). Major composition and volume changes may occur in zones of intense deformation (shear zones), as discussed below.

5.16 Regional metasomatism

Much work on metasomatism has concentrated on contact metamorphic environments, for which magma-derived or magma-controlled fluid sources can be inferred, and mineral assemblage zoning can be related to an intrusive contact. However, extensive metasomatic systems have been inferred for some regional metamorphic terranes. Some can be related to deformation zones (shear zones), but others are of broader extent. Fluid sources, mass transfer mechanisms and mineral reactions may be difficult to infer for such situations.

5.16.1 Metasomatism in shear zones

Though many rocks deformed in high-strain zones (mylonitic shear zones; see Section 6.11) show little evidence of whole-rock chemical change (e.g., Kerrich *et al.*, 1980), others show major chemical and isotopic changes, indicating substantial movement of fluid through the zones during deformation (Beach, 1973, 1976, 1980; Beach & Fyfe, 1973; Kerrich *et al.*, 1977; McCaig, 1984, 1997; Kerrich, 1986; Sinha *et al.*, 1986; Simpson, 1986; Stel, 1986; O'Hara, 1988; Losh, 1989; Jamtveit *et al.*, 1990; Dipple & Ferry, 1990; Dipple *et al.*, 1990; McCaig & Knipe, 1990; Selverstone *et al.*, 1991; Glazner & Bartley, 1992; Früh-Green, 1994; Streit & Cox, 1998; Doyle & Cartwright, 2000; Cartwright & Barnicoat, 2003; J. D. Barnes *et al.*, 2004; Keller *et al.*, 2004). The process involves a complex local interplay between contrasting processes, such as diffusion versus infiltration, continuous versus episodic effects, and channelled versus pervasive reactions (McCaig & Knipe, 1990). Later deformation and metamorphism may destroy much of the microstructural evidence (Beach, 1980; O'Hara, 1988).

Fluid flow in mylonite zones appears to occur mainly by microcracking (Brantley *et al.*, 1990; Tobisch *et al.*, 1991; Oliver, 2001). The microcracking is favoured by (1) grain-boundary sliding in aggregates of different minerals that deform at different rates, thereby tending to open grain-boundary cracks, and (2) formation of new minerals during deformation, which could lead to the same effect, or which could produce new interfaces that are potentially permeable, such as (001)mica-quartz or (001)mica-feldspar, especially when they are well aligned, as is common in shear zones; these interfaces are likely to attract fluid, owing to poor chemical bonding across them (Vernon, 1999). Though ductile processes in mylonites have been emphasized in recent years (e.g., Bell & Etheridge, 1973), crystal-plastic processes are likely to give way to mass transfer and volume change at lower metamorphic grades (Kerrich *et al.*, 1977), and recent experiments and studies of mylonites have concentrated on complex interactions between metamorphism, fluids, cataclasis and crystal-plastic deformation processes (e.g., Dixon & Williams, 1983; Knipe & Wintsch, 1985; Murrell, 1985; Rutter & Brodie, 1985).

5.16.2 Regional Na metasomatism

Evidence of regional Na and Na-Ca metasomatism is widespread, especially in Proterozoic terranes, being commonly associated with Fe oxide (Cu-U-Au-rare earth element) mineral deposits (Hitzman *et al.*, 1992; Freitsch *et al.*, 1997). The Mount Isa Inlier, north-west Queensland, Australia, and the Willyama Inlier, western New South Wales and eastern South Australia, provide good examples of this kind of metasomatic alteration (see below). Other Proterozoic examples include quartz-plagioclase metasomatic rocks in the Vaestervik area, south-eastern Sweden (Elbers & Hoeve, 1971; Hoeve, 1978).

Palaeozoic examples are also common. For example, sodic hydrothermal alteration, accompanied by Fe and Cu ± Zn minerals, is common in Late Palaeozoic to Jurassic arc-related plutonic and volcanic rocks in the western USA; the fluids responsible were isotopically heavy, moderately to highly saline, and of marine, rock formation and/or meteoric origin, with or without a magmatic component (Battles & Barton, 1995). The alteration involved addition of Na ± Ca and leaching of K ± Ca, with variable changes in Mg, Fe and Mn, resulting in assemblages of sodic plagioclase with chlorite, actinolite and diopside ± marialitic scapolite. Other Palaeozoic examples are zones of intense albitization with cassiterite, resulting from the hydrothermal leaching of Late Palaeozoic granite at Emuford, north-east Queensland (Charoy & Pollard, 1989).

Mount Isa Inlier, Australia: The Proterozoic Mount Isa Inlier, north-west Queensland, Australia, was subjected to widespread (200 km^3), locally intense Na ± Ca and alkali-chloride metasomatism during late regional metamorphism, expressed as albitization, scapolitization, and local calcic veins and plugs (Edwards & Baker, 1954; Oliver & Wall, 1987; Williams, 1988; Oliver *et al.*, 1990; Huang & Rubenach, 1995). The region exposes several albite-rich bodies, some of which are associated with Fe-(Cu-Au) orebodies (Davidson, 1994; Williams, 1994; de Jong & Williams, 1995; Oliver, 1995; Mark, 1998; Perring *et al.*, 2000; Rubenach & Lewthwaite, 2002; Oliver *et al.*, 2004; Mark *et al.*, 2004; Rubenach, 2005). The albite (which may be accompanied by magnetite) is generally inferred to be of metasomatic origin, formed by the action of saline fluids (Oliver & Wall, 1987) on metasedimentary and meta-igneous rocks. The rocks affected, with variable intensity, include metaclastic rocks, marbles/calcsilicate rocks, metadolerites, early felsic 'porphyries' and major granite intrusions (de Jong & Williams, 1995; Oliver, 1995).

Mineralogical changes involved in the Na-Ca metasomatism include albite replacing K-feldspar and oligoclase, as well as pyroxene replacing amphibole (Perring *et al.*, 2000). The assemblage: Na plagioclase + actinolite + titanite ± quartz ± magnetite ± diopsidic clinopyroxene is common (Williams, 1988; de Jong & Williams, 1995).

The evidence for metasomatism (Williams, 1994; Mark, 1998; Rubenach & Lewthwaite, 2002) consists of irregular albite veins, albite net veins, albite-rich shear zones, breccias rich in actinolite and magnetite, albite-actinolite breccia replacing metadolerite, axial-surface veins of actinolite with albite selvedges in folds, albite-diopside-magnetite replacements of laminated metasedimentary rocks, and the preservation in albite-rich aggregates of original sedimentary and igneous structures that are continuous into adjacent albite-free rocks.

The albitization, involving gain of Na and loss of Ca and K, has been inferred to have occurred during contact metamorphism (Oliver, 1995), amphibolite facies regional metamorphism (Oliver, 1995), and after the peak of metamorphism (de Jong & Williams, 1995; Oliver, 1995; Mark & de Jong, 1996; Rubenach & Barker, 1998), and is spatially and temporally related to granite intrusions (Oliver, 1995; Mark & de Jong, 1996). Most of the metasomatism post-dates the peak of metamorphism and associated penetrative tectonic structures, and occurred at a range of retrograde conditions (Williams, 1988, 1994; de Jong & Williams, 1995). Four phases of deformation-related hydrothermal activity occurred from at least 1750 to 1000 Ma ago, intense metasomatism occurring in all phases (Oliver, 1995). The alteration took place in multiple, overprinting, fluid-buffered systems, operating in fluid-overpressured brittle-ductile shear zones, fracture systems and megabreccia systems (Williams, 1988; de Jong & Williams, 1995).

The metasomatizing fluids have been inferred to be saline and hotter than 500°C, at $\geq$300–350 MPa (Mark, 1998); some may have contained CO_2 (Perring *et al.*, 2000). The fluids may have interacted with evaporite-bearing metasedimentary sequences, especially in the earlier phases of alteration (Oliver, 1995; Barton & Johnson, 1996), and may have interacted with scapolite-bearing metamorphic rocks in the later phases (Oliver, 1995). Cu-Al minerals are commonly associated with late, lower T ($<$350°C) K alteration, in the form of new biotite and/or K-feldspar with hematite $\pm$ epidote, as well as silication, which overprint earlier Na-Ca alteration, and are highly localized in dilational fractures (Williams, 1988; de Jong & Williams, 1995). The ore-forming fluids may have leached metals during the Na-Ca alteration, depositing them later in structurally controlled sites (Dilles & Einaudi, 1992; Hitzman *et al.*, 1992; Barton & Johnson, 1996; Perring *et al.*, 2000).

Scapolite-rich rocks of regional extent occur in the Cloncurry area, Mount Isa block, north-west Queensland, Australia, the scapolite occurring in calcareous and non-calcareous metapelites, felsic and mafic metavolcanic rocks and metadolerites. Edwards & Baker (1954) inferred regional metasomatic scapolitization of metasedimentary rocks, but Ramsay & Davidson (1970) inferred isochemical metamorphism of shales and marls containing evaporitic halite, the meta-igneous rocks being metasomatized by fluids derived from the evaporitic metasedimentary rocks.

Willyama Inlier, Australia: Albite-rich gneisses and granofelses are scattered through the Proterozoic Willyama Supergroup, both in the Broken Hill Block (western New South Wales) and the Olary Block (eastern South Australia). They are commonly layered (Figs. 5.16, 5.17) and contain up to 11.06% Na_2O at Broken Hill (Vernon, 1961, table 1) and 10.91% Na_2O at Olary (Clark *et al.*, 2005, table 2). The layering suggests original sedimentary bedding, and some possible cross-bedding is present, though it could

Fig. 5.16 Layered albite-rich rock, Olary area, South Australia, showing extensive late fracturing and albite veining.

be truncated folding (Section 7.2.2). The most sodic albite-rich rocks are more sodic than the keratophyric rocks or evaporitic analcite tuffs that have been suggested as possible precursors (Coombs, 1965; Plimer, 1977; Brown *et al.*, 1983; Cook & Ashley, 1992). Therefore, at least some Na metasomatism appears to have been involved in their formation (Vernon,

Fig. 5.17 Layered albite-rich rock, Stirling Vale area, Broken Hill, Australia, South Australia, showing extensive late replacement by relatively coarse-grained albite aggregates. Knife 9 cm long.

1961; Kent *et al.*, 2000). Some or all of the albitization in the Broken Hill Block appears to be post-peak metamorphism, because of (1) widespread occurrence of the lower-grade assemblage: quartz + muscovite in albite-rich rocks in granulite facies areas, and (2) partial albitization of K-feldspar in pegmatites and leucosomes in migmatitic metasedimentary rocks adjacent to the albite-rich bodies (Vernon, 1961, plate 15).

The albite-rich rocks in the Olary Block were also formed late in the metamorphic history, namely the retrograde stages of a major amphibolite facies metamorphism (Kent *et al.*, 2000). They are fine-grained and finely bedded (Fig. 5.16), and have been interpreted as (1) Na-rich, evaporitic sediments of felsic volcanic provenance (Cook & Ashley, 1992), (2) epiclastic or pyroclastic rhyolitic rocks that underwent pre- or synmetamorphic Na metasomatism (Pepper & Ashley, 1998), or (3) synmetamorphic Na-metasomatized quartzofeldspathic sediments (Kent *et al.*, 2000). They show evidence of repeated veining by albite (Fig. 5.16), as do the Broken Hill albite-rich rocks (Fig. 5.17). This repeated albitization of initially albitic rock layers during regional metamorphism may be responsible for the extremely high sodium contents of some of these rocks. Plimer (1977) ruled out a metasomatic origin for the Broken Hill albite-rich rocks on the basis that the contacts with adjacent rocks are not gradational and that original structures are preserved. However, as discussed previously, sharp contacts are the rule in metasomatic rocks, and faithfully preserved structures are also common. The source of the metasomatizing fluids in the Willyama Inlier are obscure, though oxygen isotope data suggest they may have been derived from devolatilization of deeper crustal rocks (Kent *et al.*, 2000).

5.16.3 Regional Fe metasomatism

Evidence of local Fe metasomatism in a regional metamorphic terrane is provided by grandite garnet-rich zones at the margins of marble and marl layers in the Reynolds Range Group, central Australia (Cartwright *et al.*, 1995). More extensive Fe metasomatism occurs in the form of Fe-oxide bodies, of which thousands occur globally in Proterozoic and Phanerozoic rocks (Barton & Johnson, 1996), commonly with abundant Na metasomatism. The orebodies occur with many types of igneous rocks in various geological settings. Some are strata-bound, massive Fe oxide rocks in metavolcanic and metasedimentary rocks, but most are discordant, variably brecciated and tabular to irregular in shape. They consist of abundant Ti-poor Fe oxide minerals (magnetite or hematite), and have a characteristic suite of minor elements (Cu, U, Au, Ag, Co and rare earth elements) responsible for phosphate minerals (apatite and rare earth phosphates), Cu-Fe sulphide minerals, and sporadic Au, U, Ag and Co minerals (Barton & Johnson, 1996). The orebodies typically show evidence of hydrothermal alteration and some possible magmatic connection (Hitzman *et al.*, 1992), though the abundance of hydrothermal alteration, even in rocks of mafic igneous composition, suggests non-magmatic controls of the infiltrating solutions (Barton & Johnson, 1996). Early, deeper alteration consists of magnetite ± apatite rocks, commonly with abundant sodic (albite ± scapolite + hornblende) alteration, superimposed on which may be later hydrolytic ± potassic alteration with hematite ± Cu-Fe sulphide ± REE minerals (Barton & Johnson, 1996). Mafic igneous rocks have abundant early metasomatic scapolite with hornblende, and late chlorite-carbonate alteration. Felsic rocks only rarely have scapolite, and instead show early albitic ± potassic alteration and later silicic-'sericitic' alteration. Carbonate host rocks have calcic or magnesian skarn assemblages, with hydrous assemblages superimposed on anhydrous assemblages (see Section 5.16.4).

Banded iron-formations in regional metamorphic terranes (Chapter 1) commonly show evidence of large-scale Fe concentration, but typically this is due to hydrothermal or supergene redistribution of Fe during deposition or diagenesis, and especially to supergene enrichment produced where erosion exposes existing iron-formation (Morris, 1985).

5.16.4 Regional skarns

Anhydrous pyroxene-garnet skarns, with superimposed hydrous actinolite-chlorite-carbonate assemblages, may occur in carbonate host rocks associated with regional metasomatic Fe oxide bodies, as discussed in the previous section (Barton & Johnson, 1996), and with some silver-base metal orebodies.

The South Australian Olary Block of the Proterozoic Willyama Inlier contains numerous skarns in the form of andradite-rich garnet-epidote alteration zones, and breccias with a clinopyroxene-actinolite matrix, formed in the retrograde stages (at400 °C–650 °C) of a major amphibolite facies regional metamorphism (Kent *et al.*, 2000). The breccias have a complex hydrothermal history involving fluid pressure fluctuations (Clark & James, 2003). The zones of andradite-rich garnet and epidote occur as veins, breccias, space-filling aggregates and massive replacement bodies in calcsilicate host rocks, reflecting locally high fluid pressures and high fluid:rock ratios. Addition of Ca, Mn, Cu and Fe^{3+} and loss of Na, Mg, Rb and Fe^{2+} occurred in the metasomatism, which was promoted by hypersaline, oxidized, chemically complex fluids containing Na, Ca, Fe^{3+}, Cl and $(SO_4)^{2-}$. The source of the fluids is unknown, but oxygen isotope data suggest the fluids may have been derived from devolatilization of deeper crustal rocks (Kent *et al.*, 2000).

5.16.5 Regional Si metasomatism

Evidence of local silica metasomatism in a regional metamorphic terrane is provided by marbles with up to 66 per cent of wollastonite formed by fluid infiltration during high-temperature (700 °C) regional metamorphism in the Reynolds Range Group, central Australia (Cartwright *et al.*, 1995). Because the interlayered rocks do not contain enough Si to make so much wollastonite, Cartwright *et al.* (1995) inferred that Si metasomatism occurred, contributing to the reaction: calcite + quartz = wollastonite + CO_2.

5.16.6 Carbonic metasomatism?

Phenomena involving arrested in-situ formation of granulite from adjacent amphibolite facies gneiss across southern India and Sri Lanka (Janardhan *et al.*, 1979) have been explained as the products of structurally controlled metasomatism and dehydration induced by carbonic fluids ascending from an unknown source (Newton, 1990, 1992; Jackson & Santosh, 1992). This interpretation has been questioned on the basis of field, petrological, geochemical and isotopic data (Burton & O'Nions, 1990; Raith & Srikantappa,

1993; Harley & Santosh, 1995), as the observed relationships may also be explained by partial melting (Sandiford *et al.*, 1988; Burton & O'Nions, 1990), which could have dehydrated the rocks by preferential absorption of H_2O into the melt.

A similar example involves a series of striking migmatitic structures in rectilinear networks through western Fiordland, New Zealand, involving, for the most part, narrow anorthositic dykes that cut hornblende-bearing orthogneiss (Blattner, 1976). Adjacent to the dykes, host rocks show patchy, spatially restricted recrystallization and dehydration to garnet granulite, on a decimetre scale. Major element differences between the host hornblende granulite and garnet granulite are restricted to variations in Na and Cu content (Daczko *et al.*, 2001). A variety of causal processes has been inferred, including: metasomatism due to the ingress of a carbonic, mantle-derived fluid (Bradshaw, 1989), hornblende breakdown leading to water release and limited partial melting of host rocks (Blattner, 1976), and dehydration induced by volatile scavenging by a migrating silicate melt (Daczko *et al.*, 2001). Variability in dyke assemblages, together with a correlation between dehydration structures and host-rock silica content, are consistent with open-system behaviour involving volatile scavenging by a migrating trondhjemitic liquid (Clarke *et al.*, 2005).

5.17 Metasomatism and ore deposits in metamorphic terranes

5.17.1 Metamorphism superimposed on regional metasomatism

Some metallic rocks (called 'ore deposits' if they are of economic value) in metamorphic terranes result from metamorphism of pre-existing sulphide or oxide rocks of sedimentary and/or volcanic origin (e.g., Parr *et al.*, 2004), though the origin of some stratiform sulphide-rich rocks with metamorphic structural characteristics are controversial, mainly because high-grade metamorphism and deformation tend to obscure original features.

Unusual or unexpected mineral assemblages in and adjacent to some other orebodies reflect metasomatic hydrothermal alteration before regional metamorphism. For example, if felsic rocks hydrothermally metasomatized at low temperatures (forming aluminous clay minerals or pyrophyllite, by argillic alteration) are later metamorphosed, abundant higher-grade aluminous minerals (for example, andalusite, sillimanite or cordierite) may be formed. These are minerals that would not otherwise have formed in rocks of felsic igneous composition (Gresens, 1971; Vernon *et al.*, 1987; McLelland *et al.*, 2002a,b). Such unexpected aluminous rocks may be indicators of former wall-rock alteration around potential ore deposits in metamorphosed terranes. For example, Sykes & Moody (1978) suggested that early argillic alteration was followed by regional metamorphism to form an andalusite-bearing assemblage at Hillsborough, North Carolina, USA.

An example of hydrothermally metasomatized rocks later subjected to amphibolite facies regional metamorphism is provided by the Big Bell orebody, Western Australia, inferred to be a high-grade equivalent of greenschist facies Archaean gold deposits (Phillips & Nooy, 1988). The rocks consist of gold-sulphide-bearing schists with muscovite and K-feldspar, as well as biotite schists containing a wider variety of complex metamorphic assemblages than those in the surrounding region. These assemblages reflect intense pre-metamorphic metasomatic alteration of mafic rocks, and so may act as a guide for exploring for similar ore deposits in high-grade metamorphic terranes.

The same may be said of protoliths of the Arinteiro-Bama copper deposits of Santiago de Compostela, Spain (P. J. Williams, 1983). The rocks of the ore horizon are hydrothermally altered mafic rocks. Though the original structures have been obscured by amphibolite facies regional metamorphism, the assemblage of garnet + gedrite ± biotite ± staurolite (a highly unusual Al-rich assemblage in metamorphosed mafic rocks), plus the fact that the rocks are compositionally equivalent to high-Fe, hydrothermally altered (chloritized) mafic rocks, indicate that the high-grade metamorphism was isochemical with respect to non-volatile chemical components and so was superimposed on rocks previously metasomatized at low temperatures. All the altered rocks are enriched in Si, Fe, Mg and S and are depleted in Ca and Sr, relative to unaltered equivalents, except that biotite-rich horizons are enriched in K. Garnet-gedrite rocks could be useful ore exploration guides in other areas of metamorphosed volcanogenic sulphide-bearing rocks (P. J. Williams, 1983).

Some skarn-type orebodies (including possibly the Broken Hill Lode, New South Wales, Australia, which is a Ag-Pb-Zn orebody with minerals such as rhodonite, pyroxmangite, knebelite, calcite and fluorite, among many others) may be the result of early metasomatism, on which high-grade regional metamorphism has been superimposed.

5.17.2 Metasomatism accompanying regional metamorphism

Metallic rocks and accompanying metasomatism may form in metamorphic terranes during metamorphism and deformation, through the action of fluids. The processes involved provide examples of the various types of metasomatic reactions discussed in Section 5.12, such as hydration, hydrolysis reactions and skarn formation. As fluids carry metallic elements in solution (Yardley, 2005), fluid-rock interactions are important in the formation of many metallic rocks.

Some of these fluids have been inferred to be of metamorphic origin, produced by dehydration reactions (e.g., Groves & Phillips, 1987; Barnicoat *et al.*, 1991), whereas others may be released from granite magmas intruded during metamorphism. Evidence is accumulating that CO_2-rich fluids may be involved, as suggested by metal-bearing CO_2-rich fluid inclusions in the minerals of some granite-related Au, Fe oxide-Cu-Au and Sn-W orebodies in regional metamorphic rocks (Perring *et al.*, 2000).

'Orogenic' gold deposits typically occur in terranes subjected to low- to moderate-*P*, low- to high-*T* metamorphism (Powell *et al.*, 1991; Ridley & Diamond, 2002; Ridley *et al.*, 2000; Groves *et al.*, 2003). They occur in Archaean to Tertiary rocks, and are the main types of gold deposits in metamorphic belts. They are formed in all stages of orogenic evolution, later stages overprinting earlier ones (Groves *et al.*, 2003). Several are large enough to be classed as 'giant' orebodies (>250 t Au). They are integral products of subduction-related accretionary or collisional terranes, and develop in arc, back-arc or accretionary prism terranes, consisting predominantly of mafic/ultramafic, sedimentary or granitic rocks (Groves *et al.*, 2003). They form mainly at greenschist facies conditions, but may form at up to

granulite facies conditions, with the result that minerals resulting from wall-rock alteration vary from carbonate to diopside and from muscovite to biotite/phlogopite. Variable amounts of other metals (Ag, As, B, Bi, Cu, Pb, Sb, Te, W, Zn) are associated with the gold (Groves *et al.*, 2003).

Orogenic gold deposits show evidence of structural control, in the form of veins, vein arrays, and strata-bound replacement bodies, but may also be disseminated. The fluid and metal sources are controversial, but have been attributed to large volumes of hydrous fluid released from prograde devolatilization reactions in mafic and ultramafic rocks, scavenging gold and sulphur from the rocks (Powell *et al.*, 1991). However, large volumes of granitic magma are commonly produced during the metamorphism, and these may be distally connected with the gold deposits (Groves, *et al.*, 2003), though they are likely to be mixed with and/or dominated by simultaneously generated metamorphic fluids (Powell *et al.*, 1991).

Archaean greenstone gold deposits, which are among the world's largest, have worldwide similarities, though they show a wide variety of styles. As summarized by Phillips & Groves (1983) and Groves & Phillips (1987), the deposits are epigenetic (occurring as replacement bodies), show distinctive wall-rock alteration (forming Fe sulphide, K-mica ± albite, and Ca-Mg-Fe carbonate minerals) and are characterized by distinctive metal associations involving Au, Ag, As, Sb, W and B with low proportions of base metals. The gold-bearing hydrothermal systems were synchronous with regional metamorphism and emplacement of granites. Gold deposition occurred at greenschist facies metamorphic conditions (300°C–400°C and 1–2 kbar), mainly in response to decreasing gold solubility with cooling, though many large gold orebodies formed by sulphidation of Fe-rich host rocks, with simultaneous deposition of Fe-sulphide minerals and gold (Groves & Phillips, 1987).

The low salinity, reduced, near-neutral, H_2O-CO_2 fluids carried gold as reduced sulphur complexes (Groves & Phillips, 1987), and had X_{CO_2} up to 0.3 (Mikucki & Ridley, 1993; Powell *et al*, 1991). The inferred constancy of the concentration of major ore-fluid components (e.g., CO_2 and Cl ± K) suggest fluid buffering and high fluid-rock ratios along fluid pathways (Mikucki & Ridley, 1993). The origin of the fluids is in doubt. Though the source regions have been inferred to be external to the greenstone belts themselves (Mikucki & Ridley, 1993), stable isotope data are consistent with either a magmatic or metamorphic source (Groves & Phillips, 1987; Ridley *et al.*, 2000). However, the absence of granites near some large gold orebodies favours a metamorphic origin for the fluids, possibly due to devolatilization of the deeper parts of the greenstone succession (Groves & Phillips, 1987; Powell *et al.*, 1991) – an interpretation supported by the work of Barnicoat *et al.* (1991), discussed below.

Gold deposits formed at middle to upper amphibolite facies and granulite facies conditions in the Yilgarn Block, Western Australia, involved infiltration of H_2O-CO_2 (X_{CO_2} = 0.1–0.25) fluids of low salinity that deposited gold at 250°C–400°C and 1–3 kbar within *c.* 100°C of the metamorphic peak (Barnicoat *et al.*, 1991). These deposits suggest that at least some of the gold-bearing fluids in granite-greenstone terranes originate in or below the middle crust. Though the source of the ore fluids in these terranes is a matter of debate, chemical and isotopic evidence suggests a deep crustal source for some components, the fluids being modified along conduits on the way to depositional sites in greenschist facies domains (Barnicoat *et al.*, 1991).

Quartz vein-hosted gold deposits in turbidite sequences in slate belts occur in fault- and fold-controlled dilatant fractures that are generated where fluid pressure locally exceeds the rock (lithostatic) pressure (Cox *et al.*, 1991). Wall-rock hydrothermal alteration assemblages, as well as chemical and isotopic data, indicate that large volumes of C-O-H metamorphic fluids, channelled into high-permeability fault zones, are involved. The gold is deposited where oxidized gold-bearing 'primary' fluid intersects dilational fracture networks and mixes with reduced CH_4-bearing 'secondary' fluid formed by the interaction of the primary fluid with graphitic slates (Cox *et al.*, 1991).

Intrusion-related gold deposits may be a distinct group of gold deposits in regional metamorphic terranes, as they occur in the proximal parts of magmatic-hydrothermal systems, rather than resulting from a more widespread metamorphic fluid (Groves *et al.*, 2003). They are mainly Phanerozoic and contain no 'giant' orebodies.

Silver-base metal ore deposits may be associated with metasomatic hydrothermal alteration involving extreme acid leaching (advanced argillic alteration) in solfataric volcanic systems, for example in Colorado, Nevada and Spain (Harvey & Vitaliano, 1964; Hemley *et al.*, 1980). Secondary minerals include kaolinite, diaspore, pyrophyllite and alunite. Similarly, porphyry copper deposits show a range of metasomatic alteration types, including advanced argillic alteration in higher levels of the system, producing minerals such as 'sericite', pyrophyllite and quartz, with smaller amounts of andalusite, overprinted by later, lower-temperature diaspore, pyrophyllite and alunite (Gustafson & Hunt, 1975; Meyer *et al.*, 1968; Hemley *et al.*, 1980). Deeper levels of these systems may be characterized by abundant Na-K feldspathization of plagioclase, with andalusite and rare corundum (Hemley *et al.*, 1980). Na tends to be fixed in hotter, lowest levels and K tends to be fixed higher in the system, where convective fluids are cooling, as expected from experimental work (Hemley *et al.*, 1980) and exemplified in the Yerington porphyry copper deposit, Nevada, USA (Carten, 1986).

Alteration and mineralization around some massive sulphide deposits are characterized by fixation of seawater Mg in the form of hydrolytic chloritization of feldspar, with concomitant loss of Ca and alkalies, according to the reaction: anorthite $+ 5Mg^{2+} + 8H_2O + SiO_2 =$ clinochlore (chlorite) $+ Ca^{2+} + 8H^+$ (Hemley *et al.*, 1980). The Mg metasomatism increases with increasing temperature, as indicated by experiments, and occurs in the root zones of massive sulphide convective systems. Chloritization of mafic minerals is mainly a hydration process, with minor hydrolytic base cation leaching. Reaction zones of fluids discharging from 'black smokers' at mid-ocean ridges at 300°C–400°C may involve Mg, Ca and Na fixation reactions proceeding to different degrees at different places in the submarine geothermal system; Mg fixation is characteristic of downwelling limbs of sub-seafloor convection cells, whereas Ca and Na fixation are related more to hydrothermal up-flow zones (Seyfreid *et al.*, 1988). At 250°C–450°C, the following assemblages have been produced in response to increasing seawater:rock ratios: Chl-Ab-Ep-Act, Chl-Ab-Ep-Act-Qtz, Chl-Ab-Qtz and Chl-Qtz, the chlorite becoming progressively more magnesian (Mottl, 1983).

Aluminous assemblages with andalusite or kyanite may form in some less mafic systems, owing to Si leaching (Hemley *et al.*, 1980). An example is the Cyprus style of alteration/mineralization, in which the rocks beneath pyrite-chalcopyrite orebodies are characterized by chlorite-illite-kaolinite alteration of the mafic host rocks, with a decreasing degree of argillic alteration as chloritization increases (Callaghan, 1966; Searle, 1972; Constantinou & Govett, 1973). The near-surface, high-sulphide zone is characterized by marked addition of Fe, slight loss of Al, and marked loss of Ca, Na and Mg, whereas lower levels are characterized by Mg fixation, which displaced K upwards, to form illite in higher-level rocks richer in sulphide minerals.

Some regional metamorphic orebodies are associated with the synmetamorphic formation of skarns at upper amphibolite facies conditions. For example, in the Proterozoic Cannington Ag-Pb-Zn ore deposit, northeastern Australia, skarns consisting of Ca-rich garnet ± pyroxmangite and hedenbergite, with fayalite, apatite and quartz, overprinted by sulphide minerals, were formed by initially anhydrous, then hydrous infiltration metasomatism controlled by fractures (Chapman & Williams, 1998; Roache *et al.*, 2005). The skarn-forming process involved addition of Ca, Fe, Mn, Ca, P, C and Si, and depletion in alkalies. Hydrous Fe-K-Cl-rich assemblages characterized by hornblende, biotite, pyrosmalite and dannemorite occur in veins and replacing the anhydrous skarns. Still later sulphide-magnetite-fluorite aggregates occur with more retrograde hydration, producing phyllosilicates, such as greenalite (Chapman & Williams, 1998). Ag-Pb ore is associated with pyroxmangite (Fe-Mn areas), and Zn (+ Cu, As) ore is concentrated with hedenbergite (Ca-Fe areas), implying either redistribution of sulphides during metamorphism or transport of metals to the deposition sites during a later part of the same alteration system responsible for the anhydrous skarns (Chapman & Williams, 1998).

Another example in the same region is provided by Zn-bearing skarns in at the Maramungee prospect and in small isolated outcrops along a 90-km zone with sulphide minerals in the eastern Selwyn Range area (Williams & Heinemann, 1993; Williams & Baker, 1995). Exoskarns developed preferentially in calcareous calcsilicate parts of amphibolite facies metasedimentary gneisses consist of Mn hedenbergite + Fe-Mn-rich grossular + quartz + apatite ± plagioclase (involving addition of Mn and Fe). In the metabasites and meta-intermediate igneous rocks, endoskarns consisting of Ca-rich almandine ± clinopyroxene (involving addition of Ca and loss of Mg and alkalies), are restricted to narrow zones at the contacts of less fractured plutons, but are more extensive in fractured plutons. The skarns contain some of the potentially economic minerals. Most of the skarns probably were formed after the metamorphic peak. Early anhydrous assemblages may show replacement by retrograde, variably hydrous, oxidized and Cl-rich assemblages containing epidote-clinozoisite, K + Cl-rich hornblende, muscovite and chlorite with sulphide minerals (Williams & Baker, 1995).

Fe oxide deposits are common in Proterozoic to Phanerozoic rocks, typically associated with sodic and other metasomatic alteration, as discussed in Section 5.16.3. Some Fe oxide deposits are associated with the formation of regional metasomatic skarns (Section 5.16.4). Fe-rich orebodies derived from banded iron-formation (Section 7.2.4) by hydrothermal or, more commonly, supergene secondary iron enrichment represent the largest and most concentrated secondary accumulations of any single metalliferous element in Earth's crust, rivalled only by concentrations of aluminium in some bauxitic deposits. These concentrations involve a transfer of immense amounts of material in solution, without leaving obvious evidence of the passage of these solutions (Morris, 1985, p. 785).

Chapter 6

Deformation of Metamorphic Rocks

6.1 Introduction

This chapter is concerned with the main tectonic structures of metamorphic rocks, both microscopic and mesoscopic, but does not deal with detailed structural analysis (for example, fold or fault analysis), strain analysis or detailed deformation processes, which are discussed by Ramsay (1967), Spry (1969), Hobbs *et al.* (1976), Nicolas & Poirier (1976), Vernon (1976), Passchier & Trouw (1996), Blenkinsop (2000) and Vernon (2004).

Metamorphic rocks are commonly subjected to tectonic deformation, with the result that they undergo a change of shape (*strain*). The type and amount of strain depend on both the local directed pressure (*non-hydrostatic stress; deviatoric stress*) and the mechanical properties of the rock. Deformed rocks are sometimes known as *tectonites*.

Because deformation in Earth's crust and mantle cannot be observed in progress, it is necessary to rely on experimental deformation of rocks and minerals at various temperatures and pressures, to learn about the processes involved. Deformation experiments on minerals have been possible since the development of various types of high-pressure, high-temperature deformation apparatus in the 1960s, and now a great deal of experimental information relevant to grain shapes, deformation mechanisms, recovery-recrystallization processes and preferred orientation of grains in rocks is available.

Most deformation experiments have involved single minerals, such as quartz, calcite, olivine, ice, halite, gypsum, feldspar, mica, pyroxene and sulphide minerals. Some of the results can be applied directly to the interpretation of microstructures of mineralogically simple rocks, such as quartzite, marble, dunite, pyroxenite, anorthosite, glacier ice and sulphide rocks. Some experiments have involved the deformation of two-phase materials, such as salt-mica (Means & Williams, 1974) and ice-mica (Wilson, 1984), which also help to make more realistic interpretations of deformation in natural rocks.

However, because most common rocks are chemically complex systems involving several minerals deforming at different rates, deformation of natural rocks is more complicated than in most experiments. Moreover, natural strain rates are much slower than those in most experiments, suggesting that caution should be exercised when applying experimental results to rocks. Nevertheless, the relatively few experiments on actual rocks (e.g., Shea & Kronenberg, 1993) have provided evidence on deformation contrasts between different minerals, as well as on mechanisms of strain localization. In addition, during metamorphism, chemical reactions can affect deformation processes and vice versa, as discussed in Section 6.5. Experiments on complex systems, especially with active chemical reactions and fluids, are more difficult to carry out, but are being undertaken, especially in systems involving melting (e.g., Arzi, 1978; van der Molen & Paterson, 1979; Rushmer, 1995; Wolf & Wyllie, 1991; Rutter & Neumann, 1995).

Deformation of low-melting-temperature minerals can be observed directly in experiments conducted in a deformation apparatus mounted in a microscope. These have been carried out for ice, using transmitted light (Wilson, 1986) and stibnite, using reflected light (McQueen *et al.*, 1980). Another approach, which has produced stimulating ideas for the interpretation of rock microstructures, is the use of low-temperature organic compounds as analogues of natural minerals. Because these compounds can be deformed, recrystallized and melted at room temperature, and because they are transparent (some helpfully coloured), changes in grain shape and distribution can be observed, photographed and videotaped in transmitted light in a microscope during the actual deformation of a thin section (Urai *et al.*, 1980; Means & Xia, 1981; Means & Jessell, 1986; Means, 1989; Means & Park, 1994; Park & Means, 1996). The main advantage of this 'see-through' experimental technique is that the entire microstructural history can be recorded, from the original to the deformed states, whereas in conventional experiments, only the starting material and deformed product can be examined. This is because experiments on most minerals must be conducted at high temperatures and pressures in special confined apparatus, in order to achieve ductile flow, so that direct observation has not been possible. The main disadvantage of the synkinematic microscopy technique is that very few real minerals are involved. Nevertheless, many of the microstructures produced resemble those in deformed natural rocks. Morever, several unexpected grain-scale processes have been observed, which argues for caution in the interpretation of rock microstructures.

6.2 Outline of deformation processes

Deformation mechanisms can be classified in various ways, but broadly speaking, *brittle* and *ductile* deformation can be distinguished at the microscope scale. In brittle deformation, fractures occur across and/or between grains, and the resulting fragments move relative to each other. In ductile deformation, the grains change their shapes or move relative to each other without fracturing (loss of cohesion) at the grain scale (e.g., Passchier & Trouw, 1996, p. 21). In both situations, but especially during brittle deformation, a change of shape of an aggregate may be accomplished or assisted by dissolution of minerals at some sites, transfer of dissolved chemical components in solution, and deposition at other sites in the deforming aggregate (*stress-induced solution transfer*).

6.2.1 Brittle deformation

Deformation by fracture and frictional processes is called *cataclastic flow* (*cataclasis*), and involves the formation of new surfaces, loss of cohesion by

Fig. 6.1 Fragmental deformed granite (cataclasite) from the Hunter Thrust, Mitchell's Flat, north-east of Singleton, New South Wales, Australia, showing microstructural evidence of brittle deformation in the form of angular fragments, of various sizes, of quartz (clear), feldspar (cloudy, owing to fine-grained alteration products) and quartz-feldspar aggregates. Sample kindly provided by Tanya Wilson. Plane-polarized light; base of photo 4 mm.

fracturing, and frictional sliding along the fracture surfaces. Microstructures indicative of cataclastic flow include microfractures and displacements on cleavages, as well as rotations or displacements of rigid particles without internal (crystal-plastic) deformation (Fig. 6.1). However, if microcracking is fine and pervasive, the deformation may appear to be ductile at the scale of the light microscope, although transmission electron microscopy (TEM) reveals the microfractures (Tullis & Yund, 1987; Green, 1992). Grain-scale, transient fracturing may occur in many metamorphic rocks undergoing prograde metamorphism, as it is probably the main mechanism for removing fluid produced in devolatilization reactions, though typically it leaves no microstructural evidence in the resulting rock.

Deformation controlled by fracture processes involves the formation of thoroughgoing fractures and movement along the fractures. Details of fracture processes in deformation and the various mechanisms of fracturing have been discussed by Knipe (1989, pp. 135–137) and Blenkinsop (2000). *Frictional grain-boundary sliding* involves the sliding of grains past each other, without the development of thoroughgoing fractures. This type of brittle deformation is also called 'independent particulate flow' (Borradaile, 1981). The sliding depends on loss of cohesion and the overcoming of friction between grains, and so is distinguished from high-temperature grain-boundary sliding (see below), in which cohesion between grains is maintained (i.e., no fractures are formed). Frictional grain-boundary sliding is favoured by low confining pressure, as well as high fluid pressure, which reduces the 'effective pressure'; i.e., the fluid pressure reduces grain-to-grain contacts. It is most common in slumping and faulting in unconsolidated sediments (Paterson & Tobisch, 1983; Maltman, 1981), as well as in fault gouges and cataclasites.

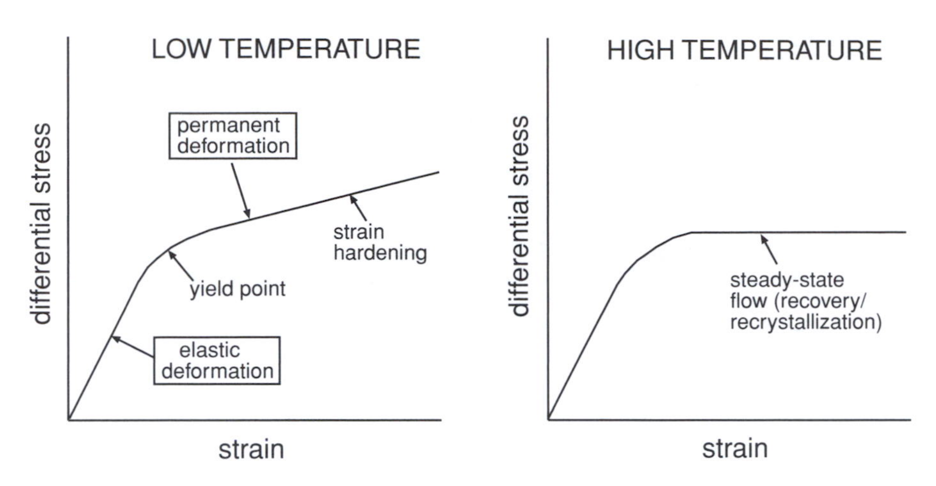

Fig. 6.2 General forms of typical stress-strain curves for experimental deformation at constant strain rate and temperature. Differential stress (stress difference) is the difference between the maximum and minimum compressive stresses exerted on the mineral sample. An initial elastic (recoverable) deformation is followed (at the yield point) by permanent (plastic) deformation. At lower temperatures, the mineral becomes progressively harder to deform (i.e., it undergoes strain strengthening or strain 'hardening'), owing to tangling of dislocations. In contrast, at higher temperatures, deformation can proceed at relatively low, approximately constant differential stress, owing to recovery and recrystallization during deformation.

In mylonitic deformation (Section 6.11), some minerals may deform by cohesive flow and others by fracture processes in the same rock at the same time. Though the overall process is one of ductile deformation (Vernon, 1974), because cohesive processes dominate over fracture processes, some of the deformation involves fracturing. Furthermore, brittle and ductile processes may alternate, even during the deformation of the same mineral (McLaren & Pryer, 2001).

6.2.2 Ductile deformation

As summarized by Paterson (2001), ductile flow of rocks can occur by the following three main mechanisms: (1) change of shape of grains by *crystal plasticity*, which is referred to as *dislocation creep*, (2) change of grain shape by diffusion through or around grains, called *diffusion creep*, and (3) relative movement of grains, referred to as *granular flow* or *grain-boundary sliding*.

In experiments on the ductile flow of minerals at constant strain rate and temperature, an initial elastic (recoverable) deformation is followed (at the 'yield point') by permanent (plastic) deformation (Fig. 6.2). For plastic flow, the strain rate depends on several factors, including the differential stress (the difference between the maximum and minimum compressive stress acting on the mineral or rock sample), the amount of accumulated strain, the temperature, the fluid pressure, the grainsize, the activity of chemical components, and the distribution of small grains of other minerals.

6.2.3 Crystal plasticity (dislocation creep)

Crystal plasticity (*crystal plastic flow*) is permanent deformation by non-cataclastic flow (*ductile flow*). The flow involves slip (*translation gliding*) and/or *deformation twinning*, without loss of cohesion on the grain scale

Fig. 6.3 Diagrams showing the general processes of slip and deformation twinning. Twinning produces a change in crystallographic orientation, which appears as a change in absorption colour and/or interference colour in thin section, whereas slip produces no such optical effects.

(Fig. 6.3). These processes enable a grain to change its shape by allowing one part of the crystal to undergo shear with respect to a neighbouring part (Hobbs *et al.*, 1976). Microstructural evidence of crystal plastic deformation includes *kink bands* (Fig. 6.4) and *deformation twins* (Fig. 6.5), which have been produced experimentally.

Fig. 6.4 Biotite porphyroblast in a schist, showing kink bands, revealed by difference in absorption colour, owing to their orientation differences. The widths and degrees of misorientation of the kink bands are variable, which distinguishes them from deformation twins, for which the misorientation is constant. From Vernon (2004, fig. 5.6). Plane-polarized light; base of photo 1.5 mm.

Fig. 6.5 Lenticular deformation twins in plagioclase, their concentration varying with local strain inside the grain. From Vernon (2004, fig. 3.89). Crossed polars; base of photo 1.5 mm.

Slip (translation gliding) is the main primary mechanism of crystal plastic deformation. It causes atomic layers of a grain to slide past each other without fracturing and without changing the orientation of the slipped portion of the grain. Therefore, it cannot be detected in thin section, though secondary structures resulting from slip, such as kink bands (Fig. 6.4) provide indirect evidence; this is in contrast to deformation twinning, in which a change of orientation is produced (Figs. 6.3 and 6.5). The shapes of grains are changed in the slip process; individual grains may become very elongated or may become converted to stretched out aggregates of much smaller new grains formed by *recrystallization* (Section 6.3.2) during the deformation.

Slip occurs on specific planes (commonly planes of dense atomic packing) and in specific directions in the crystal. A *slip system* is the combination of a slip plane and a slip direction in this plane. Slip systems have been determined for many minerals at various temperatures (e.g., Hobbs *et al.*, 1976, table 2.1; Nicolas & Poirier, 1976). Because of the crystallographic control of slip planes, ductile deformation of grain aggregates typically results in a strong *crystallographic preferred orientation*. Some minerals with relatively high crystallographic symmetry, such as quartz, olivine and calcite, have several slip systems and can deform relatively easily over a range of conditions. In contrast, many other minerals (such as mica, plagioclase, hornblende, pyroxene and kyanite) are of lower symmetry and may have only one dominant slip system. The more slip systems a mineral has, the more readily a grain of that mineral can change its shape in response to local differential stress.

Slip takes place by movement of *dislocations*, as explained by Hobbs *et al.* (1976) and Vernon (1976, 2004). Dislocations are *line defects* in crystals, by which one row of atoms is decoupled from the rest of the lattice (Fig. 6.6). Though in this sense a dislocation involves very local breaking of atomic bonds (hence the name), it isn't a fracture. Movement of a dislocation, in response to imposed stress, causes displacement of one part of a crystal

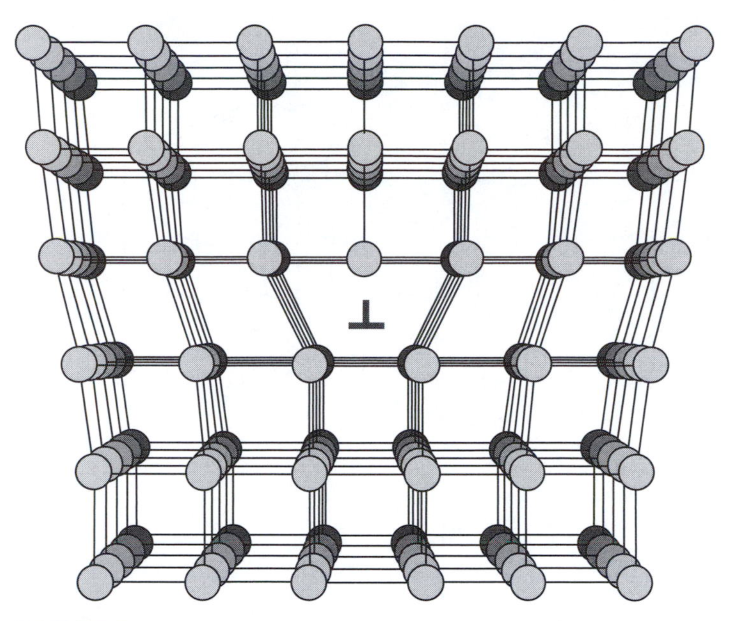

Fig. 6.6 Diagram of part of a hypothetical, simplified cubic crystal, showing an extra half-plane of atoms and the consequent point of emergence of a dislocation (line defect) represented by the inverted T symbol; i.e., the dislocation is a line pointing away from the viewer. This is an *edge dislocation* (see Fig. 6.7). Based on a diagram in Guy (1959, p. 110).

with respect to another (Fig. 6.7). The movement of dislocations along slip planes through crystals enables solid crystalline materials to change shape without breaking (Fig. 6.8A). Displacement can be achieved if all the atoms in a lattice row break at the same time, but this is fracture, and the stress required for flow has been found experimentally to be many orders of magnitude less than that required for fracture.

A good analogy is a large carpet on a floor. If you want to move it, you can take hold of one of the ends and try pulling it. However, this requires a lot of effort, owing to the weight of the carpet and the frictional forces that tend to stick it to the floor. A much easier way is to form a hump or bulge parallel to the edge of the carpet and move the hump along (Fig. 6.8B), which only requires overcoming the friction underneath the hump, not the whole carpet. As the hump is moved along, the carpet settles back onto the floor behind you and pulls away from the floor under the new position of the hump. Eventually the hump reaches the end of the carpet, which has then moved a small distance, equivalent to the size of the hump. If you make and move many of these humps parallel to each other, you will eventually move the carpet to where you want it.

A dislocation is analogous to the hump in the carpet. The force on the mineral grain causes one row of atoms at a time to break. Then the next row breaks and the one behind it joins together again. So, successive rows break, one at a time, until the 'break' (dislocation) moves right through the mineral, causing a displacement of one row of atoms (Fig. 6.8A). Many thousands of these minute displacements cause the mineral to change its visible shape. Each dislocation needs only a very small amount of energy,

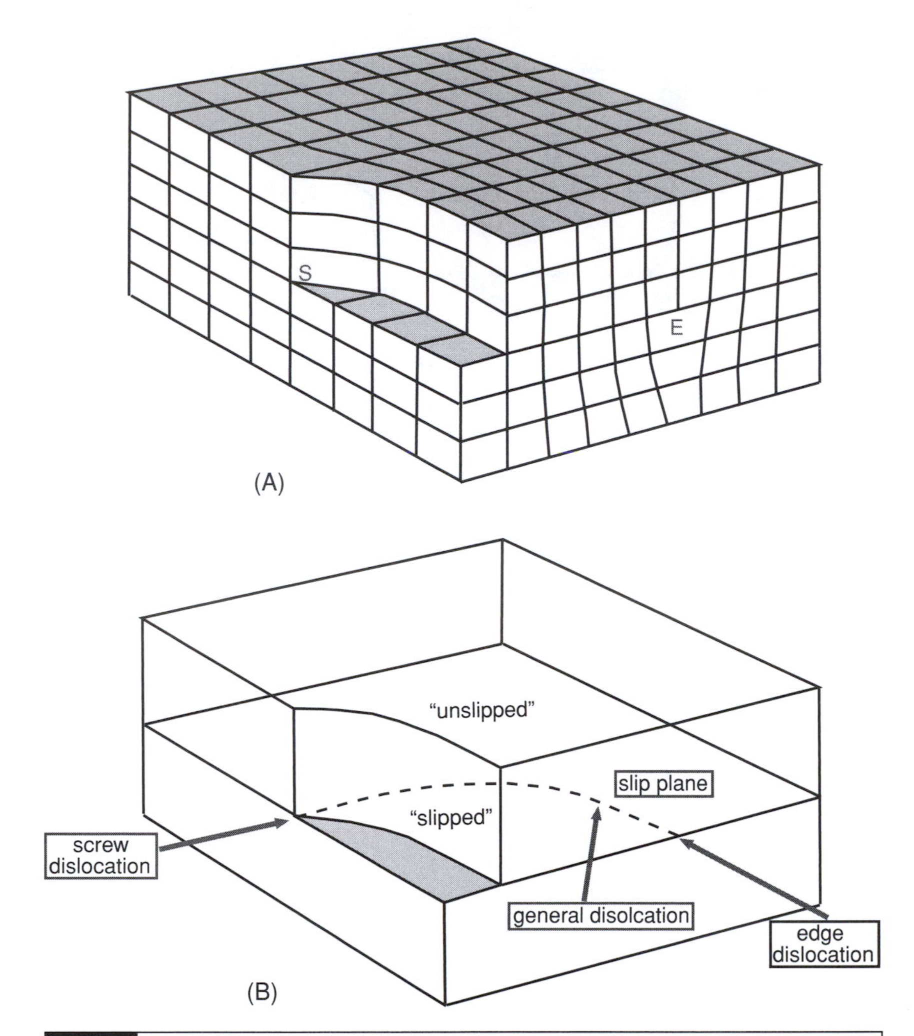

Fig. 6.7 (A) Diagram of dislocations inside portion of an idelalized cubic crystal, showing an *edge dislocation* (E) and a *screw dislocation* (S), joined by a *general dislocation*. (B) Diagram helping to visualize the situation shown in (A). The dislocation is a line separating the portions of the crystal that have been displaced (slipped) from portions that have not been displaced (unslipped).

and the process does not require the mineral to change the overall arrangement of its atoms, so that the mineral retains its identity during the deformation. The movement of dislocations is different from deformation by fracture, which involves the breaking of many rows of atoms at the same time.

Some dislocations are formed by growth mistakes as minerals crystallize. However, many more are generated by stress concentrations during deformation, especially near crystal defects, cracks and mineral inclusions. Images of dislocations are observable only in the transmission electron microscope (McLaren, 1991), as their strain effects are too small to affect light.

Broadly speaking, crystal plasticity may be divided into low- and high-temperature types (Fig. 6.2). *Low-temperature plasticity* occurs at less than roughly half the melting temperature at laboratory strain rates, and is

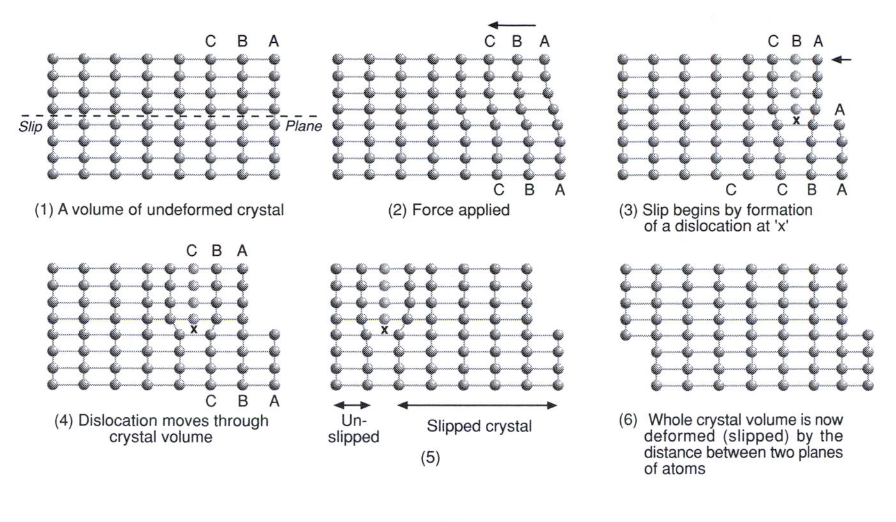

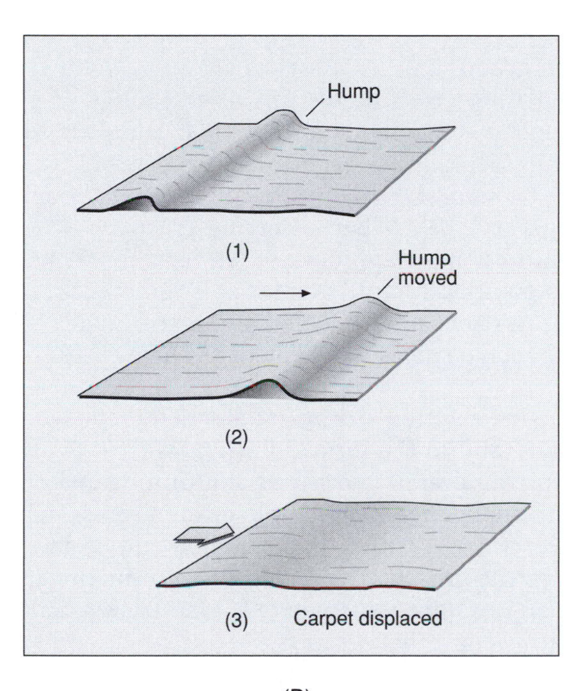

Fig. 6.8 (A) Sketch showing how a dislocation moves through part of a mineral, represented by a simplified grid of atoms joined by bonds (lines) and separated for clarity. In stage 1 the crystal is undeformed, but has a plane of atoms (a slip plane) along which deformation can occur if a force is applied. By stage 3, the atoms in plane A–A above the slip plane have linked up with plane B–B below the slip plane, forcing half of the B–B row to break away along the line X (coming straight off the page), which is called a *dislocation*. This extra half-plane of atoms then links to the C–C row of atoms below the slip plane, forcing half of the C–C plane to break away along the line X (stage 4). In this way, the dislocation X (with its extra half-plane of atoms) can move right through the volume of crystal, providing a very small permanent deformation (stage 6). The dislocation marks the boundary between deformed ('slipped') and undeformed crystal, as shown for stage 5. This process is not the same as going directly from stage 1 to stage 6, which would involve forming a fracture by breaking *all* the bonds along the slip plane simultaneously. Modified from Hobbs *et al.* (1976, fig. 2.2) and Vernon (2000b, fig. 33). (B) Sketch showing how a hypothetical carpet can be moved across a floor by forming and moving a hump. This is much easier than moving the whole carpet at once by pulling at one end, because only the frictional forces under the hump need to be overcome at any one time. Comparing this with (A), the floor is analogous to the slip plane in a crystal and the hump is analogous to the dislocation. From Vernon (2000b, fig. 32). For a colour version of this figure, please go to the plate section.

dominated by glide of dislocations in slip planes. This leads to interference, tangling and hence immobilization of dislocations, causing the mineral to increasingly resist strain (Fig. 6.2). The process is called 'strain hardening (strengthening)'. *High-temperature plasticity* is dominated by thermally activated *recovery* and *recrystallization* processes (Section 6.3), which cause 'softening (weakening).' The processes involve untangling of dislocations, and consequently the mineral is able to continue to deform ('creep') at relatively small differential stresses (Fig. 6.2). The amount of strain accumulation depends on the competition between strain hardening and recovery/dynamic recrystallization.

Plastic deformation at high temperature (dislocation creep) is probably the main deformation process in the deeper parts of Earth's crust (e.g., Yund & Tullis, 1991). The resulting grains may show undulose (undulatory) extinction and subgrains, sutured grain boundaries, and a pronounced shape and/or crystallographic preferred orientation.

6.2.4 Diffusion creep

Diffusion creep (*diffusive mass transfer*) involves change of grain shape by diffusion of chemical components, either (1) in aqueous solution ('*stress-induced solution transfer*' or '*dissolution-precipitation creep*'; also called 'pressure solution'), or (2) by solid-state diffusion along grain boundaries ('Coble creep') or through crystals ('Nabarro-Herring creep', which requires high temperatures and so is most appropriate for very hot rocks, such as peridotites in Earth's mantle). Typically material is removed from sites of high normal compressive stress and deposited at low-stress sites, with the result that a volume of rock changes its shape (e.g., Durney, 1972; Rutter, 1976, 1983).

The term 'pressure solution' (Sorby, 1908) strictly refers to the actual dissolving of minerals, and so the term 'solution-transfer' (Durney, 1972) has been proposed for the overall process of solution, transfer and redeposition of chemical components. A preferable term is 'stress-induced solution transfer' (Passchier & Trouw, 1996, p. 26), which emphasizes the necessity for deformation in the process. The terms 'dissolution-precipitation creep' or simply 'solution-precipitation creep (e.g., den Brok & Spiers, 1991) also imply a deformation-controlled process.

Stress-induced solution transfer is especially effective at low metamorphic temperatures (at which diffusion occurs more readily than dislocation creep) and produces microstructures such as truncated detrital grains (Figs. 6.9–6.10), truncated ooliths, truncated fossils, truncated pebbles, dark seams ('stylolitic surfaces') rich in fine-grained insoluble material (Fig. 6.10), tectonic overgrowths, and 'beard' structures (McClay, 1977; Cox & Etheridge, 1982; Powell, 1982), as shown in Figs. 6.10 and 6.11. However, the process may also occur in the deformation of high- and medium-grade metamorphic rocks, producing veins and 'beard' structures (Lafrance & Vernon, 1993, 1999; Wintsch & Yi, 2002).

Dislocation creep tends to swamp stress-induced solution transfer at higher temperatures (Wheeler, 1992). However, because diffusion occurs along grain boundaries, stress-induced solution transfer is accentuated by finer grainsizes, and so may dominate dislocation creep, even at the high temperatures of the lower crust and upper mantle, at which dislocation creep would otherwise predominate (Wheeler, 1992). Microstructures generally taken to indicate diffusion creep in deformed rocks include equant

Fig. 6.9 Slate from the Hill End area, New South Wales, Australia, showing slaty cleavage, composed of thin mica-chlorite-graphite folia anastomosing around clastic grains of quartz and lenticular aggregates ('stacks') of chlorite interlayered with white mica. Many of the quartz clasts and mica-chlorite aggregates are elongate parallel to the foliation and have shapes suggestive of solution processes. The cleavage of the intergrown chlorite and white mica in the 'stacks' varies from perpendicular to only slightly oblique to the slaty cleavage. From Vernon (2004, fig. 5.54A). Plane-polarized light; base of photo 1.5 mm.

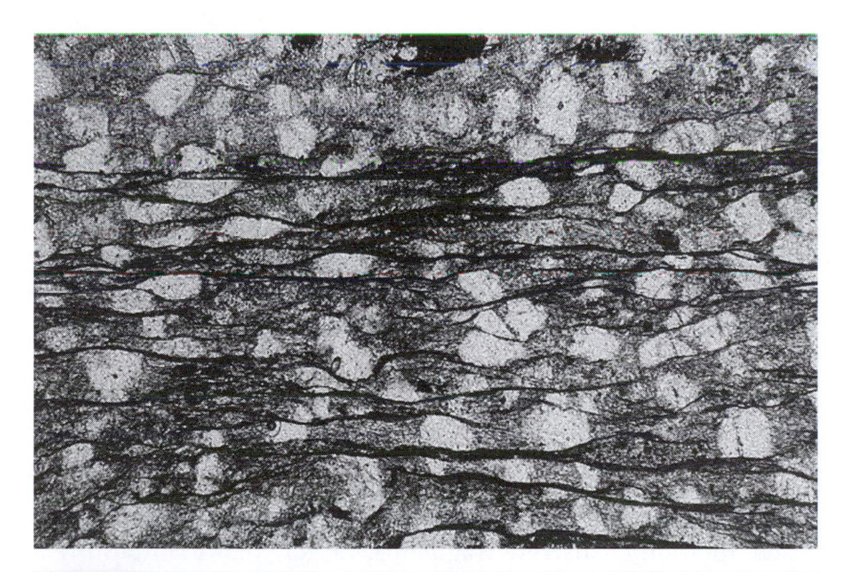

Fig. 6.10 Penetrative domainal cleavage in a clay-bearing sandstone (analogous to slaty cleavage in a slate) from the root zone of the Wildhorn nappe, Valais, Switzerland, consisting of dark folia (seams) formed by solution, anastomosing around clasts of quartz, the shapes of which suggest solution against the seams. This has resulted in somewhat rectangular grain shapes and elongation parallel to the foliation, which can be compared with pre-cleavage grain shapes at the top of the photo. Fine-grained mica and chlorite between the seams occur mainly as aggregates aligned parallel to the foliation, forming 'beards' against the quartz clasts. Section cut perpendicular to the cleavage and parallel to the lineation. Sample by courtesy of David Durney. From Vernon (2004, fig. 5.56). Plane-polarized light; base of photo 1.5 mm.

Fig. 6.11 Phyllite from the Hill End area, New South Wales, Australia, showing 'beards' (aligned intergrowths of fibrous white mica and quartz) adjacent to a plagioclase clast and extending parallel to the slaty cleavage. From Vernon (2004, fig. 5.15A). Plane-polarized light; base of photo 1.5 mm.

grain shapes, indented grains, overgrowths and a lack of crystallographic preferred orientation (Bons & den Brok, 2000).

6.2.5 Ductile grain-boundary sliding

Ductile grain-boundary sliding (*granular flow*, *grainsize-sensitive flow*, 'superplastic flow') involves relative grain movement without loss of cohesion, normally in the absence of fluid, and is an important deformation process in very fine-grained aggregates. Resulting potential gaps between rotating grains are filled by diffusive mass transfer (Ashby & Verall, 1973; Nicolas & Poirier, 1976; Schmid *et al.*, 1977), dislocation motion (Tullis, 1983), or both these processes (Kenkmann & Dresen, 2002), so that the aggregate remains coherent.

In some rocks, ductile grain-boundary sliding may be promoted by the formation of transient, fine-grained reaction products in metamorphic reactions, or by fluid, which assists diffusion and results in a kind of high-temperature 'pressure solution' (Tullis & Yund, 1991; Tullis *et al.*, 1996). This can be called *fluid-assisted diffusion creep*. Fluid-assisted ductile grain-boundary sliding should be conceptually distinguished from frictional grain-boundary sliding, which not only involves an intergranular fluid, but also rotation of discrete fragments, rather than maintaining a coherent aggregate during deformation.

6.2.6 Conditions favouring various deformation mechanisms

Different deformation mechanisms dominate at different conditions of temperature, pressure, strain rate, differential stress, grainsize, fluid content and fluid composition, though several deformation mechanisms may operate simultaneously, even if one dominates. For example, experiments have shown that the main factors favouring ductile flow of solid dry rocks are: (1) high confining pressure (which makes it difficult for the rock to expand and hence break during deformation), (2) high temperature (which allows dislocations to move freely through minerals) and (3) slow application of the deforming force (which gives the dislocations enough time to move). Therefore, in dry rocks, ductile flow tends to dominate in the deeper parts of Earth's crust and in the mantle, where the rocks are hot and under high confining pressures. Generally fracture dominates at depths of less than about 15 kilometres and flow dominates at greater depths, though many exceptions occur, and the conditions vary with the mineral concerned. For example, quartz tends to be ductile at lower temperatures than feldspars.

Higher confining pressure and lower fluid pressure tend to promote dislocation creep over cataclastic behaviour, and larger grainsizes tend to favour dislocation creep and deformation twinning, owing to greater ease of accommodation of strain produced by these processes at grain boundaries, compared with the situation in fine-grained aggregates.

6.3 Recovery and recrystallization

During deformation, dislocations moving in different slip planes may interfere with each other and form 'tangles', which inhibit their movement and hence further deformation of the mineral (strain strengthening or strain hardening). *Recovery* and *recrystallization* are processes that tend to reduce the concentration and/or tangling of dislocations, and so produce volumes of material capable of continued deformation. Thus, ductile deformation is a competition between strain strengthening ('hardening') and recovery processes (Fig. 6.2).

6.3.1 Recovery

Recovery includes all processes that attempt to return a crystal to the undeformed state without the formation of high-angle (high energy) boundaries (Hobbs *et al.*, 1976). In other words, no new grains are formed. Recovery may be *dynamic* or *static*, depending on whether it occurs during or after deformation, respectively. During recovery, dislocations are able to free themselves from tangles by dislocation 'climb' (the movement of edge dislocations out of their slip planes by the addition or loss of point defects, which is a heat-activated process) and dislocation 'cross-slip' (the movement of screw dislocations from one slip plane to another). Both these processes untangle dislocations and so reduce the amount of strain strengthening. The freed dislocations migrate to form *subgrain boundaries* (Fig. 6.12), which are ordered arrays ('walls') of dislocations (Spry, 1969; Hobbs *et al.*, 1976). This leaves relatively strain-free volumes (*subgrains*) between the subgrain boundaries, so that deformation can continue. Optically, subgrain boundaries tend to be relatively evenly spaced, and show small misorientation angles (Fig. 6.13). Subgrains have been observed in a variety of minerals, including quartz, calcite, plagioclase, olivine, garnet, cordierite, galena and pyrite.

6.3.2 Recrystallization

Recrystallization involves the formation of strain-free volumes inside deformed grains by the creation and/or movement of grain boundaries, in response to deformation (Figs. 6.13–6.15). During recrystallization, strain energy is reduced by: (1) migration of existing high-angle (high-energy, random, irrational) grain boundaries, kink-band boundaries or twin

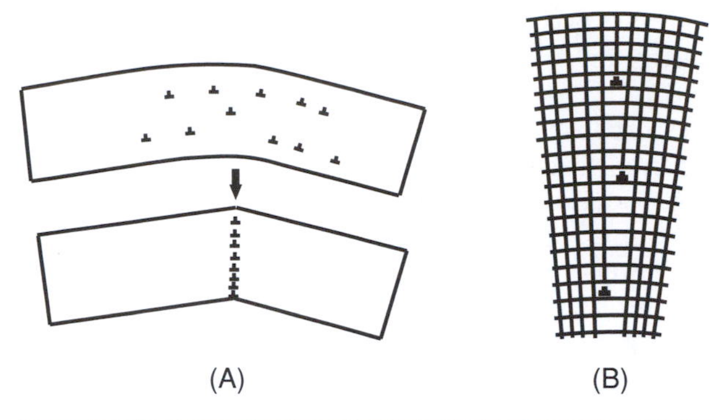

Fig. 6.12 (A) Sketch showing change from a bent grain volume with random edge dislocations in slip planes (top) to a grain with subgrain (low-angle, low-energy) boundary (ordered array of dislocations) formed by the glide and climb of the dislocations, leaving dislocation-free regions (subgrains) on either side of the boundary. (B) Diagrammatic representation of a simple subgrain boundary formed by an ordered array of evenly spaced edge dislocations.

boundaries (leaving strain-free volumes behind the migrating interface), (2) development and migration of new high-angle grain boundaries and (3) development of new low-energy crystal faces, *all in the same mineral*. A generally applicable definition of recrystallization is: 'the development and/or migration of high-energy grain boundaries or crystal faces in the solid state, in response to deformation, and in the same mineral.' If grains of a different mineral are produced (Fig. 6.16), the process is *neocrystallization*.

Fig. 6.13 Elongate and equant subgrains in a relic of deformed quartz surrounded by new (recrystallized) grains. Note the very small orientation differences between the subgrains, compared with the much larger orientation differences between most of the new grains. From Vernon (2004, fig. 5.17). Crossed polars; base of photo 1.75 mm.

Fig. 6.14 Large, strongly deformed grain of plagioclase with deformation twinning and deformation bands (kink-like features) partly recrystallized to finer-grained aggregates of polygonal, largely untwinned plagioclase, as a result of high-temperature deformation of an anorthosite in the Giles Complex, South Australia. From Vernon (2004, fig. 5.23). Crossed polars; base of photo 4.4 mm.

Recrystallization typically produces aggregates of new (recrystallized) grains that are strain-free and therefore capable of continued deformation. The new grains may be (1) polygonal, in minerals with relatively three-dimensional lattice structures, such as quartz, feldspar, calcite or olivine

Fig. 6.15 Relics (porphyroclasts) of deformed (bent and locally kinked) orthopyroxene with exsolution lamellae of clinopyroxene, surrounded by recrystallized aggregates of polygonal grains of orthopyroxene and clinopyroxene, in a deformed gabbro from the Giles Complex, South Australia. From Vernon (2004, fig. 5.25). Crossed polars; base of photo 4.4 mm.

Fig. 6.16 Kinked grain of biotite that has been partly neocrystallized to much smaller new grains of biotite and muscovite with crystal faces, which are typical of mica aggregates (Section 3.2.1), Swiss Alps. From Vernon (2004, fig. 5.26). Crossed polars; base of photo 3.5 mm.

(Figs. 6.13–6.15), (2) crystals with low-energy faces, in minerals with strongly anisotropic lattice structures, such as mica (Fig. 6.16), or (3) irregularly shaped, where grain boundary migration recrystallization is the dominant process (Fig. 6.17). As noted above, recrystallization does not involve the production of new minerals. However, small compositional changes between new and old grains commonly occur in recrystallized minerals with complex chemical compositions (e.g., plagioclase), and can be accommodated in the definition of 'recrystallization.'

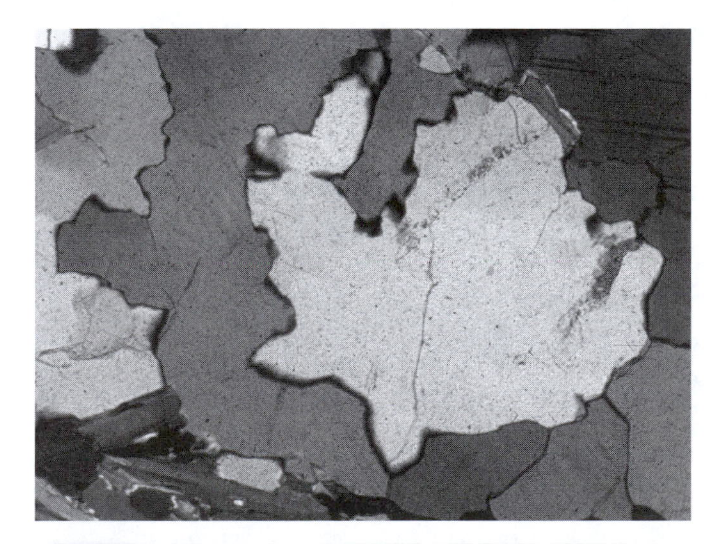

Fig. 6.17 Quartz with deeply indented boundaries formed by grain-boundary migration recrystallization, in deformed tonalite, San José pluton, Baja California, México. From Vernon (2004, fig. 5.30A). Crossed polars; base of photo 2.4 mm.

'Nucleation' of new grains during recrystallization (as opposed to neocrystallization) generally does not involve the formation of completely new grains from new nuclei developed randomly within old grains, but instead involves either (1) *subgrain rotation*, (2) *strain-induced grain boundary migration* ('bulge nucleation') or (3) *grain-boundary migration recrystallization*. The first two processes can produce similar microstructures (Lloyd & Freeman, 1994), and, to complicate matters, microfracturing can produce slightly misaligned fragments that optically resemble subgrains (e.g., Urai *et al.*, 1986; den Brok & Spiers, 1991; Lloyd & Freeman, 1994).

(1) *Subgrain rotation recrystallization* (Hobbs, 1968; Poirier & Guillopé, 1979) occurs when dislocations accumulate in subgrain boundaries, causing the boundaries to progressively increase their complexity and misorientation. By this process, subgrain boundaries become grain (high-energy, high-angle) boundaries. The progressive crystallographic misorientation involves limited grain-boundary migration, with the result that orientation relationships between old and new (recrystallized) grains may be recognizable (e.g., Hobbs, 1968).

(2) *Strain-induced grain-boundary migration* (grain-boundary bulging, bulging recrystallization, 'slow grain-boundary migration') involves differential migration of parts of a high-angle boundary, such as a grain boundary, kink-band boundary or deformation twin boundary. The migration occurs by diffusion of atoms across the boundary, which consequently moves in the opposite direction to the diffusion direction, and forms a 'bulge'. Strain-induced grain-boundary migration is driven by strain energy differences (differences in the dislocation concentration) on either side of the boundary.

(3) *Grain-boundary migration recrystallization* ('fast grain-boundary migration') produces bulges with long wavelengths, which migrate through the aggregate, continuously converting parts of it from one lattice orientation to another. These changing 'orientation domains' move through the aggregate in a complex way, leading to irregular grain shapes (Fig. 6.17). The process occurs at relatively high temperatures (amphibolite facies) in quartz and calcite (Schmid *et al.*, 1980, 1987; Fitz Gerald & Stünitz, 1993), but is also promoted by free water on the migrating boundaries (e.g., Urai *et al.*, 1986). For example, Mancktelow & Pennacchioni (2003) found that in mylonite zones in the western Alps, dynamic recrystallization in dry quartz was dominated by subgrain rotation and 'slow' (bulging) grain boundary migration (producing fine-grained granoblastic aggregates), whereas dynamic recrystallization in wet quartz was dominated by 'fast' grain boundary migration, producing much larger grains with strongly indented boundaries. The temperatures required for this process to operate in feldspar appear to be so high that no examples had been reported until Lafrance *et al.* (1995, 1998) inferred that irregular grain boundaries and dissected grains in deformed anorthosite are the result of 'fast' grain boundary migration, possibly with some subgrain rotation, occurring at about 1050°C.

6.4 Deformation contrasts between minerals in rocks

As mentioned previously, a major factor governing ductility is the number of slip systems available for deformation to occur without producing holes

Fig. 6.18 Intensely deformed granite (mylonite) from east of Armidale, New South Wales, Australia, showing marked contrast between strong, fractured and locally bent relics (porphyroclasts) of plagioclase and finely recrystallized, drawn-out aggregates of quartz and biotite. From Vernon (2000b, fig. 163). Crossed polars; base of photo 1.4 cm.

or cracks. Minerals with more slip systems are able to change their shapes in response to local stress distributions more readily than others. This can lead to localization of deformation into high-strain zones or to the opening of pores or local microcracks, which are important in localizing fluids (Chapter 5) and melts (Chapter 4).

Stronger and weaker minerals coexist in natural rocks. For example, strong feldspar and weak quartz coexist in granite deforming at relatively low temperature (<500 °C). The feldspar deforms plastically a little before it fractures (brittle deformation), whereas the quartz flows and recrystallizes in a ductile manner, commonly forming fine-grained recrystallized aggregates (Fig. 6.18). Evidence of both ductile and brittle behaviour is seen in many felsic mylonites (Section 6.11), in which feldspar deforms cataclastically, whereas quartz and mica deform mainly by dislocation creep, commonly assisted by neocrystallization (Vernon *et al.*, 1983), as shown in Figs. 6.18 and 6.19.

6.5 Chemical changes during deformation

6.5.1 General effects of water

Chemical reactions, which typically accompany deformation, can affect deformation and vice versa. Metamorphic reactions (1) provide new strain-free grains, (2) may release or consume fluid, which affects mineral deformation, and (3) may produce stronger minerals ('reaction strengthening') or weaker minerals ('reaction weakening').

Fluids can be very important in the deformation of rocks, for the following main physical and chemical reasons: (1) water promotes grain-boundary sliding (Rutter, 1972; White & Knipe, 1978; Etheridge & Wilkie, 1979), (2) water increases the rates of some chemical reactions and (3) water opens transient fractures (by hydraulic fracturing or 'hydrofracturing') as

Fig. 6.19 Strongly deformed granite, Wologorong Batholith, southern New South Wales, Australia, showing fractured relics of K-feldspar and plagioclase (some with oscillatory zoning) interspersed with elongate aggregates of recrystallized quartz and foliated aggregates of neocrystallized muscovite, new biotite and epidote formed by reactions involving the former igneous biotite. The plagioclase has resisted the deformation, in marked contrast to the quartz and mica, which recrystallized/neocrystallized to fine-gralned aggregates and flowed readily, forming an anastomosing, mylonitic foliation. From Vernon (2000b, fig. 162; 2004, fig. 5.38). Crossed polars; base of photo 2.4 cm.

the local fluid pressure exceeds the local minimum compressive stress plus the local tensile strength of the rock in the direction of fracture opening, reducing the effective mean stress, especially along grain boundaries. Fluid produced by prograde reactions must escape from the reaction sites for devolatilization reactions to proceed (Chapters 2 and 5), and process (3) could allow this to happen.

In addition, minute spaces between grains open during deformation, owing to deformation contrasts between different minerals and volume changes between reactant and product minerals in metamorphic reactions. These pockets of lower pressure 'suck in' fluid, and so increase the diffusivity of fluid, so much so that zones of strong deformation may act as fluid channels (Chapter 5), which are potentially important for (a) fluid removal from rocks undergoing prograde metamorphism, (b) fluid introduction to rocks undergoing retrograde metamorphism, (c) melt transfer in rocks undergoing partial melting during high-grade metamorphism (Chapter 4), and (d) transfer of metal-bearing hydrothermal fluids in the formation of many metallic ore deposits (Chapter 5).

6.5.2 Reaction strengthening

A good example of reaction strengthening is the formation of porphyroblasts of strong minerals, such as cordierite, K-feldspar, andalusite and garnet, at high temperatures in the prograde metamorphism of shales. This can make metashales stronger than interbedded metasandstones (Vernon & Johnson, 2000; Vernon *et al.*, 2000, 2003) at high metamorphic grades, whereas the reverse is typically true at low and medium grades, where the more abundant, fine-grained quartz-mica matrix causes metashales to be weaker.

Though dehydration reactions generally produce coarser-grained, stronger minerals, they may produce transient fine-grained reaction products that can deform readily by grain-boundary sliding (e.g., Schmid, 1976; White, 1976) and localize deformation in narrow zones. However, the process can continue only until grain coarsening removes the weakening effect (White & Knipe, 1978). This concept of transitory reaction-enhanced ductility during prograde metamorphism led White & Knipe (1978) to suggest that rocks are weakest during metamorphic reactions, and that deformation may be most intense when new minerals are being produced, before grain coarsening occurs. Rubie (1986) extended the idea by suggesting that development of equilibrium mineral assemblages and microstructures in metamorphic rocks generally occurs in relatively short time periods under transitory fluid-present conditions, caused either by devolatilization reactions and or infiltration of external fluid. Movement of fluid and consequent reaction enhancement would be assisted by deformation and vice versa.

6.5.3 Reaction weakening

Reaction weakening can operate by (1) forming weaker minerals from strong minerals (e.g., chlorite from garnet; muscovite from feldspar), (2) grainsize reduction (e.g., the formation of fine-grained reaction products in deforming plagioclase and biotite undergoing alteration during the deformation of granites), and (3) pervasive release of water in prograde dehydration reactions.

A good example of reaction weakening is the replacement of feldspar by fine-grained aggregates rich in muscovite during hydration (for example, in retrograde metamorphism). These aggregates deform much more easily than the original feldspar. Therefore, in volcanic rocks, former feldspar phenocrysts may alter to these aggregates, deform, and so become part of the flowing former groundmass, leaving unaltered quartz phenocrysts as relics (Vernon, 1986b; Williams & Burr, 1994). Such ‘quartz-eye rocks’ are common in altered volcanic sequences associated with some orebodies (Chapter 7), and the residual quartz phenocrysts have often been misinterpreted as porphyroblasts. The residual volcanic nature of the ‘quartz-eyes’ is shown by the common occurrence of embayments, which may persist even in rocks deformed at amphibolite facies conditions, owing to the relative strength of the large quartz crystals compared with the fine-grained mica-bearing matrix (Vernon, 1986b; Williams & Burr, 1994).

Not only are some new minerals inherently weaker than the original mineral, but if the new grainsize is much smaller, this may also assist deformation, for example by inducing diffusion-accommodated grain-boundary sliding (Stünitz, 1993; Stünitz & Tullis, 2001). Moreover, this need not be a transitory effect, as discussed previously for prograde reactions, but may persist throughout subsequent deformation (e.g., Etheridge & Vernon, 1981; Vernon *et al.*, 1983), provided grain growth is restricted, as is commonly the situation in retrograde metamorphism.

6.5.4 Deformation during prograde metamorphic reactions

Effects of deformation (with consequent enhanced access of water) on retrograde metamorphic reactions and vice versa have long been appreciated, but effects of deformation on prograde metamorphic reactions and vice versa are equally important. As water is pervasively released at the grain scale by dehydration reactions, it has a potential weakening effect, and this is likely to occur through relatively large rock volumes, compared with the relatively local effect of water in retrograde metamorphism.

In low-P/high-T regional metamorphism, granite intrusions may cause metamorphism to commence prior to foliation-forming deformation (Vernon *et al.*, 1993a). The heating and pervasive release of water in the resulting dehydration prograde reactions, together with initial small grain-sizes of product minerals, may promote the deformation, which in turn may assist reactions.

6.5.5 Deformation during retrograde metamorphic reactions

Fluids are particularly important in retrograde metamorphism, and are closely related to deformation. The importance of microfractures also applies to retrograde metamorphism, in which the devolatilization reaction style of prograde metamorphism is reversed (e.g., Vernon & Ransom, 1971; Vernon, 1976). In retrograde metamorphism, reaction progress depends on the supply of fluid (mainly water or carbon dioxide, depending on the reactions concerned) from an external source. The fluid must reach local reaction sites, and the most effective way for it to do this at realistic rates is to penetrate along transient microfractures that open during deformation.

Because retrograde metamorphic reactions require addition of volatile components (for example, H_2O), the development of reaction products is commonly patchy and incomplete, except in high-strain zones, through which water can pass. For example, Max (1970) described schistose amphibolite zones anastomosing around pods of unmetamorphosed dolerite. Also, at Broken Hill, Australia, granulite facies metamorphism uniformly dehydrated the rocks, after which they fractured during cooling, allowing water back into the rocks to form shear zones with lower amphibolite facies mineral assemblages. The shear-zone rocks are fine-grained, neocrystallized schists and mylonites, with few or no high-grade relics, but elsewhere the retrograde reactions are patchy, incomplete and pseudomorphous, owing to restricted deformation and access of water (Vernon, 1969; Vernon & Ransom, 1971; Corbett & Phillips, 1981).

6.6 Foliations and lineations

Tectonic foliations are planar structures (compositional layering and/or parallel alignment of minerals) and *lineations* are linear structures (compositional rods and/or linear alignment of minerals) produced by deformation. These structures are abundant on the outcrop scale, but are also commonly seen with the microscope. Examples of different types of foliations and lineations have been illustrated by Turner & Weiss (1963) and Passchier & Trouw (1996).

Despite these superimposed tectonic foliations, sedimentary bedding may be preserved as compositional layering (Chapter 7), especially in relatively weakly deformed and low-grade metamorphic rocks, but also in many high-grade terranes, both in outcrop (Fig. 6.20) and thin section (Figs. 6.21, 6.22). In outcrop, its identification may be assisted by the presence of other sedimentary structures (Chapter 7). True bedding is oblique to the axial surfaces of microfolds in their *hinge areas*, but intense

Fig. 6.20 Granulite facies calcsilicate rock, Cadney Creek, Arunta block, central Australia, showing isoclinally folded bedding. Knife 9 cm long.

deformation may *transpose* the bedding into a new orientation parallel to newer foliations in the *limbs* of tight folds. In strongly deformed and higher-grade metamorphic rocks, bedding may be largely removed or even obliterated by such *transposition*, as shown in Fig. 6.23, and/or recrystallization/neocrystallization.

Fig. 6.21 Schist from the Picuris Range, New Mexico, USA, with (1) inferred bedding (shown by broad compositional variations: quartz-rich versus mica-rich) trending from upper left towards lower right of photo, (2) an early, fine-grained, tectonic foliation (slaty cleavage) crenulated by later folds, trending from lower left to upper right, and (3) a crenulation cleavage (more spaced, anastomosing, strongly compositionally differentiated foliation) trending steeply from upper left to lower right. Also shown are porphyroblasts of staurolite, biotite and garnet. From Vernon (2004, fig. 5.53). Crossed polars: base of photo 1.8 cm.

Fig. 6.22 Slate from the Hill End trough, central-western New South Wales, Australia, showing residual bedding (trending from bottom-left to top-right in the photo) oblique to a prominent slaty cleavage consisting of dark folia rich in phyllosilicate minerals and (?)graphite anastomosing around lenticular quartz clasts and micaceous patches (possibly former feldspar grains). Plane-polarized light; base of photo 1.75 mm.

6.7 Bedding-plane foliation

Bedding-parallel mineral alignments are common in shales that have not been strongly deformed (Williams, 1972; Etheridge & Lee, 1975; Holeywell & Tullis, 1975; Knipe & White, 1977; Beutner, 1978; Maltman, 1981; Bennett *et al.*, 1981, 1991; Morritt et al., 1982; Baker *et al.*, 1993). The alignment is called *bedding-plane foliation* (e.g., Morritt *et al.*, 1982), *diagenetic foliation* (Passchier & Trouw, 1996) or 'shaly cleavage'. The aligned minerals are typically detrital clay minerals or mica, and the detailed structures have been

Fig. 6.23 Transposition foliation (parallel to the long edge of the photo) with an earlier foliation preserved as intrafolial folds in less strongly deformed parts of a Precambrian quartz-rich rock, Weekeroo Inlier, Olary area, South Australia. Knife 9 cm long.

revealed by transmission electron microscope (TEM) and scanning electron microscope (SEM) investigations (Bennett *et al.*, 1981, 1991). Moore & Geigle (1974) described bedding-parallel foliation formed by alignment of platy and elongate minerals, in Pleistocene mudstones from the Aleutian Trench and the Gulf of Mexico. The alignment is caused mainly by mechanical rotation in response to compaction (Baker *et al.*, 1993).

In slates, clastic mica grains are commonly kinked, especially where they are oblique to the bedding (Morritt *et al.*, 1982), as a result of compaction or incipient folding. The detrital nature of the mica is indicated by frayed ends and a very high length-to-width ratio; these features are caused by abrasion and splitting along the mineral cleavage during transport. The typical detrital shapes of quartz grains in the same rock confirm that the mica is not metamorphic (Morritt *et al.*, 1982).

Compaction alone generally does not produce a slaty cleavage (see Section 6.8), unless it continues during the early stages of folding, during which axial-surface cleavages develop. In some slates, the bedding-plane foliation is so strong that the first recognizable folds are crenulations, which typically develop in finely foliated rocks (Williams, 1972; Morritt *et al.*, 1982).

6.8 Slaty cleavage

Low-grade metamorphosed shales and siltstones typically have one dominant tectonic foliation, referred to as *slaty cleavage* (Figs. 6.9, 6.10, 6.22, 6.24–6.27). The foliation is typically developed parallel to the axial surfaces of folds (Fig. 6.24), and may also be referred to as '*first-generation cleavage*' (Durney & Kisch, 1994). If the folds are tight (isoclinal), the cleavage tends to be sub-parallel to bedding in the limbs and slightly oblique to bedding in the hinges of the folds. The analogous cleavage in deformed low-grade sandstones (Fig. 6.10) is called *penetrative domainal cleavage* or 'rough cleavage' (Durney & Kisch, 1994).

Slaty cleavage is typically *compositionally differentiated* into narrow, dark folia of fine-grained, well-aligned phyllosilicate minerals (especially white mica and chlorite) and/or graphite (P or M domains) anastomosing around stronger domains containing detrital fragments, mainly of quartz (Q domains), as shown in Figs. 6.9 and 6.10. In some slates and schists, the Q domains are relatively unmodified, showing detrital microstructures (e.g., Williams, 1972; Etheridge *et al.*, 1983, fig. 2), whereas in other slates and some metasandstones, the quartz grains in the Q domains show overgrowths of quartz or of quartz-phyllosilicate fringes or 'beards' of fibrous grains that extend parallel to the cleavage (Powell, 1969, figs. 3, 5; Powell, 1982; Williams, 1972), as shown in Figs. 6.10 and 6.11. In many slates, the quartz clasts in Q domains are weakly or not deformed, with slight to no undulose extinction and little or no evidence of recrystallization. Therefore, the strong tendency towards rectangular grain shapes and elongation of the quartz grains parallel to the slaty cleavage (Fig. 6.9) is due largely to solution of those parts of clasts that are adjacent to cleavage folia (*stress-induced solution* or 'pressure solution'), rather than internal deformation of grains (crystal-plastic deformation), as discussed by Williams (1972), Gray (1978), Bell (1978), Gregg (1985) and Vernon (1998, fig. 6). Some P domains

Fig. 6.24 Folded Cambrian amphibolite facies metapelite, Harvey's Return, Kangaroo Island, South Australia, showing sedimentary bedding (prominent dark (biotite-rich) and light (quartz-rich) layers and a strong slaty cleavage roughly parallel to the axial surfaces of the folds. Coin 2.8 cm diameter. Photo by Scott Paterson.

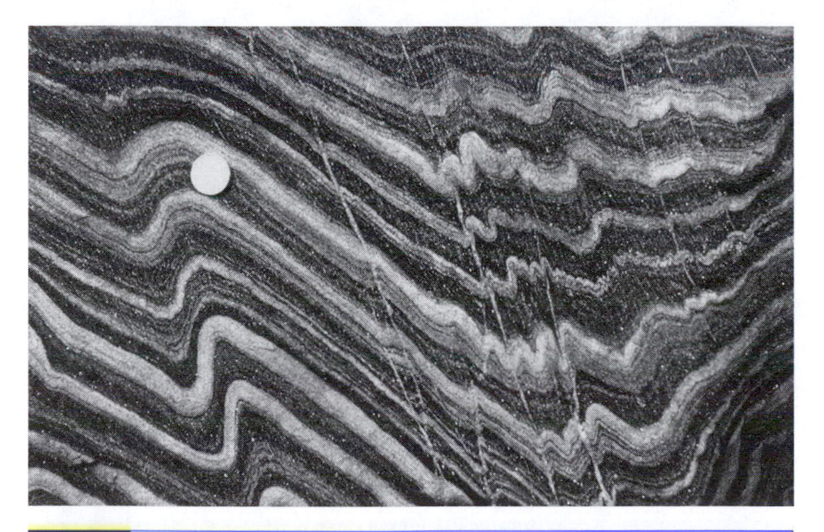

Fig. 6.25 Tectonic foliation (slaty cleavage) oblique to folded bedding in Cambrian amphibolite facies schist, Harvey's Return, Kangaroo Island, South Australia. Coin 2.8 cm diameter.

Fig. 6.26 Complex tectonic foliations oblique to folded bedding in Palaeozoic greenschist facies phyllite, Yarra area, central-western New South Wales, Australia. Plane-polarized light; base of photo 5 cm. Photo by Scott Paterson.

Fig. 6.27 Slate from the Hill End area, New South Wales, Australia, showing local development of crenulation cleavage (top-right of photo). The crenulation cleavage has been formed only in the most micaceous layer (bed), but even there its development is patchy. The slaty cleavage in the non-crenulated bed shows micaceous folia anastomosing around lenticular clasts, mainly of quartz. From Vernon (2004, fig. 5.55). Plane-polarized light; base of photo 11 mm.

rich in fine-grained, dark material in some slates (Figs. 6.9, 6.10) resemble stylolite solution seams (Williams, 1990).

Some slates show evidence of grain rotation, at least in the early stages of cleavage formation (Knipe & White, 1977; Bell, 1978; Roy, 1978; Gray, 1979; Knipe, 1981, 1989; White & Johnston, 1981; van der Pluijm & Kaars-Sijpesteijn, 1984). This mechanism for forming slaty cleavage was suggested by Sorby (1853). Moreover, weak foliations have been observed in unconsolidated sediments (Paterson *et al.*, 1985), especially in slumps (Williams *et al.*, 1969; Maltman, 1981; Paterson & Tobisch, 1993), the foliation being formed mainly by rotation of existing phyllosilicate grains.

Several studies have shown that cleavages may initiate mechanically during diagenesis and continue to develop by chemical-mechanical processes during low-grade metamorphism (Williams *et al.*, 1969; Spang *et al.*, 1979; Mackenzie *et al.*, 1987; Hammond, 1987; Boyer, 1984; Tranter, 1992). Most of the phyllosilicate alignment and cleavage development occurs by solution of old grains (of phyllosilicate minerals, as well as quartz and/or carbonate) and neocrystallization, as well as physical grain rotation, but after most or all of the dewatering and lithification of the sediment (Williams, 1972; Wood, 1974; Boulter, 1974; Etheridge & Lee, 1975; Geiser, 1975; Holeywell & Tullis, 1975; Groshong, 1976; Knipe & White, 1977; Bell, 1978; Beutner, 1978; Gray, 1978; Knipe, 1981, 1989; Knipe, 1981; Williams, 1983b; Gregg, 1985; Woodland, 1985; Lee *et al.*, 1986; Kanagawa, 1991). The compositional domains of differentiated slaty cleavages require redistribution of chemical components during cleavage formation.

Some slates have lenticular to ellipsoidal intergrowths ('stacks') of interlayered chlorite and white mica, with their cleavages mainly oriented obliquely to the slaty cleavage, though they vary from perpendicular to only slightly oblique to the slaty cleavage, and a few 'stacks' may have their cleavages approximately parallel to the slaty cleavage (Fig. 6.9). Several workers have interpreted the 'stacks' as primary clasts (for example, of biotite) modified by fracturing and growth of secondary chlorite and white mica (Beutner, 1978; Roy, 1978; van der Pluijm & Kaars-Sijpesteijn, 1984; Milodowski & Zalasiewicz, 1991; Clark & Fisher, 1995). This could account for the scarcity of 'stacks' with cleavage parallel to the slaty cleavage, because these would probably would have deformed by slip on the phyllosilicate cleavage, thereby becoming part of the slaty cleavage itself. However, the 'stacks' do not resemble typical unmodified detrital mica clasts (Gregg, 1986), which tend to be very elongate, as mentioned previously. In contrast, many 'stacks' are much thicker perpendicular to their cleavage than parallel to it, which could be taken to imply growth in Q domains during development of the slaty cleavage. Alternatively, solution and consequent truncation by the slaty cleavage of formerly much longer mica clasts could have occurred.

Most observations are consistent with the general inferences that (1) slaty cleavage may begin to develop at sub-metamorphic temperatures, and continue to develop with prograde metamorphism, and (2) mechanical rotation and/or kinking of detrital grains may predominate in the earlier stages, whereas solution and neocrystallization predominate in the later, probably most important, stages of cleavage development. Though best developed in low-grade slates and phyllites, slaty cleavage may persist, in modified form, in medium-grade schists (Fig. 6.21) and high-grade schists and gneisses.

Fig. 6.28 Mica schist from Japan, consisting mainly of muscovite and quartz with a coarser grainsize than that of phyllite. The muscovite is well aligned, forming a foliation (schistosity). Crossed polars; base of photo 1.8 cm.

6.9 Crenulation cleavage

The foliation that is characteristic of medium- to high-grade regional metamorphic rocks (schists) is generally called *schistosity* or *schistose foliation* (Figs. 6.28, 6.29). The minerals defining schistose foliations typically are coarser-grained than in slates and phyllites. Many of these foliations represent the advanced stages of a *crenulation cleavage* (Fig. 6.21), which may also occur in slates and phyllites (Fig. 6.27). The origin of crenulation cleavage has been discussed by Williams (1972, 1990), Cosgrove (1976), Gray (1977a, 1978, 1979), Gray & Durney (1979a,b), Marlow & Etheridge (1977), Hanmer (1979), Bell & Rubenach (1983). Some schistosities (possibly those shown in Figs. 6.28 and 6.29) may represent crenulation folding developed to such an extent that residual hinges have been removed by transposition and neocrystallization (Bell & Rubenach, 1983), as mentioned in Section 3.6.5 and illustrated in Fig. 3.49.

A previous slaty cleavage, an earlier crenulation cleavage or a strong depositional bedding-parallel foliation or fissility ('shaly cleavage') is required before crenulations can form. The new cleavage typically initiates at a high angle to the earlier cleavage (Figs. 6.21, 6.27), as observed in experiments on the kinking and crenulation of layer silicate aggregates (Etheridge *et al.*, 1973; Williams *et al.*, 1977).

Crenulation cleavages are compositionally differentiated into domains that are generally known as P or M (phyllosilicate-rich or mica-rich) and Q (quartz-rich). The term 'crenulation cleavage' covers a large variety of crenulation-type foliations, ranging from *zonal* (with gradational layering) to *discrete* (with abrupt layering) types (Gray 1977a), as shown in Fig. 6.21. Zonal crenulation foliations may pass into discrete crenulation foliations. Some schists show a gradation from open zonal crenulations, through

Fig. 6.29 Amphibole schist consisting of abundant hornblende and less abundant plagioclase and quartz, the amphibole being strongly aligned in a foliation. Plane-polarized light; base of photo 4.4 mm.

progressively tighter differentiated crenulations, into thoroughly recrystallized rocks from which all vestiges of former microfolds have been removed (Bell & Rubenach, 1983).

Discrete crenulation cleavages (Fig. 6.21) consist of limb-zones (P or M domains) rich in phyllosilicate minerals and/or sillimanite alternating with or anastomosing around layers or elongate lenses, respectively, of lower-strain hinge-zones (Q domains, 'microlithons') richer in other minerals (notably quartz, with or without feldspar), which preserve the folded pre-existing foliation. The P domains accommodate most of the non-coaxial (shearing) strain, whereas the Q domains appear to deform mainly coaxially (Bell, 1981, 1985).

The P (M) domains commonly are interpreted as insoluble residues resulting from the solution and removal of quartz, feldspar or carbonate, generally with evidence of neocrystallization of phyllosilicate minerals (or sillimanite at highest grades of metamorphism) parallel to the foliation. Though some curvature of phyllosilicate grains at the edges of Q domains into P domains is common, old phyllosilicate grains statistically parallel to an earlier foliation or bedding typically appear to have been dissolved against the developing P domains (e.g., Lee *et al*., 1986). Increased deformation in the high-strain limbs of crenulations causes solution and removal of minerals such as quartz, feldspar and calcite from these zones, leaving them enriched in minerals with a layered or strongly prismatic structure, such as biotite, white mica, chlorite or graphite at low-to-intermediate grades of metamorphism or sillimanite at high grades (Vernon, 1987b), as shown in Fig. 6.30. These minerals appear to be relatively stable in an environment of strong, non-coaxial strain accumulation, either because they have relatively low solubility under high normal stress or because they are able to slip on cleavages or grain boundaries, without much accumulation of crystal-plastic strain (Bell & Rubenach, 1983; Bell *et al*., 1986; Vernon,

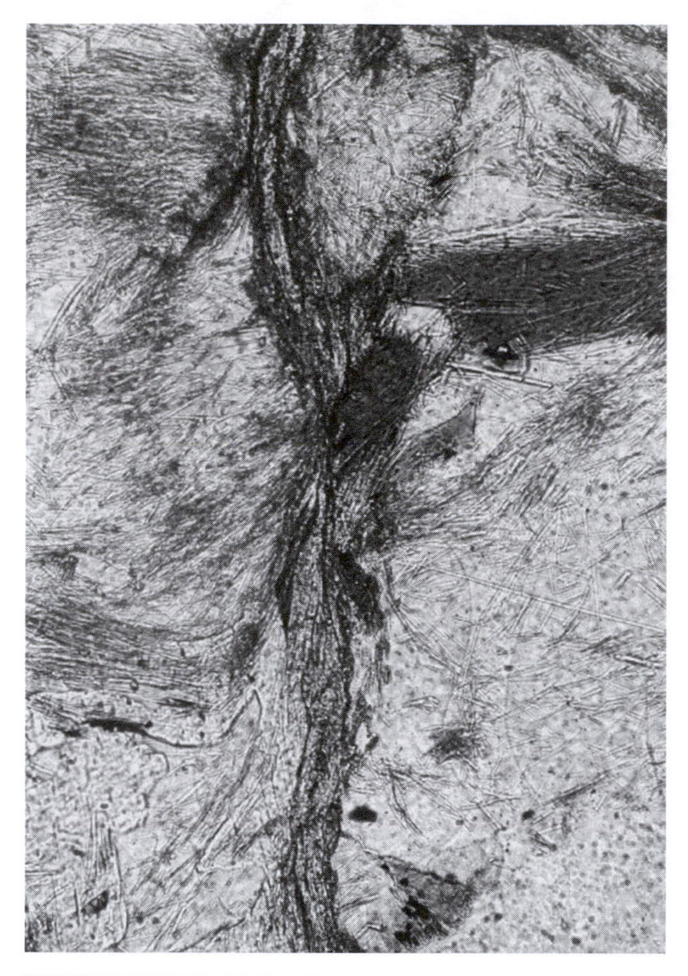

Fig. 6.30 High-grade schist from the Placitas-Juan Tabo area, New Mexico, USA, showing a folium (P domain) of fibrous sillimanite in the limb zone of a crenulation cleavage. The adjacent Q-domains show the previous (crenulated) foliation, delineated by fibrous sillimanite and biotite, curving into the P domain. However, the sillimanite in both domains belongs to the same mineral assemblage. From Vernon (2004, fig. 5.58). Plane-polarized light; base of photo 1.1 mm.

1987b). This is probably related to the strongly anisotropic crystal structures of the minerals concerned, resulting in relatively weak bonding across rational crystal boundaries, such as (001) planes in layer silicates, (0001) planes in graphite and (110) planes in sillimanite.

Compositional mapping with the electron microprobe has revealed untwinned plagioclase coexisting with quartz in the Q-domains of some crenulated schists (Williams *et al.*, 2001). These maps have also shown that non-crenulated areas undergo compositional alteration and structural modification during the crenulation of adjacent areas, and so cannot be assumed to represent the original rock prior to the formation of the crenulations (Williams *et al.*, 2001).

The components of minerals dissolved from the P domains are inferred to either be (a) concentrated in hinge zones of the crenulations or

(b) removed from the local observable system (e.g., on the scale of a large thin section, a hand specimen or even an outcrop), to be deposited in veins nearby or further afield. A controversy rages about which interpretation is correct, and possibly both are applicable in different local circumstances, as discussed Section 5.13.

Worley *et al.* (1997) found that the mechanism of formation of crenulation cleavage varies with metamorphic grade. At chlorite-biotite zone conditions (around 520°C), the formation of crenulation cleavage is dominated by local solution-deposition, resulting in distinct compositional differences between the mica of different structural domains. However, at garnet zone temperatures, the mica compositions are homogeneous across domains, reflecting increasing importance of diffusion through mineral grains ('volume diffusion') and longer grain-boundary diffusion distances, with consequent equilibration of mineral compositions.

Multiple crenulation cleavages are typical of schists, owing either to repeated, separate deformation events or to continuous deformation affected by mechanical heterogeneities, such as porphyroblasts. Even where the schistosity appears superficially to be a single foliation, detailed examination of protected, low-strain sites (for example, adjacent to porphyroblasts) commonly reveals residual crenulation microfolds that have been obliterated from unprotected areas of the matrix (Figs. 3.41, 3.42).

6.10 Gneissic foliation (gneissic layering)

Foliations in high-grade metamorphic rocks typically are less continuous than in low- and medium-grade metamorphic rocks, owing to a coarser grainsize. In high-grade metapelitic gneisses, folia rich in biotite and/or sillimanite commonly anastomose around large grains of garnet, cordierite or K-feldspar (Figs. 3.20, 4.2), the large grains and folia belonging to the same compatible assemblage (Chapter 3). Gneisses of felsic igneous composition typically consist of xenoblastic aggregates of quartz and feldspar, with or without garnet or orthopyroxene, interspersed with discontinuous, anastomosing folia rich in biotite or hornblende. Many mafic gneisses are non-foliated or only weakly foliated, consisting of polygonal (granoblastic) aggregates of calcic plagioclase, pyroxene and hornblende (Figs. 3.2, 6.31).

High-grade metamorphic rocks may also show a compositional layering, which may represent bedding in relatively weakly deformed metasedimentary gneisses, but is more likely to be transposed bedding and/or new compositional layering of tectonic origin in more strongly deformed rocks. Compositional layering may occur in mafic gneisses, commonly formed by the deformation and transposition of original igneous layering (Myers, 1978). Compositional layering is less common in felsic gneisses, formed by the deformation and metamorphism of granites.

Some high-grade gneisses contain elongate layers or rods of quartz (commonly very coarse-grained) alternating with layers of granoblastic feldspar (Fig. 6.32). The quartz layers are generally attributed to strong deformation, coupled with recrystallization, of primary quartz grains. However, some have suggested that similar quartz layers may originate as veins during intense deformation (Vollbrecht *et al.*, 1997; Williams *et al.*, 2000). Very elongate, vein-like grains and aggregates of quartz, ilmenite,

Fig. 6.31 Two-pyroxene granofels (granulite facies metabasalt), Anmatjira Range, central Australia, showing extensive recrystallization and neocrystallization, but preserving a few plagioclase grains with elongate igneous crystal shapes and even a concentrically zoned plagioclase crystal (right of centre). From Vernon (2004, fig. 4.79). Crossed polars; base of photo 10 mm.

titanite and scapolite may be formed by microfracturing and coupled mass transfer/deposition during mylonitic deformation (see below) of gabbro at high-temperature amphibolite facies conditions (Lafrance & Vernon, 1993, 1999).

Fig. 6.32 Pegmatite from central Australia that has been strongly deformed at high temperature (granulite facies), showing aggregates of recrystallized feldspar (with a few elongate porphyroclasts) alternating with elongate aggregates ('platten') of coarsely recrystallized quartz. These aggregates probably are distorted derivatives of coarse-grained quartz and feldspar in the original rock. From Vernon (2004, fig. 5.43). Crossed polars; base of photo 4.8 mm.

Fig. 6.33 Mylonite formed by intense deformation of granite, showing strong lenticular compositional foliation, dark folia being mainly biotite, light folia mainly quartz with some feldspar. Original K-feldspar phenocrysts have partly resisted the deformation and remain as lenticular, partly recrystallized porphyroclasts ('augen'). Knife 9 cm long.

6.11 Composite (transposed) foliations in mylonitic rocks

Mylonite zones (ductile shear zones) are discrete, elongate zones (on the regional, outcrop and microscope scales) with larger strain accumulation and finer grainsize than in adjacent rocks (Johnson, 1960, 1967; Christie, 1960, 1963; Bell & Etheridge, 1973; White *et al.*, 1980; Passchier & Trouw, 1996; Snoke *et al.*, 1999; Vernon, 2004). They are characterized by thin compositional folia (Fig. 6.33), strong mineral-elongation ('stretching') lineations (Fig. 6.34), and asymmetrical structures developed especially around larger residual grains ('porphyroclasts'), as discussed by Passchier & Trouw (1996) and Vernon (2004) and shown in Figs. 6.35 and 6.36.

Progressive deformation of mylonites commonly produces complicated, very localized fold and foliation sequences that are closely related in time. For example, in mylonite zones in deformed granitoid and similar rocks, complex refolding, truncation of foliations, isolated (truncated) fold hinges and divergent lineations are common (Fig. 6.36), and all may be formed by progressive deformation in the same general deformation event (e.g., Bell & Hammond, 1986; Vernon, 2004). Mylonitic deformation is also discussed in Section 6.13.

Meneilly (1983), Williams (1985) and Tobisch & Paterson (1988) have shown that elucidation of deformation sequences in such rocks is commonly difficult and may even be impossible. 'Progressive deformation' in this context implies that structures involved have similar orientations, senses of movement and metamorphic grades, and that they are produced as a relatively continuous sequence in a geologically short time (Tobisch

Fig. 6.34 Vertical stretching lineation in mylonite formed by intense deformation of granite, Wyangala Dam, near Cowra, New South Wales, Australia.

& Paterson, 1988). The resulting *composite foliations* are so complex and locally variable that conventional time-sequence terms, such as S_1, S_2, etc. and F_1, F_2, etc., which are useful in lower-strain terrains (Chapter 3), are inappropriate. Though the usual temporal correlation of folds and foliations may be applicable for a small area (e.g., part of a thin section) in such zones, correlation cannot be made with similar sequences in other small areas. Therefore, Tobisch & Paterson (1988) emphasized that temporal notations should be used only if independent evidence indicates that structural elements can be related in time.

6.12 Layering in migmatites

Partial melting (anatexis) during metamorphism can produce compositional layering in the form of leucosomes segregated into layers ('stromata') parallel to a tectonic foliation in *stromatic migmatites* or *metatexites* (Figs. 4.3, 4.5, 4.22A, 4.23, 4.24), as discussed in detail in Chapter 4. Some stromata may be due to solid-state deformation of previously formed

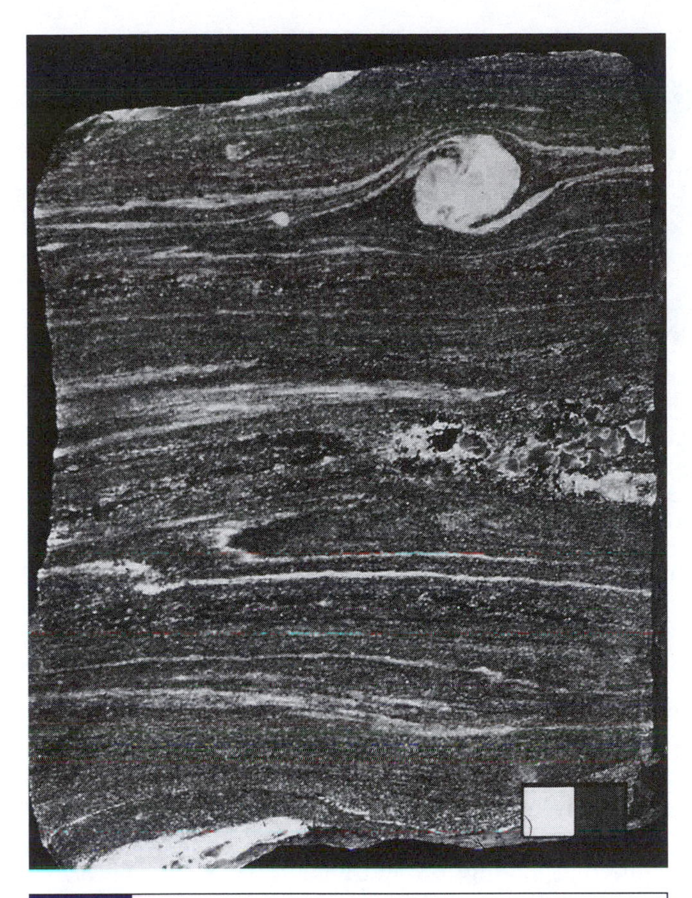

Fig. 6.35 Mylonite, Grenville Province, Ontario, Canada, showing lenticular foliation, rootless fold hinges, and a K-feldspar porphyroclast (top) with asymmetrical 'tails' of finely recrystallized feldspar. Scale (bottom-right) in cm.

Fig. 6.36 Mylonite from the deformed Wyangala Batholith, New South Wales, Australia, showing asymmetrical microfolds and isolated ('rootless') fold hinges. From Vernon (2004, fig. 5.42). Plane-polarized light; base of photo 4 mm.

Fig. 6.37 Strongly deformed migmatite, Maine, USA, showing boudinage of coarse-grained leucosome.

leucosomes (e.g., Vernon *et al.*, 2000, 2003). In thin section, leucosomes are generally seen to consist mainly of aggregates of polygonal quartz and feldspar (commonly with irregular grain boundaries), with less abundant mafic minerals, though igneous microstructures (for example, crystal faces in feldspar or cordierite) may be preserved in less deformed leucosomes (Vernon & Collins, 1988; Vernon & Johnson 2000), as shown in Fig. 4.9. Leucosomes may be lenticular, owing to boudinage during later deformation (Fig. 4.22A, 4.25, 6.37), and may also commonly show ptygmatic folding (Section 4.17), as shown in Figs. 4.34. In weakly deformed migmatites, leucosomes may remain in the stronger metapelitic beds, forming *bedded migmatites* (Section 4.3), as shown in Figs. 4.11–4.13.

6.13 Interpretation of the origin of compositional layering (original or metamorphic?)

Many metamorphic rocks show compositional layering, and the problem is to interpret its origin, using the available structural evidence. Original (primary) compositional layering may be sedimentary bedding or igneous layering, but it may be modified during metamorphism and deformation. For example, bedding may be attenuated, a bedding-parallel foliation (fissility, 'shaly cleavage') may develop as low-temperature metamorphic minerals grow during compaction (Morritt *et al.*, 1982; Paterson *et al.*, 1985), and quartzofeldspathic segregations may grow parallel to bedding during partial melting at high temperatures (e.g., Passchier *et al.*, 1990, p. 25; Vernon & Johnson, 2000). Moreover, as discussed below, new (tectonic) compositional layering can form during deformation. So it should not be assumed that all relatively continuous layering in metamorphic rocks is residual sedimentary bedding. The following are some of the ways that compositional layering can be formed.

(1) Lenticular compositional layering, which can superficially resemble lenticular sedimentary bedding (Pettijohn & Potter, 1964, plates 13B, 14,

Fig. 6.38 Strongly foliated metasedimentary rock, near Ulverstone, Tasmania, Australia. Intrafolial rootless fold hinges of former quartz veins (centre) reveal that this is not bedding, but a tectonic foliation, despite its regularity. Coin diameter 2.8 cm.

15A, 17, 18; Weiss, 1972, figs. 29, 41, 45, 46), can be formed by *mechanical processes*, such as extreme flattening and extension of large grains, grain aggregates, veins, pillows, fragments or pebbles in the original rock (Myers, 1978), as shown in Fig. 6.38. This occurs especially in high-strain domains, and the macroscopic products of extreme deformation at amphibolite facies metamorphic conditions are uniformly layered gneisses (Myers, 1978). The formation of an effective new foliation can be formed by deformation and rotation of an existing foliation or bedding into a new orientation (*transposition*), as shown in Fig. 6.23, the new foliation being referred to as a *transposition foliation* (Turner & Weiss, 1963, p. 92; Dietrich, 1960; Williams, 1967, 1983a; Hobbs *et al.*, 1976).

(2) Tectonic compositional lamination is also produced by the high-strain (mylonitic) deformation of relatively massive, coarse-grained igneous and high-grade metamorphic rocks, such as gabbros, granites, anorthosites and granulite-facies granofelses (Evans & Leake, 1960; Ramsay & Graham, 1970; Vernon & Ransom, 1971; Moore, 1973; Vernon, 1974; Myers, 1978), as shown in Figs. 6.39 and 6.40. The layering, which typically is compositional and finely lenticular, results from solid-state deformation and recrystallization of different minerals or rock-types (Vernon, 1974; Myers, 1978).

The resulting compositional layering can be superficially simple (Moore, 1973; Passchier *et al.*, 1990, fig. 3.9), perhaps suggesting residual bedding, as in the mylonite shown in Fig. 6.41A and the gneiss shown in Fig. 6.41B. However, close examination of intensely deformed outcrops may reveal residual structures indicating a non-sedimentary origin, such as intrafolial folds, rootless isoclinal fold hinges (Fig. 6.42A) or lenses of different rock-types (Fig. 6.42B). For example, strongly layered rocks have been produced during isoclinal folding of mafic dykes in granitoids, as on Eyre Peninsula, South Australia and in the Grenville Province, Ontario, Canada (Fig. 6.43). If only the fold limbs are viewed in a relatively small outcrop, the deformed igneous pseudo-layering can resemble sedimentary bedding (Fig. 6.43B), which suggests that special care should be taken in areas of poor outcrop, particularly where the rocks are of igneous chemical composition. Very strong deformation of various rock-types, such as layered gneisses, dykes and granite veins, tends to rotate (transpose) them all into parallelism or

Fig. 6.39 Precambrian mylonitic augen gneiss, resulting from strongly deformed granite with K-feldspar megacrysts, Eyre Peninsula, South Australia. The K-feldspar megacrysts have been converted into rounded to lenticular, partly recrystallized/neocrystallized aggregates (augen), and the other minerals (quartz, plagioclase and biotite) have been changed into recrystallized/neocrystallized lenses and laminae. The intensity of deformation is illustrated by the isoclinal fold hinges shown in the centre of the photo. Photo by Scott Paterson.

approximate parallelism (Bridgewater *et al.*, 1973a,b, fig. 2c), as shown in Fig. 6.44.

(3) Compositional layering can also develop by *metamorphic differentiation*, (Fig. 6.45A,B), as discussed in Section 5.14. This process involves the redistribution of chemical and hence mineralogical components, mainly

Fig. 6.40 More intensely deformed granite from the same locality as for Fig. 6.39, showing narrow, continuous compositional layering, but still preserving small lenticular relics of the former K-feldspar megacrysts. Scale in centimetres. Photo by Scott Paterson.

(A)

(B)

Fig. 6.41 (A) Remarkably regular, fine compositional layering in mylonite, Bear Mountains Fault Zone, Guadalupe Complex, central Sierra Nevada, California, USA. The layering was formed by intense deformation of interspersed mafic-felsic intrusive igneous rocks (Vernon *et al.*, 1989). (B) Very regularly layered gneiss from Nepal, showing no evidence of its former nature. The mineral assemblage is consistent with an igneous origin, but also with sedimentary rocks of similar felsic composition.

in response to differential strain accumulation, and commonly in conjunction with active deformation (e.g., Turner, 1941; Dietrich, 1960; Kretz, 1966b; Talbot & Hobbs, 1968; Ghaly, 1969; Williams, 1972; Hobbs *et al.*, 1976; Gray, 1977; Marlow & Etheridge, 1977; Gray & Durney, 1979a,b; Bell & Rubenach, 1983; Vernon, 1998, 2004). Diffusion of chemical components may also modify pre-existing layering, such as bedding (Fig. 6.45C).

(4) Relatively continuous primary compositional layering occurs in some igneous rocks, especially mafic-ultramafic igneous complexes (Hess, 1960; Wadsworth, 1961; Wager & Brown, 1968, Moore, 1973; McBirney & Noyes, 1979; Myers, 1981; Conrad & Naslund, 1989; Cawthorn, 1996). Where such rocks are metamorphosed and deformed, this original layering

(A)

(B)

Fig. 6.42 (A) Finely laminated Precambrian amphibolitic gneiss resulting from intense deformation of high-grade metamorphic gneiss, involving extreme transposition of former compositional layering. The tectonic origin of the compositional layering is evidenced by rootless, intrafolial, isoclinal fold hinges (bottom-left) and small, lenticular, dark boudins (top-centre). Also shown are strongly folded, distorted, thin, transgressive felsic veins (top half of photo). Skerry Bay, between Tongue and Bettyhill, Sutherland, Scotland. (B) Layered Precambrian amphibolitic gneiss, the mafic-felsic layering resulting from intense deformation of high-grade metamorphic gneiss, involving extreme transposition of former compositional layering. The lenticular nature of the layers and the presence of extremely attenuated isoclinal folds testify to the intensity of the deformation. Road-cut opposite Eilan Donan Castle on Loch Duich, Sutherland, Scotland.

may be locally preserved, even if pseudomorphed by metamorphic minerals (e.g., Frost, 1975), but is commonly distorted, transposed and variably obscured. Therefore, the problem is to distinguish it from imposed, tectonometamorphic layering, if present. Tectonic layering in deformed

(A)

(B)

Fig. 6.43 (A) Isoclinal folding of Precambrian mafic dykes (dark) that previously had intruded felsic (light) igneous rocks, Grenville Province, Ontario, Canada. Knife 9 cm long. (B) Isoclinally folded mafic (dark) and felsic (light) Proterozoic meta-igneous rocks, Grenville Province, Ontario, Canada. The tight folding has produced repeated, relatively regular layering, which could be mistaken for bedding in small outcrops without the intrafolial fold shown here (centre). Knife 9 cm long.

(A)

(B)

Fig. 6.44 (A) Strongly deformed Precambrian metasedimentary rocks with former quartz and pegmatite veins, Eyre Peninsula, South Australia. The veins have been tightly folded and dismembered (boudinaged). The former bedding and the veins all show a strong tendency to rotate into parallelism with the new foliation, regardless of their original orientations. The former orientation and thickness of sedimentary bedding cannot be determined from exposures like this. Lens cap for scale. (B) Proterozoic mylonite in the Anmatjira Range, central Australia, showing very regular layering formed by intense isoclinal folding of former igneous rocks of both mafic (dark) and felsic (light) composition. Knife 9 cm long. Smaller outcrops may show only the layering, but larger exposures, on the scale shown, reveal the isoclines.

(A)

(B)

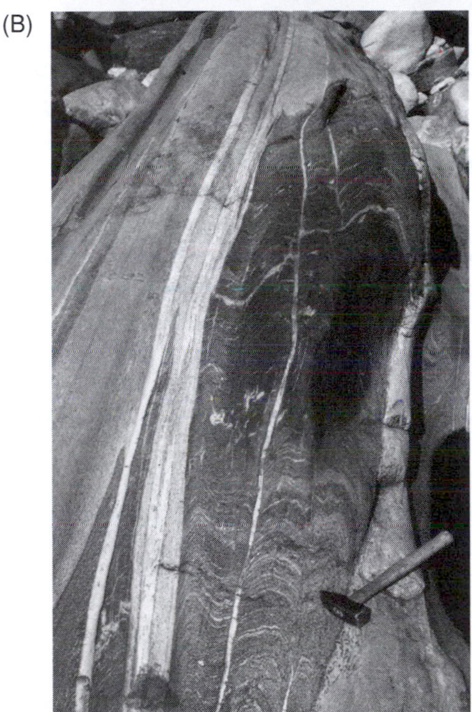

Fig. 6.45 (A) Strong compositional layering in mica schist, inferred to have been formed by metamorphic differentiation (Turner, 1941), Otago area, New Zealand. The light layers are composed of quartz and albite. (B) Spectacular compositional layering formed by metamorphic differentiation superimposed on earlier folded layering, Lavertezzo, Alps. (C) Bedding in calcsilicate rock, the composition of individual beds having been modified by diffusion during contact metamorphism, King Island, Australia (Edwards *et al.*, 1956).

igneous rocks may be remarkably continuous (Moore, 1973, fig. 17), but is generally lenticular and commonly shows augen of formerly coarse-grained minerals (Moore, 1973, fig. 15), even where residual isoclinal fold hinges have been obliterated. The whole-rock chemical composition of

(C)

Fig. 6.45 *(cont.)*.

these rocks enables distinction between the layering and sedimentary bedding.

Compositional layering may also be formed by distortion of microgranitoid enclave ('mafic' enclave) magma globules caused by flow in granite magmas (Fig. 6.46A). Generally this type of layering is characterized by elongation of euhedral plagioclase, K-feldspar or hornblende crystals, without plastic deformation of the crystals (Vernon *et al.*, 1988; Vernon 2000a) as shown in Fig. 6.46B.

(5) Layering may be caused by flow sorting in diatremes (Wilshire, 1961) and schlieren formed by in situ cyclic crystallization, gravitational accumulation or flow sorting in granites (e.g., Vernon & Paterson, in press). Conceivably, relics of such layering could be preserved in metamorphic equivalents of these rocks.

(6) Layering generally attributed to partial melting (*anatexis*) commonly develops in migmatites during metamorphism at conditions of the upper amphibolite and granulite facies (Chapter 4). The melted material plus new crystals (*neosome*) or melt segregated from the new crystals (*leucosome*) tends to segregate into layers ('stromata') separated by largely (though commonly not entirely) melt-free rock (*mesosome*) parallel to a tectonic foliation; the rocks are called *stromatic migmatites* or *metatexites*. Isoclinally folded leucosomes in deformed migmatites may also be mistaken for sedimentary bedding, especially if viewed in outcrops that do not show fold hinges. Layer-parallel ('lit-par-lit') injection of magma can also produce comparitionally layered rocks (Dietrich, 1960).

(7) Layering can be caused by grainsize variations brought about by pinning of grain boundaries, for example of quartz, feldspar or carbonate by small, commonly aligned, dispersed grains of minerals such as mica, graphite, chlorite or sillimanite (Voll, 1960; Vernon, 1968, 1976). The layering does not involve metamorphic differentiation, but simply reflects existing small compositional differences, either bedding or a tectonic layering.

The following features are indicative of (or at least consistent with) an interpretation of layering as sedimentary bedding, though confirmatory

(A)

(B)

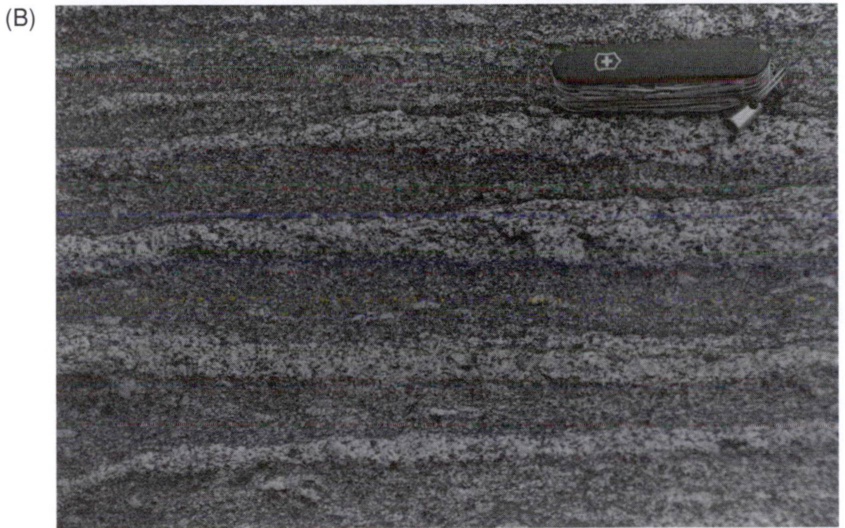

Fig. 6.46 (A) Igneous compositional layering in granite, Ardara, Ireland, formed by elongation of microgranitoid ('mafic') enclave magma globules during flow of the granite magma.
(B) Magnified view of the same kind of layering as in (A), at the same locality, showing elongation of euhedral plagioclase crystals in the more mafic layers, caused by suspension (magmatic) flow.

evidence (especially other sedimentary structures) should always be sought: (a) distinct compositional differences between layers; (b) markedly variable layer thickness; (c) regular thickness over relatively wide area, such as a large outcrop (Pettijohn & Potter 1964, plates 1–4, 7, 8), especially of distinctive metasedimentary rock-types, such as marble or metaquartzite (Turner & Weiss, 1963, p. 93); (d) thick and massive bedded units, with weak or no internal bedding (Pettijohn & Potter 1964, plates 10B, 25); (e) characteristic sedimentary patterns of repetition (Pettijohn &

Potter 1964, plates 1–4, 7, 8; Weiss 1972, fig. 48), in a way that cannot be interpreted as being due to isoclinal folding; (f) pebbles in some layers; (g) fossils or trace fossils in some layers; (h) layering that does not cross or truncate any other layering, except for cross-bedding, and (i) gross compositional layering oblique to the main tectonic foliation (Turner & Weiss, 1963, p. 93).

As emphasized by Hobbs *et al.* (1976, p. 152), workers in deformed metamorphic terrains must try to prove that layering is bedding, rather than assuming it to be bedding. The most effective way of doing this is to observe undoubted sedimentary structures that are consistent with a sedimentary origin for the layering. However, this can also be difficult, because some tectonic structures resemble sedimentary structures, as discussed below. Moreover, even when the interpretation of layering as sedimentary bedding is relatively clear, it should not be assumed that the bedding is undistorted or that it maintains original orientations, because bedding and layering of other origins are commonly strongly distorted and transposed into new orientations by intense deformation in regional metamorphic terranes (Turner & Weiss, 1963; Hobbs *et al.*, 1976).

6.14 Non-tectonic deformation of sedimentary structures: Structures produced by soft-sediment deformation

Soft-sediment deformation (e.g., Matland & Kuenen, 1951; Hills, 1963, pp. 132–3) is distinguished by definition from the solid-state deformation that characterizes penetrative deformation of metamorphic tectonites, but may be difficult to distinguish from tectonic structures in metamorphosed sedimentary rocks (e.g., Williams *et al.*, 1969; Hobbs *et al.*, 1976). Bedding may be intricately and even isoclinally folded by slumping, and currents may produce convolute lamination and folded cross-bedding, all these processes taking place in unconsolidated sediment (Pettijohn & Potter, 1964; Williams *et al.*, 1969; Conybeare & Crook, 1968; Hobbs *et al.*, 1976). Weak to moderately strong foliations have been observed in unconsolidated sediments (Williams *et al.*, 1969; Hobbs *et al.*, 1976, p. 153; Morritt *et al.*, 1982; Paterson *et al.*, 1985), especially in slumps (Williams *et al.*, 1969; Maltman, 1981; Paterson & Tobisch, 1983), the foliation being formed mainly by rotation and alignment of existing phyllosilicate grains. Moreover, primary bedding may be lenticular, pod-like, bifurcating, discontinuous and disrupted, as a result of local current disturbances in soft sediment (e.g., Pettijohn & Potter, 1964, plates 13B, 14, 15A).

Examples of strongly lenticular to 'flaser' bedding (Pettijohn & Potter 1964, plates 17, 18) may superficially resemble lenticular foliations in metasedimentary tectonites. A particularly spectacular example is shown in Pettijohn & Potter (1964, plate 18C), which looks very much like a mylonitic rock. If such a rock were metamorphosed while maintaining a relatively fine grainsize, it could be mistaken for a mylonite or augen gneiss. However, examination of the microstructure should reveal evidence of a mylonitic origin, for example, partial recrystallization or neocrystallization of formerly larger gains (e.g., Passchier & Trouw, 1996; Vernon, 2004).

Therefore, if soft-sediment structures are preserved in metamorphic rocks, they may be difficult to distinguish from tectonometamorphic structures, and the distinction may be somewhat artificial, since soft-sediment deformation may be triggered by tectonic processes, and may pass continuously into tectonometamorphic deformation. The problem has been discussed in detail by Hobbs *et al.*, 1976, pp. 157–9. Many criteria that have been suggested are ambiguous, but the following appear to be the most reliable.

1. Relatively undeformed trace fossils (e.g., worm burrows) indicate that folds formed in soft sediment.
2. Bending of metamorphic foliations and/or metamorphic mineral grains around fold closures indicates tectonometamorphic folding.
3. Deformation of fossils related to their position in a fold indicates tectonometamorphic folding.
4. Tectonometamorphic cleavages should transect slump folds (e.g., Wright & Henderson, 1992), provided they form much later.

6.15 Heterogeneous deformation ('deformation partitioning')

Deformation of rocks is typically heterogeneous on all scales (Wynne-Edwards, 1969; Bridgewater *et al.*, 1973a; Coward, 1973a,b; Watson, 1973; Escher *et al.*, 1976; Cobbold, 1977; Carreras *et al.*, 1980; Poirier, 1980; Bell, 1981; Lister & Williams, 1983). Volumes of rock undergoing weak or no deformation may coexist with zones undergoing strong deformation (Cobbold, 1977). The deformation histories in high-strain zones may be coaxial or non-coaxial, and volumes of rock undergoing strong, non-coaxial deformation may anastomose, on all scales (millimetre to kilometre), around pods of rock undergoing weaker coaxial deformation (Max, 1970; Bridgewater *et al.*, 1973a, fig. 2; Bell, 1981). In metasedimentary rocks, this pattern (often referred to as '*deformation partitioning*') may be independent of bedding. Furthermore, the pattern of deformation partitioning into high- and low-strain zones may change with time (Bell & Rubenach, 1983). The result of deformation partitioning is that original sedimentary or igneous structures may be preserved in one place while being obliterated in adjacent volumes. In some areas, enough bedding may be preserved to enable mapping of local form surfaces and to distinguish and correlate stratigraphic units (Turner & Weiss, 1963, p. 91), at least over limited distances, as at Broken Hill, Australia, though in other places strong transposition of bedding has been well documented (P. F. Williams, 1967, 1983a).

Though deformation partitioning may be independent of bedding (e.g., Bell, 1981), it may also be controlled by bedding. For example, in interbedded sandstone-shale successions, the shaly beds commonly undergo strong, typically non-coaxial deformation (involving development of prominent foliations), whereas the sandy beds may undergo weak, locally non-coaxial deformation, involving minimal distortion of sedimentary structures (Weiss & McIntyre, 1957; Lister & Williams, 1983, p. 15), though the reverse can be true in high-grade metamorphic rocks, as discussed in Section 7.2.2. This could well apply to the least deformed beds in disharmonic folds (Hobbs *et al.*, 1976, fig. 7.18). Even in areas of strong isoclinal folding, involving extreme distortion and transposition, sedimentary structures may be preserved in fold limbs, as in macroscopic folds at Broken Hill, Australia (Hobbs *et al.*, 1976, p. 114, fig. 4–21). Furthermore, reactivation of bedding or slaty cleavage in later episodes of deformation may be restricted to selected, commonly more micaceous beds (Bell, 1986).

Therefore, it should not be assumed that (1) because sedimentary or igneous structures can be recognized in some units, no unit of the terrane has been appreciably deformed, or (2) because one unit or part of an area has been strongly deformed, the whole terrane must have been equally strongly deformed. Situations in which some beds may be deformed non-coaxially, while adjacent beds are deformed coaxially (and so are much more likely to preserve original structures) are common (Lister & Williams, 1983).

A clear example of heterogeneous deformation was described by Weiss & McIntyre (1957, p. 575) in their classic study of the Loch Leven area, Scotland, as follows. 'The interiors of some of the more massive bodies of quartzite have suffered so little deformation that they hardly fall into the category of tectonites. However, thin layers of quartzite enclosed within schist are strongly deformed, and the schists are intensely strained and internally reconstructed.' Ripple marks and cross-bedding are excellently preserved in the metaquartzites (Section 7.2.2). As noted by Weiss & McIntyre (1957, p. 575), 'most of the strain has been concentrated in the kinematically susceptible schist'.

Another striking example of deformation partitioning is provided by the occurrence of ignimbrite with well-preserved pumice lenses (fiamme) preserved as non-deformed relics in multiply deformed schists in the Olary area, South Australia (Section 7.3.4). Evidently, the massive ignimbrite unit was much less susceptible to deformation than the adjacent, foliated, mica-rich schists.

Another example is provided by the Archaean Yilgarn Block, Western Australia, for which Binns *et al.* (1976) observed that primary igneous structures of basalts and komatiites are best preserved in non-deformed or weakly deformed parts of the terrane ('static-style domains', in which the metamorphic grade varies from prehnite-pumpellyite to middle amphibolite facies), but are obliterated in strongly deformed parts ('dynamic-style domains', in which the grade varies from middle to upper amphibolite facies). This study also suggests that, though the grade tends to be higher in the more strongly deformed parts of the area, deformation appears to exert a larger control than metamorphism on the preservation of primary structures. Similarly, deformation and metamorphism in Caledonian metadolerite dykes in County Mayo, Ireland, resulted in foliated zones anastomosing around spheroidal zones with residual doleritic microstructures and minerals (Max, 1970).

Another good example of the preservation of sedimentary and/or igneous structures in high-grade (amphibolite to granulite facies) rocks is the Mount Stafford area, central Australia (Vernon *et al.*, 1990), where graded and cross bedding are commonly preserved in metasediments, even where extensively melted, and doleritic to gabbroic structures are preserved in metamorphosed mafic igneous rocks (Section 7.3.3), except in local high-strain zones.

Even in the most strongly deformed terranes, small residual pods may show relic structures, such as <1 m pods of anorthositic gabbro with

Fig. 6.47 Residual pods of eclogite (E) partly and locally converted to blueschist (B) in more strongly deformed zones that provided access for water, Port Macquarie, eastern New South Wales, Australia.

original igneous microstructures in the Archaean North Atlantic craton (Bridgewater *et al.*, 1973b, p. 501).

Deformation appears to provide pathways for seawater penetration in the oceanic crustal metamorphism of gabbros, with the result that alteration occurs preferentially in deformed zones (Ito & Anderson, 1983). Similar situations may apply in the transformation of eclogite to blueschist (Fig. 6.47) and in retrograde metamorphism generally (Chapter 2), all these settings requiring introduction of volatiles, especially water.

Chapter 7

Parent Rocks

7.1 The general problem

The determination and interpretation of *P-T-X* conditions and their relationship to deformation and partial melting events (Chapters 2–5) are at the very heart of metamorphic geology. However, metamorphic geologists must decipher the structures of deformed rocks in the field (Chapter 6) before meaningful detailed laboratory work can be undertaken. Furthermore, the geological history of a metamorphic region is incomplete without an understanding of the parent rocks ('protoliths'). In addition, because many major orebodies occur in regional metamorphic terranes, ore exploration geologists need to gain as much information as possible on the nature of the parent rocks, in order to formulate realistic ore-search models. The aim is to try to 'see through' the obscuring complexities of metamorphic mineral assemblages and imposed structures, and infer premetamorphic minerals and structures.

In strongly deformed, high-temperature metamorphic terranes, much structural evidence of parent rocks may be obliterated or obscured, because of metamorphic reactions, deformation, recrystallization/neocrystallization and grain growth, especially in rocks heated for long periods of time or repeatedly heated. If so, bulk chemical composition may be the only indicator of parentage. Unfortunately, bulk chemical composition points to only the broadest rock categories, such as 'felsic igneous,' 'mafic igneous,' or 'pelitic.' Some structurally different rocks – for example, volcanic and plutonic felsic rocks – are commonly indistinguishable on chemical grounds, necessitating a search for structural evidence.

Ideally, structural and metamorphic changes should be traced from unmetamorphosed or weakly metamorphosed parts of a rock unit or terrane to more strongly deformed and metamorphosed parts (e.g., King & Rast, 1955, 1956; Knill, 1960). However, commonly this is not possible. Fortunately, in some terranes, even in high-grade metamorphic zones up to granulite facies conditions, original mesoscopic and/or microscopic structures are preserved well enough to be confidently identified, so that more reliable inferences about parent rocks can be made. For example, many metamorphic regions have zones of relatively low strain between higher-strain zones, and residual sedimentary and igneous structures may be preserved in the low-strain zones, as discussed in Section 6.15.

Some of the clearest primary structures are found in rocks that remain fine-grained, owing to finely dispersed particles of unreactive minerals, such as graphite, which 'pin' mobile grain boundaries and so slow grain growth (Voll, 1960; Vernon, 1976). Residual structures are also commonly preserved in contact metamorphism, in which the duration of heating is relatively short and the effects of deformation may be small. However, residual structures may also occur in relatively coarse-grained, even partly melted rocks (e.g., Vernon *et al.*, 1990, 2003; Greenfield *et al.*, 1996; Vernon & Johnson, 2000), as discussed below. In summary, conditions favouring the preservation of original structures are: (1) short duration of metamorphism, (2) low strain and (3) small, dispersed grains of unreactive or stable minerals that keep the grainsize relatively small and inhibit grain growth, which would tend to obscure original microstructures.

In this chapter, we outline the main chemical characteristics of the main metamorphic rock groups, describe examples of residual sedimentary and igneous structures in metamorphic rocks, and discuss the recognition of rocks that have undergone metasomatism prior to or during metamorphism, together with their potential importance for mineral exploration. Emphasis will be placed on the Broken Hill region, New South Wales, Australia, as an example of problems of the interpretation of parent rocks in an amphibolite-granulite facies terrane. This area has been intensively studied from the viewpoint of determining the nature of parent rocks, owing to its importance for economic mineral exploration.

7.2 Metasedimentary rocks

7.2.1 Broad chemical characteristics of metasedimentary rocks

The use of bulk chemical composition to infer parental metamorphic rocks has been discussed by, among others, Bastin (1909) and Leith & Mead (1915), and average compositions of the main sedimentary rock-types have been calculated by Clarke (1924) and listed by Pettijohn (1949, table 19). The chemical compositions of most sedimentary rocks are more variable than those of igneous rocks, owing to the chemical and physical sorting processes involved in weathering, transport and deposition, which eliminate certain minerals and develop others. For example, SiO_2 may reach more than 99 wt% in some quartz sandstones and cherts, Al_2O_3 may reach almost 70 wt% in bauxite, Fe_2O_3 may reach 75 wt%, in iron-rich deposits, MgO may reach 20 wt% in dolomite, and CaO may reach 56 wt% in pure calcite limestone (Mason, 1966). However, many metasediments have less extreme compositions, and mechanically concentrated and rapidly deposited detrital sediments, such as feldspathic sandstones ('greywackes') may have very similar compositions to those of many igneous rocks (Pettijohn, 1949, p. 82), as shown in Fig. 7.1A, which shows the main broad compositional ranges of metasedimentary rocks. The main minerals resulting from these compositional restrictions are shown in Fig. 7.1B.

Claystones (shales, pelites) are formed by accumulations rich in clay minerals, with the result that metapelites show broadly similar chemical characteristics, namely high Al_2O_3, $K_2O > Na_2O$, and $MgO > CaO$, reflecting abundance of the clay minerals, illite and montmorillonite in claystones (Mason, 1966). K generally dominates over Na, because of its retention in many clay minerals, whereas Na tends to be dissolved in seawater (Pettijohn, 1949, p. 83). Metapelites of all grades represented in the

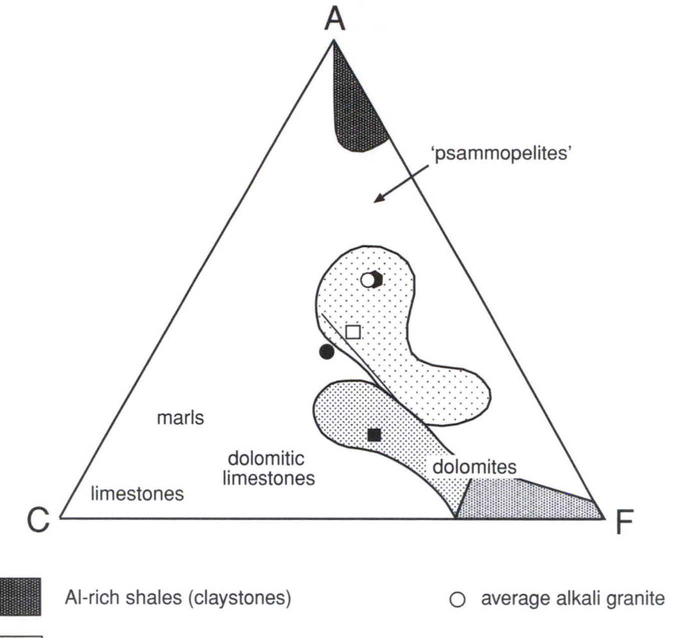

Al-rich shales (claystones)

feldspathic sandstones ('greywackes')

basaltic & andesitic rocks

ultramafic rocks

○ average alkali granite

□ average calc-alkali granite

● average granodiorite

■ average basalt

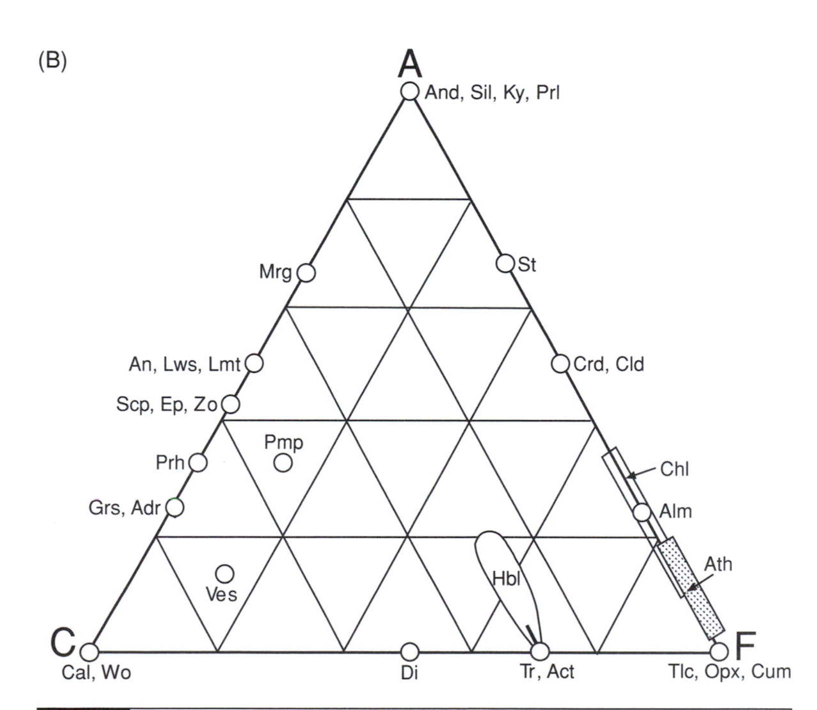

Fig. 7.1 (A) Chemical compositions of common broad rock-types plotted on an ACF diagram, for which A = $Al_2O_3 + Fe_2O_3 - (Na_2O + K_2O)$, C = $CaO - 3.3\ P_2O_5$, and F = $MgO + FeO + MnO$, after Winkler (1979, fig. 5.7) and Stroud *et al.* (1983, fig. 6). (B) Chemical compositions of common metamorphic minerals plotted on an ACF diagram, after Winkler (1979, fig. 5.5). Comparison with (A) shows the effect of bulk-rock chemical composition on the metamorphic minerals produced. See list in the Appendix for mineral name abbreviations.

Broken Hill regional metamorphic terrane, Australia, contain over 20 wt% Al_2O_3 and 55–60 wt% SiO_2, with 4–5 wt% K_2O (Willis *et al.*, 1983, table 2), reflecting an abundance of biotite, with muscovite, andalusite and cordierite at lower grades and sillimanite, cordierite and garnet at higher grades.

However, though aluminous minerals, such as cordierite, garnet, andalusite and sillimanite are usually indicators of a metapelitic parentage, especially where abundant, they are common in small amounts as primary minerals in peraluminous ('S-type') granites (Chappell & White 1974; Clemens & Wall, 1981) and equivalent volcanic rocks (Zeck, 1970; Wyborn *et al.*, 1981; Clemens & Wall, 1984). They also occur in felsic igneous rocks that have been enriched in Al by metasomatism (e.g., in the 'porphyry copper' environment) in the form of base-cation leaching (Gustafson & Hunt, 1975; Lowder & Dow, 1978; Vernon, 1979; Carpenter & Allard, 1980; Valiant *et al.*, 1983; Watanabe & Hasegawa, 1986). These aluminous rocks may be later metamorphosed, resulting in Al-rich minerals that do not reflect a sedimentary parentage (Espenshade & Potter, 1960; Sykes & Moody, 1978; Carpenter & Allard, 1980; Allard & Carpenter, 1981; Vernon *et al.*, 1987).

In addition, the whole-rock composition of pelitic rocks may change during prograde metamorphism, especially by removal in solution of the more soluble components, such as Ca, Na, K and especially Si, and consequent concentration of the less soluble components, such as Al and Ti (Ague, 1991). Nevertheless, the essential pelitic chemical composition is retained.

Quartz-rich metasandstones (metapsammites) are enriched in Si, and plagioclase-rich (i.e., less mature) metasandstones are also relatively rich in Na and/or Ca. For example, metapsammites of all grades represented in the Broken Hill area, Australia, contain around 80 wt% SiO_2 and 8–10 wt% Al_2O_3 (in mica, aluminosilicate, cordierite, garnet and feldspar), with 7.8 wt% CaO and 0.9–7.4 wt% Na_2O (Willis *et al.*, 1983, table 2), reflecting the presence of plagioclase. In feldspathic metapasmmites, CaO and Na_2O may be as high as 4 and 2.5 wt%, respectively (Willis *et al.*, 1983, table 2). The K_2O content (in biotite and K-feldspar) is around 7.2 wt%. Psammopelitic rocks have compositions varying between these extremes (Willis *et al.*, 1983, table 3).

Metamorphosed limestones and marls are enriched in Ca and/or Mg. For example, in metamorphosed impure carbonate rocks in the Broken Hill area, Australia (Stroud *et al.*, 1983, table 5), CaO varies from 6 to 46 wt% (most commonly 10–20 wt%) and MgO varies from 0.5 to 15 wt% (most commonly 7–10 wt%), both components reflecting an abundance of carbonate and/or calcsilicate minerals. Al_2O_3 varies from 0.9 to 19 wt% (most commonly 9–12 wt%), SiO_2 varies from 42 to 68 wt% (most commonly 50–60 wt%) and the Na_2O and K_2O contents combined are generally less than 1 wt%.

Concentrations of rare elements are generally not marked or diagnostic in sedimentary rocks, apart from phosphates, borates and nitrates, some Mn deposits, and accumulations of Cu, V and U with organic matter (Krauskopf, 1967, p. 592).

The most difficult problem with inferring parental sedimentary rock-types from their chemical compositions concerns the origin of metamorphic quartzofeldspathic rocks. The problem is to distinguish immature (feldspar-rich) metasedimentary rocks (e.g. volcanic sandstones or arkoses)

(A)

(B)

Fig. 7.2 (A) Bedding is preserved as quartz-rich layers, despite strong isoclinal folding and later crenulation folding, in this schist-metaquarzite block from the Gordon River Dam site, Tasmania, Australia. Knife 9 cm long. (B) Bedding in strongly deformed, amphibolite facies, Dalradian metasedimentary rocks, Connemara, western Ireland. The metapelites (bottom) are much more strongly deformed than the metapasmmites (top), and also have abundant lenticular leucosomes indicative of partial melting.

from volcanic and granitic parent rocks. A generally reliable guide is that reworked quartzofeldspathic sediments tend to be richer in quartz, and hence Si, than felsic igneous rocks (Pettijohn, 1949, p. 260).

Another problem concerns inferences about the parentage of migmatitic rocks formed by partial melting, owing to the very common redistribution and removal of melt. Even stromatic migmatites (Section 4.7.2 commonly have lost melt from the leucosomes, so that bulk chemical compositions may not be reliable indicators of parent rock-types.

7.2.2 Residual structures of metapelitic and metapsammitic rocks

Metamorphosed metapelitic and metapsammitic rocks commonly preserve characteristic primary structures. For example, residual bedding is

Fig. 7.3 Finely bedded metapelite with andalusite porphyroblasts, Broken Hill area, western New South Wales, Australia, with a clastic dyke that also has some porphyroblasts, indicating its pre-metamorphic origin. Photo by Nick Cook.

common in metamorphosed sedimentary rocks, even where strongly folded (Fig. 7.2). Fine bedding laminations (e.g., in pelite, chert or ironstone) may be preserved, even where metamorphism has produced new minerals, including porphyroblasts (Oftedahl, 1967; Talbot & Hobbs, 1968), as shown in Figs. 7.3 and 7.4, though the thickness and continuity of bedding may be altered by the deformation. Residual bedding is typically oblique to tectonic foliations (Figs. 6.21, 6.22, 6.24).

Cross-bedding and ripple marks may be well preserved in metamorphic terranes, for example, in amphibolite facies metaquartzites in the Scottish Highlands (Bailey, 1930; Vogt, 1930; Wilson *et al.*, 1953; Weiss & McIntyre, 1957; Tobisch, 1955), as shown in Fig. 7.5. Graded bedding is preserved in And-St-Grt-Bt zone (low-pressure amphibolite facies) metasediments in the Coos Canyon area, Maine (Guidotti, 2000, fig. 7.4), as shown in Fig. 7.6A,

(A)

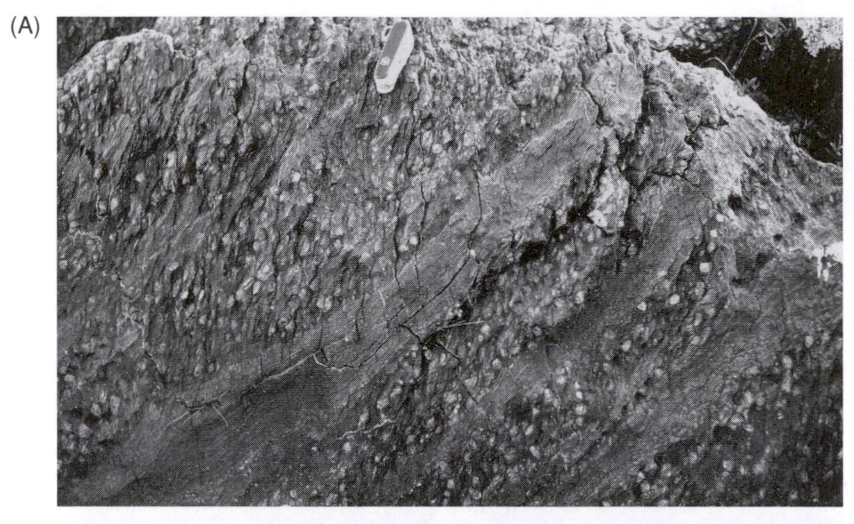

(B)

Fig. 7.4 (A) Bedding preserved in amphibolite facies schist with abundant andalusite porphyroblasts, Weekeroo Inlier, Olary district, South Australia. Knife 9 cm long. (B) Bedding preserved in granulite facies schist with abundant sillimanite delineating a foliation in pelitic beds (lighter grey), Southern Cross Mine area, Broken Hill district, western New South Wales, Australia. Knife 9 cm long.

and in granulite facies metasedimentary rocks of the Broken Hill area (Glen & Laing, 1975; Laing, 1980, 1996a; Willis *et al.*, 1983), as shown in Fig. 7.6B. Graded and cross bedding are locally preserved in low-pressure amphibolite to granulite facies metasediments in the Cooma area, south-east Australia (Hopwood, 1976; Granath, 1976; Johnson *et al.*, 1994), the Mount Stafford area, central Australia (Vernon *et al.*, 1990), the Olary Block, South Australia (Clarke *et al.*, 1986), and the Rio Mora area, New Mexico, USA (Grambling & Codding, 1982). Graded bedding may be preserved even in highest-grade metamorphic zones, in which partial melting has been extensive (Glen & Laing, 1975; Laing, 1980; Stevens & Willis, 1983; Clarke

(A)

(B)

Fig. 7.5 (A) Cross-bedding in Dalradian metaquartzite, Loch Leven, Scotland. This outcrop occurs very close to intensely, multiply deformed schists interbedded with the quartzite (Weiss & McIntyre, 1957). (B) Ripple marks on a bedding surface at the same locality.

et al., 1986; Vernon *et al.*, 1990, 2003; Greenfield *et al.*, 1996; Vernon & Johnson, 2000), as shown in Figs. 7.7 and 7.8.

Generally, graded bedding appears to be more reliable as a younging indicator than cross-bedding in metamorphic terranes, but only if complete Bouma sequences are preserved, as grading can be reversed in Bouma A horizons (Paul F. Williams, personal communication). Preferential growth of porphyroblastic minerals in the more pelitic parts of graded beds may produce a so-called 'reverse' grading (Shrock, 1948; Hobbs *et al.*, 1976), as shown in Fig. 7.8. This occurs especially in relatively weakly deformed parts of low-pressure/high-temperature (LPHT) areas, such as the Palaeozoic Cooma and Cambalong Complexes of south-eastern Australia (Vernon & Johnson, 2000; Vernon *et al.*, 2003), the Mount Stafford area of the Proterozoic Arunta block, central Australia (Vernon *et al.*, 1990), and

(A)

(B)

Fig. 7.6 (A) Graded bedding in crenulated schists with staurolite porphyroblasts in selected (probably relatively iron-rich) beds, Coos Canyon, western Maine, USA. Knife 9 cm long. (B) Graded bedding in sillimanite–biotite–garnet–K-feldspar (granulite facies) metasedimentary rocks, Broken Hill, New South Wales, Australia. The foliation, delineated mainly by elongate pods of sillimanite, is oblique to the bedding.

the Proterozoic Broken Hill region (Glen & Laing, 1975, p. 16). Nevertheless, such 'reverse grading' may be used as a stratigraphic younging indicator, provided the quartz-rich part of the bed is regarded as the older fraction, regardless of its finer grainsize.

Moreover, the growth of large porphyroblasts of strong metamorphic minerals, such as garnet, cordierite and K-feldspar in the pelitic tops of graded beds increases their strength, so that most of the strain is accumulated in the less pelitic bases of the beds (Vernon *et al.*, 2003;

(A)

(B)

Fig. 7.7 (A) Graded bedding preserved in high metamorphic grade migmatitic rocks, the quartz-rich bases of beds grading up into sillimanite-rich, partly melted metapelitic tops of beds (upside down) at The Hawk, Cape Sable Island, Nova Scotia, Canada. The foliation is oblique to the bedding. (B) Bedding preserved in psammitic layers, despite strong crenulation folding in metapelitic layers at the same locality. The metapelitic layers show light-coloured felsic leucosomes, which are evidence of partial melting.

Vernon, 2004), as shown in Fig. 7.8. In fact, the increased strength of the tops of the beds may help to preserve graded bedding as a residual structure in some regional metamorphic terranes.

Graded bedding may occur in some layered igneous intrusions (Parsons, 1987; Myers, 1981), but no confusion with sedimentary grading should arise, because of the basic to ultrabasic mineral assemblages and chemical composition of the igneous material. Truncated curved layering resembling cross-bedding also occurs in some plutonic igneous rocks, such as mafic-ultramafic layered complexes (Parsons & Butterfield, 1981; Parsons, 1987; Parsons & Becker, 1987; Irvine, 1987; Upton, 1987), and a

Fig. 7.8 Graded bedding in Ordovician amphibolite facies rocks in the Cambalong Complex, SE Australia. Cordierite porphyroblasts grew preferentially in the metapelitic tops of the beds, which also contain leucosome (light coloured) as a result of partial melting. This is an example of metamorphic 'reverse grading.' Because of the strength of the abundant large porhyroblasts, the metapelitic parts of the beds became stronger than the metapsammitic parts, which consequently flowed more and developed a foliation in response to later deformation. The leucosome migrated locally, but most remained within the host beds. Knife 9 cm long.

structure of similar appearance is formed by truncated schlieren layering in granitoids (Vernon & Paterson, in press). However, though precautions should be taken when inferring parent rocks of meta-igneous rocks of these compositions, they can cause no confusion in the interpretation of potential cross-bedding in most metasedimentary rocks, except possibly arkoses, and then the bulk chemical composition may help distinguish them from metamorphosed felsic igneous rocks.

Distorted pebbles are commonly preserved in metamorphosed conglomerates (Flinn, 1956; Smith *et al.*, 1969; Windley & Bridgewater, 1971; Weiss, 1972, figs. 175–109; Hobbs *et al.*, 1976; James, 1976; Till & Snee, 1995, fig. 7.4d; Vernon, 2002b, fig. 117) as shown in Fig. 7.9, though care should be taken to avoid misinterpreting dismembered quartz veins (Fig. 7.10) and large porphyroblasts (Fig. 7.11) as deformed pebbles (e.g., Ramsay, 1956; Naha & Majumdar, 1971).

Many other sedimentary and organic structures may be seen in less strongly foliated metamorphic rocks of variable grade. Some may be useful as younging indicators, especially in low-grade metamorphic areas, but potentially in higher grade (amphibolite facies) areas as well (Hobbs *et al.* 1976, pp. 148–151). These structures include: *ripple marks* (Hobbs & Talbot, 1966; Glen & Laing, 1975; Eriksson, 1981; Clarke *et al.*, 1986; Huang & Rubenach, 1995, fig. 7.5b), as shown in Fig. 7.5B; *worm burrows*; *clastic dykes* (Powell, 1969), as shown in Fig. 7.3; *sedimentary breccias* (Bernoulli &

Fig. 7.9 Deformed conglomerate, showing pebbles of various rock-types elongated parallel to a strongly foliated matrix. Ruler 15 cm long. Photo by Scott Paterson.

Weissert, 1985; Cook & Ashley, 1992, fig. 7.4), as shown in Fig. 7.12; *sole structures (sole markings)*, such as load casts, sandstone balls, flame structures and rip-up clasts (Hobbs & Talbot, 1966; Glen & Laing, 1975; Hobbs *et al.*, 1976, pp. 149–50; Kukla *et al.*, 1990); *scour surfaces* (Kukla *et al.*, 1990); *ball-and-pillow structures* (Hobbs *et al.*, 1976, pp. 148–151; Clarke *et al.*, 1986; Kukla *et al.*, 1990); *slump structures* and *convolute laminations* (Oftedahl, 1967; Hobbs *et al.*, 1976, pp. 156–7; Horwitz & Ramanaidou, 1993); *slump rolls* (Hobbs & Talbot, 1966); *oolitic structures* (Chapman, 1893; Cloos, 1947; James, 1992); *mud (desiccation) cracks* (Hobbs *et al.*, 1976, pp. 150–1; Hälbich & Altermann, 1991); *calcareous concretions* (Wright & Henderson, 1992; Stevens, 1998), as shown in Fig. 7.13; and *evaporitic structures*, such as gypsum rosettes (Hälbich & Altermann, 1991) or pseudomorphs of such structures (Cook & Ashley, 1992). Features such as sand volcanoes, sand dykes and slumped horizons, which are typical of near-seafloor compaction processes, have been observed in low-grade rocks of the Meguma Group, Nova Scotia, Canada (Wright & Henderson, 1992, p. 286).

Fossils are preserved in some metamorphic rocks, especially those of low metamorphic grade (e.g. greenschist facies), but they also occur in some high-grade, even migmatitic rocks (Vogt, 1930; Bucher, 1953; Boucot & Thompson, 1963; Vance, 1968, p. 304; Grew *et al.*, 1970; Termier & Termier, 1970; Boucot & Rumble, 1978, 1980, 1986; Rumble *et al.*, 1982; Hill, 1985; Hiroi *et al.*, 1987; Franz *et al.*, 1991; Hanel *et al.*, 1999).

However, residual sedimentary structures and fossils are commonly distorted in deformed in regional metamorphic rocks (e.g. Chapman, 1893;

(A)

(B)

Fig. 7.10 (A) Deformed metasediment showing elongate lensoids that superficially resemble deformed pebbles, but which are actually dismembered quartz veins, Ben Hutig, west of Tongue, Sutherland, Scotland. (B) Dismembered quartz veins in schist, near Ulverstone, Tasmania, Australia, showing an earlier stage of the deformation process than that shown in (A).

Cloos, 1947; Flinn, 1956; Hills, 1963, pp. 122–7; Turner & Weiss, 1963, p. 96; Hobbs *et al.*, 1976). A potential problem is that some tectonic structures can resemble sedimentary structures. For example, intersecting foliations can look like cross-bedding (Bowes & Jones, 1958) and so can truncated folds, especially truncations produced by transposition during tight tectonic folding (e.g. Hobbs *et al.*, 1976, fig. 5.33d). Many examples are so problematic that no confident inferences of cross-bedding can be made (Fig. 7.14). In addition, small crenulation folds may resemble ripple marks (Ingerson, 1940), and disrupted quartz veins can resemble pebbles (Fig. 7.10), as mentioned previously. Therefore, care needs to be taken when interpreting structural evidence – for example, contrasting interpretations of sedimentary structures in metamorphic rocks at Broken Hill by Condon (1959) and Williams (1959) – and doubtful structures should not be used for stratigraphic interpretations.

Fig. 7.11 Andalusite porphyroblasts superficially resembling pebbles, Reynolds Range, central Australia. Base of photo approximately 30 cm.

Sedimentary microstructures are generally less clearly preserved, but can be detected in some low- to medium-grade metamorphic rocks, for example *quartz and feldspar clasts* in metapsammitic rocks (Muffler & White, 1969, plates 1, 2; Ernst, 1971, figs. 3, 4; Bishop, 1972, fig. 4; Read & Eisbacher, 1974, fig. 11; Kamineni *et al.*, 1991, fig. 4; Vernon, 2004, fig. 4.74), as shown in Fig. 7.15; *rock fragments* in various rock-types (Dunlop & Buick, 1981), especially *pebbles* in metaconglomerates; *distorted ooids* (Fig. 7.16); *fossils or fossil fragments* (Fig. 7.17); *trace fossils*; and *vitroclastic microstructures* in low-grade (e.g., zeolite or greenschist facies) rocks (Coombs, 1954; LaBerge, 1966; Raam, 1968; Hunahashi *et al.*, 1972; Read & Eisbacher, 1974, fig. 11a; Dimroth & Lichtblau, 1979; Lowe, 1999a,b; Ransom *et al.*, 1999). In addition, many shales and some sandstones have relatively strong alignment of detrital mica and clay grains parallel to bedding (Hobbs *et al.*, 1976, fig. 3.8; Maltman, 1981; Morritt *et al.*, 1982; Paterson *et al.*, 1985), and this alignment can be preserved during metamorphism (Hobbs *et al.*, 1976, p. 153; Maltman, 1981).

(A)

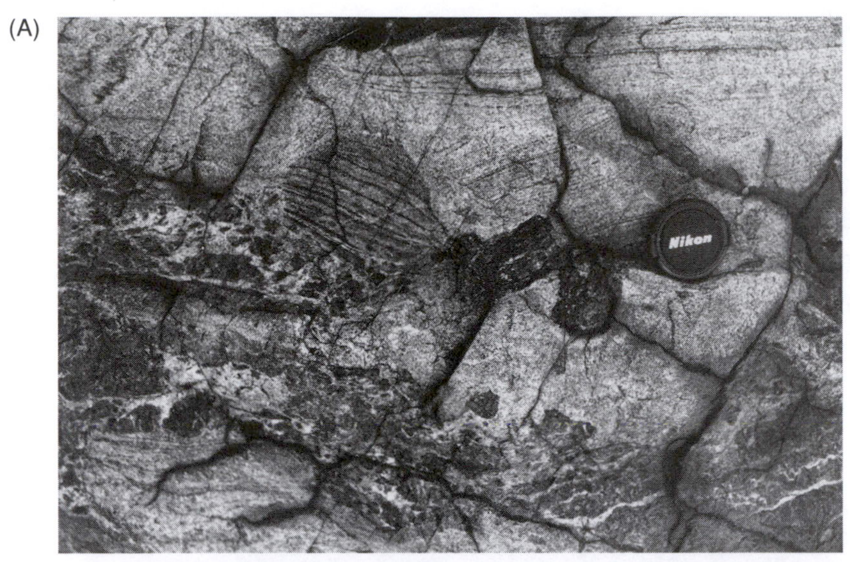

(B)

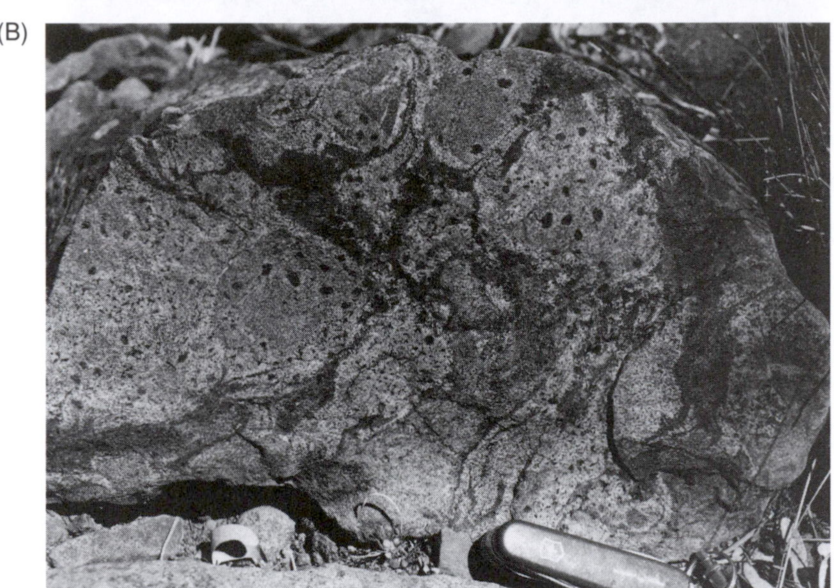

Fig. 7.12 (A) Sedimentary brecciation structures preserved in amphibolite facies rocks at Mount Stafford, central Australia (Vernon *et al.*, 1990, fig. 17.9, with permission of Unwin Hyman Ltd). The metapelitic fragments contain dark cordierite porphyroblasts. Lens cap for scale. (B) Similar structures preserved in granulite facies rocks with garnet and cordierite porphyroblasts, also in the Mount Stafford area (Vernon *et al.*, 1990, fig. 17.10, with permission of Unwin Hyman Ltd). Knife 9 cm long. (C): Calcsilicate breccia, Olary area, South Australia, consisting of angular fragments of calcsilicate rock separated by clinopyroxene-actinolite aggregates of metasomatic origin (Kent *et al.*, 2000), as discussed in Section 5.16.4.

7.2.3 Residual structures of calcareous metasediments

Calcareous metasedimentary rocks may be strongly layered (Fig. 7.18), the layering generally being interpreted as bedding (e.g., Stroud *et al.*, 1983; Cook & Ashley, 1992). However, bedding may be thinned and distorted by deformation, and consequently the origin of layering may be doubtful in strongly deformed rocks. Chemical evidence generally can be used to

(C)

Fig. 7.12 (*cont.*).

Fig. 7.13 Epidote-rich calcsilicate nodule in upper amphibolite facies rocks at The Monuments, north of Broken Hill, western New South Wales, Australia. The structure has been interpreted as a former calcareous nodule (Stevens, 1998).

(A)

(B)

Fig. 7.14 (A) Moine metaquartzite, with truncated folds resembling cross-bedding, Strathan, west of Tongue, Sutherland, Scotland. (B) Strongly deformed, partly melted, feldspathic metasedimentary rock, Harcuvar Mountains, California, USA. Though some of the structures could represent distorted cross-bedding, some could be due to foliation truncations caused by the deformation. Knife 9 cm long.

distinguish calcareous rocks from layered amphibolites of igneous origin; for example, metamorphosed calcareous rocks tend to have higher Ca, Mg and Si and lower Al and Fe than mafic igneous rocks (Edwards, 1958; Stroud *et al.*, 1983, p. 260). Generally limestones and dolomites plot separately from mafic igneous rocks on ACF diagrams (Fig. 1A), though some overlap of mafic rocks with marls and calcareous shales may occur (Fig. 1A), as exemplified by calcareous rocks in the Broken Hill area, Australia (Stroud *et al.*, 1983, fig. 6). Also, mafic rocks tend to have much higher Cr and Ni contents (Evans & Leake, 1960).

Calcareous concretions are common in sedimentary rocks, and may give rise to calcsilicate pods in metasediments. For example, Wright & Henderson (1992, p. 286) described calcareous concretions in slates of the Meguma Group, Nova Scotia, Canada; silt beds pass through the

Fig. 7.15 Angular to subangular clasts of quartz (clear) and feldspar (clouded) with metamorphic biotite and epidote (high relief), in greenschist facies metasandstone, Hill End trough, central-western New South Wales, Australia. Plane-polarized light; base of photo 4.8 mm. From Vernon (2004, fig. 4.74).

concretions, showing that the concretions are not detrital. Zoned calc-silicate ellipsoids (containing minerals such as quartz, calcic plagioclase, amphibole, epidote, diopside, garnet and titanite) in metasedimentary rocks in the Broken area Hill, Australia (Fig. 7.13), are probably the higher-grade metamorphosed equivalents of diagenetic carbonate concretions (Stroud *et al.*, 1983; Stevens, 1998).

Fig. 7.16 Chlorite ooids in a weakly metamorphosed calcareous shale from the Swiss Alps. The ooids have been distorted and elongated by deformation, and many have had their shapes truncated by solution seams. Sample by courtesy of David Durney. Plane-polarized light; base of photo 4 mm. From Vernon (2004, fig. 2.26).

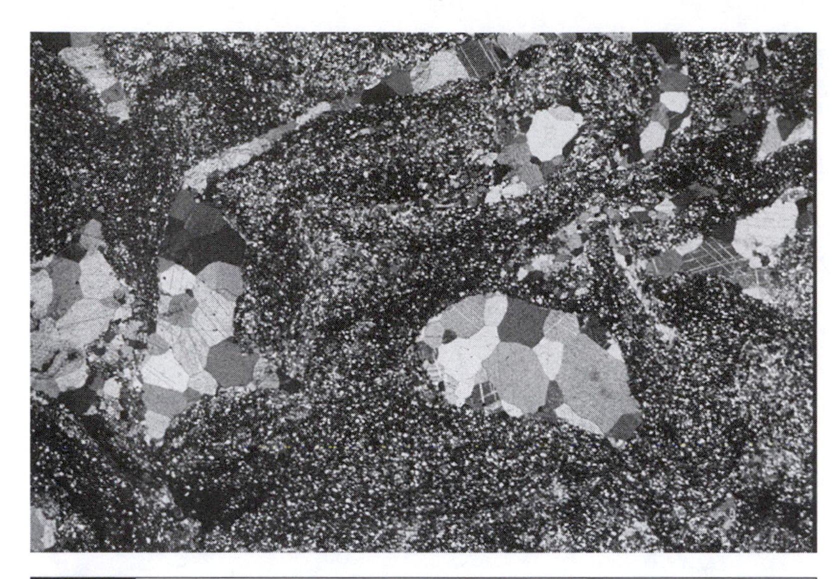

Fig. 7.17 Metamorphosed impure limestone (marl), Cox's River, Hartley area, New South Wales, Australia, showing shell fragments that have been recrystallized to coarse-grained polygonal aggregates of calcite. Crossed polars; base of photo 7.8 cm.

Calcsilicate beds or layers interlayered with marble are generally stronger than the marble, and so fracture and undergo boudinage while the marble flows coherently during strong deformation.

7.2.4 Chemical metasediments

Metasediments of predominantly chemical origin include iron-rich rocks, cherts and manganese-rich rocks. These rocks are typically fine-grained, and may remain relatively fine-grained during metamorphism, commonly showing bedding laminations (Oftedahl, 1967; Goodwin, 1973; Stanton, 1976; Trendall, 1983; Barnes *et al.*, 1983; James, 1992; Peter, 2003).

Common examples are layered iron-rich rocks, known as 'iron-formation', 'banded iron-formation (BIF)' or 'jaspilite', for example, quartz-magnetite-rock and quartz-hematite rock. Other minerals, such as siderite, garnet, apatite, pyroxene, riebeckite, stilpnomelane, minnesotaite, chlorite and sulphide minerals, may also be present.

Banded iron-formations occur across all grades of regional metamorphism. Prograde metamorphism of iron-formations worldwide, inferred to be isochemical except for H_2O and CO_2, produces amphiboles, then pyroxenes and, at highest grades, olivine-bearing assemblages (Klein, 2005). At lowest metamorphic grades in the Archaean of the Yilgarn Block, Western Australia, the BIF minerals are quartz, magnetite, greenalite, stilpnomelane, siderite and ankerite, whereas at highest grades the minerals are quartz, magnetite, clinopyroxene and orthopyroxene (Gole, 1981). At lowest grades in Fe metashales interbedded with the Yilgarn banded iron-formations the minerals are chamosite, stilpnomelane, siderite and pyrite, whereas at highest grades the minerals are hornblende and almandine, with or without biotite (Gole, 1981). Mafic rocks associated with iron-formations in the Hamersley Basin, Western Australia, contain prehnite-pumpellyite to greenschist facies assemblages produced by burial metamorphism (Smith *et al.*, 1982).

(A)

(B)

Fig. 7.18 (A) Continuous layering, inferred to be bedding, in Proterozoic calcsilicate rock, Cloncurry area, north Queensland. (B) Folded bedding in Proterozoic calcsilicate rock, Cadney Creek, Arunta Block, central Australia. Knife 9 cm long.

Compositional layers in banded iron-formations, invariably interpreted as sedimentary bedding, vary from continuous to lenticular and wavy (James, 1992), and may be isoclinally folded (Barnes *et al.*, 1983, photo 4). The layers are typically 0.5–3 cm thick, and are commonly internally laminated on a scale of millimetres or less (James & Sims, 1973; Gole, 1981). The layers consist of chert alternating with iron-rich rock consisting of siderite, siderite plus Fe silicates, magnetite, magnetite plus hematite or hematite (James & Sims, 1973). Stratigraphic iron-formation units are up to hundreds of metres thick and hundreds or even thousands of kilometres in lateral extent, in the cratonic shields of all continents (James & Sims, 1973; Gole, 1981). Millimetre-scale laminations in some iron-formations can be laterally continuous over many kilometres (Trendall, 1983). For example, 'macrobands' in the Dales Gorge Member are continuous over the whole of the 60 000 km^2 of the exposed outcrop of the of the Hamersley Group, Western Australia (Trendall, 1983). As well as bedding laminations, other sedimentary structures may be preserved in ironstones (some reflecting current action), for example, graded bedding, cross-bedding, scour-and-fill structures, mudcracks and raindrop imprints (Hitzman *et al.*, 1992, pp. 256–7), pisolites (Trendall, 1973a), and algal structures, ooliths and stylolitic seams (Bayley & James, 1973; James, 1992). The laminations in some iron-formations may be residual glacial varves (Trendall, 1973b).

Banded iron-formations are probably the most common chemical sedimentary rocks in the Early and Middle Precambrian (LaBerge, 1973). Their most common age range is 1900–2500 Ma, though examples as old as 3800 Ma and as young as 600 Ma have been reported (Klein, 2005). Ironstones and cherts in younger rocks typically occur in separate units, rather than being interlaminated (LaBerge, 1973), though Schultz (1966) described interlayered chert and ironstone of Lower Carboniferous age.

The largest iron-rich orebodies result from secondary iron concentration, involving hydrothermal or supergene redistribution of iron in existing BIFs. Some of this iron concentration may occur during deposition or diagenesis, but most occurs by supergene enrichment produced where erosion exposes existing iron-formations during slow uplift that keeps pace with erosion and enables large ore bodies to form over millions to hundreds of millions of years (Morris, 1985). Such supergene enrichment probably began when the atmosphere became sufficiently rich in oxygen about 2000 Ma ago (Morris, 1985).

A large variety of metamorphosed chemical sedimentary rock-types occurs interbedded with metaclastic and metavolcanic rocks in the amphibolite-granulite facies complex (Willyama Supergroup) of the Broken Hill area, Australia, namely 'banded iron-formation' (Richards, 1966; Williams, 1967; Stanton, 1976; Barnes *et al.*, 1983), quartz-gahnite rock, garnet-quartz rock, and tourmaline-rich rocks (Vernon, 1969; Barnes *et al.*, 1983; Plimer, 1984).

7.3 Meta-igneous rocks

7.3.1 Broad chemical characteristics of metamorphosed igneous rocks

The broad parentage of most metamorphosed rocks of mafic igneous composition can be inferred on the basis of bulk chemical composition

(Mason 1966; Le Maitre 1976), as shown in Fig. 7.1A, provided metasomatic changes have not occurred. For example, amphibolites and pyroxene-hornblende gneisses interlayered with metasediments in the Broken Hill area, Australia, have 47–50 wt% SiO_2, 13–14 wt% Al_2O_3, 7–8 wt% MgO, 8–11 wt% CaO and 14–22 wt% total Fe as Fe_2O_3, and have been interpreted as Fe-rich tholeiitic basalts (Stroud *et al.*, 1983, table 2). However, undoubted dolerites and gabbros in the same area have very similar chemical compositions (Stroud *et al.*, 1983, table 11). Therefore, the distinction between metamorphosed volcanic and intrusive mafic rocks cannot be confidently made on the basis of bulk chemical compositions alone.

Most metamorphosed ultramafic rocks are distinctively rich in Mg and poor in Si (Mason, 1966; Le Maitre, 1976), and serpentinized peridotites are very rich in H_2O. Metamorphosed ultramafic rocks in the Broken Hill area, Australia, have around 14–26 wt% MgO, 14–18 wt% total Fe as Fe_2O_3, and 41–42 wt% SiO_2 (Stroud *et al.*, 1983, table 11).

Metamorphosed felsic igneous rocks (Fig. 7.1A) commonly can be matched relatively closely with average rhyolite to dacite compositions (Le Maitre, 1976). For example, felsic gneisses in the Broken Hill area, Australia, can be matched in this way (Brown *et al.*, 1983, table 2, fig. 2), and have been interpreted as felsic volcanic to pyroclastic rocks. However, their compositions are very similar to those of average granites to granodiorites (Le Maitre, 1976), and so, as for mafic igneous rocks, distinction between volcanic and intrusive felsic rocks cannot be confidently made on the basis of bulk chemical compositions alone.

7.3.2 Residual structures in metamorphosed igneous rocks

A large range of mesoscopic (outcrop-scale) primary (magmatic) structures may be preserved in weakly or non-deformed, metamorphosed igneous rocks, for example, *pillows* (McCall, 1971; Windley & Bridgewater, 1971; Pearce & Birkett, 1974; Binns *et al.*, 1976; Dimroth & Lichtblau, 1979; Condie, 1981, pp. 77, 87; Boardman, 1986; Schaefer & Morton, 1991; Ransom *et al.*, 1999; Komiya *et al.*, 1999), as shown in Fig. 7.19; *amygdales* (Harker, 1932, figs. 39A, 43A,B; Coleman & Lee, 1963, fig. 4; Condie, 1981, p. 87; Sivell & Waterhouse, 1984, p. 20; Boardman, 1986; Schaefer & Morton, 1991; Till & Snee, 1995, fig. 4c); *lapilli* (Boardman, 1986; Schaefer & Morton, 1991; Ransom *et al.*, 1999); chilled *mafic dyke margins* (Davidson, 1990, fig. 22); *igneous brecciation* (Wadsworth, 1961; Condie, 1981, p. 82; Boardman, 1986); *abundant, large xenoliths* (Wintsch *et al.*, 1990); *microgranitoid enclaves* (Prestvik, 1980, p. 559, Vernon & Williams, 1988; Vassallo & Vernon, 2000), as shown in Fig. 7.20; *aplite dykes and veins in metagranites* (Vernon & Williams, 1988; Vassallo & Vernon, 2000), especially where tectonic foliations pass from the granite into the aplite (Fig. 7.21); *cumulate layering*, some with grading (Myers, 1981); *comb layering* (Myers, 1981); and *elongate, skeletal to dendritic shapes of olivine and clinopyroxene ('spinifex structure')* in komatiites and komatiitic basalts in less deformed parts of Archaean greenstone terranes, even where the primary minerals are pseudomorphed by metamorphic minerals of the prehnite-pumpellyite, greenschist or even lower amphibolite facies (Viljoen & Viljoen, 1969; Naldrett & Gasparrini, 1971; Nesbitt, 1971; Pyke *et al.*, 1973; Arndt *et al.*, 1977; Donaldson, 1982; Hill *et al.*, 1990, 1995). Care should be taken to distinguish such dendritic olivine of igneous origin from elongate, but less obviously dendritic olivine of metamorphic origin, which may occur in the same region. Not only is the

(A)

(B)

Fig. 7.19 (A) Pillows in Archaean greenschist facies metabasalt, Pilbara region, Western Australia, showing metasomatic alteration zones related to the edges of the pillows. Knife (below and to left of centre) 9 cm long. (B) Distorted pillows in greenschist facies metabasalt, path to Monviso, Piedmont region, Italy. Photo by Guido Gosso.

Fig. 7.20 Elongate microgranitoid enclave ('mafic enclave') containing a K-feldspar megacryst in deformed granite (augen gneiss), Wyangala Dam, central-western New South Wales, Australia. Coin diameter 2.8 cm.

(A)

(B)

(C)

Fig. 7.21 (A) Aplite veins in deformed, foliated granite, Ardara area, Donegal, Ireland. The obliqueness of the foliation to the boundaries of the aplite, as well as the distortion and incipient boudinage of the aplite veins, indicate that the aplite was present before the deformation. (B) Aplite vein oblique to gneissic foliation in deformed granite, Wyangala Dam, central-western New South Wales, Australia. The gneissosity passes from the deformed granite through the aplite, showing that the aplite pre-dates the deformation. (C) Aplite layers in deformed megacrystic charnockitic gneiss, Anmatjira Range, central Australia. The gneissic foliation, delineated partly by leucosomes formed by partial melting during granulite facies metamorphism, transects the aplite layers, indicating that the aplite was present before the deformation.

(A)

(B)

Fig. 7.22 (A) Elongate plagioclase lath shapes preserved in strongly neocrystallized, granulite facies metadolerite, despite extensive grain boundary migration of the edges of the laths, Baja California, Mexico. Crossed polars; base of photo 7.7 mm. From Vernon (2004, fig. 4.78). (B) Metadolerite, in which metamorphic amphibole has pseudomorphed former pyroxene, preserving poikilitic microstructure (right) and has also grown across primary pyroxene-plagioclase grain boundaries, thereby partly obscuring igneous grain shapes, Foothills terrane, central Sierra Nevada, California, USA. Crossed polars; base of photo 4.4 mm.

chemical composition different (low Ni and Ca), but the habit of the metamorphic olivine is simpler, the crystals being either needle-like or bladed, but not skeletal or feathery (Evans & Trommsdorff, 1974).

Residual igneous microstructures are also commonly preserved, especially in contact metamorphic rocks (Binns, 1965, plates VA,B; Andrew, 1984, figs. 2, 5), but also in many regionally metamorphosed igneous and tuffaceous rocks of up to granulite facies. Examples are *elongate plagioclase laths* (Harker, 1932, figs. 39B, 41A, B, 42A; Walker *et al.*, 1960, fig. 1A; Binns, 1964, p. 291; Joplin, 1968, figs. 59A,B; Smith, 1968; Reed & Morgan,

Fig. 7.23 Residual plagioclase phenocrysts in neocrystallized, amphibolite facies metabasalt, Arunta Block, central Australia. Crossed polars; base of photo 11 mm.

1971, fig. 3; Vallance, 1974; Binns *et al.*, 1976, figs. 2A,B; Condie, 1981, figs. 3–19; Vernon, 2004, figs. 4.78, 4.79), as shown in Fig. 7.22; *plagioclase phenocrysts*, especially those with complex oscillatory zoning (Harker, 1932, figs. 23C, 40B; Walker *et al.*, 1960, figs. 1A; Vogel & Spence, 1969; Vernon, 1976, p. 91; Perfit *et al.*, 1980; Condie, 1981, p. 108; Boardman, 1986; Allibone, 1991, figs. 7; Vernon, 2004, figs. 4.77, 4.79), as shown in Figs. 7.23–7.25; *K-feldspar phenocrysts*, variably altered (Lentz & Goodfellow, 1993); *K-feldspar megacrysts* (Vernon, 1986b, 1999, 2004); *embayed quartz phenocrysts*

Fig. 7.24 Residual zoned plagioclase phenocryst in felsic metavolcanic or volcaniclastic rock, Mount Lofty Ranges, South Australia. The general shape of the phenocryst was preserved in spite of grain boundary movement at its margins. The internal compositional zoning was preserved by being protected from the grain boundary adjustment. Crossed polars; base of photo 4.4 mm. From Vernon (2004, fig. 4.76).

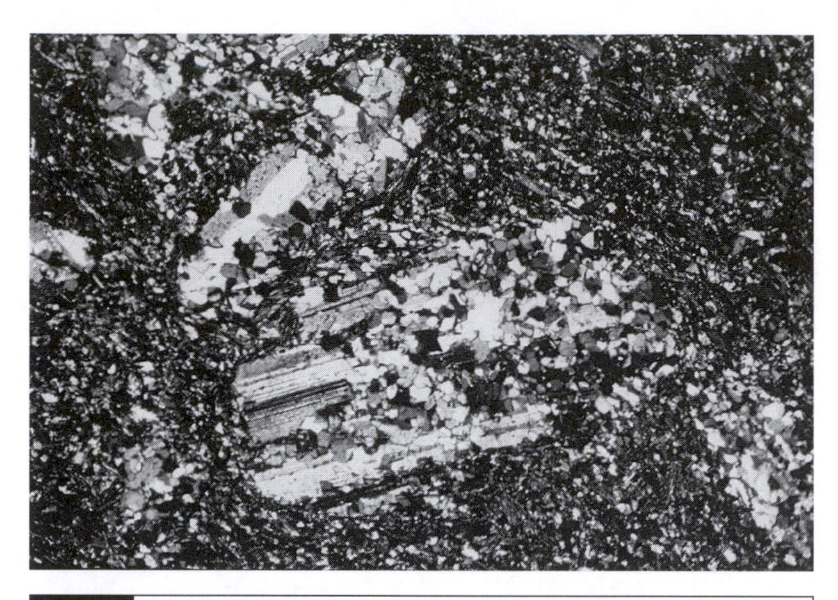

Fig. 7.25 Despite partial recrystallization to fine-grained polygonal aggregates, the general shapes of plagioclase phenocrysts are preserved in this amphibolite facies metabasalt from the Snowy Mountains, south-eastern New South Wales, Australia. Crossed polars; base of photo 4.4 mm. From Vernon (2004, fig. 4.77).

(Vernon, 1986b, 1999, 2004; Williams & Burr, 1994; Stevens & Barron, 2002), as shown in Fig. 7.26; *cumulate structures* (Condie, 1981, p. 106; Hill *et al.*, 1990); *comb layering* Furnes, 1973; Condie, 1981, p. 103); *variolitic structure* (Furnes, 1973, Condie, 1981, p. 87); *ophitic, subophitic* and other *poikilitic structures* (Glassley & Sørensen, 1980, fig. 4; Boardman, 1986; Carney *et al.*, 1991, fig. 8a; Vernon, 2004, fig. 4.78), as shown in Figs 7.22B and 7.27; *gabbroic microstructure* (Bloxham, 1954; Ito & Anderson, 1983); *perlitic cracking*

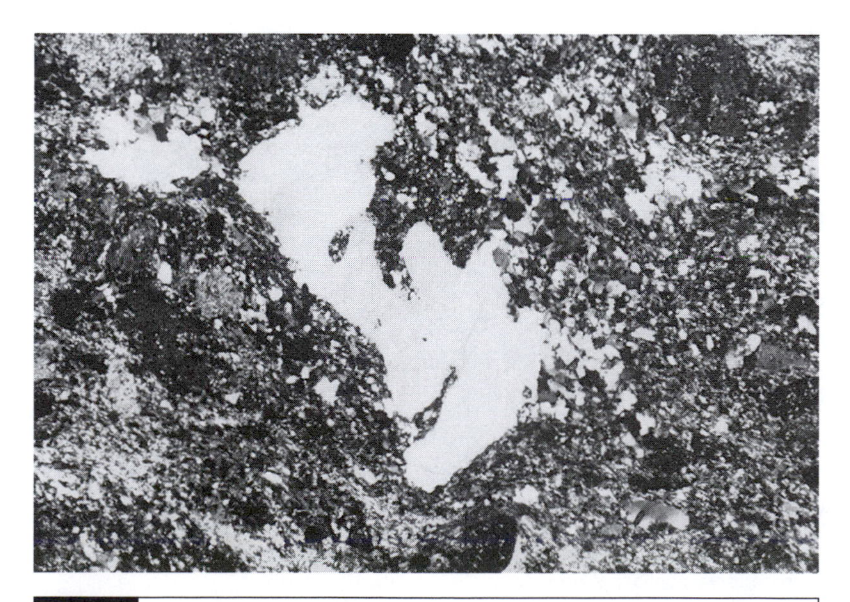

Fig. 7.26 Residual embayed quartz phenocryst in an amphibolite facies metarhyolite from the Picuris Range, New Mexico, USA. Crossed polars; base of photo 4 mm. From Vernon (2004, fig. 4.75).

Fig. 7.27 Ophitic microstructure variably preserved in strongly neocrystallized, granulite facies metagabbro, despite extensive grain boundary migration, Mount Stafford, Arunta Block, central Australia. Crossed polars; base of photo 4.4 mm. From Vernon *et al.* (1990, fig. 17.15A, with permission of Unwin Hyman Ltd).

(Hunahashi *et al.*, 1972; Sivell & Waterhouse, 1984, p. 20); *'spinifex' structure* (Condie, 1981, pp. 77, 87; Hill *et al.*, 1990), *amygdaloidal structures* (Dimroth & Lichtblau, 1979, figs. 17, 18); *microgranitoid enclaves* (Vassallo & Vernon, 2000), and *dimensional preferred orientation of euhedral feldspar crystals* (Paterson *et al.*, 1989; Vernon, 2000a, 2004), which normally indicates magmatic flow, and may be preserved even in partly recrystallized/neocrystallized rocks (Fig. 7.28).

Fig. 7.28 Extensively neocrystallized amphibolite facies metabasic rock, central Sierra Nevada, California, USA, showing predominantly polygonal plagioclase grain shapes, despite which, many grains are elongate in a residual magmatic flow foliation. This interpretation is supported by the alignment of twin interfaces parallel to the grain elongation. Crossed polars; base of photo 2.7 mm.

The presence of crystal faces in feldspar or quartz is one of the best indicators of igneous parentage (Vernon, 1999, 2004), because these minerals typically form crystal faces where they grow in fluids, whereas they almost invariably form random (high-energy) interfaces in metamorphic rocks (Kretz, 1966a; Vernon, 1968, 1975, 1976; Kehlenbeck, 1972). Rare exceptions may reflect growth in a fluid film in the metamorphic environment (Vernon, 2004).

Fine, complex oscillatory zoning in plagioclase is generally an indicator of magmatic crystallization (e.g., Vernon 1976, 2004, pp. 266–7), except possibly for rare crystallization in cavities (Vernon, 1976, p. 91). Most zoning in metamorphic plagioclase is relatively simple (Binns, 1964; Cannon, 1966).

Residual igneous grain shapes may be preserved even where original minerals have been pseudomorphed by metamorphic minerals and in metasomatic rocks, for example, former elongate plagioclase crystals pseudomorphed by minerals such as albite and pumpellyite in metabasalts (Smith, 1968; Reed & Morgan, 1971; Jolly & Smith, 1972).

7.3.3 Residual structures in metamorphosed mafic igneous rocks

Elongate plagioclase grains (Fig. 7.22) favour an igneous origin (Paterson *et al.*, 1989; Vernon, 2000a, 2004). Residual phenocrysts suggest a volcanic origin, as in the Anmatjira Range, central Australia (Fig. 7.23), and locally at Broken Hill, Australia (Brown *et al.*, 1983, p. 237), whereas coarse- or medium-grained residual ophitic microstructures suggest an intrusive origin (Max, 1970), as at Mount Stafford, central Australia (Fig. 7.22), and locally at Broken Hill (Brown *et al.*, 1983, p. 242; Stevens *et al.*, 1988, p. 309). Alignment of elongate plagioclase grains, especially where accompanied by a parallel alignment of twin interfaces (Vernon *et al.*, 1989; Vernon, 2004), is consistent with the presence of a magmatic flow foliation prior to metamorphism; this situation applies even where recrystallization/neocrystallization has obliterated all or nearly all original crystal faces, producing irrational grain boundaries (Figs. 6.31, 7.28, 7.29). Parallel alignment of olivine phenocrysts also indicates magmatic flow in metamorphosed mafic-ultramafic volcanic or shallow intrusive rocks (e.g., Vernon, 2004). Elongation of plagioclase grains in coarse-grained, thoroughly recrystallized, high-grade, mafic granofelses may be only local (Figs. 6.31, 7.22B) or slight to absent (Fig. 7.29), though rare zoned plagioclase phenocrysts may be preserved, even in these rocks (Figs. 6.31, 7.24).

In high-grade metamorphic terranes, mafic gneisses and granofelses are chemically distinguishable from most calcsilicate rocks, as discussed above. However, intense recrystallization makes distinction between extrusive and intrusive origins of former mafic igneous rocks difficult. Some mafic gneisses and granofelses contain abundant, large elongate plagioclase grains, despite intense recrystallization and the formation of irrational grain boundaries between the high-grade metamorphic minerals (including orthopyroxene and brown hornblende), for example in the Arunta Block, central Australia (Figs. 7.22, 7.27), suggesting a gabbroic or doleritic precursor. Such rocks can be distinguished from post-metamorphic rocks with igneous microstructures by the evidence of solid-state adjustment of grain boundaries and the absence of crystal faces on the residual plagioclase grains. Intrusive contacts (Brown *et al.*, 1983, p. 137; Stevens *et al.*, 1988, p. 309) generally suggest a non-volcanic origin,

Fig. 7.29 Strongly recrystallized, granulite facies metagabbro, with some residual elongate plagioclase grain shapes, though not as obvious as for Fig. 7.28. Musgrave Ranges, central Australia. Crossed polars; base of photo 3.5 mm.

although local intrusive contacts may be formed where subaqueous mafic lava lows enter unconsolidated sediments.

Ishizuka (1985) described prograde metamorphism of mafic igneous rocks, in which low-grade (zeolite facies) rocks with well-preserved igneous structures pass through medium-grade rocks with partly preserved structures, into high-grade (hornblende granulite facies) granoblastic rocks without residual structures. Similarly, Mason (1978, pp. 164–6) described progressive obliteration of igneous microstructures in prograde ocean-floor metamorphism of basalt from low-grade to amphibolite facies conditions. In addition, Dimroth & Dressler (1978, fig. 3) depicted progressive destruction of gabbro microstructures, from greenschist to amphibolite facies conditions; ductile deformation was insufficient to promote recrystallization at low grades.

Amygdales may be preserved in non-deformed mafic to intermediate volcanic rocks (e.g., Vernon, 2004, figs. 3.97, 3.98). Skeletal and dendritic microstructure ('spinifex structure'), hyaloclastic brecciation, and polyhedral jointed flow tops may be preserved in non-deformed komatiites metamorphosed at greenschist facies conditions (Viljoen & Viljoen, 1969). Residual tuffaceous structures have been observed in mafic and ultramafic rocks in the Archaean Yilgarn Block of Western Australia (Binns *et al.*, 1976).

7.3.4 Residual structures in metamorphosed felsic igneous rocks

The precursors of felsic (quartzofeldspathic) gneisses may difficult to determine in the absence of reliable structural indicators, because chemically these rocks could be volcanic or granitic. The sheet-like shapes of many of them may suggest a volcanic or volcaniclastic origin, but many granitoids in the middle crust are of sheeted form, for example, in Proterozoic low-pressure/high-temperature (LPHT) terranes, such as the Arunta Block, central Australia, and the Willyama Inlier, Broken Hill, western New South Wales (Vernon *et al.*, 1990; Collins *et al.*, 1991; Collins & Vernon, 1991; Vernon & Williams, 1988; Vassallo & Vernon, 2000). Granites have also intruded as sheets in other Precambrian terranes, such as western Greenland (e.g. Bridgewater *et al.*, 1974; Myers, 1976), and Arizona, USA (Karlstrom *et al.*, 1993, p. 224, fig. 7.10; Karlstrom & Williams, 1995). Moreover, very elongate plutons have been emplaced as sheets parallel to the regional foliation, (with syn-intrusion or post intrusion regional deformation) in parts of the Palaeozoic Lachlan Fold Belt, south-eastern Australia (Vernon *et al.*, 1983; Paterson *et al.*, 1990; Healy *et al.*, 2004). Other younger examples are the Palaeozoic Main Donegal Granite of Donegal, Ireland (Pitcher & Berger, 1972; Hutton, 1982; 1992), the Great Tonalite Sill of southeastern Alaska, USA, and British Columbia, Canada (Hutton & Ingram, 1992), and Mesozoic granite sheets in the Cascade Range, Washington, USA (Paterson & Miller, 1998; Miller & Paterson, 2001).

Wintsch *et al.* (1990) attempted to distinguish between extrusive and intrusive origins for felsic gneisses in eastern Connecticut, USA, on the basis of chemical and structural criteria. They suggested that (1) evidence for an extrusive origin includes interlayered contacts with metasediments, metarhyolites and metabasalts, as well as a 20–50 metre-scale chemical cyclicity, typical of modern ash-flow tuffs, and (2) evidence for an intrusive origin includes abundant amphibolite and calcsilicate xenoliths, as well as kilometre-scale chemical zoning or homogeneity. They also suggested that a massive structure lacking compositional layering is evidence of an intrusive origin, but cumulate layering is relatively common in granites (Vernon & Paterson, in press), and so this criterion is of limited value.

Phenocrysts versus porphyroblasts

Identification of metamorphosed porphyritic igneous rocks requires reliable criteria for distinguishing phenocrysts from porphyroblasts, especially with regard to K-feldspar. Deformation of both porphyritic igneous rocks and porphyroblastic metamorphic rocks can produce augen gneisses, in which it may be difficult to determine the origin of K-feldspar augen (porphyroclasts). Furthermore, porphyroblasts commonly grow in deformed rocks (Chapter 3), so that the problem is expanded to include the distinction between residual phenocrysts, residual porphyroblasts and new porphyroblasts.

Phenocrysts of K-feldspar are characterized by euhedral shape, oscillatory zoning (Fig. 7.30A,B) and simple twinning (Vernon, 1986a, 1990b, 1999, 2004), whereas porphyroblasts of feldspar generally do not show these features. In addition, inclusions in K-feldspar phenocrysts are commonly euhedral and arranged zonally (parallel to former crystal faces of the growing phenocryst), as shown in Fig. 7.30A,B, whereas inclusions in K-feldspar porphyroblasts are typically rounded (for quartz and feldspar inclusions) or more elongate with rounded corners (for biotite inclusions),

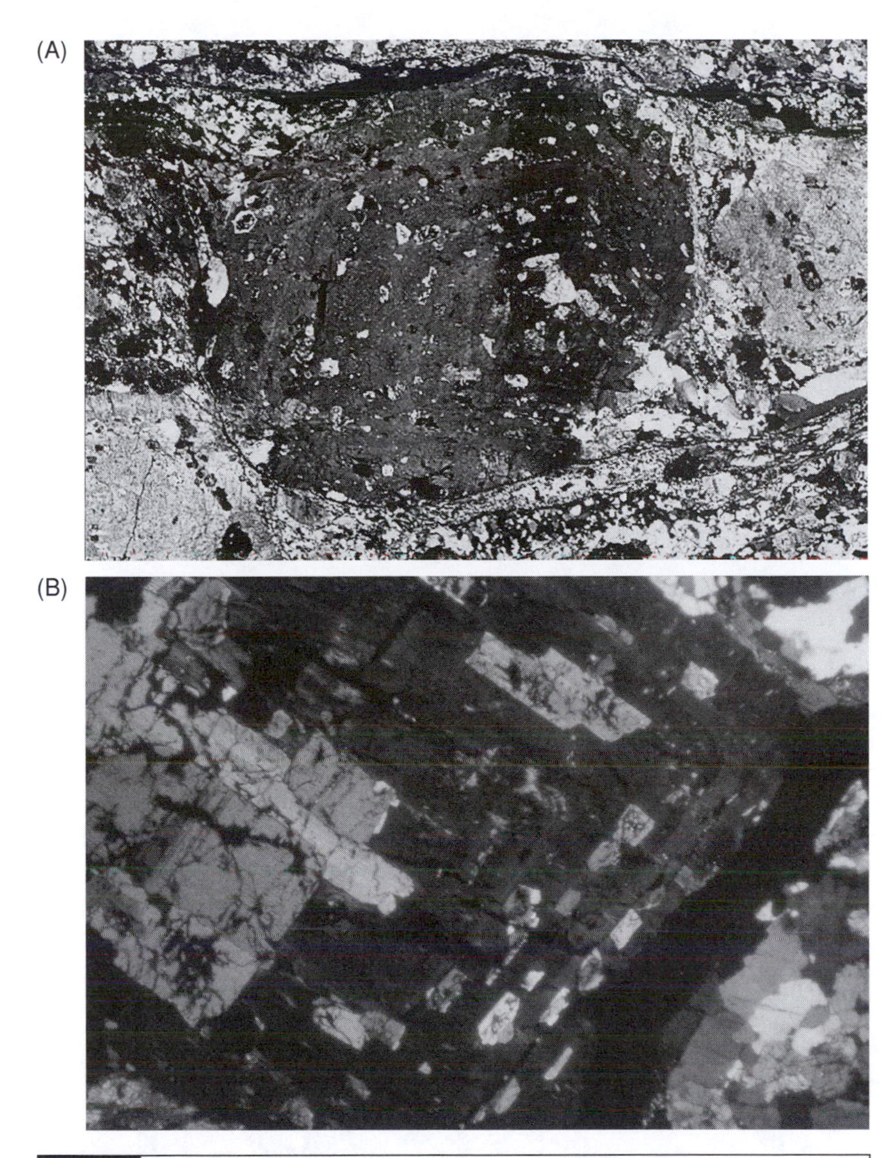

Fig. 7.30 (A) Auge of K-feldspar with oscillatory zoning and zonally (crystallographically) arranged euhedral plagioclase inclusions, both features indicating a phenocrystic origin, in a strongly deformed (mylonitic) granite, Papoose Flat, California, USA. From Vernon (2000a, fig. 10) and Vernon & Paterson (2002, fig. 8). Crossed polars; base of photo 3 cm. (B) K-feldspar auge from the same locality as the rock depicted in A, showing zonally arranged inclusions of euhedral plagioclase, indicative of a origin as a phenocryst. Crossed polars; base of photo 4 mm. (C) Porphyroblast of K-feldspar in an amphibolite facies metapelite, Cooma Complex, SE Australia, showing random, rounded inclusions of quartz and elongate inclusions of biotite. Also shown are very narrow albite exsolution lamellae and thicker, more lenticular, replacive albite 'flames' in the K-feldspar. Crossed polars; base 3.5 mm.

as shown in Fig. 7.30C, and either random (Fig. 7.30C) or arranged in inclusions trails that reflect a pre-existing foliation pattern (Chapter 3).

Megacrysts (augen) of K-feldspar in augen gneisses and mylonites are generally residual phenocrysts (i.e., porphyroclasts), rather than porphyroblasts, because they show the twinning, zoning and inclusion relationships mentioned above (Vernon 1986a, 1990b; Vernon & Paterson, 2002; Vernon,

(C)

Fig. 7.30 (cont.).

2004). Many felsic mylonites and augen gneisses contain spectacular examples of residual phenocrysts that have persisted almost undeformed (apart from microcline twinning and slight marginal recrystallization), complete with compositional zoning and/or zonally-arranged inclusions of igneous plagioclase (Figs. 7.30–7.32). This is possible because of the much greater strength of a large single crystal of feldspar, compared with the relatively weak, fine-grained matrix, which undergoes dynamic recrystallization and neocrystallization, possibly with grain-boundary sliding (Vernon, 2004).

Fig. 7.31 Augen gneiss, resulting from the deformation of a megacrystic granite, Arunta Block, central Australia. Some of the K-feldspar augen are rounded to irregular in shape, whereas others have survived the deformation and recrystallization/neocrystallization as euhedral crystals. Most of the strain has accumulated in the former groundmass, which now consists of thin, lenticular folia rich in biotite (dark) and quartz or quartz/feldspar (light). Knife 9 cm long.

Fig. 7.32 Residual K-feldspar megacrysts, one of which is almost euhedral, in a fine-grained mylonite formed by the intense deformation of megacrystic granite, east of Armidale, New South Wales, Australia. From Vernon (2004, fig. 5.51). Crossed polars; base of photo 4.9 cm.

Zoning and inclusion patterns in former K-feldspar phenocrysts are commonly truncated by tectonic folia in augen gneisses and mylonites (Vernon & Paterson 2002, fig. 8), as shown in Fig. 7.30A. Truncation of igneous plagioclase with oscillatory zoning by sillimanite folia has been described by Vernon *et al.* (1987).

Structural indicators of felsic volcanic precursors

The best structural indicators of metavolcanic rocks are embayed quartz phenocrysts (Fig. 7.26) and rectangular feldspar phenocrysts (Figs. 7.23, 7.24, 7.32). In addition, fiamme (pumice lenses) indicate former welded tuff (ignimbrite), as shown in Fig. 7.33.

Embayed quartz phenocrysts (Fig. 7.26) may survive deformation and metamorphism well (Vernon, 1986b, 1996b, 1999, 2004; Page & Laing, 1992; Williams & Burr, 1994; Laing, 1996b; Pepper & Ashley, 1998; Stevens & Barron, 2002), even at moderately high grades of metamorphism (Vernon, 2004, fig. 4.75), because they are large single crystals, compared with a fine-grained, commonly micaceous groundmass that is much more easily deformed (Etheridge & Vernon, 1981). The former groundmass undergoes intense recrystallization/neocrystallization, owing to the fine grainsize, which promotes fluid flow and metamorphic reactions, with consequent 'reaction softening' and increased strain rate, whereas the unreactive, stronger quartz phenocrysts are preserved intact (Etheridge & Vernon, 1981).

Though phenocrysts of feldspar may be preserved in metamorphosed volcanic rocks (Pepper & Ashley, 1998; Vernon, 2004, figs. 4.76, 4.77; Stevens & Barron, 2002, photograph 10b), they are less likely to be preserved in the most altered rocks, owing to pre-metamorphic and/or metamorphic reactions converting the feldspar to fine-grained micaceous aggregates. These

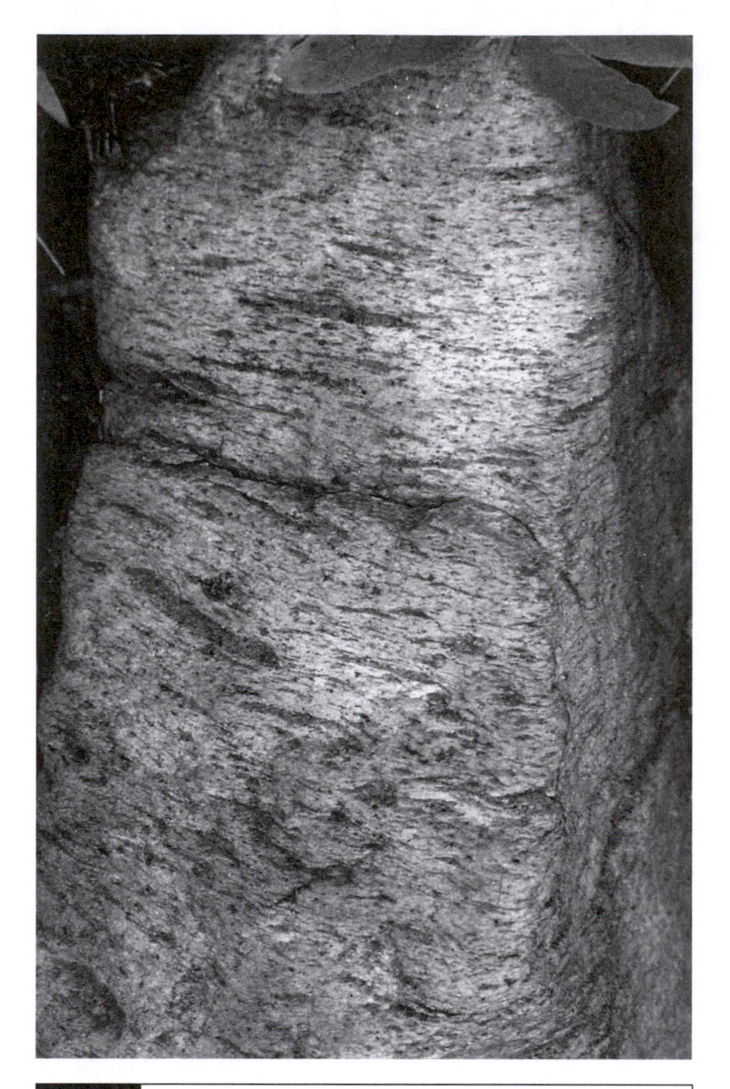

Fig. 7.33 Ignimbrite with fiamme (pumice lenses) occurring amid strongly deformed amphibolite facies schists, Olary area, South Australia. The original pyroclastic structure is remarkably well preserved, despite some deformation.

aggregates deform readily and so become part of the foliated deforming matrix (Frater, 1983), leaving the strong quartz phenocrysts as 'quartz-eyes' in a schistose rock (Vernon, 1986b).

Aggregates of 'decussate' biotite (Fig. 7.34) commonly represent former biotite phenocrysts that have been deformed and recrystallized (Vernon *et al.*, 1983; Stevens & Barron, 2002, photographs 8, 17; Kruhl & Vernon, 2005, fig. 7.9). Because minerals tend to be dispersed in metamorphic rocks, single-mineral clusters require specific circumstances, such as the pseudomorphing of pre-existing larger grains (Vernon, 2004, pp. 277–279). The recrystallized aggregates may be equant in contact metamorphic rocks (Fig. 7.34A,B) and may be elongate parallel to a foliation in regional metamorphic rocks (Fig. 7.34C).

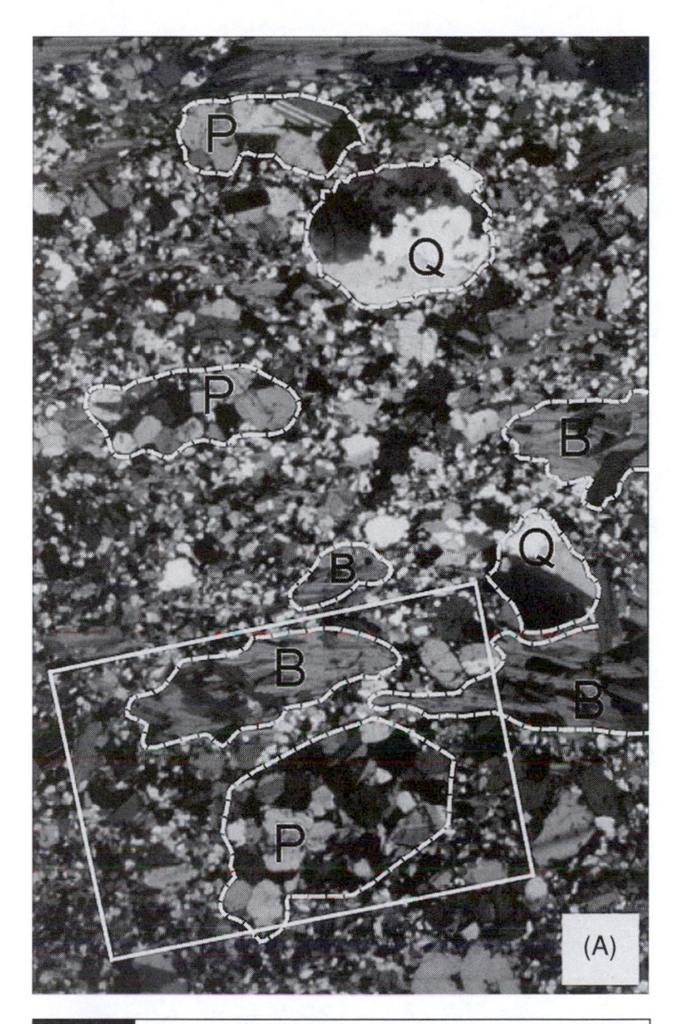

Fig. 7.34 (A) Weakly deformed, contact metamorphosed porphyritic microtonailte, Monte'e Senes, north-east Sardinia, Italy, showing residual, partly recrystallized quartz crystals (Q), coarsely recrystallized aggregates replacing former plagioclase crystals (P) and recrystallized ('decussate') aggregates of biotite (B) marking former primary biotite crystals. Crossed polars; base of photo 7 mm. After Kruhl & Vernon, 2005, fig. 9A; with permission of the *Canadian Mineralogist*. (B) Enlarged view of part of the area shown in A. Crossed polars; base of photo 3.1 mm. After Kruhl & Vernon, 2005, fig. 9B; with permission of the *Canadian Mineralogist*. (C) 'Decussate' aggregates representing recrystallized former phenocrysts of biotite in a felsic tuff metamorphosed at low-pressure granulite facies conditions in the Broken Hill area, New South Wales, Australia. The recrystallized aggregates show a tendency to elongation parallel to the foliation. Also shown (top-right) is a radial aggregate similar to that shown in Fig. 7.35. From Stevens & Barron (2002, photograph 17), with permission of the Geological Survey of New South Wales. Plane-polarized light; base of photo 3.5 mm.

Fig. 7.34 (*cont.*).

In contrast to quartz phenocrysts, fiamme are easily distorted and obscured by deformation and recrystallization/neocrystallization. Some elongate, ovoid aggregates in deformed, high-grade metamorphic rocks have been interpreted as former fiamme (e.g., Brown *et al.*, 1983; Pepper & Ashley, 1998), but generally such interpretations are equivocal. Nevertheless, local felsic rocks in low-strain areas in the Proterozoic Olary Block, eastern South Australia, have well preserved fiamme and quartz phenocrysts (Fig. 7.33), indicating a former welded tuff; such occurrences are rare.

Former spherulites may be preserved in low-grade metamorphic rocks, but have rarely been reported in higher-grade terranes. However, Stevens & Barron (2002) described albite-rich spheroidal structures in metavolcanic rocks in the Broken Hill area, Australia (Fig. 7.35). They consist of radial aggregates resembling spherulites, but have been replaced by blocky albite

(A)

(B)

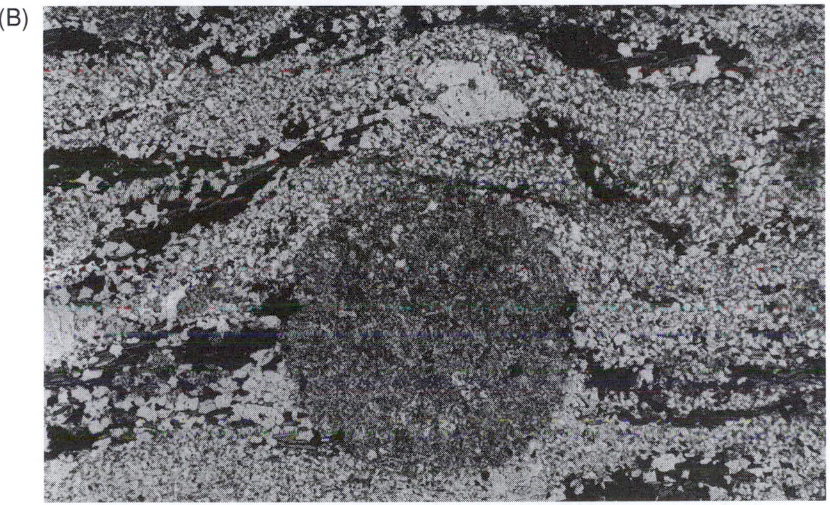

Fig. 7.35 (A) Spherical feldspar aggregate in a felsic tuff metamorphosed at low-pressure granulite facies conditions in the Broken Hill area, New South Wales, Australia. The spherical structure may represent a former spherulite, as suggested by poorly preserved radial internal structure, though this interpretation is in some doubt. Plane-polarized light; base of photo 10 mm. (B) Same rock, showing the foliation anastomosing around a spherical aggregate, which evidently was stronger than the fine-grained neocrystallized/recrystallized matrix during the deformation. Plane-polarized light; base of photo 10 mm. (C) Same field of view as in B, showing that the spherical aggregate consists of an aggregate of blocky grains of albite, which may have permitted preservation of the spherical structure during deformation and high-grade metamorphism. Crossed polars; base of photo 10 mm. All photos from Stevens & Barron (2002, photograph 15), with permission of the Geological Survey of New South Wales.

grains. This replacement by coarser-grained aggregates appears to have preserved the spherulitic shapes by making them stronger than the matrix, which is deflected around them (Fig. 7.35). The structures may represent neocrystallized spherulites, or may be former lithophysae, though these are generally larger than the Broken Hill examples (Stevens & Barron, 2002, p. 11).

(C)

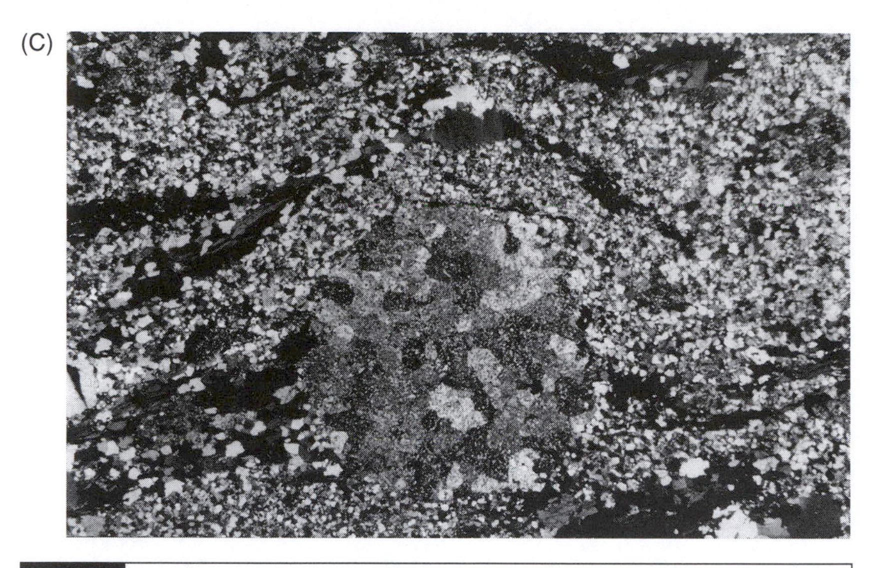

Fig. 7.35 (cont.).

Structural indicators of granitic precursors

Structural features indicating a granitic origin include: (1) euhedral K-feldspar megacrysts – especially with simple twinning, oscillatory compositional (especially Ba) zoning, zonally arranged inclusions (not inclusion trails), and euhedral plagioclase inclusions (Vernon, 1986a), (2) K-feldspar megacrysts aligned in a magmatic foliation or lineation (Vernon,

Fig. 7.36 Deformed, metamorphosed (granulite facies) megacrystic granite ('Alma Gneiss'), Broken Hill, western New South Wales, Australia, showing alignment of elongate, lenticular to rectangular porphyroclasts (augen) of K-feldspar, preserving evidence of a former magmatic foliation. Deformed enclaves are also aligned parallel to this foliation, suggesting that at least some of the elongation may have taken place during magmatic flow. However, the lenticular shapes of most of the former megacrysts and the anastomosing, lenticular foliation of the rock indicate that considerable solid-state deformation occurred during metamorphism.

1986a), as shown in Fig. 7.36, (3) aplite veins (restricted to plutonic rocks and therefore very reliable indicators), commonly with superimposed tectonic foliations (Fig. 7.21), (4) micrograniṭoid (igneous, 'mafic') enclaves, implying magma mingling (Vernon, 1983, 1990a, 1991a), as shown in Fig. 7.20, (4) metasedimentary xenoliths, especially large ones (Wintsch *et al.*, 1990), and (5) intrusive contacts (Brown *et al.*, 1983, p. 137; Vassallo & Vernon, 2000). Microgranitoid enclaves typically contain K-feldspar megacrysts in megacrystic host rocks, owing to magma mixing (Vernon, 1990a), and so may preserve them in deformed equivalents (Fig. 7.20). K-feldspar megacrysts are much more common in granites than in volcanic rocks, and survive relatively well, even in strongly deformed rocks, such as augen gneisses and mylonites (Vernon, 1986a; 1990b; 2004), as shown in Figs. 7.20, 7.31, 7.32 and 7.36.

The Broken Hill granite/volcanic controversy

A good example of the controversy that can attend the interpretation of the origin of felsic gneisses in high-grade regional metamorphic terranes is provided by the high-grade (amphibolite to lower granulite facies) Proterozoic Broken Hill Block, western New South Wales, Australia. Felsic gneisses, occurring as sheet-like bodies, are common (Brown *et al.*, 1983, fig. 1). The felsic gneisses have been interpreted by different people as: (1) metasedimentary (arkoses, volcaniclastic or granitized sediments), (2) volcanic flows and/or ignimbrites, and (3) metagranitoids. The correct interpretation of their origin is important, because stratigraphic interpretations and ore search models depend on whether the precursors of these rocks are regarded as intrusive or extrusive/clastic rocks.

The first geologists to work on the felsic gneisses around Broken Hill. (Andrews, 1922; Browne, 1922; Stillwell, 1922) interpreted most of them as deformed granites, and this interpretation continued to be in vogue for many years (Gustafson *et al.*, 1950; Binns, 1964; Vernon, 1969). In contrast, most recent workers have favoured a volcanic/volcaniclastic origin (Shaw, 1973; Plimer, 1975; Johnson & Klinger, 1976; Stanton, 1976; Brown *et al.*, 1983; Willis *et al.*, 1983; Hobbs *et al.*, 1984; Chenhall & Phillips, 1985; Stevens *et al.*, 1988; Page & Laing, 1992; Laing, 1996a,b) or an epiclastic origin (King & Thompson, 1953; Haydon & McConachy, 1987). These interpretations, if correct, would enable the felsic gneisses to be used as stratigraphic units, as has been done (Stevens & Willis, 1983; Stevens *et al.*, 1988). The inference of King & Thompson (1953) that the rocks represent granitized sediments was discredited by the foundering of the granitization hypothesis itself.

The main reason for the change in interpretation appears to have been the sheet-like nature of many of the felsic gneisses, which, taken at face value, tended to fit best with the stratigraphic approach to ore search in the Broken Hill area that became popular around 40 years ago. However, sheet-like granites are common, as discussed previously, and so this form of occurrence should not be used as an argument against an intrusive granitic origin.

An important piece of evidence for a volcanic or volcaniclastic origin is the local occurrence of what appear to be embayed quartz phenocrysts (e.g. Stanton, 1976; Laing, 1996b, fig. 7.3a, p. 59; Pepper & Ashley, 1998) and possible pumice lenses or 'fiamme' (Brown *et al.*, 1983, photo 3a,

p. 164). Embayed quartz phenocrysts constitute good evidence of a volcanic or subvolcanic precursor (Vernon, 1986b, 1999, 2004; Williams & Burr, 1994), and so the examples at Broken Hill, especially the ones described by Laing (1996b) and Stevens & Barron (2002), can reasonably be taken to indicate the presence of volcanic, volcaniclastic or high-level intrusive felsic precursor rocks. Less common feldspar phenocrysts (Stevens & Barron, 2002, photo 10b) and elongate decussate biotite aggregates that appear to have replaced deformed biotite phenocrysts (Stevens & Barron, 2002, photos 8, 17) are also consistent with volcanic precursors.

Spherical albite aggregates (Fig. 7.35) described by Stevens & Barron (2002) may be recrystallized spherulites or lithophysae, as discussed above. The fiamme illustrated by Brown *et al.* (1983, fig. 7.15) are less convincing, although excellent fiamme occur in amphibolite facies rocks of broadly similar age in the Olary Block, South Australia; these are surprisingly preserved in undeformed felsic rocks amid strongly deformed schists (Fig. 7.33). Therefore, at least some of the Broken Hill felsic gneisses appear to be of volcanic, volcaniclastic or high-level intrusive origin.

However, recent investigations have shown that many of the Broken Hill felsic gneisses are former granites, especially those containing K-feldspar megacrysts (Vernon, 1996b; Vernon & Williams, 1988; Vassallo & Vernon, 2000). The evidence includes the presence of euhedral to subhedral K-feldspar megacrysts with oscillatory zoning, aplite dykes, metasedimentary xenoliths, large microgranitoid enclaves (some with K-feldspar megacrysts, as is typical of such enclaves in megacrystic granites, as shown in Fig. 7.20) and, above all, intrusive contacts. Some of these contacts involve intrusive breccia formed by the intricate injection of megacrystic granite into an earlier massive granite (Vassallo & Vernon, 2000, fig. 10). In places, the felsic gneisses show aligned K-feldspar megacrysts, which typically reflect magmatic flow (Paterson *et al.*, 1989; Vernon, 2000a), as shown in Fig. 7.36.

The parentage of felsic gneisses without residual megacrysts, embayed phenocrysts or fiamme may be impossible to reliably identify without field evidence of contact relationships. The same applies to mylonitic (Fig. 7.37) and other multiply deformed (Fig. 7.38) rocks in the Broken Hill area. For example, sedimentary structures, such as cross-bedding and graded bedding, may be preserved in some units (Glen & Laing, 1975; Laing, 1980, 1996a,b; Willis *et al.*, 1983), whereas extremely strong deformation produces transposition of earlier-formed layering into a new, lenticular foliation in other units (Williams, 1967; Vernon, 1969; Vassallo & Vernon, 2000). In the same area, igneous microstructures are preserved in the interiors of large units of gabbro, whereas complete recrystallization to polygonal aggregates (pyroxene-hornblende granofelses and gneisses) occurs at the edges of the gabbro bodies and in thinner mafic units throughout the area (Binns, 1964). Consequently, the parentage (flows, sills, dykes, tuffs?) of these thinner units is not as clear as that of the undoubted metagabbros. Some of the felsic gneisses have been so strongly deformed that mylonites free of obvious megacryst remnants have resulted (Fig. 7.37), making it difficult or impossible to infer the rocks' parentage. Still others have been multiply deformed (Fig. 7.38), making outcrop-scale identification difficult or impossible without field evidence of contact relationships.

Fig. 7.37 Mylonite formed by the intense deformation in a shear zone of a felsic rock of indeterminate parentage, Broken Hill-Adelaide Road, Thackaringa area, Broken Hill district, Australia. The original rock may have been a granite, but could equally well have been a felsic volcanic rock. The larger feldspar patches may represent former pegmatite or leucosome, suggesting that the rock may have been partly melted (i.e., a migmatite) before the deformation. The smaller feldspar lenses may represent former K-feldspar megacrysts in a granite, but could also represent severely dismembered leucosomes in a felsic migmatite.

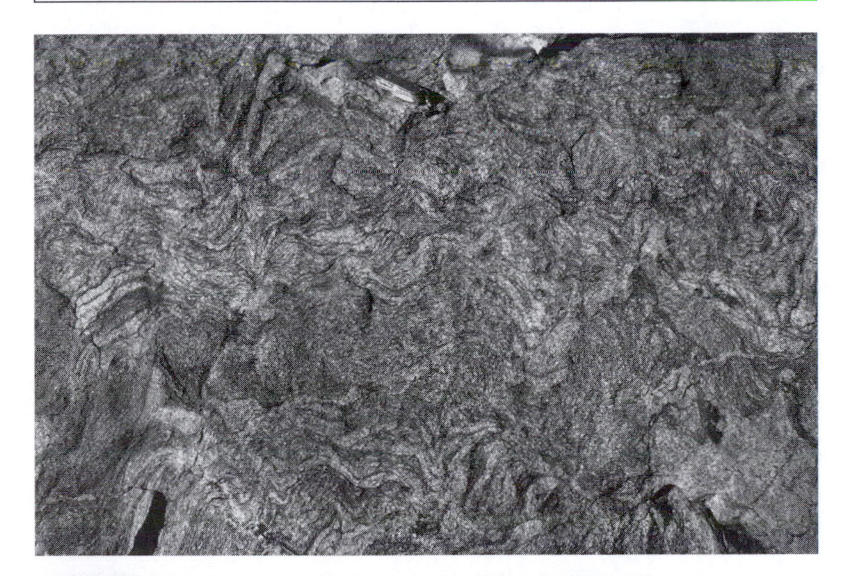

Fig. 7.38 Felsic gneiss that has undergone tight isoclinal folding, followed by crenulation folding, Southern Cross Mine area, Broken Hill district, western New South Wales, Australia. The parent rock was probably an intrusive leucogranite, on the basis of field relationships, but this interpretation could not be made with confidence from this outcrop alone.

Fig. 7.39 Folded layered sulphide rock, consisting mainly of galena (shiny layers) and sphalerite (dark layers), Mount Isa, north-west Queensland.

7.4 Sulphide-rich rocks

Sulphide-rich rocks in many regional metamorphic terranes have undergone the same metamorphic-deformation history as the silicate, carbonate and oxide rocks (Fig. 7.39), and so should be considered as metamorphic rocks, rather than being referred to only as 'ore deposits' and treated differently from the adjacent rocks.

Some sulphide-rich rocks appear to have undergone partial melting and mobilization in high-grade metamorphic terranes, for example, Broken Hill, Australia and Hemlo, Canada (Lawrence, 1967; Mavrogenes *et al.*, 2001; Tomkins & Mavrogenes, 2001, 2002; Frost *et al.*, 2002, 2005; Tomkins *et al.*, 2004, 2007; Sparks & Mavrogenes, 2005), as discussed in Section 4.19.

7.5 Metasomatized parent rocks

As recognized by Harker (1893), a metamorphic mineral assemblage depends not only on the physical conditions of metamorphism, but also on the bulk chemical composition of the rock, regardless of whether or not the rock undergoes pre-metamorphic alteration (Chapter 1). For example, metasomatic hydrothermal alteration, such as occurs in geothermal areas, wall-rock alteration around hydrothermal ore deposits, alteration in porphyry copper deposits, and deuteric alteration of igneous rocks, all of which change a rock's chemical composition, may occur prior to the metamorphism responsible for a rock's present mineral assemblage and structure. The metamorphism simply operates on the chemical system available to it, and produces an appropriate mineral assemblage, with or without residual structures that may help to identify the parent rock-type. This fundamental principle applies, regardless of the geological environment in which the rock acquired its present chemical composition (Vernon, 1976, p. 18).

Thus, if pre- or syn-metamorphic chemical alteration, known as *metasomatism* or *allochemical metamorphism* (see Chapter 5) considerably alters the bulk chemical composition of a rock, unexpected minerals may be produced when later metamorphosed. Clear examples of this are basaltic and andesitic rocks subjected to hydrothermal metamorphism (for example, geothermal, ocean-floor or burial metamorphism), during which glass is converted to chlorite plus quartz, which are later regionally or contact metamorphosed to produce cordierite-anthophyllite rocks (Vallance, 1967, 1969; Irving & Ashley, 1976; Reinhardt & Rubenach, 1989). These rocks may show structural evidence of their igneous origins, despite anomalously high magnesium contents.

Another example is provided by base-cation leaching (hydrogen metasomatism) caused by the action of acidic hydrothermal solutions on felsic volcanic or shallow intrusive rocks at subvolcanic (e.g., 'porphyry copper') conditions, which produces aluminous minerals such as kaolinite or pyrophyllite, less commonly andalusite (Gustafson & Hunt, 1975; Lowder & Dow, 1978; Valiant *et al.*, 1983; Watanabe & Hasegawa, 1986). The leaching process involves hydrogen metasomatism (Hemley & Jones, 1964; Vernon, 1987), as discussed in Chapter 5. Later metamorphism of such altered rocks produces aluminous minerals, such as andalusite, sillimanite or cordierite, which would not be expected to grow in metamorphosed felsic volcanic or plutonic rocks (Espenshade & Potter, 1960; Sykes & Moody, 1978; Carpenter & Allard, 1980; Allard & Carpenter, 1981; Vernon *et al.*, 1987), as mentioned in Section 7.2.1.

Unexpected mineral assemblages, especially anomalously aluminous assemblages in metamorphosed felsic igneous or volcaniclastic rocks, are potential indicators of former hydrothermal alteration zones around orebodies, and so may be guides to undetected orebodies in metamorphic terranes (see Section 5.17).

Appendix: Mineral Symbols Used in This Book

After Kretz (1983), extended by Bucher and Frey (1994)

Act	actinolite
Ab	albite
Alm	almandine
Als	aluminosilicate (Al_2SiO_5)
Am	amphibole
Ann	annite
Adr	andradite
Anl	analcite (analcime)
And	andalusite
An	anorthite
Ath	anthophyllite
Bt	biotite
Brc	brucite
Cal	calcite
Chl	chlorite
Cld	chloritoid
Cpx	clinopyroxene
Crd	cordierite
Crn	corundum
Czo	clinozoisite
Cum	cummingtonite
Di	diopside
Dol	dolomite
En	enstatite
Ep	epidote
Fa	fayalite
Fo	forsterite
Fs	ferrosilite
Grt	garnet
Gln	glaucophane
Gr	graphite
Grs	grossular
Hbl	hornblende
Hd	hedenbergite
Ilm	ilmenite
Jd	jadeite
Kfs	K-feldspar
Ky	kyanite
Lmt	laumontite
Lws	lawsonite
Mrg	margarite
Mgs	magnesite
Mag	magnetite
Ms	muscovite
Ol	olivine
Ne	nepheline
Omp	omphacite
Opx	orthopyroxene
Or	orthoclase
Pg	paragonite
Per	periclase
Phl	phlogopite
Pg	paragonite
Pgt	pigeonite
Pl	plagioclase
Prh	prehnite
Pmp	pumpellyite
Prl	pyrophyllite
Prg	pargasite
Prp	pyrope
Qtz	quartz
Rt	rutile
Spr	sapphirine
Scp	scapolite
Ser	'sericite'
Srp	serpentine
Sil	sillimanite
Sps	spessartine
Spl	spinel
St	staurolite
Stp	stilpnomelane
Tlc	talc
Ts	tschermakite
Ttn	titanite (sphene)
Toz	topaz
Tur	tourmaline
Tr	tremolite
Ves	vesuvianite (idocrase)
Wo	wollastonite
Zrn	zircon
Zo	zoisite
Apy	arsenopyrite
Ccp	chalcopyrite
Chr	chromite
Cv	covellite
Gn	galena
Py	pyrite
Sp	sphalerite

Glossary of Microstructural and Other Terms

This glossary uses definitions that conform to those in glossaries in standard texts, modified to take into account our own experiences and preferences as little as possible. Glossaries of metamorphic terms are in Joplin (1968), Shelley (1993), Best (2003) and Vernon (2004), and glossaries of deformation terms are in Shelley (1993), Best (2003) and Vernon (2004).

Acicular: Crystal with needle-like habit.

Amphibolite: Compositional term for a metamorphic rock (typically regional) composed largely of amphibole (especially hornblende) and plagioclase.

Amygdaloidal: Microstructure of a rock containing amygdules (*qv*).

Amygdule, amygdale: Gas vesicle (*qv*) in former volcanic rock, now filled with secondary minerals, either deuteric (precipitated from hydrothermal solutions derived from the magma) or metamorphic.

Anatexis: Partial melting of a solid rock at high-temperature metamorphic conditions.

Annealing: Technological process involving removal of strain by heating after deformation; unsatisfactory term for rocks, because absence of deformation during the heating commonly cannot be determined, even in contact metamorphic environments.

Augen: Residual large crystals (*porphyroclasts; qv*) in strongly deformed (generally gneissic or mylonitic) rocks. The augen typically tend to be lenticular or eye-shaped, and the finer-grained mylonitic foliation is deflected around them. The most common augen are of K-feldspar in deformed granites. The singular is *auge* (German for "eye").

Augen gneiss: Gneissic rock containing augen (*qv*), commonly, not necessarily of K-feldspar.

Augen mylonite: Mylonite (*qv*) with augen (*qv*), commonly but not necessarily of K-feldspar; see *augen gneiss*.

Beard structure: Fibrous intergrowths ('beards') projecting out from clasts or porphyroblasts parallel to the foliation in deformed rocks.

Bedded migmatite: Migmatite (*qv*) in which the leucosome (*qv*) remains in the parent rock and bedding is preserved.

Blasto, blastic: Prefix and suffix, respectively, referring to growth in the solid state (i.e., in metamorphic rocks); for example, *blastophitic* (*qv*), *blastoporphyritic* (*qv*), *blastopsammitic (qv)* and *porphyroblastic (qv)*.

Blastomylonite: Mylonite (*qv*) in which grain growth has occurred, either during or after the deformation.

Blastophitic: Metamorphosed igneous rock with residual ophitic (*qv*) microstructure.

Blastoporphyritic: Metamorphosed igneous rock with residual porphyritic (*qv*) microstructure.

Blastopsammitic: Residual clastic (*qv*) microstructure of metamorphosed sandstone (psammite, arenite).

Boudinage: The formation of boudins (*qv*) by deformation.

Boudins: Lines of tablet or sausage-shaped mineral grains or rock fragments enclosed in more ductile, deformed material; formed by the stretching and breaking of a stronger layer in weaker material that is extending in a ductile manner.

Breccia: Rock composed of mineral and/or rock fragments, generally formed by explosive volcanic activity (*volcanic breccia*) or deformation in fault zones (*fault breccia*).

Burial metamorphism: Low-temperature regional metamorphism without obvious deformation, owing to deep burial; grades into diagenesis (*qv*); residual igneous and sedimentary microstructures are typically preserved; similar minerals and microstructures can be produced by metamorphism in geothermal areas.

Calcsilicate rock: Metamorphosed marl or impure (clay and/or quartz-bearing) limestone or dolomite.

Cataclastic: Microstructure of rocks (cataclasites) formed by brittle deformation, involving fracturing of grains and relative movement of fragments.

Chondrite: Stony meteorite with spherical globules ('chondrules') containing dendritic crystals, the shapes of which suggest that the globules were once droplets of melted rock; inferred to have accreted to form the inner planets of the solar system.

Clastic (detrital, fragmental) microstructure: Microstructure dominated by detrital fragments (clasts) in sedimentary rocks.

Contact aureole granite: Granite intrusion, around which metamorphism is localized, as in *contact metamorphism* (*qv*).

Contact metamorphism: Metamorphism localized around an igneous intrusion.

Corona, reaction rim: Rim or mantle resulting from incomplete reactions between two minerals or between a mineral and components in a fluid in a metamorphic rock; coronas are commonly, (though not necessarily) composed of symplectic intergrowths (*qv*). Reaction rims may also occur in igneous rocks as a result of incomplete magmatic reactions between crystals and melt.

Crenulation cleavage: Foliation formed by alignment of the strongly deformed limbs of microfolds of a pre-existing foliation, commonly slaty cleavage (*qv*).

Cryptocrystalline: Very fine-grained, crystalline material that is incompletely resolved by the microscope.

Crystallite: Minute, embryonic crystal in some glassy volcanic or impact rocks, typically with acicular (needle-like), dendritic, and intricately curved or hair-like shapes. Crystallites are too small to show birefringence.

Crystalloblastic: Term implying growth of minerals in solid metamorphic rocks.

Cumulate microstructure: Microstructure of cumulates, which typically are mafic-ultramafic rocks with more crystals than could be precipitated from the inferred parent magma; the accumulation of the crystals can be physical (e.g., by crystal settling or floating) or by crystallization processes (e.g., crystallization of early minerals and removal of the interstitial liquid).

Decussate: Microstructure characterized by criss-cross (random) arrangement of elongate mineral grains dominated by crystal faces (rational-impingement boundaries); common in sheet silicate minerals (e.g., biotite, muscovite) and wollastonite, especially in contact metamorphic aureoles.

Dendritic: Shape of single crystals with spiky or branched habit precipitated (typically from fluid) at conditions of strong supersaturation.

Deuteric minerals: Secondary minerals precipitated from aqueous fluids concentrated at the late magmatic stage, typically replacing primary (magmatic) minerals, the process being known as *deuteric alteration*.

Devitrification: Replacement of volcanic glass by crystalline aggregates, as in the formation of *spherulites* (*qv*).

Diagenetic: Processes involving alteration (reactions between clastic minerals and aqueous fluid) and lithification of sediment, grading into burial metamorphism (*qv*).

Diatexite: Magma containing melt, solid reaction products and restite (*qv*); a migmatite characterized by obliteration of former structural features.

Differentiated layering: Compositional metamorphic layering formed from an originally more compositional homogeneous rock by segregation of chemical components (and consequently minerals), typically during deformation; also see *metamorphic differentiation*.

Dihedral angle: Term used in two ways: (1) angle between two meeting grain boundaries, typically used for solid aggregates (e.g., metals, metamorphic rocks); also the angle between meeting crystal/liquid interfaces; (2) interfacial angle of one mineral against two grains of another mineral or the angle formed where liquid meets two mineral grains, reserving 'interfacial angle' for boundaries between grains of the same mineral.

Dislocations: Line defects in crystalline materials that assist crystal growth and deformation.

Ductile grain-boundary sliding, granular flow: Movement of adjacent grains without loss of cohesion during deformation. The process should be distinguished from frictional grain boundary sliding, which involves loss of cohesion at the boundaries.

Ejecta: Volcanic fragments ejected in an explosive eruption; also called *tephra* and *pyroclasts*.

Enclave: Body of external origin in an igneous rock. The main varieties are *xenoliths* (fragments of solid rocks) and *microgranitoid enclaves* (*qv*; former magma globules derived by mingling of magmas).

Epitaxial: Adjective of *epitaxy* (*qv*).

Epitaxy: Nucleation and growth of a mineral in another with a systematic relationship between the two crystal structures.

Equigranular, equant: Even-grained.

Fault rock: Rock formed by deformation in a fault zone; includes *cataclasite* (*qv*) and *mylonite* (*qv*).

Fissility: Tendency of rocks to split into thin sheets along bedding or cleavage planes, owing mainly to strong alignment of sheet silicate minerals.

Foliation: Penetrative planar structure of metamorphic-deformation origin.

Frictional grain boundary sliding, independent particulate flow: The sliding of grains past each other, without the development of thoroughgoing fractures.

Glass: Material of volcanic, frictional or impact origin formed by the rapid cooling (quenching) of melt (especially felsic melt, but also mafic melt, less commonly) on Earth's surface; glass has the rigidity of a solid, but the atomic arrangement more akin to that of a silicate melt, characterized by short-range order, as opposed to the long-range order of atomic arrangements in crystalline materials, which makes glass optically isotropic; because this less ordered arrangement of atoms represents a higher-energy situation than a compositionally equivalent crystalline material, glass is metastable and tends to crystallize ('devitrify') with time, especially in the presence of water, commonly forming spherulitic aggregates.

Gneiss: Coarse-grained, generally high-grade regional metamorphic rock with a gneissic foliation (*qv*). Orthogneiss is derived from an igneous parent rock such as granite, paragneiss is derived from sedimentary parent rock.

Gneissic foliation, gneissosity: Crudely layered, commonly discontinuous foliation typical of coarse-grained metamorphic rocks and deformed coarse-grained rocks, characterized by folia (rich in minerals such as biotite, sillimanite or hornblende) anastomosing around lenticular pods of granoblastic aggregates (rich in minerals such as quartz, feldspar, cordierite or garnet).

Grain growth: Solid-state changes in the sizes and shapes of grains and crystals (without changing the minerals), to minimize the total interfacial free energy of the system.

Granitoid: Loose term for any coarse-grained intrusive rock of felsic to intermediate composition.

Granoblastic: Aggregate of xenoblastic (*qv*), equant, predominantly polygonal (*qv*) grains in metamorphic rocks; common in aggregates of quartz (in quartzite), calcite (in marble) and olivine (in peridotite).

Granofels: Useful, non-genetic term to describe a granular metamorphic rock (contact or regional), commonly, though not necessarily, of high-grade. It can often be used to replace 'granulite', which has metamorphic grade connotations.

Helicitic: Microstructure characterized by curved trails of inclusions in porphyroblasts (commonly of garnet, but also albite, staurolite and chloritoid) formed by overgrowth of pre-existing microfolded structures; may be difficult to distinguish from rotational (snowball) microstructure (*qv*).

Hornfels: Typically granoblastic rock, generally non-foliated ('massive'), occurring in contact metamorphic aureoles; may be foliated where the contact metamorphism overprints earlier regional metamorphism. Vartieties are indicated by mineralogical or compositional adjectives, for example: biotite hornfels, calc-silicate hornfels, mafic hornfels.

Hornfelsic: Not very useful term to describe the structure of a hornfels (contact metamorphic rock), which tends to be relatively even and fine-grained.

Hybridization, magma mixing: Mixing of two different magmas; the liquid components of both magmas mix, even though the solid crystals may only partly react in an attempt to equilibrate with the mixed melt; contrast with *magma mingling* (*qv*).

Hydrogen metasomatism, base-cation leaching: Leaching of cations by acid solutions; may produce unexpected Al-rich minerals, such as cordierite, andalusite and sillimanite, in metamorphosed felsic igneous rocks.

Idioblastic: Crystals with well-formed (low-energy) crystal faces in metamorphic rocks.

Imbrication: See *tiling*.

Inclusion: Mineral enclosed in a larger grain of another mineral; the term is sometimes used for enclaves (*qv*), but the former meaning is very widely accepted, and so should be reserved for it.

Inclusion trails: Lines of included mineral grains (inclusions) in porphyroblasts, resulting from overgrowth of pre-existing foliations, though not necessarily of the mineral grains originally delineating those foliations.

Isograd: Boundary (shown as a line on a geological map) representing the first appearance of an index mineral (or assemblage) for a metamorphic zone.

Leucosome: Light-coloured (leucocratic), felsic material, commonly mobile, produced in migmatization (commonly involving partial melting or anatexis) when separated from solid mafic products of the melting reaction. Leucosomes consist of melt or melt products, such as crystals precipitated from the melt or from residual melt separated from precipitated crystals; igneous microstructures are common in undeformed or weakly deformed leucosomes.

Lineation: Linear structure (compositional rods and/or linear alignment of minerals) produced by deformation.

M domain: See *P domain*.

Magma mingling: Intimate interpenetration of two or more different magmas, without hybridization (*qv*), forming relatively fine-grained globules (microgranitoid enclaves, magmatic enclaves) of the magma with the higher liquidus temperature in the other; though the magmas may be potentially miscible, relatively rapid cooling (and consequent increase in viscosity) of one of the magmas inhibits mixing; the magmatic enclaves may show chilled rims, which contrasts with liquid immiscibility, for which both liquids are in equilibrium and so do not quench against each other.

Magma mixing, hybridization: See *hybridization*.

Marble: Metamorphic rock composed entirely or almost entirely of carbonate minerals, that is, metamorphosed limestone or dolomite.

Marl: Impure (clay and/or quartz-bearing) limestone or dolomite. The metamorphosed equivalents are called *calcsilicate rocks* (*qv*).

Matrix: Aggregate that is distinctly finer-grained than the porphyroblasts in a metamorphic rock. The usage is similar to that of *groundmass* in an igneous rock.

Megacryst: Very large phenocryst, usually referring to K-feldspar phenocrysts in a granitic rock. The adjective is *megacrystic*.

Melanosome: Dark selvedge (mainly of mafic minerals, especially biotite) adjacent to the leucosome (*qv*) in some migmatites (*qv*); typically consists of the solid products of the melting reaction that segregate from the leucosome.

Mesosome: Darker volumes between light-coloured leucosomes (*qv*) in a migmatite (*qv*); not necessarily representing the original rock, as it may be chemically modified during the migmatization process, for example, during anatexis (*qv*), and may contain small amounts of former melt and/or crystals produced in the melting reaction.

Meta-: Prefix denoting a metamorphic rock; for example, metapelite (= metamorphosed pelite), metapsammite, metabasalt, metagabbro, metadolerite.

Metabasite: Metamorphic rock of basic (mafic) chemical composition.

Metamorphic differentiation: The formation of compositional layering during metamorphism in an originally uniform rock; see *differentiated layering*.

Metamorphic facies: A grouping of all mineral assemblages in a set of rock compositions formed under the same broad *P-T* conditions. Most metamorphic facies are named after characteristic metabasic assemblages (greenschist, amphibolite, blueschist, eclogite). However, a metamorphic facies does not refer to a single rock-type, but embraces a number of rock-types, all formed under the same conditions.

Metamorphic grade: General term referring to the intensity of metamorphism. Rocks are referred to as being of 'low grade', 'medium grade' or 'high grade' informally, with no well defined *P-T* boundaries between them. Grade is commonly related mostly to temperature (e.g., in low-*P* metamorphism), though pressure also has an effect, and dominates over *T* in high-*P* metamorphism.

Metamorphic zone: Volume of metamorphic rock (shown as an area on a geological map) characterized by a particular mineral or mineral assemblage.

Metapelite: Metamorphosed pelite (shale, claystone, argillite).

Metapsammite: Metamorphosed psammite (sandstone, arenite).

Metasomatism allochemical metamorphism: Change of bulk chemical composition of a rock during metamorphism; also see *hydrogen metasomatism*.

Microboudinage: Boudinage (*qv*) on the microscope scale.

Microcrystalline: Microstructure of a rock with crystal and grain shapes detectable at the microscope scale.

Microgranitoid enclave, igneous enclave, mafic enclave: Enclave with igneous microstructure in granitoid rocks, typically finer-grained and more mafic than the host rock (though generally felsic or intermediate, not mafic). Some people use 'microgranular enclave', but this conveys little meaning, as it just implies multiple grains on the microscope scale, and many very different rocks can be described in this way. 'Microgranitoid' signifies a microstructure like than of typical granitoids (*qv*), but finer-grained.

Microkrystite: Partly or completely crystallized spherule (*qv*).

Microtektite: Tektite (*qv*) with a diameter equal to or less than 1 mm.

Migmatite: Metamorphic rock characterized by light (felsic) aggregates or layers interspersed with darker material. Most migmatites are inferred to have been formed by partial melting (*anatexis, qv*), though not all, and the distinction can be difficult. *Anatectic migmatites* form by partial melting during high-grade metamorphism, and consist of melt-rich material (called *neosome*) and darker, melt-free or melt-poor rock (called *mesosome*). The neosome is commonly separated into light, quartzofeldspathic, melt-rich material (*leucosome*) and dark, mafic crystal-rich material (*melanosome*) composed largely of peritectic minerals (*qv*). Most light-coloured layers (*stromata*) in migmatites are leucosomes. Many observed leucosomes have lost part of their original melt. Igneous microstructures occur in the leucosomes of some migmatites.

'Millipede' structure: Microstructure of porphyroblasts with inclusion trails that curve into the matrix foliation by making opposite concave-outwards patterns at each end of the porphyroblast.

Mimetic crystallization: Growth of new minerals in bedding planes or foliations, thereby preserving or intensifying existing microstructures.

Mineral assemblage (paragenesis): Assemblage of minerals inferred to have grown simultaneously.

Mylonite: Thinly foliated and strongly lineated rock, commonly with porphyroclasts (*qv*), formed by intense deformation in elongate, relatively restricted zones.

Myrmekite: Vermicular (worm-like, symplectic) intergrowth of quartz and sodic plagioclase (generally oligoclase), formed by replacement of K-feldspar, typically in deformed granitic rocks.

Neoblast: Term sometimes used for a grain or crystal of a new mineral that has grown in the solid state.

Neocrystallization: Solid-state replacement of a mineral by grains of new minerals; contrast with *recrystallization* (*qv*).

Neosome: Mobile material (potential leucosome and melanosome; *qv*) formed in a migmatite (qv), typically by anatexis (partial melting).

Ophitic: Microstructure in which euhedral crystals of plagioclase are completely enclosed by much larger crystals of pyroxene; typical of the mafic intrusive igneous rock *dolerite*.

Order of crystallization: Order of first crystallization of minerals, especially in igneous rocks; it is determined by melting experiments on the magma composition concerned, and generally cannot be determined by inspection with the microscope. Minerals belonging to the same mineral assemblage or pargenesis (*qv*) crystallize together, so that the concept of an order of crystallization is commonly inappropriate for metamorphic rocks, except where reliable microstructural relationships (such as partial pseudomorphism, corona-type reaction rims, and inclusion assemblages occurring in porphyroblasts but not in the matrix) are available.

Orthogneiss: See *gneiss*.

Ostwald ripening: Solution of minute crystalline particles and corresponding growth of larger particles, reflecting the greater solubility of very small particles compared to larger ones, in response to the tendency to reduce the total surface free energy. It applies to particles crystallized from liquid or solid solution, and occurs at the nanometre to micrometre scales (e.g., at the scale of transmission electron microscope observation). Crystals observable optically are probably too large to be controlled by this process, in view of their inherent stability and the prohibitively large diffusion distances (Jackson, 1967). Therefore, it appears to be most applicable to the nucleation and early growth stages, for which diffusion distances are very short, so that surface energy can drive diffusion. It occurs in synthetic materials, and evidence for the process has been observed in experiments on felsic and mafic magmatic systems. Consequently, it may be applicable to the very earliest stages of crystallization of magmas, though it is more doubtful for porphyroblasts in metamorphic rocks, owing to the low driving force of surface energy, which is many orders of magnitude smaller than the chemical potentials driving crystal growth controlled by normal nucleation and growth processes (Carlson, 1999, 2000). Normal solid-state grain growth (Section 3.2.1) generally is excluded from the definition (e.g., Martin & Doherty, 1976, p. 234; Solomatov, 2002, p. 1446). This is justified on the grounds that solution and re-precipitation of minute grains is not involved in normal grain growth, just the modification and elimination of relatively large grains by local grain-boundary migration. However, Shelley (1993) and Best (2003) have referred to normal grain growth as 'Ostwald ripening'.

P domain, M domain: Elongate, relatively high-strain domain rich in mica or sillimanite, separated by lower-strain Q domains (rich in quartz or quartz + feldspar) in a slate or crenulated schist.

Palaeosome: Part of a migmatite that is unaffected by the migmatization process, thereby reflecting the original rock composition; however, commonly the unmelted or melt-poor parts of migmatites are chemically altered in the process, so that the term *mesosome* (*qv*) is generally used instead.

Paragenesis (mineral assemblage): Assemblage of minerals inferred to have grown simultaneously.

Paragneiss: See *gneiss*.

Pelite, claystone, argillite, shale: Fine-grained sedimentary rock. The metamorphosed equivalent is *metapelite*.

Penetrative domainal cleavage, 'rough cleavage': Equivalent of slaty cleavage (*qv*) in a deformed sandy rock.

Peritectic minerals: Solid products of a melting reaction, such as cordierite and/or garnet in partly melted metapelites.

Petrogenetic grid (*P-T* grid): *P-T* diagram showing univariant reaction lines and invariant points, indicating stability of mineral assemblages for all whole-rock compositions in the chosen chemical system, in contrast to *P-T* pseudosections (*qv*), which show phase relationships for a particular whole-rock chemical composition.

Phase: A chemically and physically homogeneous part of a *system* (e.g., a body of magma) that is bound by an interface with other phases (e.g., the melt phase and other crystal phases in a magma).

Phyllite: Low-grade metamorphic rock similar to *slate* (qv), but with larger chlorite and white mica grains, resulting in a less perfect cleavage.

Phyllonite: Mylonite (*qv*) rich in mica.

Plastic deformation, (crystal-plastic deformation): Permanent deformation of crystalline materials in response to applied differential stress.

Poikiloblast: Porphyroblast (*qv*) with many inclusions.

Poikiloblastic: Structure of a porphyroblast with many inclusions, and of a rock with poikiloblasts.

Polygonal aggregate: 'Foam-like' aggregate of grains with smoothly curved interfaces, formed by adjustment of grain boundaries in the solid state to minimize interfacial free energy; common in minerals with three-dimensional atomic structures, such as quartz, calcite, olivine, feldspar, galena, pyrite and ice.

Polymetamorphic: Rocks that have undergone polymetamorphism (*qv*). Care must be taken (with detailed microstructural analysis and geochronology) to ensure that separate events, rather than pulses of a single event, have taken place.

Polymetamorphism: More than one metamorphic event.

Porphyroblast: Large crystal, relative to the grainsize of the matrix (*qv*), in a metamorphic rock.

Porphyroblastic: Structure of a rock with porphyroblasts.

Porphyroclast: Large clast, relative to the grainsize of the matrix, in a mylonite (*qv*), being a relic of a formerly large grain, such as a phenocryst or a porphyroblast (*qv*).

Porphyroclastic: Deformed rock with porphyroclasts (*qv*).

Preferred orientation: Statistical alignment of mineral grains; may apply to shape (*dimensional preferred orientation*) and/or crystal axes (*crystallographic preferred orientation*); synonym of *texture* in materials science.

Pressure-shadow, pressure, fringe, strain shadow, 'beard': Elongate aggregate of granoblastic or fibrous minerals (especially quartz and calcite) occupying dilated zones adjacent to a porphyroblast, extending parallel to the foliation.

Pressure-solution: See *stress-induced solution transfer*.

Prograde metamorphism: Metamorphism resulting from progressively increasing *T* and/or *P*.

Psammite, sandstone, arenite: Clastic sedimentary rock composed of sand sized fragments. The metamorphosed equivalent is *metapsammite*.

Pseudomorph: A mineral or an aggregate of minerals replacing an older mineral grain or crystal, preserving the original size and shape.

Pseudomorphous: Adjective of *pseudomorph* (*qv*).

Pseudosection: Phase diagram, commonly with intensive variables defining axes, illustrating the dependency of phase equilibria on the chosen intensive variables in a model chemical system. *P-T* pseudosections show phase relationships for a particular whole-rock chemical composition, in contrast to *P-T* ('petrogenetic') grids (*qv*), which show stability of assemblages for all whole-rock compositions in the chosen system. *T-X* or *P-X* pseudosections show how phase relationships vary for a particular whole-rock composition at a specified *T* or *P*, respectively, illustrating how variation in the mole fraction of the chosen *X* component (commonly H_2O or CO_2) affects phase relationships (e.g., Elmer *et al.*, 2006).

Pseudotachylite, pseudotachylyte: Glassy rock occurring in veins with sharp contacts with the original rock, formed by rapid brittle sliding between rock volumes or by meteorite/comet impacts, commonly in the absence of water; frictional heat builds up at a fast enough rate to cause partial melting; characterized by extremely rapid fragmentation, frictional heating, partial melting and rapid cooling (quenching of the melt).

***P-T* grid:** See *petrogenetic grid*.

Q domain: Elongate, relatively low-strain domain rich in quartz or quartz + feldspar, separated by higher-strain P (M) domains in a slate or crenulated schist.

Quartzite: Metamorphosed pure or nearly pure quartz-rich rock, such as quartz sandstone (orthoquartzite) or chert; the term 'metaquartzite' is technically preferable, but not always used.

Reaction rim: Mineral layer formed by incomplete reaction between minerals, exemplified by *coronas* (*qv*) in metamorphic rocks.

Recovery: Term embracing all processes involved in returning a plastically deformed crystal to the undeformed state, without the formation of high-angle (high-energy) grain boundaries; that is, no new grains are formed. The result is typically the formation of *subgrains* (*qv*), the optical appearance of which is known as *undulose extinction*, especially in quartz.

Recrystallization: Conversion of a mineral grain into an aggregate of smaller new grains of the same mineral, as a result of deformation. Recrystallization may be either *dynamic* (during deformation) or *static* (after deformation), but it can be difficult to distinguish between the two in the light microscope. Most natural recrystallization is probably dynamic, as deformation typically accompanies metamorphism. Some change in composition between old and new grain is common (e.g., in plagioclase), but the growth of different new minerals is referred to as *neocrystallization*. The fact that many recrystallized aggregates consist of polygonal grains has led some people to use 'polygonization' incorrectly as a term for recrystallization, but the term refers to *recovery* (*qv*) which is a process that does not involve the formation of new grains.

Regional aureole granite: Granite intrusion, around which occur metamorphic aureoles of regional extent.

Regional metamorphism: Metamorphism on a regional scale, in contrast to localized metamorphism around intrusions (*contact metamorphism*; *qv*).

Relic (residual structure): Structure remaining after a deformation or metamorphic event, such as a porphyroclast in a mylonite, a phenocryst in a metamorphosed volcanic rock or a partly replaced porphyroblast in a retrograde metamorphic rock. 'Relict' is sometimes used as a synonym of 'residual'.

Resistate, resister: Unmelted rock in a migmatite terrane.

Restite: Solid material that either results from a partial melting reaction (peritectic minerals) or is unaffected in the same rock, and is either (1) left behind in a migmatite from which mobile melt has been removed, or (2) carried with the melt in a diatexite (*qv*).

Retrograde metamorphism: Metamorphism resulting from progressively decreasing *T* and/or *P*.

Ribbon structure (plattern structure): Elongate aggregates of coarse-grained minerals typically quartz, in high-grade, granoblastic metamorphic rocks; formed either by intense ductile deformation or crystallization in veins.

Rodingite: Light-coloured, tough, commonly flinty hydrous calc-silicate rock representing former gabbro associated with serpentinite; rodingite is formed by Ca metasomatism, involving replacement of former plagioclase by grossular ± vesuvianite and other calcsilicate minerals, commonly preserving gabbroic microstructures.

Schist: Regional metamorphic rock, coarser-grained than *phyllite* (*qv*), with *schistose foliation* (*qv*) and commonly (though not necessarily) with porphyroblasts (*qv*).

Schistose: Structure of a metamorphic rock with schistosity (*qv*).

Schistosity; schistose foliation: Foliation consisting of elongate minerals (e.g., mica, sillimanite, amphibole), commonly anastomosing around porphyroblasts; characteristic of schists.

Schlieren: Layers or streak-like concentrations, typically of relatively coarse-grained mafic minerals in granites and diatexites (*qv*), mostly formed by physical sorting during magmatic flow.

Schreinemakers analysis: Geometric technique used in constructing phase diagrams to establish the number and relative location of univariant curves, invariant points and multi-variant fields for equilibrium between a chosen set of phases in a model system (Niggli, 1954; Korzhinskii, 1959; Zen, 1966). The technique was originally used to predict relationships between univariant reactions on a *P-T* projection, but the principles are commonly needed in constructing any phase diagram involving axes with intensive variables. For a model system involving C components and C+2 phases, all phases can coexist at an invariant point. This lies at the intersection a series of univariant reactions involving C+1 phases, which is labelled by convention with the *absent* phase, shown in square brackets. A system with more than C+2 phases has more than one invariant point, with a cluster of interconnected univariant reaction lines about each.

Sheaf structure: Term (relatively uncommon these days) used for sheaf-shaped aggregates of fibrous minerals, such as anthophyllite, hornblende or wollastonite, in metamorphic rocks; amphibole sheaves are characteristic of *garbenschiefer*, 'garben' being the German for sheaves.

Skarn (tactite): Coarse-grained, metasomatically zoned silicate rock developed between carbonate-rich rocks (metamorphosed limestones, dolomites and Mn-rich carbonate rocks) and Al- and Si-rich rocks (granitic or other intrusive rocks, shales or schists), especially in contact metamorphic environments, but also in some regional metamorphic situations.

Skeletal crystal: Single crystal with a skeletal habit, formed at conditions of strong supersaturation; common in volcanic rocks and meteorites.

Slate: Fine-grained, low-grade regional metamorphic rock with a slaty cleavage (*qv*).

Slaty cleavage: Foliation characterized by narrow dark folia of fine-grained, well-aligned phyllosilicate minerals (especially white mica and chlorite) and/or graphite (P or M domains), anastomosing around stronger domains containing detrital fragments, mainly of quartz (Q domains); typical of slates (metamorphosed shales); the analogous cleavage in deformed sandy rocks is called *penetrative domainal cleavage* (*qv*) or 'rough cleavage'.

Snowball structure: Spiral inclusion trails in garnet porphyroblasts.

Spherules: Former silicate melt droplets up to a few millimetres across (now glassy or devitrified) formed by melting of Earth rocks at meteorite/comet impact sites and scattered widely. 'Spherule' is a general term covering *tektite* (*qv*), *microtektite* (*qv*) and *microkrystite* (*qv*).

Spherulite: Spherical aggregate of radiating fibres (mainly K-feldspar, with or without silica minerals, especially quartz) that mainly grows in volcanic glass during devitrification (*qv*), though some may grow in strongly undercooled melt.

Spherulitic: Structure of a rock with spherulites; also refers to an aggregate of radiating fibres, not necessarily spherical.

Strain shadow: Region adjacent to a clast or porphyroblast that is protected from deformation, so that it may preserve earlier microstructures that have been obliterated from the rest of the matrix.

Stress-induced solution transfer, dissolution-precipitation creep, 'pressure solution': Change of grain shape by diffusion of chemical components in aqueous solution as material is dissolved from sites of high normal compressive stress and deposited at sites of low stress. Mass transfer in fluid is a major mechanism for the development of foliations, especially slaty cleavage and discrete crenulation cleavage in low-grade metamorphic rocks.

Stretched pebble: Pebble in a conglomerate that has been elongated by solid-state deformation.

Stromatic migmatite: Migmatite in which the leucosome (*qv*) has segregated into layers (stromata).

Stylolite: Curved, irregular or toothed surface, typically containing fine-grained insoluble residues, formed by stress-induced solution transfer (*qv*).

Subgrains: Regions of small misorientation within an individual grain, formed by *recovery* (*qv*) in response to deformation; optically appearing as extinction differences, such as undulose extinction in quartz. Elongate subgrains may be referred to as 'extinction bands.'

Subophitic: Microstructure in which euhedral crystals of plagioclase are partly enclosed in larger grains of pyroxene (compare with *ophitic*).

Symplectite, symplectic intergrowth: Worm-like (vermicular) intergrowth of minerals. *Myrmekite* (*qv*) is the most common example.

Tektites: Spherical, ellipsoidal, disk-like, dumbbell-shaped, and irregularly shaped blobs of silicate glass formed by bolide (meteorite or comet) impact, typically showing sculptured external forms that reflect spinning, and commonly with some modification by ablation during transport in Earth's atmosphere. Many show contorted flow layering, and some are vesicular. Tektites occur as uncrystallized glass objects in four widespread 'strewnfields' on Earth's surface.

Texture: Used synonymously with 'microstructure' by many English-writing petrologists, but used for 'preferred orientation' by materials scientists and in some non-English languages. The IUGS Subcommission on the Systematics of Metamorphic Rocks has recommended that the term be replaced by 'microstructure' (Brodie *et al.*, 2002).

Tiling, imbrication: Rotational alignment in magma of elongate euhedral crystals that are not internally deformed or of fragments of phenocrysts that have undergone boudinage.

Transposition: Strong modification and realignment of a foliation or bedding into a new orientation.

Ultramylonite: Mylonitic rocks fewer than 10% of porphyroclasts (assuming the original rock was sufficiently coarse-grained to produce porphyroclasts).

Variolitic: Microstructure of glassy basaltic rocks characterized by radiating (spherulitic, fan-like or sheaf-like) aggregates (varioles) of fibrous crystals, for example, of plagioclase and clinopyroxene, or of plagioclase with interfibrillar clinopyroxene.

Vesicle: Spherical or ellipsoidal cavity caused by formation and expansion of gas, typically in a volcanic rock or shallow intrusion.

Vesicular: Structure of a rock with vesicles.

Vugh: Cavity filled or partly filled with minerals that precipitate from hydrothermal solution.

Xenoblastic: Irregular grains showing no crystal-face boundaries in a metamorphic rock.

Xenocryst: Solid fragment of a foreign mineral caught up in a magma; if they are out of equilibrium with the melt, xenocrysts commonly show partial alteration or resorption (solution), indicating reaction with the magma.

Xenolith: Solid fragment of a foreign rock caught up in a magma, and which may show evidence of reaction with it; contrast with *microgranitoid enclave* (*qv*).

References

Abbott, R. N. & Clarke, D. B. 1979: Hypothetical liquidus relationships in the subsystem $Al_2O_3 - FeO - MgO$ projected from quartz, alkali feldspar and plagioclase for $a(H_2O) \leq 1$. *Canadian Mineralogist*, **17**, 549–60.

Addison, W. D., Brumpton, G. R., Vallini, D. A., McNaughton, N. J., Davis, D. W., Kissin, S. A., Fralick, P. W. & Hammond, A. L. 2005: Discovery of distal ejecta from the 1850 Ma Sudbury impact event. *Geology*, **33**, 193–6.

Agard, P., Goffé, B., Touret, J. L. R. & Vidal, O. 2000: Retrograde mineral and fluid evolution in high-pressure metapelites (Schists lustrés unit, Western Alps). *Contributions to Mineralogy and Petrology*, **140**, 296–316.

Agostino, P. N. 1971: Theoretical and experimental investigations on ptygmatic structures. *Bulletin of the Geological Society of America*, **82**, 2651–60.

Ague, J. J. 1991: Evidence for major mass transfer and volume strain during regional metamorphism of pelites. *Geology*, **19**, 855–8.

Ague, J. J. 1994: Mass transfer during Barrovian metamorphism of pelites, south-central Connecticut; II, Channelized fluid flow and the growth of staurolite and kyanite. *American Journal of Science*, **294**, 1061–134.

Ague, J. J. 1997: Crustal mass transfer and index mineral growth in Barrow's garnet zone, northeast Scotland. *Geology*, **25**, 73–6.

Ague, J. J. 2002: Gradients in fluid composition across metacarbonate layers of the Wepawug Schist, Connecticut, USA. *Contributions to Mineralogy and Petrology*, **143**, 38–56.

Ague, J. J. 2003: Fluid infiltration and transport of major, minor and trace elements during regional metamorphism of carbonate rocks, Wepawug Schist, Connecticut, USA. *American Journal of Science*, **303**, 753–816.

Ague, J. J. & Rye, D. 1999: Simple models of CO_2 release from metacarbonates with implications for interpretation of directions and magnitudes of fluid flow in the deep crust. *Journal of Petrology*, **40**, 1443–62.

Aitchison, J. C., Clarke, G. L., Meffre, S. & Cluzel, D. 1995. Eocene arc-continent collision in New Caledonia and implications for regional southwest Pacific tectonic evolution. *Geology*, **23**, 161–4.

Alias, G., Sandiford, M., Hand, M. & Worley, B. 2002: The P-T record of synchronous magmatism, metamorphism and deformation at Petrel Cove, southern Adelaide Fold Belt. *Journal of Metamorphic Geology*, **20**, 351–63.

Allard, G. O. & Carpenter, R. H. 1981: Alteration in the Otake geothermal field - a model for interpreting certain metamorphic assemblages in metamorphosed volcanic terranes. 1981 IAVCEI Symposium: Arc Volcanism. Tokyo: Hakone, pp. 1–2.

Allibone, A. 1991: Volcanogeneic and granitoid rocks form northwest Stewart Island. *New Zealand Journal of Geology and Geophysics*, **34**, 35–50.

Allibone, A. H. & Norris, R. J. 1992: Segregation of leucogranite microplutons during syn-anatectic deformation: an example from the Taylor Valley, Antarctica. *Journal of Metamorphic Geology*, **10**, 589–600.

Alvarez, W., Engelder, J. T. & Lowrie, W. 1976: Formation of spaced cleavage and folds in brittle limestone by dissolution. *Geology*, **4**, 698–701.

Andrew, A. S. 1984: P-T-X(CO_2) conditions in mafic and calc-silicate hornfelses from Oberon, New South Wales, Australia. *Journal of Metamorphic Geology*, **2**, 143–63.

Andrews, E. C. 1922: The geology of the Broken Hill district. *Memoirs of the Geological Survey of New South Wales*, **8**, 295 pp.

Anhaeusser, C. R. 1992: Structures in granitoid gneisses and associated migmatites close to the granulite boundary of the Limpopo Belt, South Africa. *Precambrian Research*, **55**, 81–92.

Annen, C., Blundy, J. D. & Sparks, R. S. J. 2006: The genesis of intermediate and silicic magmas in deep crustal hot zones. *Journal of Petrology*, **47**, 505–39.

Arai, S. 1975: Contact metamorphosed dunite-harzburgite complex in the Chugoku district, western Japan. *Contributionns to Mineralogy and Petrology*, **52**, 1–16.

Arndt, N. T., Naldrett, A. J. & Pyke, D. R. 1977: Komatiitic and iron-rich tholeiitic lavas of Munro Township, northeast Ontario. *Journal of Petrology*, **18**, 319–69.

Arzi, A. A. 1978: Critical phenomena in the rheology of partially-molten rocks. *Tectonophysics*, **44**, 173–84.

Ashby, M. F. & Verall, R. A. 1973: Diffusion-accommodated flow and superplasticity. *Acta Metallurgica*, **21**, 149–63.

Ashworth, J. R. 1976: Petrogenesis of migmatites in the Huntly–Portsoy area, north-east Scotland. *Mineralogical Magazine*, **40**, 661–82.

Ashworth, J. R. 1979: Comparative petrography or deformed and undeformed migmatites from the Grampian Highlands of Scotland. *Geological Magazine*, **116**, 445–56.

Ashworth, J. R. 1985: Introduction. *In*: Ashworth, J. R. (ed.): *Migmatites*. Glasgow: Blackie, pp. 1–35.

Ashworth, J. R. & McLellan, E. L. 1985: Textures. *In*: Ashworth, J. R. (ed.): *Migmatites*. Glasgow: Blackie, pp. 80–203.

Aslund, T., Oliver, N. H. S. & Cartwright, I. 1995: Metasomatism of the Revenue Granite and aureole rocks, Mt Isa Inlier, Queensland: syndeformational fluid flow and fluid-flow interaction. *Australian Journal of Earth Sciences*, **42**, 291–9.

Atherton, M. P. 1977: The metamorphism of the Dalradian rocks of Scotland. *Scottish Journal of Geology*, **13**, 331–70.

Ayres, M., Harris, N. & Vance, D. 1997: Possible constraints on anatectic melt residence times from accessory mineral dissolution rates; an example from Himalayan leucogranites. *Mineralogical Magazine*, **61**, 29–36.

Ayrton, S. 1969: On the origin of gneissic banding. *Eclogae Geologica Helvetica*, **62**, 567–70.

Baba, S. 1998: Proterozoic anticlockwise P-T path of the Lewisian Complex of South Harris, Outer Hebrides, NW Scotland. *Journal of Metamorphic Geology*, **16**, 819–41.

Baba, S. 1999: Sapphirine-bearing orthopyroxene-kyanite/sillimanite granulites from South Harris, NW Scotland: evidence for Proterozoic UHT metamorphism in the Lewisian. *Contributions to Mineralogy and Petrology*, **136**, 33–47.

Bailey, E. B. 1930: New light on sedimentation and tectonics. *Geological Magazine*, **67**, 77–92.

Baker, D. W., Chawla, K. S. & Krizek, R. J. 1993: Compaction fabrics of pelites: experimental consolidation of kaolinite and implications for analysis of strain in slate. *Journal of Structural Geology*, **15**, 1123–37.

Baker, G. 1959: Rodimgite in nickeliferous serpentinite, near Beaconsfield, northern Tasmania. *Journal of the Geological Society of Australia*, **6**, 21–35.

Baker, G. 1963: Form and sculpture of tektites. *In*: O'Keefe, J. A. (ed.): *Tektites*. Chicago: University of Chicago Press, pp. 1–24.

Baker, T. & Lang, J. R. 2003: Reconciling fluid inclusion types, fluid processes, and fluid sources in skarns: an example from the Bismark Deposit, Mexico. *Mineralium Deposita*, **38**, 474–95.

Banno, S. 1998: Pumpellyite-actinolite facies of the Sanbagawa metamorphism. *Journal of Metamorphic Geology*, **16**, 117–28.

Barbey, P., Brouand, M., Le Fort, P. & Pecher, A. 1996: Granite-migmatite genetic link: example of the Manaslu granite and Tibetan Slab migmatites in central Nepal. *Lithos*, **38**, 63–79.

Barker, A. J. 1990: *Introduction to Metamorphic Textures and Microstructures*. Glasgow: Blackie.

Barker, A. J. 1994: Interpretation of porphyroblast inclusion trails: limitations imposed by growth kinetics and strain rates. *Journal of Metamorphic Geology*, **12**, 681–84.

Barker, A. J., Bennett, D. G., Boyce, A. J. & Fallick, A. E. 2000: Retrogression by deep infiltration of meteoric fluids into thrust zones during late-orogenic rapid unroofing. *Journal of Metamorphic Geology*, **18**, 307–18.

Barker, A. J. & Zhang, Y. 1998: The role of microcracking and grain-boundary dilation during retrograde reactions. *In*: Treloar, P. J. & O'Brien, P. (eds.): What drives metamorphism and metamorphic reactions? *Geological Society, London, Special Publications*, **138**, 247–68.

Barker, D. S. & Burmester, R. F. 1970: Leaching of quartz from Precambrian hypabyssal rhyolite porphyry, Llano County, Texas. *Contributions to Mineralogy and Petrology*, **28**, 1–8.

Barnes, C. G., Dumond, G., Yoshinobu, A. S. & Prestvik, T. 2004: Assimilation and crystal accumulation in a mid-crustal magma chamber; the Sausfjellet pluton, north-central Norway. *Lithos*, **75**, 389–412.

Barnes, J. D., Selverstone, J. & Sharp, Z. D. 2004: Interactions between serpentine devolatilisation, metasomatism and strike-slip localization during deep crustal shearing in the Eastern Alps. *Journal of Metamorphic Geology*, **22**, 283–300.

Barnes, R. G., Stevens, B. P. J., Stroud, W. J., Brown, R. E., Willis, I. L. & Bradley, G. M. 1983: Zinc, manganese, and iron-rich rocks and various minor rock types. *In*: Stevens, B. P. J. & Stroud, W. J. (eds.): *Rocks of the Broken Hill Block: Their Classification, Nature, Stratigraphic Distribution, and Origin. Records of the Geological Survey of New South Wales*, **21**, 289–323.

Barnett, D. E. & Chamberlain, C. P. 1991: Relative scales of thermal- and fluid infiltration-driven metamorphism in fold nappes, New England, U.S.A. *American Mineralogist*, **76**, 713–27.

Barnicoat, A. C., Fare, R. J., Groves, D. I. & McNaughton, N. J. 1991: Synmetamorphic lode-type gold deposits in high-grade Archean settings. *Geology*, **19**, 921–4.

Barr, D. 1985: Migmatites in the Moines. *In*: Ashworth, J. R. (ed.): *Migmatites*. Glasgow: Blackie, pp. 225–64.

Barrière, M. 1981: On curved laminae, graded layers, convection currents and dynamic crystal sorting in the Ploumanac'h (Brittany) subalkaline granite. *Contributions to Mineralogy and Petrology*, **77**, 217–24.

Barrow, G. 1893: On an intrusion of muscovite-biotite gneiss in the southeastern Highlands of Scotland, and its accompanying metamorphism. *Quarterly Journal of the Geological Society of London*, **49**, 330–58.

Barrow, G. 1912: On the geology of the lower Deeside and the southern Highland border. *Proceedings of the Geologists' Association*, **23**, 274–90.

Barth, T. W. F. 1948: Oxygen in rocks: a basis for petrographic calculations. *Journal of Geology*, **56**, 50–60.

Barton, M. D., Ilchik, R. P. & Marikos, M. A. 1991: Metasomatism. *In*: Kerrick, D. M. (ed.): *Contact Metamorphism. Mineralogical Society of America, Reviews in Mineralogy*, **26**, 321–50.

Barton, M. D. & Johnson, D. A. 1996: Evaporitic-source model for igneous-related Fe oxide-(REE-Cu-Au-U) mineralization. *Geology*, **24**, 259–62.

Bastin, E. S. 1909: Chemical composition as a criterion in identifying metamorphosed sediments. *Journal of Geology*, **17**, 445–72.

Battles, D. A. & Barton, M. D. 1995: Arc-related sodic hydrothermal alteration in the western United States. *Geology*, **23**, 913–6.

Baumgartner, L. P. & Ferry, J. M. 1991: A model for coupled fluid-flow and mixed-volatile mineral reactions with applications to regional metamorphism. *Contributions to Mineralogy and Petrology*, **106**, 273–86.

Baumgartner, L. P. & Rumble, D. 1988: Transport of stable isotopes: I. Development of kinetic continuum theory for stable isotope transport. *Contributions to Mineralogy and Petrology*, **98**, 417–30.

Baxter, E. F., Ague, J. J. & DePaolo, D. J. 2002: Prograde temperature-time evolution in the Barrovian type-locality constrained by Sm/Nd garnet ages from Glen Clova, Scotland. *Journal of the Geological Society*, **159**, 71–82.

Bayley, R. W. & James, H. L. 1973: Precambrian iron-formations of the United States. *Economic Geology*, **68**, 934–59.

Bea, F., Montero, P., Garuti, G. & Zacharini, F. 1997: Pressure-dependence of rare earth element distribution in amphibolite- and granulite-grade garnets. A LA-ICP-MS study. *Geostandards Newsletter*, **21**, 253–70.

Bea, F., Pereira, M. D. & Stroh, A. 1994: Mineral/leucosome trace element partitioning in a peraluminous migmatite (an ablation ICP-MS study). *Chemical Geology*, **117**, 291–312.

Beach, A. 1973: The mineralogy of high temperature shear zones at Scourie, N.W. Scotland. *Journal of Petrology*, **14**, 213–48.

Beach, A. 1974: A geochemical investigation of pressure solution and the formation of veins in a deformed greywacke. *Contributions to Mineralogy and Petrology*, **46**, 61–8.

Beach, A. 1976: The interrelation of fluid transport, deformation, geochemistry, and heat flow in early Proterozoic shear zones in the Lewisian complex. *Philosophical Transactions of the Royal Society of London*, **A280**, 596–604.

Beach, A. 1977: Vein arrays, hydraulic fractures and pressure-solution structures in a deformed flysch sequence, S.W. England. *Tectonophysics*, **40**, 201–26.

Beach, A. 1980: Retrograde metamorphic processes in shear zones with special reference to the Lewisian complex. *Journal of Structural Geology*, **2**, 257–63.

Beach, A. & Fyfe, W. S. 1973: Fluid transport and shear zones at Scourie, Sutherland: Evidence of overthrusting? *Contributions to Mineralogy and Petrology*, **36**, 175–80.

Beard, J. S., Ragland, P. C. & Crawford, M. L. 2005: Reactive bulk assimilation: A model for crust-mantle mixing in silicic magmas. *Geology*, **33**, 681–4.

Bebout, G. E. & Barton, M. D. 1989: Fluid flow and metasomatism in a subduction zone hydrothermal system, Catalina Schist Terrane, California. *Geology*, **17**, 976–80.

Bebout, G. E. & Barton, M. D. 1993: Metasomatism during subduction: products and possible paths in the Catalina Schist, California. *Chemical Geology*, **108**, 61–92.

Beck, R. 1899: Neus von den Afrikanischen Diamantlagerstatten. *Zeitschrift für Praktische Geologie*, 417–19.

Becker, H., Jochum, K. P. & Carlson, R. W. 1999: Constraints from high-pressure veins in eclogites on the composition of hydrous fluids in subduction zones. *Chemical Geology*, **160**, 291–308.

Bell, T. H. 1978: The development of slaty cleavage across the Nackara Arc of the Adelaide Geosyncline. *Tectonophysics*, **51**, 171–201.

Bell, T. H. 1981: Foliation development – The contribution, geometry and significance of progressive bulk inhomogeneous shortening. *Tectonophysics*, **75**, 273–96.

Bell, T. H. 1985: Deformation partitioning and porphyroblast rotation in metamorphic rocks: a radical reinterpretation. *Journal of Metamorphic Geology*, **3**, 109–18.

Bell, T. H. 1986: Foliation development and refraction in metamorphic rocks: reactivation of earlier foliations and decrenulation due to shifting patterns of deformation partitioning. *Journal of Metamorphic Geology*, **4**, 421–44.

Bell, T. H. & Etheridge, M. A. 1973: Microstructure of mylonites and their descriptive terminology. *Lithos*, **6**, 337–48.

Bell, T. H., Forde, A. & Wang, J. 1995: A new indicator of movement direction during orogenesis: measurement technique and application to the Alps. *Terra Nova*, **7**, 500–8.

Bell, T. H., Ham, A. P. & Kim, H. S. 2004: Partitioning of deformation along an orogen and its effects on porphyroblast growth during orogenesis. *Journal of Structural Geology*, **26**, 825–45.

Bell, T. H. & Hammond, R. L. 1986: On the internal geometry of mylonite zones. *Journal of Geology*, **92**, 667–86.

Bell, T. H. & Hayward, N. 1991: Episodic metamorphic reactions during orogenesis: the control of deformation partitioning on reaction sites and reaction duration. *Journal of Metamorphic Geology*, **9**, 619–40.

Bell, T. H. & Hickey, K. A. 1999: Complex microstructures preserved in rocks with a simple matrix: significance for deformation and metamorphic processes. *Journal of Metamorphic Geology*, **17**, 521–35.

Bell, T. H., Hickey, K. A. & Upton, G. J. G. 1998: Distinguishing and correlating multiple phases of metamorphism across a multiply deformed region using the axes of spiral, staircase and sigmoidal inclusion trails in garnet. *Journal of Metamorphic Geology*, **16**, 767–94.

Bell, T. H. & Johnson, S. E. 1989: Porphyroblast inclusion trails: the key to orogenesis. *Journal of Metamorphic Geology*, **7**, 279–310.

Bell, T. H. & Kim, H. S. 2004: Preservation of Acadian deformation and metamorphism through intense Alleghanian shearing. *Journal of Structural Geology*, **26**, 1591–613.

Bell, T. H. & Mares, V. M. 1999. Correlating deformation and metamorphism around orogenic arcs. *American Mineralogist*, **84**, 1727–40.

Bell, T. H. & Rubenach, M. J. 1980: Crenulation cleavage development - evidence for progressive, bulk inhomogeneous shortening from millipede microstructures in the Robertson River Metamorphics. *Tectonophysics*, **68**, T9–T15.

Bell, T. H. & Rubenach, M. J. 1983: Sequential porphyroblast growth and crenulation cleavage development during progressive deformation. *Tectonophysics*, **92**, 171–94.

Bell, T. H., Rubenach, M. J. & Fleming, P. D. 1986: Porphyroblast nucleation, growth and dissolution in regional metamorphic rocks as a function of deformation partitioning during foliation development. *Journal of Metamorphic Geology*, **4**, 37–67.

Bence, A. E. & Albee, A. L. 1968: Empirical correction factors for the electron microanalysis of silicates and oxides. *Journal of Geology*, **76**, 382–403.

Bennett, R. H., Bryant, W. R. & Keller, G. H. 1981: Clay fabric of selected submarine sediments; fundamental properties and models. *Journal of Sedimentary Petrology*, **51**, 217–32.

Bennett, R. H., O'Brien, N. R. & Hulbert, M. H. 1991: Determinants of clay and shale microfabric signatures: processes and mechanisms. *In*: Bennett, R. H., Bryant, W. R. & Hulbert, M. H. (eds.): *Microstructure of Fine-grained Sediments*. New York: Springer, pp. 5–32.

Bergantz, G. W. 1989: Underplating and partial melting: Implications for melt generation and extraction. *Science*, **33**, 267–309.

Bergantz, G. W. & Dawes, R. 1994: Aspects of magma generation and ascent in continental lithosphere. *In*: Ryan, M. P. (ed.): *Magmatic Systems*. San Diego: Academic Press, pp. 291–317.

Berger, A. & Rosenberg, C. L. 2002: Preservation of chemical-melt equilibria in natural anatexite: the effects of deformation and rapid cooling. *Contributions to Mineralogy and Petrology*, **144**, 416–27.

Berlenbach, J. W. & Roering, C. 1992: Sheath-like structures in pseudotachylytes. *Journal of Structural Geology*, **14**, 847–56.

Berman, R. G. 1988: Internally consistent data for minerals in the stem Na_2O-K_2O-CaO-MgO-FeO-Fe_2O_3-Al_2O_3-SiO_2-TiO_2-H_2O-CO_2. *Journal of Petrology*, **29**, 445–522.

Bernoulli, D. & Weissert, H. 1985: Sedimentary fabrics in Alpine ophicalcites, South Pennine Arosa zone, Switzerland. *Geology*, **13**, 755–8.

Best, M. G. 2003: *Igneous and Metamorphic Petrology*. Second edition. Malden, Massachusetts: Blackwell.

Beus, A. 1983: On the possible mechanism of formation of euhedral crystals in metasomatic processes. *Bulletin de Minéralogie*, **106**, 411–6.

Beutner, E. C. 1978: Slaty cleavage and related strain in Martinsburg Slate, Delaware Water Gap, New Jersey. *American Journal of Science*, **278**, 1–23.

Bhattacharya, A., Mazumdar, A. C. & Sen, S. K. 1988: Fe-Mg mixing in cordierite: constraints from natural data and implications for cordierite-garnet geothermometry in granulites. *American Mineralogist*, **73**, 338–44.

Bingen, B., Demaiffe, D. & Hertogen, J. 1996: Redistribution of rare earth elements, thorium, and uranium over accessory minerals in the course of amphibolite to granulite facies metamorphism: the role of apatite and monazite in orthogneisses from southwestern Norway. *Geochimica et Cosmochimica Acta*, **60**, 1341–54.

Binns, R. A. 1964: Zones of progressive regional metamorphism in the Willyama Complex, Broken Hill district, New South Wales. *Journal of the Geological Society of Australia*, **11**, 283–330.

Binns, R. A. 1965: Hornblendes from some basic hornfelses in the New England region, New South Wales. *Mineralogical Magazine*, **34**, 52–65.

Binns, R. A., Gunthorpe, R. J. & Groves, D. I. 1976: Metamorphic patterns and development of greenstone belts in the eastern Yilgarn Block, Western Australia. *In*: Windley, B. F. (ed.): *The Early History of the Earth*. Chichester: Wiley, pp. 303–13.

Bishop, D. G. 1972: Authigenic pumpellyite and other metamorphic effects in the Kyeburn Formation, central Otago. *New Zealand Journal of Geology and Geophysics*, **15**, 243–50.

Black, L. P., Harley, S. L., Sun, S. S. & McCulloch, M. T. 1987: The Rayner Complex of East Antarctica: complex isotopic systematics within a Proterozoic mobile belt. *Journal of Metamorphic Geology*, **5**, 1–26.

Black, P. M. 1977: Regional high-pressure metamorphism in New Caledonia: phase equilibria in the Ouégoa district. *Tectonophysics*, **43**, 89–107.

Blackburn, W. H. 1968: The spatial extent of chemical equilibrium in some high-grade metamorphic rocks from the Grenville of southeastern Ontario. *Contributions to Mineralogy and Petrology*, **31**, 275–99.

Blake, D. H., Elwell, R. W. D., Gibson, I. L., Skelhorn, R. R. & Walker, G. P. L. 1965: Some relationships resulting from the intimate association of acid and basic magmas. *Quarterly Journal of the Geological Society of London*, **121**, 31–50.

Blattner, P. 1976: Replacement of hornblende by garnet in granulite facies assemblages near Milford Sound, New Zealand. *Contributions to Mineralogy and Petrology*, **55**, 181–90.

Blenkinsop, T. G. 2000: *Deformation Microstructures and Mechanisms in Minerals and Rocks*. Dordrecht: Kluwer.

Bloch, S. & Hofmann, A. W. 1978: Magnesium metasomatism during hydrothermal alteration of new oceanic crust. *Geology*, **6**, 275–7.

Bloxham, T. W. 1954: Rodingite from the Girvan-Bellantrae complex, Ayrshire. *Mineralogical Magazine*, **30**, 525–8.

Boardman, S. J. 1986: Early Proterozoic bimodal volcanic rocks in central Colorado, U.S.A., Part I: Petrography, stratigraphy and depositional history. *Precambrian Research*, **34**, 1–36.

Bodorkos, S., Sandiford, M., Oliver, N. H. S. & Cawood, P. A. 2002: High-*T*, low-*P* metamorphism in the Palaeoproterozoic Halls Creek orogen, northern Australia: the middle crustal response to a mantle-related thermal pulse. *Journal Metamorphic Geology*, **20**, 217–37.

Boger, S. D. & Hansen, D. 2004: Metamorphic evolution of the Georgetown Inlier, northeast Queensland, Australia; evidence for an accreted Palaeoproterozoic terrane? *Journal of Metamorphic Geology*, **22**, 511–27.

Bohlen, S. R. & Boettcher, A. L. 1982: The quartz-coesite transformation: a precise determination and the effects of other components. *Journal of Geophysical Research*, **87**, 7073–8.

Bohlen, S. R., Dollase, W. A. & Wall, V. J. 1986: Calibration and application of spinel equilibria in the system FeO-Al_2O_3-SiO_2. *Journal of Petrology*, **27**, 1143–56.

Bohlen, S. R. & Liotta, J. J. 1986: A barometer for garnet amphibolites and garnet granulites. *Journal of Petrology*, **27**, 1025–56.

Bohlen, S. R., Wall, V. J. & Boettcher, A. L. 1983: Experimental investigations and geologic applications of equilibria in the system FeO-TiO_2-Al_2O_3-SiO_2-H_2O. *American Mineralogist*, **68**, 1049–58.

Bohor, B. F., Foord, E. E., Modreski, P. J. & Triplehorn, D. M. 1984: Mineralogic evidence for an impact event at the Cretaceous-Tertiary boundary. *Science*, **224**, 867–9.

Bonin, B. 2004: Do coeval mafic and felsic magmas in post-collisional to within-plate regimes necessarily imply two contrasting, mantle and crustal, sources? A review. *Lithos*, **78**, 1–24.

Bonney, T. G. 1899: The parent rock of the diamond in South Africa. *Proceedings of the Royal Society of London*, **65**, 223–36.

Bons, P. D. 2007: From leucosome to magma chamber – just a matter of scale? Sixth Hutton Symposium, Stellenbosch, South Africa, Abstract Volume, pp. 53–4.

Bons, P. D., Arnold, J., Elburg, M. A., Kalda, J., Soesoo, A. & van Milligen, B. P. 2004: Melt extraction and accumulation from partially molten rocks. *Lithos*, **78**, 25–42.

Bons, P. D. & den Brok, B. 2000: Crystallographic preferred orientation development by dissolution-precipitation creep. *Journal of Structural Geology*, **22**, 1713–22.

Bons, P. D., Dougherty-Page, J. & Elburg, M. A. 2001: Stepwise accumulation and ascent of magmas. *Journal of Metamorphic Geology*, **19**, 627–33.

Borradaile, G. J. 1981: Particulate flow of rocks and the formation of cleavage. *Tectonophysics*, **72**, 305–321.

Bosworth, T. O. 1910: Metamorphism around the Ross of Mull Granite. *Quarterly Journal of the Geological Society of London*, **66**, 376–96.

Bottrell, S. H., Greenwood, P. B., Yardley, B. W. D., Shepherd, J. J. & Spiro, B. 1990: Metamorphic and post-metamorphic fluid flow in the low-grade rocks of the Harlech Dome, north Wales. *Journal of Metamorphic Geology*, **8**, 131–43.

Boucot, A. J. & Rumble, D. 1978: Devonian brachiopods from the sillimanite zone, Mount Moosilauke, New Hampshire. *Science*, **201**, 348–9.

Boucot, A. J. & Rumble, D. 1980: Regionally metamorphosed (high sillimanite zone, granulite facies) Early Devonian brachiopods from the Littleton Formation, New Hampshire. *Journal of Palaeontology*, **54**, 188–95.

Boucot, A. J. & Rumble, D. 1986: Comment on "Remarkable fossil locality: Crinoid stems from migmatite of the Coast Plutonic Complex, British Columbia". *Bulletin of the Geological Society of America*, **74**, 1313–32.

Boucot, A. J. & Thompson, J. B. 1963: Metamorphosed Silurian brachiopods from New Hampshire. *Geology*, **14**, 631.

Boulter, C. A. 1974: Tectonic deformation of soft sedimentary clastic dikes from the Precambrian rocks of Tasmania, Australia, with particular reference to their relations with cleavages. *Bulletin of the Geological Society of America*, **85**, 1413–20.

Bowen, N. L. 1940: Prograde metamorphism of siliceous limestone and dolomite. *Journal of Geology*, **48**, 225–74.

Bowes, D. R. & Jones, K. A. 1958: Sedimentary features and tectonics in the Dalradian of Western Perthshire. *Transactions of the Royal Society of Edinburgh*, **17**, 135–7.

Boyer, S. E. 1984: Origin and significance of compositional layering in Late Precambrian sediments, Blue Ridge Province, North Carolina, USA. *Journal of Structural Geology*, **6**, 121–33.

Bradshaw, J. Y. 1989: Early Cretaceous vein-related garnet granulite in Fiordland, Southwest New Zealand: a case for infiltration of mantle-derived CO_2-rich fluids. *Journal of Geology*, **97**, 697–717.

Brady, J. B. 1977: Metasomatic zones in metamorphic rocks. *Geochimica et Cosmochimica Acta*, **41**, 113–26.

Brady, J. B. 1988: The role of volatiles in the thermal history of metamorphic terranes. *Journal of Petrology*, **29**, 1187–213.

Brantley, S. L., Evans, B., Hickman, S. H. & Crerar, D. A. 1990: Healing of microcracks in quartz; implications for fluid flow. *Geology*, **18**, 136–39.

Braun, I., Raith, M. & Ravindra Kumar, G. R. 1996: Dehydration–melting phenomena in leptynitic gneisses and the generation of leucogranites: a case study from the Kerala Khondalite Belt, Southern India. *Journal of Petrology*, **37**, 1285–305.

Brearley, A. J. 1987: A natural example of the disequilibrium breakdown of biotite at high temperature – TEM observations and comparison with experimental kinetic data. *Mineralogical Magazine*, **51**, 93–101.

Bridgewater, D., Escher, A. & Watterson, J. 1973a: Tectonic displacements and thermal activity in two contrasting Proterozoic mobile belts from Greenland. *Philosophical Transactions of the Royal Society of London*, **A273**, 513–33.

Bridgewater, D., McGregor, V. R. & Myers, J. S. 1974: A horizontal tectonic regime in the Archaean of Greenland and its implications for early crustal thickening. *Precambrian Research*, **1**, 179–97.

Bridgewater, D., Watson, J. & Windley, B. F. 1973b: The Archaean craton of the North Atlantic region. *Philosophical Transactions of the Royal Society of London*, **A273**, 493–512.

Brimhall, G. H. & Dietrich, W. E. 1987: Constitutive relationships between chemical composition, volume, density, porosity and strain in metasomatic hydrothermal systems. *Geochimica et Cosmochimica Acta*, **51**, 567–87.

Brodie, K. H., Fettes, D., Harte, B. & Schimid, R. 2002: Towards a unified nomenclature in metamorphic petrology. Structural terms, including fault rocks. A proposal on behalf of the IUGS Subcommission on the Systematics of Metamorphic Rocks. Provisional recommendations; web version of 31 October, 2002. http://www.bgs.ac.uk/SCMR/scmr_products.html.

Brooks, W. E. 1986: Distribution of anomalously high K_2O volcanic rocks in Arizona: Metasomatism at the Picacho Peak detachment fault. *Geology*, **14**, 339–42.

Brown, E. H. 1996: High-pressure metamorphism caused by magma loading in Fiordland, New Zealand. *Journal of Metamorphic Geology*, **14**, 441–52.

Brown, G. M. & Fyfe, W. S. 1972: The transition from metamorphism to melting: status of the granulite and eclogite facies. *24th International Geological Congress, Section* **2**, 27–34.

Brown, M. 1973: The definition of metatexis, diatexis and migmatite. *Proceedings of the Geologists' Association*, **84**, 371–82.

Brown, M. 1979: The petrogenesis of the St. Malo migmatite belt, Amorican Massif, France, with particular reference to the diatexites. *Neues Jahrbuch für Mineralogie Abhandlungen*, **135**, 48–74.

Brown, M. 1994: The generation, segregation, ascent and emplacement of granite magma: the migmatite-to-crustally derived granite connection in thickened orogens. *Earth-Science Reviews*, **36**, 83–130.

Brown, M. 2001: Orogeny, migmatites and leucogranites: A review. *Proceedings of the Indian Academy of Sciences (Earth and Planetary Science)*, **110**, 313–36.

Brown, M. 2002: Retrograde processes in migmatites and granulites revisited. *Journal of Metamorphic Geology*, **20**, 25–40.

Brown, M. 2004: The mechanism of melt extraction from lower continental crust of orogens. *Transactions of the Royal Society of Edinburgh: Earth Sciences*, **95**, 35–48.

Brown M. 2005a: Synergistic effects of melting and deformation: an example from the Variscan Belt, western France. *In*: Gapais, D., Brun, J. P. & Cobbold, P. R. (eds.): *Deformation Mechanism, Rheology and Tectonics: From Minerals to the Lithosphere. Geological Society of London Special Publication*, **243**, 205–26.

Brown, M. 2005b: Comment. *In*: Brown, M., Pitcher, W. S. & Vernon, R. H.: Invited comments on Clemens's 'Granites and granitic magmas'. *Proceedings of the Geologists' Association*, **116**, 23–5.

Brown, M. 2007: Crustal melting and melt extraction, ascent and emplacement in orogens: mechanisms and consequences. *Journal of the Geological Society of London*, **164**, 709–30.

Brown, M., Averkin, Y. A., McLellan, E. L. & Sawyer, E. W. 1995a: Melt segregation in migmatites. *Journal of Geophysical Research*, **100**, 15655–79.

Brown M. & Earle, M. M. 1983: Cordierite-bearing schists and gneisses from Timor, eastern Indonesia: *P-T* conditions of metamorphism and tectonic implications. *Journal of Metamorphic Geology*, **1**, 183–203.

Brown M. & Rushmer, T. 1997: The role of deformation in the movement of granitic melt: views from the laboratory and the field. *In*: Holness M. B. (ed.): *Deformation-enhanced Fluid Transport in the Earth's Crust and Mantle. Mineralogical Society Series*, **8**, London: Chapman and Hall, pp. 111–14.

Brown M., Rushmer, T. & Solar, G. S. 1995b: Introduction to special section: Mechanisms and consequences of melt segregation from crustal protoliths. *Journal of Geophysical Research*, **100**, 15551–63.

Brown M. & Solar, G. S. 1999: The mechanism of ascent and emplacement of granite magma during transpression: a syntectonic granite paradigm. *Tectonophysics*, **312**, 1–33.

Brown, M. A., Brown, M., Carlson, W. D. & Denison, C. 1999: Topology of syntectonic melt-flow networks in the deep crust: inferences from three-dimensional images of leucosome geometry in migmatites. *American Mineralogist*, **84**, 1793–818.

Brown, R. E., Stevens, B. P. J., Willis, I. L., Stroud, W. J., Bradley, G. M. & Barnes, R. G. 1983: Quartzo-feldspathic rocks. *In*: Stevens, B. P. J. & Stroud, W. J. (eds.): *Rocks of the Broken Hill Block: Their Classification, Nature, Stratigraphic Distribution, and Origin. Records of the Geological Survey of New South Wales*, **21**, 127–226.

Browne, W. R. 1922: Report on the petrology of the Broken Hill region, excluding the Great Lode and its immediate vicinity. *Memoirs of the Geological Survey of New South Wales*, **8**, 295–353.

Bucher, H. W. 1953: Fossils in metamorphic rocks: a review. *Bulletin of the Geological Society of America*, **64**, 275–300.

Bucher, K. & Frey, M. 1994: *Petrogenesis of Metamorphic Rocks.* Berlin: Springer.

Buick, I. S. & Cartwright, I. 1996: Fluid-rock interaction during low-pressure polymetamorphism of the Reynolds Range Group, central Australia. *Journal of Petrology*, **37**, 1097–124.

Buick, I. S. & Cartwright, I. 2000: Stable isotope constraints on the mechanism of fluid flow during contact metamorphism around the Marulan Batholith, NSW, Australia. *Journal of Geochemical Exploration*, **69–70**, 291–5.

Buick, I. S., Gibson, R. L., Cartwright, I., Maas, R., Wallmach, T. & Uken, R. 2000: Fluid flow in metacarbonates associated with emplacement of the Bushveld Complex, South Africa. *Journal of Geochemical Exploration*, 69–70, 391–5.

Burnham, C. W. 1997: Magmas and hydrothermal fluids. *In*: Barnes, H. L. (ed.): *Geochemistry of Hydrothermal Ore Deposits*. New York: Wiley, pp. 71–136.

Burt, D. M. 1974: Metasomatic zoning in Ca-Fe-Si exoskarns. *In*: Hofmann, A. W., Giletti, B. J., Yoder, H. S. & Yund, R. A. (eds.): *Geochemical Transport and Kinetics. Carnegie Institution of Washington Publication*, **634**, 287–93.

Burton, K. W. & O'Nions, R. K. 1990: The timescale and mechanism of granulite formation at Kurunegala, Sri Lanka. *Contributions to Mineralogy and Petrology*, **106**, 66–89.

Butler, B. C. M. 1965: A chemical study of some rocks of the Moine Series of Scotland. *Quarterly Journal of the Geological Society of London*, **121**, 163–208.

Butler, P. 1969: Mineral compositions and equilibria in the metamorphosed iron formation of the Gagnon region, Quebec, Canada. *Journal of Petrology*, **10**, 56–101.

Butler, R. W. H., Harris, N. B. W. & Whittington, A. G. 1997: Interactions between deformation, magmatism and hydrothermal activity during active crustal thickening; a field example from Nanga Parbat, Pakistan Himalayas. *Mineralogical Magazine*, **61**, 37–52.

Byerly, G. R. & Lowe, D. R. 1994: Spinels from Archean impact spherules. *Geochimica et Cosmochimca Acta*, **58**, 3469–86.

Callaghan, E. 1966: Emplacement of massive cupriferous pyrite orebodies, Skouriotissa, Cyrpus. *Geological Society of America Special Paper*, **87**, 25–6.

Camacho, A., Vernon, R. H. & FitzGerald, J. D. 1995: Large volumes of pseudotachylyte in the Woodroffe Thrust, eastern Musgrave Ranges, Australia. *Journal of Structural Geology*, **17**, 371–83.

Cannon, R. T. 1966: Plagioclase zoning and twinning in relation to the metamorphic history of some amphibolites and granulites. *American Journal of Science*, **264**, 526–42.

Cardenas, A. A., Girty, G. H., Hanson, A. D., Lahren M. M., Knaack, C. & Johnson, D. 1996: Assessing differences in composition between low metamorphic grade mudstones and high-grade schists using logratio techniques. *Journal of Geology*, **104**, 279–93.

Carlisle, D. B. & Bradman, D. R. 1991: Nanometre-size diamonds in the Cretaceous-Tertiary boundary clay of Alberta. *Nature*, **352**, 708–9.

Carlson, W. D. 1999: The case against Ostwald ripening of porphyroblasts. *Canadian Mineralogist*, **37**, 403–413.

Carlson, W. D. 2000: The case against Ostwald ripening of porphyroblasts: reply. *Canadian Mineralogist*, **38**, 1029–31.

Carlson, W. D. 2002: Scales of disequilibrium and rates of equilibration during metamorphism. *American Mineralogist*, **87**, 185–204.

Carmichael, D. M. 1969: On the mechanism of prograde metamorphic reactions in quartz-bearing pelitic rocks. *Contributions to Mineralogy and Petrology*, **20**, 244–67.

Carmichael, D. M. 1970: Intersecting isograds in the Whetstone Lake area, Ontario. *Journal of Petrology*, **11**, 147–217.

Carmichael, D. M. 1978: Metamorphic bathozones and bathograds: a measure of the depth of post-metamorphic uplift and erosion on the regional scale. *American Journal of Science*, **278**, 769–97.

Carney, J. N., Treloar, P. J., Barton, C. M., Crow, M. J., Evans, J. A. & Simango, S. 1991: Deep-crustal granulites with migmatitic and mylonitic fabrics from the Zambezi Belt, northeastern Zimbabwe. *Journal of Metamorphic Geology*, **9**, 461–79.

Carpenter, R. H. & Allard, G. O. 1980: Mineralization, alteration and volcanism in the Lincolnton-McCormick district, Georgia and South Carolina. *Geological Society of America Abstracts with Programs*, **7**, 398–99.

Carreras, J., Julivert, M. & Santanach, P. 1980: Hercynian mylonite belts in the eastern Pyrenees: an example of shear zones associated with late folding. *Journal of Structural Geology*, **2**, 5–9.

Carrington, D. P. & Harley, S. L. 1995: Partial melting and phase relations in high-grade metapelites: an experimental petrogenetic grid in the KFMASH system. *Contributions to Mineralogy and Petrology*, **120**, 270–91.

Carson, C. J., Clarke, G. L. & Powell, R. 2000: Hydration of eclogite, Pam Peninsula, New Caledonia. *Journal of Metamorphic Geology*, **18**, 79–90.

Carson, C. J., Powell, R., & Clarke, G. L. 1999: Calculated mineral equilibria for the eclogite facies in Na_2O-CaO-MgO-FeO-Al_2O_3-SiO_2-H_2O: application to the Pouébo Terrane, Pam Peninsula, New Caledonia. *Journal of Metamorphic Geology*, **17**, 9–24.

Carson, C. J., Powell, R., Wilson, C. J. L. & Dirks, P. H. G. M. 1997: Partial melting during tectonic exhumation of a granulite terrain: an example from the Larsemann Hills, east Antarctica. *Journal of Metamorphic Geology*, **15**, 105–26.

Carswell, D. A. 1989: Eclogites and the eclogite facies: definitions and classification. *In*: Carswell, D. A. (ed.): *Eclogite Facies Rocks*. Glasgow & London: Blackie, pp. 1–13.

Carswell, D. A., Curtis, C. D. & Kanaris-Sotiriou, R. 1974: Vein metasomatism in peridotite at Kalskaret near Tafjord, south Norway. *Journal of Petrology* **15**, 383–402.

Carswell, D. A. & Harley, S. L. 1989: Mineral barometry and thermometry. *In*: Carswell, D. A. (ed.): *Eclogite Facies Rocks*. Glasgow: Blackie, pp. 83–110.

Carten, R. B. 1986: Sodium-calcium metasomatism: chemical, temporal, and spatial relationships at the Yerington, Nevada, porphyry copper deposit. *Economic Geology*, **81**, 1495–519.

Cartwright, I. 1999: Regional oxygen isotope zonation at Broken Hill, New South Wales, Australia: large-scale fluid flow and implications for Pb-Zn-Ag mineralization. *Economic Geology*, **94**, 357–74.

Cartwright, I. & Barnicoat, A. C. 1999: Stable isotope geochemistry of alpine ophiolites: a window to ocean floor hydrothermal alteration and constraints on fluid-rock interaction during high-pressure metamorphism. *International Journal of Earth Sciences*, **88**, 219–36.

Cartwright, I. & Barnicoat, A. C. 2003: Geochemical and stable isotope resetting in shear zones from Täschalp: constraints on fluid flow during exhumation in the Western Alps. *Journal of Metamorphic Geology*, **21**, 143–61.

Cartwright, I. & Buick, I. S. 1995: Formation of wollastonite-bearing marbles during late regional metamorphic channelled fluid flow in the Upper Calcsilicate Unit of the Reynolds Range Group, central Australia. *Journal of Metamorphic Geology*, **13**, 397–417.

Cartwright, I. & Buick, I. S. 1999: The flow of surface-derived fluids through Alice Springs age middle-crustal ductile shear zones, Reynolds Range, central Australia. *Journal of Metamorphic Geology*, **17**, 397–414.

Cartwright, R. B., Buick, I. S., Weaver, T. R., Vry, J. K. & Oliver, N. H. S. 1995: Patterns of fluid flow in Proterozoic calc-silicates: fluid channelling and variations in fluid fluxes and intrinsic permeabilities. *Australian Journal of Earth Sciences*, **42**, 259–65.

Cartwright, I. & Oliver, N. H. S. 1992: Direction of fluid flow during contact metamorphism around the Burstall Granite, Australia. *Journal of the Geological Society of London*, **149**, 693–6.

Cartwright, I., Power, W. L., Oliver, N. H. S., Valenta, R. K & McLatchie, G. S. 1994: Fluid migration and vein formation during deformation and greenschist facies metamorphism at Ormiston Gorge, central Australia. *Journal of Metamorphic Geology*, **12**, 373–86.

Cartwright, I. & Valley, J. W. 1990: Fluid–rock interaction in the north-west Adirondack Mountains, New York State. *In*: Ashworth, J. R. & Brown, M. (eds.): *High-temperature Metamorphism and Crustal Anatexis*. London: Unwin Hyman, pp. 180–97.

Castro, A., Patiño-Douce, A. E., Corretgé, L. G., de la Rosa, J. D., El-Biad, M. & El-Hmidi, H. 1999: Origin of peraluminous granites and granodiorites, Iberian massif, Spain. An experimental test of granite petrogenesis. *Contributions to Mineralogy and Petrology*, **135**, 255–76.

Cawthorn, R. G. 1996. *Layered Intrusions*. Amsterdam: Elsevier.

Chamberlain, C. P. & Rumble, D. 1988: Thermal anomalies in a regional metamorphic terrane: an isotopic study of the role of fluids. *Journal of Petrology*, **29**, 1215–32.

Chapman, F. 1893: On oölitic and other limestones with sheared structures from Ilfracombe. *Geological Magazine*, **20**, 100–4.

Chapman, L. H. & Williams, P. J. 1998: Evolution of pyroxene-pyroxenoid-garnet alteration at the Cannington Ag-Pb-Zn deposit, Cloncurry district, Queensland, Australia. *Economic Geology*, **93**, 1390–405.

Chappell, B. W. & White, A. J. R. 1974: Two contrasting granitic types. *Pacific Geology*, **8**, 173–4.

Chappell, B. W. & White, A. J. R. 1991: Restite enclaves and the restite model. *In*: Barbarin, M. B. & Didier J. (eds.): *Enclaves and Granite Petrology*. Amsterdam: Elsevier, pp. 375–81.

Charoy, B. & Pollard, P. J. 1989: Albite-rich, silica-depleted metasomatic rocks at Emuford, northeast Queensland: mineralogical, geochemical, and fluid inclusion constraints on hydrothermal evolution and tin mineralization. *Economic Geology*, **84**, 1850–74.

Chenhall, B. E. & Phillips, E. R. 1985: The petrology and geochemistry of quartzofeldspathic gneisses, Broken Hill mines area, Broken Hill, New South Wales. *Proceedings of the Linnean Society of New South Wales*, **108**, 1–21.

Chinner, G. A. 1960: Pelitic gneisses with varying ferrous/ferric ratios from Glen Clova, Angus. *Journal of Petrology*, **1**, 178–217.

Chinner, G. A. 1961: The origin of sillimanite in Glen Clova, Angus. *Journal of Petrology*, **2**, 312–23.

Chopin, C. 1984: Coesite and pure pyrope in high-grade blueschists of the western Alps: a first record and some consequences. *Contributions to Mineralogy and Petrology*, **86**, 107–18.

Chopin, C. 2003: Ultrahigh-pressure metamorphism: tracing continental crust into the mantle. *Earth and Planetary Science Letters*, **212**, 1–14.

Christensen, J. N., Selverstone, J., Rosenfeld, J. L. & De Paolo, D. J. 1994: Correlation by Rb-Sr geochronology of garnet growth histories from different structural levels within the Tauern Window, Eastern Alps. *Contributions to Mineralogy and Petrology*, **118**, 1–12.

Christie, J. M. 1960: Mylonitic rocks of the Moine thrust zone in the Assynt region, northwest Scotland. *Transactions of the Edinburgh Geological Society*, **18**, 79–93.

Christie, J. M. 1963: The Moine thrust zone in the Assynt region, northwest Scotland. *University of California Publications in Geological Sciences*, **40**, 345–419.

Cihan, M. 2002: The pitfalls of sectioning rocks perpendicular to foliations and lineations in the matrix. *Geological Society of Australia Abstracts*, **67**, 162.

Claoué-Long, J. C., Compston, W., Roberts, J. & Fanning, C. M. 1995: Two Carboniferous ages: a comparison of SHRIMP zircon dating with conventional zircon ages and $^{40}Ar/^{39}Ar$ analysis. *In*: Berggren, W. B., Kent, D. V., Aubry, M.-P. & Hardenbol, J. (eds.): *Geochronology Time Scales and Global Stratigraphic Correlation: SEPM (Society for Sedimentary Geology) Special Publication*, **54**, 3–21.

Clark, C., Hand, M., Faure, K. & Schmidt Mumm, A. 2006: Up-temperature flow of surface-derived fluids in the mid crust: the role of pre-orogenic burial of hydrated fault rocks. *Journal of Metamorphic Geology*, **24**, 367–87.

Clark, C. & James, P. 2003: Hydrothermal brecciation due to fluid pressure fluctuations: examples from the Olary Domain, South Australia. *Tectonophysics*, **366**, 187–206.

Clark, C., Schmidt Mumm, A. & Faure, K. 2005: Timing and nature of fluid flow and alteration during Mesoproterozoic shear zone formation, Olary Domain, South Australia. *Journal of Metamorphic Geology*, **23**, 147–64.

Clark, M. B. & Fisher, D. M. 1995: Strain partitioning and crack-seal growth of chlorite-muscovite aggregates during progressive noncoxial strain: an example from the slate belt of Taiwan. *Journal of Structural Geology*, **17**, 46–74.

Clarke, D. B., Halliday, A. N. & Hamilton, P. J. 1988: Neodymium and strontium isotopic constraints on the origin of the peraluminous granitoids of the South Mountain Batholith, Nova Scotia, Canada. *Chemical Geology*, **73**, 15–24.

Clarke, F. W. 1924: Data of geochemistry. *United States Geological Survey Bulletin* **770**.

Clarke, G. L. 1988: Structural constraints on the Proterozoic reworking of Archaean crust in the Rayner Complex, MacRobertson and Kemp Land coast, East Antarctica. *Precambrian Research*, **40/41**, 137–56.

Clarke, G. L., Aitchison, J. C. & Cluzel, D. 1997: Eclogites and blueschists of the Pam Peninsula, NE New Caledonia: a reappraisal. *Journal of Petrology*, **38**, 843–76.

Clarke, G. L., Burg, J.-P. & Wilson, C. J. L. 1986: Stratigraphic and structural constraints on the Proterozoic tectonic history of the Olary Block, South Australia. *Precambrian Research*, **34**, 107–37.

Clarke, G. L., Collins, W. J. & Vernon, R. H. 1990: Successive overprinting granulite facies metamorphic events in the Anmatjira Range, central Australia. *Journal of Metamorphic Geology*, **8**, 65–88.

Clarke, G. L., Daczko, N. R., Klepeis, K. A. & Rushmer, T. 2005: Roles for fluid and/or melt advection in forming high-P mafic migmatites, Fiordland, New Zealand. *Journal of Metamorphic Geology*, **23**, 557–67.

Clarke, G. L., Daczko, N. R. & Nockolds, C. 2001: A method for applying matrix corrections to X-ray intensity maps using the Bence-Albee algorithm and Matlab. *Journal of Metamorphic Geology*, **19**, 635–44.

Clarke, G. L., Klepeis, K. A. & Daczko, N. R. 2000: Cretaceous granulites at Milford Sound, New Zealand; metamorphic history and emplacement in a convergent margin setting. *Journal of Metamorphic Geology*, **18**, 359–74.

Clarke, G. L. & Powell, R. 1991: Decompressional coronas and symplectites in granulites of the Musgrave Complex, central Australia. *Journal of Metamorphic Geology*, **9**, 441–50.

Clarke, G. L., Powell, R. & Fitzherbert, J. A. 2006: The lawsonite paradox: a comparison of field evidence and mineral equilibria modelling. *Journal of Metamorphic Geology*, **24**, 715–25.

Clarke, G. L., White, R. W., Fitzherbert, J. A., Lui, S. & Pearson, N. J. 2007: Contrasting behaviour of rare earth and major element during partial melting in granulite facies migmatites, Wuluma Hills, Arunta Block, central Australia. *Journal of Metamorphic Geology*, **25**, 1–18.

Clemens, J. D. 1984: Water contents of silicic to intermediate magmas. *Lithos*, **17**, 273–87.

Clemens, J. D. 1988: Volume and composition relationships between granites and their lower crustal source regions: an example from central Victoria Australia. *Australian Journal of Earth Sciences*, **35**, 445–49.

Clemens, J. D. 1998: Observations on the origins and ascent mechanisms of granitic magmas. *Journal of the Geological Society of London*, **155**, 843–51.

Clemens, J. D. 2003: S-type granitic magmas – petrogenetic issues, models and evidence. *Earth-Science Reviews*, **61**, 1–18.

Clemens, J. D. 2005a: Granites and granitic magmas: strange phenomena and new perspectives on old problems. *Proceedings of the Geologists' Association*, **116**, 9–16.

Clemens, J. D. 2005b: Rejoinder to 'Granite and granitic magmas'. *Proceedings of the Geologists' Association*, **116**, 25–9.

Clemens, J. D. & Holness, M. B. 2000: Textural evolution and partial melting of arkose in a contact aureole: a case study and implications. *Electronic Geosciences*, **5**:4.

Clemens, J. D. & Mawer, C. K. 1992: Granitic magma transport by fracture propagation. *Tectonophysics*, **20**, 339–60.

Clemens, J. D., Petford, N. & Mawer, C. K. 1997: Ascent mechanisms of granitic magmas: causes and consequences. *In*: Holness, M. B. (ed.): *Deformation-enhanced Fluid Transport in the Earth's Crust and Mantle. Mineralogical Society Series*, **8**, London: Chapman and Hall, pp. 145–72.

Clemens, J. D. & Vielzeuf, D. 1987: Constraints on melting and magma production in the crust. *Earth and Planetary Science Letters*, **86**, 287–306.

Clemens, J. D. & Wall, V. J. 1981: Origin and crystallization of some peraluminous (S-type) granitic magmas. *Canadian Mineralogist*, **19**, 111–31.

Clemens, J. D. & Wall, V. J. 1984: Origin and evolution of a peraluminous silicic ignimbrite suite: the Violet Town Volcanics. *Contributions to Mineralogy and Petrology*, **88**, 354–71.

Clemens, J. D. & Watkins, J. M. 2001: The fluid regime of high-temperature metamorphism during granitoid magma genesis. *Contributions to Mineralogy and Petrology*, **140**, 600–6.

Cloos, E. 1947: Oolite deformation in the South Mountain Fold, Maryland. *Bulletin of the Geological Society of America*, **58**, 843–917.

Cloos, M. 1985. Thermal evolution of convergent margins: thermal modeling and reevaluation of isotopic Ar-ages for blueschists in the Franciscan Complex of California. *Tectonics*, **4**, 421–33.

Cobbold, P. R. 1977: Description and origin of banded deformation structures. I. Regional strain, local perturbations, and deformation bands. *Canadian Journal of Earth Sciences*, **14**, 1721–31.

Cocherie A., Legendre, O., Peucat, J. J. & Kouamelan, A. N. 1998: Geochronology of polygenetic monazites constrained by in situ electron microprobe Th-U-total lead determination: implications for lead behaviour in monazite. *Geochimica et Cosmochimica Acta*, **62**, 2475–97.

Coleman, R. G. 1963: Serpentinites, rodingites, and tectonic inclusions in alpine-type mountain chains. *Geological Society of America Special Paper*, **73**, 74–86.

Coleman, R. G. & Lee, D. E. 1963: Glaucophane-bearing metamorphic rock types of the Cazadero area, California. *Journal of Petrology*, **4**, 260–301.

Coleman, R. G., Lee, D. E., Beatty, L. B. & Brannock, W. W. 1965: Eclogites and eclogites: their differences and similarities. *Bulletin of the Geological Society of America*, **76**, 483–507.

Collins, W. J. 1996: Lachlan Fold Belt granitoids: products of three-component mixing. *Transactions of the Royal Society of Edinburgh: Earth Sciences*, **87**, 171–81.

Collins, W. J. 1998: Evaluation of petrogenetic models for Lachlan Fold Belt granitoids: implications for crustal architecture and tectonic models. *Geology*, **30**, 535–8.

Collins, W. J. 2002: Hot orogens, tectonic switching, and creation of continental crust. *Geology*, **30**, 535–38.

Collins, W. J., Flood, R. H., Vernon, R. H. & Shaw, S. E. 1989: The Wuluma granite, Arunta Block, central Australia: an example of in situ, near-isochemical granite formation in a granulite-facies terrane. *Lithos*, **23**, 63–83.

Collins, W. J. & Richards, S. W. 2001: To migma add magma: relation of the Cooma Complex to the S-type Murrumbidgee Batholith. *In*: Chappell, B. W. & Fleming, P. D.

(eds.): *S-Type Granites and Related Rocks. Australian Geological Survey Organization Record* **2001/02**, 33–4.

Collins, W. J., Richards, S. W. Healy, B. & Ellison, P. I. 2000: Origin of heterogeneous mafic enclaves by two-stage hybridisation in magma conduits (dykes) below and in granitic magma chambers. *Transactions of the Royal Society of Edinburgh: Earth Sciences*, **91**, 27–45.

Collins, W. J. & Sawyer, E. 1996: Pervasive granitoid magma transfer through the lower-middle crust during non-coaxial compressional deformation. *Journal of Metamorphic Geology*, **14**, 565–79.

Collins, W. J. & Shaw, R. D. 1995: Geochronological constraints on orogenic events in the Arunta Inlier: a review. *Precambrian Research*, **71**, 315–46.

Collins, W. J. & van Kranendonk, M. J. 1999: Model for the development of kyanite during partial convective overturn of Archean granite-greenstone terranes: the Pilbara Craton, Australia. *Journal of Metamorphic Geology*, **17**, 145–56.

Collins, W. J. & Vernon, R. H. 1991: Orogeny associated with anticlockwise P-T-t paths: evidence from low-P, high-T metamorphic terranes in the Arunta Inlier, central Australia. *Geology*, **19**, 835–8.

Collins, W. J. & Vernon, R. H. 1992: Palaeozoic arc growth, deformation and migration across the Lachlan Fold Belt, southeastern Australia. *Tectonophysics*, **214**, 381–400.

Collins, W. J., Vernon, R. H. & Clarke, G. L. 1991: Discrete Proterozoic structural terranes associated with low-P, high-T metamorphism, Anmatjira Range, Arunta Inlier, central Australia: tectonic implications. *Journal of Structural Geology*, **13**, 1157–71.

Compston, W., Williams, I. S. & Clements, S. W. J. 1982: U-Pb ages with single zircons using the sensitive high mass-resolution ion microprobe. *30th Annual Conference on Mass Spectrometry and Allied Topics. Honolulu: American Society for Mass Spectrometry*, pp. 593–5.

Compston, W., Williams, I. S. & Meyer, C. 1984: U-Pb geochronology of zircons from lunar breccia 73217 using the sensitive high mass-resolution ion microprobe. *Journal of Geophysical Research*, **89** Supplement, B525–34.

Compton, R. R. 1960: Contact metamorphism in Santa Rosa Range, Nevada. *Bulletin of the Geological Society of America*, **71**, 1383–416.

Condie, K. C. 1981: Archean Greenstone Belts. *In*: Windley, B. F. (ed.): *Developments in Precambrian Geology 3*. New York: Elsevier.

Condie, K. C. & Allen, P. 1980: Origin of Archean migmatites from the Gwenoro Dam area, Zimbabwe-Rhodesia. *Contributions to Mineralogy and Petrology*, **74**, 35–43.

Condon, M. A. 1959: Sedimentary structures in the metamorphic rocks and orebodies of Broken Hill. *Proceedings of the Australasian Institute of Mining and Metallurgy*, **189**, 47–65.

Connolly, J. A. D. 1990: Multi-variable phase diagrams: an algorithm based on generalized thermodynamics. *American Journal of Science*, **290**, 666–718.

Connolly, J. A. D., Holness, M. B., Rubie, D. C. & Rushmer, T. 1997: Reaction-induced microcracking: an experimental investigation of a mechanism for enhancing anatectic melt extraction. *Geology*, **25**, 591–4.

Connolly, J. A. D. & Kerrick, D. M. 1987: An algorithm and computer program for calculating composition phase diagrams. *Calphad*, **11**, 1–55.

Conrad, M. E. & Naslund, H. R. 1989: Modally-graded rhythmic layering in the Skaergaard intrusion. *Journal of Petrology*, **30**, 251–69.

Conrad, W. K., Nicholls, I. A. & Wall, V. J. 1988: Water-saturated and -undersaturated melting of metaluminous and peraluminous crustal compositions at 10 kb: evidence for the origin of silicic magmas in the Taupo volcanic zone, New Zealand, and other occurrences. *Journal of Petrology*, **29**, 765–803.

Constantinou, G. & Govett, G. J. S. 1973: Geology, geochemistry, and genesis of the Cyrpus sulfide deposits. *Economic Geology*, **68**, 843–58.

Conybeare, C. E. B. & Crook, K. A. W. 1968: Manual of sedimentary structures. *Australian Bureau of Mineral Resources, Geology and Geophysics Bulletin*, **102**, 1–327.

Cook, N. D. J. & Ashley, P. M. 1992: Meta-evaporite sequence, exhalative chemical sediments and associated rocks in the Proterozoic Willyama Supergroup, South Australia: implications for metallogenesis. *Precambrian Research*, **56**, 211–26.

Coombs, D. S. 1954: The nature and alteration of some Triassic sediments from Southland, New Zealand. *Transactions of the Royal Society of New Zealand*, **82**, 65–109.

Coombs, D. S. 1960: Lower grade mineral facies in New Zealand. *Reports of the International Geological Congress, 21st Session, Norden*, **13**, 339–51.

Coombs, D. S. 1961: Some recent work on the lower grades of metamorphism. *Australian Journal of Science*, **24**, 203–15.

Coombs, D. S. 1965: Sedimentary analcime rocks and sodium-rich gneisses. *Mineralogical Magazine*, 144–58.

Coombs, D. S. 1974: On the mineral facies of spilitic rocks and their genesis. *In*: Amstutz, G. C. (ed.): *Spilites and Spilitic Rocks*. Berlin: Springer, pp. 373–85.

Corbett, G. J. & Phillips, G. N. 1981: Regional retrograde metamorphism of a high-grade terrain: the Willyama Complex, Broken Hill, Australia. *Lithos*, **14**, 59–73.

Cosgrove, J. W. 1976: The formation of crenulation cleavage. *Journal of the Geological Society of London*, **132**, 155–78.

Coward, M. P. 1973a: Heterogeneous deformation in the development of the Laxfordian complex of South Uist, Outer Hebrides. *Journal of the Geological Society of London*, **129**, 137–60.

Coward, M. P. 1973b: The structure and origin of areas of anomalously low intensity finite deformation in the basement complex of the Outer Hebrides. *Tectonophysics*, **16**, 117–40.

Cox, S. F. & Etheridge, M. A. 1982: Fiber development in deformed hydrothermally altered acid volcanic rock. *In*: Borradaile, G. J., Bayly, M. B. & Powell, C. M. (eds.): *Atlas of Deformational and Metamorphic Rock Fabrics*. New York: Springer, pp. 304–5.

Cox, S. F. & Etheridge, M. A. 1989: Coupled grain-scale dilatancy and mass transfer during deformation at high fluid pressure: examples from Mount Lyell, Tasmania. *Journal of Structural Geology*, **11**, 147–62.

Cox, S. F., Wall, V. J., Etheridge, M. A. & Potter, T. F. 1991: Deformational and metamorphic processes in the formation of mesothermal vein-hosted gold deposits – examples from the Lachlan Fold Belt in central Victoria, Australia. *Ore Geology Reviews*, **6**, 391–423.

Craig, J. R. 1990: Ore textures and paragenetic studies – some modern case histories and sources of comparative data. *In*: Jambor, J. L. & Vaughan, D. J. (eds.): *Advanced Microscopic Studies of Ore Minerals. Mineralogical Association of Canada Short Course Handbook*, **17**, 263–317.

Craig, J. R., Vokes, F. M., & Simpson, C. 1991: Rotational fabrics in pyrite from Ducktown, Tennessee. *Economic Geology*, **86**, 1737–46.

Criss, R. E. & Taylor, H. P. 1986: Meteroic-hydrothermal systems. *In*: Valley, J. W., Taylor, H. P. & O'Neil, J. R. (eds.): *Stable Isotopes in High Temperature Geological Processes. Mineralogical Society of America, Reviews in Mineralogy*, **16**, 373–424.

Crowley J. L. & Ghent E. D. 1999: An electron microprobe study of the U-Th-Pb systematics of metamorphosed monazite: the role of Pb diffusion versus overgrowth and recrystallization. *Chemical Geology*, **157**, 285–302.

Crowley, J. L., Ghent, E. D., Carr, S. D., Simony, P. S. & Hamilton, M. A. 2000: Multiple thermotectonic events in a continuous metamorphic sequence, Mica Creek area, southeastern Canadian Cordillera. *Geological Materials Research*, **2/6**, 45 pp.

Curewitz, D. & Karson, J. A. 1999: Ultracataclasis, sintering, and frictional melting in pseudotachylytes from East Greenland. *Journal of Structural Geology*, **21**, 1693–1713.

Currie, K. L. & Ferguson, J. 1971: A study of fenitization around the alkaline carbonatite complex at Callander Bay, Ontario, Canada. *Canadian Journal of Earth Sciences*, **8**, 481–97.

Daczko, N. R., Clarke, G. L. & Klepeis, K. A. 2001. The transformation of two-pyroxene hornblende granulite to garnet granulite: simultaneous melting and fracturing of the lower crust, Fiordland, New Zealand. *Journal of Metamorphic Geology*, **19**, 547–60.

Daines, M. J. & Kohlstedt, D. L. 1997: Influence of deformation on melt topology in peridotites. *Journal of Geophysical Research*, **102**, 10257–71.

Dallmeyer, R. D. 1979: $^{40}Ar/^{39}Ar$ ages of biotite and hornblende from a progressively remetamorphosed basement terrane: their bearing on interpretation of release spectra. *Geochimica et Cosmochimica Acta*, **39**, 1655–99.

Dalrymple, G. B. & Lanphere, M. A. 1971: $^{40}Ar/^{39}Ar$ technique of K/Ar dating: a comparison with the conventional technique. *Earth and Planetary Science Letters*, **12**, 300–8.

Dalrymple, G. B. & Lanphere, M. A. 1974: $^{40}Ar/^{39}Ar$ age spectra of some undisturbed terrestrial samples. *Geochimica et Cosmochimica Acta*, **38**, 715–38.

Daniel, C. G. & Spear, F. S. 1999: The clustered nucleation and growth processes of garnet in regional metamorphic rocks from north-west Connecticut, USA. *Journal of Metamorphic Geology*, **17**, 503–20.

Daniels, J. L. 1974: The geology of the Blackstone region, Western Australia. *Geological Survey of Western Australia Bulletin*, **123**.

Davidson, A. 1990: Two transects across the Grenville Front, Killarney and Tyson Lake areas, Ontario. *In*: Salisbury, M. H. & Fountain, D. M. (eds.): *Exposed Cross-sections of the Continental Crust*. Dordrecht: Kluwer, pp. 343–400.

Davidson, C., Schmid, S. M. & Hollister, L. S. 1994: Role of melt during deformation in the deep crust. *Terra Nova*, **6**, 133–42.

Davidson, G. J. 1994: Hostrocks to the stratabound iron-formation–hosted Starra gold-copper deposit, Australia. I. Sodic lithologies. *Mineralium Deposita*, **29**, 237–49.

Davidson, J. P., De Silva, S. L., Holden, P. & Halliday, A. N. 1990. Small-scale disequilibrium in a magmatic inclusion and its more silicic host. *Journal of Geophysical Research*, **95**, 17661–75.

Davidson, J. P., Ferguson, K. M., Colucci, M. T. & Dungan, M. A. 1988: The origin and evolution of magmas from the San Pedro-Pellado volcanic complex, S. Chile. Multicomponent sources and open system evolution. *Contributions to Mineralogy and Petrology*, **100**, 429–55.

Davidson, J. P., Tepley, F. J. F., Palacz, Z. & Meffan-Main, S. 2001: Magma recharge, contamination and residence times revealed by in situ laser ablation analysis of feldspar in volcanic rocks. *Earth and Planetary Science Letters*, **184**, 427–42.

Dawson, J. B. & Carswell, D. A. 1990: High-temperature and ultra-high pressure eclogites. *In*: Carswell, D. A. (ed.): *Eclogite Facies Rocks*. Glasgow & London: Blackie, pp. 315–49.

Deer, W. A., Howie, R. A., & Zussmann, J. 1992: *An Introduction to the Rock-Forming Minerals*. Second edition. Essex: Longman.

de Jong, G. & Williams, P. J. 1995: Giant metasomatic system formed during exhumation of mid-crustal Proterozoic rocks in the vicinity of the Cloncurry Fault, northwest Queensland. *Australian Journal of Earth Sciences*, **42**, 281–90.

den Brok, S. W. J. & Spiers, C. J. 1991: Experimental evidence for water weakening of quartzite by microcracking plus solution-precipitation creep. *Journal of the Geological Society of London*, **147**, 541–48.

den Tex, E. 1963: A commentary on the correlation of metamorphism and deformation in space and time. *Geologie en Mijnbouw*, **42**, 170–6.

Denison, C. & Carlson, W. D. 1997: Three-dimensional quantitative textural analysis of metamorphic rocks using high-resolution computed X-ray tomography: Part II. Application to natural samples. *Journal of Metamorphic Geology*, **15**, 45–57.

DePaolo, D. J. 1988: *Neodymium Isotope Geochemistry: An Introduction*. New York: Springer.

DePaolo, D. J., Perry, F. V. & Baldridge, W. S. 1992: Crustal vs. mantle sources of granitic magmas: a two parameter model based on Nd isotopic studies. *Transactions of the Royal Society of Edinburgh: Earth Sciences*, **83**, 439–46.

Dickerson, R. P. & Holdaway, M. J. 1989: Acadian metamorphism associated with the Lexington batholith, Bingham, Maine. *American Journal of Science*, **289**, 945–74.

Dietrich, R. V. 1960: Banded gneisses. *Journal of Petrology*, **1**, 99–120.

Dilles, J. H. & Einaudi, M. T. 1992: Wall-rock alteration and hydrothermal flow paths about the Ann-Mason porphyry copper deposit, Nevada; a 6-km vertical reconstruction. *Economic Geology*, **87**, 1963–2001.

Dimroth, E. & Dressler, B. 1978: Metamorphism in the Labrador Trough. *In*: Fraser, J. A. & Heywood, W. W. (eds.): *Metamorphism in the Canadian Shield. Geological Survey of Canada Paper* **78–10**, 215–36.

Dimroth, E. & Lichtblau, A. P. 1979: Metamorphic evolution of Archean hyaloclasites, Noranda area, Quebec, Canada. Part I: Comparison of Archean and Cenozoic sea-floor metamorphism. *Canadian Journal of Earth Sciences*, **16**, 1315–40.

Dipple, G. M. & Ferry, J. M. 1990: Identification of the scales of differential element mobility in a ductile fault zone. *Journal of Metamorphic Geology*, **8**, 645–61.

Dipple, G. M. & Ferry, J. M. 1992a: Fluid flow and stable isotopic alteration in rocks at elevated temperatures with applications to metamorphism. *Geochimica et Cosmochimica Acta*, **56**, 3539–50.

Dipple, G. M. & Ferry, J. M. 1992b: Metasomatism and fluid flow in ductile fault zones. *Contributions to Mineralogy and Petrology*, **112**, 149–64.

Dipple, G. M., Wintsch, R. P. & Andrews, M. S. 1990: Identification of the scales of differential element mobility in a ductile fault zone. *Journal of Metamorphic Geology*, **8**, 645–61.

Dixon, J. & Williams, G. 1983: Reaction softening in mylonites from the Arnaboll thrust, Sutherland. *Scottish Journal of Geology*, **19**, 157–68.

Dodson, M. H. & Mclelland-Brown, E. 1985: Isotopic and palaeomagnetic evidence for rates of cooling, uplift and erosion. *In*: Snelling, N. J. (ed.): *The Chronology of the Geologic Record*. Geological Society of London Memoir, **10**, 315–25.

Donaldson, C. H. 1982: Spinifex-textured komatiites: a review of textures, mineral compositions and layering. *In*: Arndt, N. T. & Nisbet, E. G. (eds.): *Komatiites*. London: Allen & Unwin, pp. 1721–6.

Dontsova, Y. I. 1970: Oxygen isotope exchange in rock-forming processes. *Geochemistry International*, **7**, 624–36.

Doyle, C. & Cartwright, I. 2000: The role of fluids in retrograde shearing at Broken Hill, Australia. *Journal of Geochemical Exploration*, 69–70, 575–6.

Droop, G. T. R. 1985: Alpine metamorphism in the south-east Tauern window. *Journal of Metamorphic Geology*, **3**, 371–402.

Droop, G. T. R. & Bucher-Nurminen, K. 1984: Reaction textures and metamorphic evolution of sapphirine-bearing granulites from the Gruf Complex, Italian Central Alps. *Journal of Petrology*, **25**, 766–803.

Dumond, G., Mahan, K. H., Williams, M. L. & Karlstrom, K. E. 2007: Crustal segmentation, composite looping pressure-time paths and magma-enhanced metamorphic field gradients: Upper Granite Gorge, Grand Canyon, USA. *Bulletin of the Geological Society of America*, **119**, 202–20.

Dunkley, D., Clarke, G. L. & Harley, S. L. 1999: Diffusion metasomatism in silica-undersaturated sapphirine-bearing granulite from Rumdoodle Peak, Framnes Mountains, east Antarctica. *Contributions to Mineralogy and Petrology*, **134**, 264–76.

Dunlap, J. W., Teyssier, C., McDougall, I. & Baldwin, S. 1991: Ages of deformation from K/Ar and $^{40}Ar/^{39}Ar$ dating of white micas. *Geology*, **19**, 1213–6.

Dunlop, J. S. R. & Buick, R. 1981: Archaean epiclastic sediments derived from mafic volcanics North Pole, Pilbara Block, Western Australia. *Special Publication of the Geological Society of Australia*, **7**, 225–33.

Durney, D. W. 1972: Solution-transfer, an important geological deformation mechanism. *Nature*, **235**, 315–17.

Durney, D. W. & Kisch, H. J. 1994: A field classification and intensity scale for first-generation cleavages. *AGSO Journal of Australian Geology and Geophysics*, **15**, 257–95.

Durney, D. W. & Ramsay, J. G. 1973: Incremental strains measured by syntectonic crystal growths. *In*: de Jong, K. A. & Scholten, R. (eds.): *Gravity and Tectonics*. New York: Wiley, 67–96.

Earle, M. M. 1980: A note on the relationship between inclined isothermal surfaces and subduction-zone metamorphism. *Bulletin of the Geological Society of America*, **68**, 313–24.

Eckert, J. O., Newton, R. C. & Kleppa, O. J. 1991: The ΔH of reaction and recalibration of garnet-clinopyroxene-plagioclase-quartz geobarometers in the CMAS system by solution calorimetry. *American Mineralogist*, **76**, 148–60.

Edwards, A. B. 1958: Amphibolites from the Broken Hill district. *Journal of the Geological Society of Australia*, **5**, 1–32.

Edwards, A. B. & Baker, G. 1954: Scapolitisation in the Cloncurry district of northwestern Queensland. *Journal of the Geological Society of Australia*, **1**, 1–33.

Edwards, A. B., Baker, G. & Callow, K. J. 1956: Metamorphism and metasomatism at King Island scheelite mine. *Journal of the Geological Society of Australia*, **3**, 55–98.

Ehlers, C., Londroos, A. & Selonen, O. 1993: The late Svecofennian granite-migmatite zone of southern Finland – a belt of transpressive deformation and granite emplacement. *Precambrian Research*, **64**, 295–309.

Einaudi, M. T. & Burt, D. M. 1982: Introduction. Terminology, classification, and composition of skarn deposits. *Economic Geology*, **77**, 745–63.

Einaudi, M. T., Meinert, L. D. & Newberry, R. J. 1981: Skarn deposits. *Economic Geology 75th Anniversary Volume*, 317–91.

Elbers, F. J. & Hoeve, J. 1971: Quartz-plagioclase metasomatic rocks in the Vaestervik area, Southeastern Sweden. *Geologische Rundschau*, **60**, 1426–41.

Elburg, M. A. 1996a: Evidence of isotopic equilibration between microgranitoid enclaves and host granodiorite, Warburton Granodiorite, Lachlan Fold Belt, Australia. *Lithos*, **38**, 1–22.

Elburg, M. A. 1996b: Genetic significance of multiple enclave types in a peraluminous ignimbrite suite, Lachlan Fold Belt, Australia. *Journal of Petrology*, **37**, 1385–408.

Elburg, M. A. 1996c: U-Pb ages and morphologies of zircon in microgranitoid enclaves and peraluminous host granite: evidence for magma mingling. *Contributions to Mineralogy and Petrology*, **123**, 177–89.

Elburg, M. A. 2001: Generation of microgranitoid enclaves in S-type granites by magma mingling and chemical equilibration. *In*: Chappell, B. W. & Fleming, P. D. (eds.): *S-Type Granites and Related Rocks. Australian Geological Survey Organization Record*, **2001/02**, 37–8.

Elburg, M. A. & Nicholls, I. A. 1995: Origin of microgranitoid enclaves in the S-type Wilson's Promontory Batholith, Victoria: evidence for magma mingling. *Australian Journal of Earth Sciences*, **42**, 423–5.

Ellis, D. J. 1980: Osumilite-sapphirine-quartz granulites from Enderby Land, Antarctica: P-T conditions of metamorphism, implications for garnet-cordierite equilibria and the evolution of the deep crust. *Contributions to Mineralogy and Petrology*, **74**, 201–10.

Ellis, D. J. & Green, D. H. 1979: An experimental study on the effect of Ca upon garnet-clinopyroxene Fe-Mg exchange equilibria. *Contributions to Mineralogy and Petrology*, **71**, 131–22.

Ellis, D. J. & Obata, M. 1992: Migmatite and melt segregation at Cooma, New South Wales. *Transactions of the Royal Society of Edinburgh: Earth Sciences*, **83**, 95–106.

Ellis, D. J., Sheraton, J. W., England, R. N. & Dallwitz, W. B., 1980. Osumilite-sapphirine-quartz granulites from Enderby Land, Antarctica – Mineral assemblages and reactions. *Contributions to Mineralogy and Petrology*, **72**, 123–43.

Elmer, F. L. 2004: *Mineral Equilibria in Low-grade Carbonate-bearing Rocks*. Unpublished PhD thesis, University of Melbourne.

Elmer, F. L., White, R. W. & Powell, R. 2006: Devolatilization of metabasic rocks during greenschist-amphibolite facies metamorphism. *Journal of Metamorphic Geology*, **24**, 497–513.

Engel, A. E. & Engel, G. C. 1960: Prograde metamorphism and granitization of the major paragneiss, northwest Adirondack Mountains, New York. *Bulletin of the Geological Society of America*, **71**, 1–58.

England, P. C. & Richardson, S. W. 1977: The influence of erosion upon the mineral facies of rocks from different metamorphic environments. *Journal of the Geological Society of London*, **134**, 201–13.

England, P. C. & Thompson, A. B. 1984a: Pressure-temperature-time paths of regional metamorphism. I. Heat transfer during the evolution of regions of thickened continental crust. *Journal of Petrology*, **25**, 894–928.

England, P. C. & Thompson, A. B. 1984b: Pressure-temperature-time paths of regional metamorphism. II. Their inference and interpretation using mineral assemblages in metamorphic rocks. *Journal of Petrology*, **25**, 929–55.

Eriksson, K. A. 1981: Archaean platform-to-trough sedimentation, east Pilbara Block, Australia. *Special Publication of the Geological Society of Australia*, **7**, 235–44.

Ernst, G. W. 1971: Petrologic reconnaissance of Franciscan metagraywackes from the Diablo Range, central California Coast ranges. *Journal of Petrology*, **12**, 413–37.

Ernst, W. G. 1973; Interpretative synthesis of metamorphism in the Alps. *Geological Society of America Bulletin*, **84**, 2053–78.

Escher, A., Sorensen, K. & Zeck, H. P. 1976: Nagssugtoquidian mobile belt in West Greenland. *In*: Escher, A. & Watt, W. J. (eds.): *Geology of Greenland*. Geological Survey of Greenland, pp. 76–95.

Eskola, P. 1915: On the relation between chemical and mineralogical composition in the metamorphic rocks of the Orijävi region in south-western Finland. *Bulletin de la Commission géologique de Finlande*, **40**, 279 pp.

Eskola, P. 1921: The mineral facies of rocks. *Norsk Geologisk Tidsskrift*, **6**, 142–94.

Eskola, P. 1932: On the principles of metamorphic differentiation. *Bulletin de la Commission géologique de Finlande*, **97**, 68–77.

Espenshade, G. H. & Potter, D. B. 1960: Kyanite, sillimanite and andalusite deposits of the southeastern States. *United States Geological Survey Professional Paper*, **336**, 121 pp.

Etheridge, M. A. 1983: Differential stress magnitudes during regional deformation and metamorphism: Upper bound imposed by tensile fracturing. *Geology*, **11**, 231–4.

Etheridge, M. A., Hobbs, B. E. & Paterson, M. S. 1973: Experimental deformation of single crystals of biotite. *Contributions to Mineralogy and Petrology*, **38**, 21–36.

Etheridge, M. A. & Lee, M. F. 1975: Microstructure of slate from Lady Loretta, Queensland, Australia. *Bulletin of the Geological Society of America*, **86**, 13–22.

Etheridge, M. A. & Vernon, R. H. 1981: A deformed polymictic conglomerate - the influence of grain size and composition on the mechanism and rate of deformation. *Tectonophysics*, **79**, 237–54.

Etheridge, M. A., Wall, V. J., Cox, S. F. & Vernon, R. H. 1984: High fluid pressures during regional metamorphism and deformation. *Journal of Geophysical Research*, **89**, 4344–58.

Etheridge, M. A., Wall, V. J. & Vernon, R. H. 1983: The role of the fluid phase during regional metamorphism and deformation. *Journal of Metamorphic Geology*, **1**, 205–26.

Etheridge, M. A. & Wilkie, J. C. 1979: Grain size reduction, grain boundary sliding and the flow strength of mylonites. *Tectonophysics*, **58**, 159–78.

Eugster, H. P. 1959: Reduction and oxidation in metamorphism. *In*: Abelson, P. H. (ed.): *Researches in Geochemistry. Volume 1*. New York: Wiley, pp. 397–421.

Eugster, H. P. 1970: Thermal and ionic equilibria among muscovite, K-feldspar and aluminosilicate assemblages. *Fortschritte Mineralogie*, **47**, 106–23.

Evans, B. W. & Berti, J. W. 1986: A revised metamorphic history for the Chiwaukum Schist, Washington Cascades. *Geology*, **14**, 695–98.

Evans, B. W. & Leake, B. E. 1960: The composition of the striped amphibolites of Connemara, Ireland. *Journal of Petrology*, **1**, 337–63.

Evans, B. W. & Trommsdorff, V. 1974: On elongate olivine of metamorphic origin. *Geology*, **2**, 131–2.

Evans, K. A. & Bickle, M. J. 1999: Determination of time-integrated fluid fluxes from the reaction progress of multivariant assemblages. *Contributions to Mineralogy and Petrology*, **134**, 277–93.

Evans, K. A. & Bickle, M. J. 2005: An investigation of the relationship between bulk composition, inferred reaction progress and fluid-flow parameters for layered micaceous carbonates from Maine, USA. *Journal of Metamorphic Geology*, **23**, 181–97.

Evans, T. P. 2004: A method for calculating effective bulk composition modification due to crystal fractionation in garnet-bearing schist: implications for isopleth thermobarometry. *Journal of Metamorphic Geology*, **22**, 547–57.

Fagan, R. A. 1979: *Deformation, Metamorphism and Anatexis of an Early Palaeozoic Flysch Sequence in Northeastern Victoria*. Unpublished PhD thesis, University of New England, Australia.

Faulhaber, S. & Raith, M. 1991: Geothermometry and geobarometry of high-grade rocks: a case study on garnet-pyroxene granulites in southern Sri Lanka. *Mineralogical Magazine*, **55**, 33–56.

Faure, G. 1986: *Principles of Isotope Geology*. Second edition. New York: Wiley.

Ferguson, C. C. & Harte, B. 1975: Textural patterns at porphyroblast margins and their use in determining the time relations of deformation and crystallization. *Geological Magazine*, **112**, 467–80.

Ferry, J. M. 1983a: Regional metamorphism of the Vassalboro Formation, south-central Maine, USA: a case study of the role of fluid ion metamorphic petrogenesis. *Journal of the Geological Society of London*, **140**, 551–76.

Ferry, J. M. 1983b: Mineral reactions and element migration during metamorphism of calcareous sediments from the Vassalboro Formation, south-central Maine. *American Mineralogist*, **68**, 334–54.

Ferry, J. M. 1984: A biotite isograd in south-central Maine, U.S.A.: mineral reactions, fluid transfer, and heat transfer. *Journal of Petrology*, **25**, 871–93.

Ferry, J. M. 1987: Metamorphic hydrology at 13-km depth and 400–500°C. *American Mineralogist*, **72**, 39–58.

Ferry, J. M. 1991: Dehydration and decarbonation reactions as a record of fluid infiltration. *In*: Kerrick, D. M. (ed.): *Contact Metamorphism. Mineralogical Society of America, Reviews in Mineralogy*, **26**, 351–93.

Ferry, J. M. 1994a: A historical review of metamorphic fluid flow. *Journal of Geophysical Research*, **99**, 15487–98.

Ferry, J. M. 1994b: Overview of the petrologic record of fluid flow during regional metamorphism in northern New England. *American Journal of Science*, **294**, 905–88.

Ferry, J. M. 2001: Patterns of mineral occurrence in metamorphic rocks. *American Mineralogist*, **85**, 1573–88.

Ferry, J. M. & Dipple, G. M. 1991: Fluid flow, mineral reactions and metasomatism. *Geology*, **19**, 211–4.

Ferry, J. M. & Spear, F. S. 1978: Experimental calibration of the partitioning of Fe and Mg between biotite and garnet. *Contributions to Mineralogy and Petrology*, **66**, 113–7.

Finger, F., Broksa, I., Roberts, M. P. & Schermaier, A. 1998: Replacement of primary monazite by apatite–allanite–epidote coronas in an amphibolite facies granite gneiss from the eastern Alps. *American Mineralogist*, **83**, 248–58.

Finger, F. & Clemens, J. D. 1995: Migmatization and "secondary" granitic magmas: effects of emplacement of "primary" granitoids in Southern Bohemia, Austria. *Contributions to Mineralogy and Petrology*, **120**, 311–26.

Fitz Gerald, J. G. & Stünitz, H. 1993: Deformation of granitoids at low metamorphic grade. I: Reactions and grain size reduction. *Tectonophysics*, **221**, 269–97.

Fitzherbert, J. A., Clarke, G. L. & Powell, R. 2003: Lawsonite-omphacite bearing metabasalts of the Pam Peninsula, NE New Caledonia: evidence for disrupted blueschist- to eclogite-facies conditions. *Journal of Petrology*, **44**, 1805–31.

Fitzsimons, I. C. W. 1996: Metapelitic migmatites from Brattstrand Bluffs, east Antarctica – metamorphism, melting and exhumation of the mid crust. *Journal of Petrology*, **37**, 395–414.

Fleck, R. J. & Criss, R. E. 1985: Strontium and oxygen isotopic variations in Mesozoic and Tertiary plutons of central Idaho. *Contributions to Mineralogy and Petrology*, **90**, 291–308.

Fleming, P. D. & Offler, R. 1968: Pre-tectonic metamorphic crystallization in the Mt. Lofty Ranges, South Australia. *Geological Magazine*, **105**, 356–9.

Fleming, P. D. & White, A. J. R. 1984: Relationships between deformation and partial melting in the Palmer migmatites. *Australian Journal of Earth Sciences*, **31**, 351–60.

Flinn, D. 1956: On the deformation of the Funzie Conglomerate, Fetlar, Shetland. *Journal of Geology*, **64**, 480–505.

Flood, R. H. & Vernon, R. H. 1988: Microstructural evidence of orders of crystallization in granitoid rocks. *Lithos*, **21**, 237–45.

Foden, J. D., Elburg, M. A., Turner, S. P., Sandiford, M. A., O'Callaghan, J. & Mitchell, S. 2002: Granite production in the Delamerian orogen, South Australia. *Journal of the Geological Society of London*, **159**, 557–75.

Forbes, C. J., Betts, P. G., Weinberg, R. & Buick, I. S., 2005: A structural study of the Broken Hill Block, NSW, Australia. *Journal of Metamorphic Geology*, **23**, 745–70.

Foster, C. T. 1977: Mass transfer in sillimanite-bearing pelitic schists near Rangeley, Maine. *American Mineralogist*, **62**, 727–46.

Foster, C. T. 1981: A thermodynamic model of mineral segregations in the lower sillimanite zone near Rangeley, Maine. *American Mineralogist*, **66**, 260–77.

Foster, C. T. 1983: Thermodynamic models of biotite pseudomorphs after staurolite. *American Mineralogist*, **68**, 389–97.

Foster, C. T. 1986: Thermodynamic models of reactions involving garnet in a sillimanite/staurolite schist. *Mineralogical Magazine*, **50**, 427–39.

Foster, C. T. 1999: Forward modeling of metamorphic textures. *Canadian Mineralogist*, **37**, 415–29.

Foster, G., Kinny, P. D., Vance, D., Prince, C. & Harris, N. B. W. 2000: The significance of monazite U-Th-Pb age data in metamorphic assemblages; a combined study of monazite and garnet chronometry. *Earth and Planetary Science Letters*, **181**, 327–40.

Franz, G., Andrehs, G. & Rhede, D. 1996: Crystal chemistry of monazite and xenotime from Saxothuringian–Moldanubian metapelites, NE Bavaria, Germany. *European Journal of Mineralogy*, **8**, 1097–118.

Franz, G., Mosbrugger, V. & Menge, R. 1991: Carbo-Permian pteridophyll leaf fragments from an amphibolite facies basement, Tauern Window, Austria. *Terra Nova*, **3**, 137–41.

Frantz, J. D. & Mao, H. K. 1976: Bimetasomatism resulting from intergranular diffusion: I. A theoretical model for monomineralic reaction zone sequences. *American Journal of Science*, **276**, 817–40.

Frantz, J. D. & Mao, H. K. 1977: Bimetasomatism resulting from intergranular diffusion: II. Prediction of multimineralic zone sequences. *American Journal of Science*, **279**, 302–23.

Fraser, G. L., Pattison, D. R. M. & Heaman, L. M. 2004: Age of the Ballachulish and Glencoe Igneous Complexes (Scottish Highlands), and paragenesis of zircon, monazite and baddeleyite in the Ballachulish Aureole. *Journal of Geological Society of London*, **161**, 447–62.

Frater, K. M. 1983: Effects of metasomatism and development of quartz 'eyes' in intrusive and extrusive rocks at Golden Grove Cu-Zn deposits, Western Australia. *Transactions of the Institution of Mining and Metallurgy*, **92**, B121–31.

Freitsch, R., Tuisku, P., Martinsson, O. & Perdahl, J.-A. 1997: Early Proterozoic Cu-(Au) and Fe ore deposits associated with regional Na-Cl metasomatism in northern Fennoscandia. *Ore Geology Reviews*, **12**, 1–34.

Fricke, H. C., Wickham, S. M. & O'Neil, J. R. 1992: Oxygen and hydrogen isotope evidence for meteoric water infiltration during mylonitization and uplift in the ruby

Mountains-East Humbolt Range core complex, Nevada. *Contributions to Mineralogy and Petrology*, **111**, 203–21.

Frost, B. R. 1975: Contact metamorphism of serpentinite, chloritic blackwall and rodingite at Paddy-Go-Easy Pass, Central cascades, Washington. *Journal of Petrology*, **16**, 272–313.

Frost, R., Mavrogenes, J. A. & Tomkins, A. G. 2002: Partial melting of sulfide ore deposits during medium- and high-grade metamorphism. *Canadian Mineralogist*, **40**, 1–18.

Frost, R., Swapp, S. M. & Gregory, R. W. 2005: Prolonged existence of sulfide melt in the Broken Hill orebody, New South Wales, Australia. *Canadian Mineralogist*, **43**, 479–93.

Früh-Green, G. L. 1994: Interdependence of deformation, fluid infiltration and reaction progress recorded in eclogitic metagranitoids (Sesia Zone, Western Alps). *Journal of Metamorphic Geology*, **12**, 327–43.

Fukuyama, M., Nishiyama, T., Urata, K. & Mori, Y. 2006: Steady-diffusion modelling of a reaction zone between a metamorphosed basic dyke and a marble from Hirao-dai, Fukuoka, Japan. *Journal of Metamorphic Geology*, **24**, 153–68.

Furnes, H. 1973: Variolitic structure in Ordovician pillow lava and its possible significance as an environmental indicator. *Geology*, **1**, 27–30.

Fyfe, W. S. & Kerrich, R. 1985: Fluids and thrusting. *Chemical Geology*, **49**, 353–362.

Fyfe, W. S., Price, N. J. & Thompson, A. B. 1978. *Fluids in the Earth's Crust*. Amsterdam: Elsevier.

Fyfe, W. S., Turner, F. J. & Verhoogen, J. 1958. Metamorphic reactions and metamorphic facies. *Geological Society of America Memoir*, **73**.

Fyson, W. K. 1975: Fabrics and deformation of Archean metasedimentary rocks, Ross Lake-Gordon Lake area, Slave Province, Northwest Territories. *Canadian Journal of Earth Sciences*, **12**, 765–76.

Fyson, W. K. 1980: Fold fabrics and emplacement of an Archean granitoid pluton, Cleft Lake, Northwest Territories. *Canadian Journal of Earth Sciences*, **17**, 325–32.

Fyson, W. K. & Frith, R. A. 1979: Regional deformations and emplacement of granitoid plutons in the Hackett River greenstone belt, Slave Province, Northwest Territories. *Canadian Journal of Earth Sciences*, **16**, 1187–95.

Gagnevin, D., Daly, J. S., Poli, G. & Morgan, D. 2005a: Microchemical and Sr isotopic investigation of zoned K-feldspar megacrysts: insights into the petrogenesis of a granitic system and disequilibrium crystal growth. *Journal of Petrology*, **46**, 1689–724.

Gagnevin, D., Daly, J. S., Waight, T. E., Morgan, D. & Poli, G. 2005b: Pd isotopic zoning of K-feldspar megacrysts determined by laser ablation multi-collector ICP-MS: insights into granite petrogenesis. *Geochimica et Cosmochimica Acta*, **69**, 1899–915.

Gamble, J. A. 1979: Some relationships between coexisting granitic and basaltic magmas and the genesis of hybrid rocks in the Tertiary central complex of Slieve Gullion, northeast Ireland. *Journal of Volcanology and Geothermal Research*, **5**, 297–316.

Gapais, D., Lagarde, J.-L., Le Corre, C., Audren, C., Jégouzo, P., Casas Sainz, A. & van der Driessche, J. 1993: La zone de cisaillement de Quiberon: témoin d'extension de al chaîne varisque en Bretagne méridionale au Carbonifère, *Comptes Rendus de l'Acadamie des Sciences, Paris*, **316**, 1123–9.

Garcia-Casco, A. & Torres-Roldán, F. L. 1996: Disequilibrium induced by fast decompression in St-Bt-Grt-Ky-Sill-And metapelites from the Betic Belt (southern Spain). *Journal of Petrology*, **37**, 1207–39.

Garlick, G. D. & Epstein, S. 1967: Oxygen isotope ratios in coexisting minerals of regionally metamorphosed rocks. *Geochimica et Cosmochimica Acta*, **31**, 181–214.

Gebauer, D. 1990: Isotopic systems – geochronology of eclogites. *In*: Carswell, D. A. (ed.): *Eclogite Facies Rocks*. Glasgow & London: Blackie, pp. 141–59.

Geiser, A. 1975: Slaty cleavage and the dewatering hypothesis – an examination of some critical evidence. *Geology*, **3**, 717–20.

Gerdes, M. L. & Valley, J. W. 1994: Fluid flow and mass transport at the Valentine wollastonite mine, Adirondack Mountains, New York. *Journal of Metamorphic Geology*, **12**, 589–608.

Ghaly, T. S. 1969: Metamorphic differentiation in some Lewisian rocks of north west Scotland. *Contributions to Mineralogy and Petrology*, **22**, 276–289.

Gibbs, J. W. 1961: *The Scientific Papers of J. Willard Gibbs. Volume One. Thermodynamics*. New York: Dover.

Gibson, R. L. 1992: Sequential, syndeformational porphyroblast growth during Hercynian low-pressure/high-temperature metamorphism in the Canigou massif, Pyrenees. *Journal of Metamorphic Geology*, **10**, 637–50.

Gibson, R. L. 2002; Impact-induced melting of Archean granulites in the Vredefort Dome, South Africa. I: anatexis of metapelitic granulites. *Journal of Metamorphic Geology*, **20**, 57–70.

Gilotti, J. A. & McClelland, W. C. 2005: Leucogranites and the time of extension in the East Greenland Caledonides. *Journal of Geology*, **113**, 399–417.

Glass, B. P. 1967: Microtektites in deep-sea sediments. *Nature*, **214**, 372–4.

Glass, B. P. & Burns, C. A. 1988: Microkrystites, a new term for impact-produced glassy spherules containing primary crystallites. *Proceedings of Lunar and Planetary Science Conference XVIII*, 455–8.

Glass, B. P. & Heezen, B. C. 1967: Tektites and geomagnetic reversals. *Nature*, **214**, 372.

Glass, B. P. & Wu, J. 1993: Coesite and shocked quartz discovered in the Australasian and North American microtektite layers. *Geology*, **21**, 435–8.

Glass, B. P. & Zwart, M. J. 1979: North American microtektites in Deep Sea Drilling Project cores from the Caribbean Sea and Gulf of Mexico. *Bulletin of the Geological Society of America*, **90**, 595–602.

Glassley, W. E. & Sørensen, K. 1980: Constant P_S-T amphibolite to granulite facies transition in Agto (West Greenland) metadolerites: implications and applications. *Journal of Petrology*, **21**, 69–105.

Glazner, A. F. 1988: Stratigraphy, structure, and potassic alteration of Miocene volcanic rocks in the Sleeping Beauty area, central Mojave Desert, California. *Bulletin of the Geological Society of America*, **100**, 414–35.

Glazner, A. F. & Bartley, J. M. 1991: Volume flow, fluid loss and state of strain in extensional mylonites from the central Mojave Desert, California. *Journal of Structural Geology*, **13**, 587–94.

Gleason, G. C., Bruce, V. & Green, H. W. 1999: Experimental investigation of melt topology in partially molten quartz-feldspathic aggregates under hydrostatic and non-hydrostatic stress. *Journal of Metamorphic Geology*, **17**, 705–22.

Glen, R. A. & Laing, W. P. 1975: The significance of sedimentary structures in the Willyama Complex, New South Wales. *Proceedings of the Australasian Institute of Mining and Metallurgy*, no. **256**, 15–20.

Glikson, A. Y. 2004: Early Precambrian asteroid impact-triggered tsunami: excavated seabed, debris flows, exotic boulders, and turbulence features associated with 3.47–2.47 Ga-old asteroid impact fallout units, Pilbara Craton, Western Australia. *Astrobiology*, **4**, 1–32.

Glikson, A. Y. 2005a: Geochemical and isotopic signatures of Archaean to Palaeoproterozoic extraterrestrial impact eject/fallout units. *Australian Journal of Earth Sciences*, **52**, 785–98.

Glikson, A. Y. 2005b: Geochemical signatures of Archean to Early Proterozoic Maria-scale oceanic impact basins. *Geology*, **33**, 125–8.

Glikson, A. Y. 2006: Asteroid impact ejecta units overlain by iron-rich sediments in 3.5–2.4 Ga terrains Pilbara and Kaapvaal cratons: Accidental or cause–effect relationships? *Earth and Planetary Science Letters*, **246**, 149–60.

Glikson, A. Y. 2007: Siderophile element patterns, PGE nuggets and vapor condensation effects in Ni-rich, quench chromite-bearing microkrystite spherules, ~3.24 Ga S3 impact unit, Barberton greenstone belt, Kaapvaal Craton, South Africa. *Earth and Planetary Science Letters*, **253**, 1–16.

Glikson, A. Y. & Allen, C. 2004: Iridium anomalies and fractionated siderophile element patterns in impact ejecta, Brockman Iron Formation, Hamersley Basin, Western

Australia: evidence for a major asteroid impact in simatic crustal regions of the early Proterozoic earth. *Earth and Planetary Science Letters*, **220**, 247–64.

Glikson, A. Y., Allen, C. & Vickers, J. 2004: Multiple 3.47-Ga-old asteroid impact fallout units, Pilbara Craton, Western Australia. *Earth and Planetary Science Letters*, **221**, 383–96.

Glikson, A. Y. & Vickers, J. 2006: The 3.26 – 3.24 Ga Barberton asteroid impact cluster: Tests of tectonic and magmatic con sequences, Pilbara Craton, Western Australia. *Earth and Planetary Science Letters*, **241**, 11–20.

Goldschmidt, V. M. 1911: Die Kontaktmetamorphose im Kristianiagebiet. *Kristiania Videnskabelig Skrifter. I. Mathematisk–Naturvitenskap*, **11**, 483 pp.

Goldschmidt, V. M. 1922: On the metasomatic processes in silicate rocks. *Economic Geology*, **17**, 105–23.

Gole, M. J. 1981: Archean banded iron-formations, Yilgarn Block, Western Australia. *Economic Geology*, **76**, 1954–74.

Goodwin, A. M. 1973: Archean iron-formation and tectonic Basins of the Canadian Shield. *Economic Geology*, **68**, 915–33.

Gottschalk, M. 1997: Internally consistent thermodynamic data for rock forming minerals. *European Journal of Mineralogy*, **9**, 175–223.

Graham, C. M. & Powell, R. 1984: A garnet-hornblende geothermometer: calibration, testing, and application to the Pelona schist, southern California. *Journal of Metamorphic Geology*, **2**, 13–31.

Grambling, J. A. 1990: Internally-consistent geothermometry and H_2O barometry in metamorphic rocks: the example garnet-chlorite-quartz. *Contributions to Mineralogy and Petrology*, **105**, 617–28.

Grambling, J. A. & Codding, D. B. 1982: Stratigraphic and structural relationships of multiply deformed Precambrian metamorphic rocks in the Rio Mora area, New Mexico. *Bulletin of the Geological Society of America*, **93**, 127–37.

Grambling, J. A. & Williams, M. L. 1985: The effect of Fe^{3+} and Mn^{3+} on aluminium silicate phase relations in north-central New Mexico, USA. *Journal of Petrology*, **26**, 324–54.

Granath, J. W. 1976: *Petrogenesis of Metamorphically Layered Tectonites at Cooma, New South Wales*. Unpublished PhD thesis, University of Sydney.

Grant, J. A. 1983: Phase equilibria in partial melting of pelitic rocks. *In*: Atherton, M. P. & Gribble, C. D. (eds.): *Migmatites, Melting and Metamorphism*. Nantwich: Shiva, pp. 86–144.

Grant, J. A. 1985a: Phase equilibria in low-pressure partial melting of pelitic rocks. *American Journal of Science*, **285**, 409–35.

Grant, J. A. 1985b: Phase equilibria in partial melting of pelitic rocks. *In*: Ashworth, J. A. (ed.): *Migmatites*. Glasgow: Blackie, pp. 86–144.

Grant, J. A. 1986: The isochon diagram – a simple solution to Gresens' equation for metasomatism. *Economic Geology*, **81**, 1976–82.

Grant, J. A. & Frost, B. R. 1990: Contact metamorphism and partial melting of pelitic rocks in the aureole of the Lamarie anorthosite Complex, Morton Pass, Wyoming. *American Journal of Science*, **290**, 425–7.

Grant, S. M. 1988: Diffusion models for corona formation in metagabbros from the Western Grenville Province, Canada. *Contributions to Mineralogy and Petrology*, **98**, 49–63.

Gray, C. M. 1971: Strontium Isotopic Studies on Granulites. Unpublished PhD thesis, Australian National University, Canberra.

Gray, C. M. 1977: The geochemistry of central Australian granulites in relation to the chemical and isotopic effects of granulite facies metamorphism. *Contributions to Mineralogy and Petrology*, **65**, 79–89.

Gray, C. M. 1990: A strontium isotopic traverse across the granitic rocks of southeastern Australia: petrogenetic and tectonic implications. *Australian Journal of Earth Sciences*, **37**, 331–49.

Gray, C. M. & Kemp, A. I. S. 2001: Two-component model for southeastern Australian granitic rocks – reiteration and development. *In*: Chappell, B. W. & Fleming, P. D. (eds.): *S-Type Granites and Related Rocks. Australian Geological Survey Organization Record*, **2001/02**, 47–8.

Gray, D. R. 1977a: Differentiation associated with discrete crenulation cleavages. *Lithos*, **10**, 89–101.

Gray, D. R. 1977b: Morphologic classification of crenulation cleavage. *Journal of Geology*, **85**, 229–35.

Gray, D. R. 1978: Cleavages in deformed psammitic rocks from southeastern Australia: Their nature and origin. *Bulletin of the Geological Society of America*, **89**, 577–90.

Gray, D. R. 1979: Microstructure of crenulation cleavages: an indicator of cleavage origin. *American Journal of Science*, **279**, 97–128.

Gray, D. R. 1981: Compound tectonic fabrics in singly folded rocks from southwest Virginia, U.S.A. *Tectonophysics*, **78**, 229–48.

Gray, D. R. 1997: Volume loss and slaty cleavage development. *In*: Sengupta, S. (ed.): *Evolution of Geological Structures in Micro- to Macro-scales*. London: Chapman & Hall, pp. 273–91.

Gray, D. R. & Durney, D. W. 1979a: Investigations on the mechanical significance of crenulation cleavage. *Tectonophysics*, **58**, 35–79.

Gray, D. R. & Durney, D. W. 1979b: Crenulation cleavage differentiation: implications of the solution-redeposition process. *Journal of Structural Geology*, **1**, 73–80.

Gray, D. R. & Willman, C. E. 1991: Thrust-related strain gradients and thrusting mechanisms in a chevron-folded sequence, southeastern Australia. *Journal of Structural Geology*, **13**, 691–710.

Gray, D. R. & Wright, T. O. 1984. Problems of volume loss, fabric development, and strain determination in low-grade pelitic rocks, Martinsburg Formation, U.S.A. *Journal of Structural Geology*, **7**, 492.

Green, D. H. & Ringwood, A. E. 1967: An experimental investigation of the gabbro-to-eclogite transformation and its petrological applications. *Geochimica et Cosmochimica Acta*, **31**, 767–833.

Green, H. W. 1992: Analysis of deformation in geological materials. In Buseek, P. (ed.): *Minerals and Reactions at the Atomic Scale: Transmistion Electron Microscopy. Mineralogical Society of America, Reviews in Mineralogy*, **27**, 425–54.

Greenfield, J. E., Clarke, G. L., Bland, M. & Clark, D. J. 1996: *In-situ* migmatite and hybrid diatexite at Mt Stafford, central Australia. *Journal of Metamorphic Geology*, **14**, 413–26.

Greenfield, J. E., Clarke, G. L. & White, R. W. 1998: A sequence of partial melting reactions at Mt Stafford, central Australia. *Journal of Metamorphic Geology*, **16**, 363–78.

Greenwood, H. J. 1961: The system $NaAlSi_3O_8$-H_2O-argon: total pressure and water pressure in metamorphism. *Journal of Geophysical Research*, **66**, 3923–46.

Greenwood, H. J. 1962: Metamorphic reactions involving two volatile components. *Annual Report of the Director of the Geophysical Laboratory, Carnegie Institution of Washington Yearbook*, **61**, 82–6.

Greenwood, H. J. 1967: Mineral equilibria in the system MgO-SiO_2-H_2O-CO_2. *In*: Abelson, P. H. (ed.): *Researches in Geochemistry, Volume II*. New York: Wiley, pp. 542–67.

Greenwood, H. J. 1975: The buffering of pore fluids by metamorphic reactions. *American Journal of Science*, **275**, 573–93.

Gregg, W. J. 1985: Microscopic deformation mechanisms associated with mica film formation in cleaved psammitic rocks. *Journal of Structural Geology*, **7**, 45–56.

Gregg, W. J. 1986: Deformation of chlorite-white mica aggregates in cleaved psammitic and pelitic rocks from Islesboro, Maine, U.S.A. *Journal of Structural Geology*, **8**, 59–68.

Gresens, R. L. 1967a: Composition-volume relationships of metasomatism. *Chemical Geology*, **2**, 47–66.

Gresens, R. L. 1967b: Tectonic-hydrothermal pegmatites. I. The model. *Contributions to Mineralogy and Petrology*, **43**, 111–24.

Gresens, R. L. 1971: Application of hydrolysis equilibria to the genesis of pegmatite and kyanite deposits in northern New Mexico. *Mountain Geologist*, **8**, 3–16.

Grew, E. S., Mamay, S. H. & Barghoorn, E. S. 1970: Age of plant fossils from the Worcester Coal Mine, Worcester, Massachusetts. *American Journal of Science*, **268**, 113–26.

Gribble, C. D. 1968: The cordierite-bearing rocks of the Haddo House Norite in Aberdeenshire. *Contributions to Mineralogy and Petrology*, **17**, 315–30.

Griffin, W. L. 1971: Genesis of coronas in anorthosites of the upper Jotun Nappe, Indre Sogn, Norway. *Journal of Petrology*, **12**, 219–43.

Griffin, W. L. 1972: Formation of eclogites and the coronas in anorthosites, Bergen Arcs, Norway. *Geological Society of America Memoir*, **135**, 37–62.

Griffin, W. L. & Heier, K. S. 1973: Petrological implications of some corona structures. *Lithos*, **6**, 315–35.

Grocott, J. 1981: Fracture geometry of pseudotachylyte generation zones: a study of shear fractures formed during seismic events. *Journal of Structural Geology*, **3**, 169–78.

Groshong, R. H. 1976: Strain and pressure solution in the Martinsburg Slate, Delaware Water Gap, New Jersey. *American Journal of Science*, **276**, 1131–46.

Groshong, R. H. 1988: Low-temperature deformation mechanisms and their interpretation. *Bulletin of the Geological Society of America*, **100**, 1329–1360.

Grove, T. L., Gerlach, D. C. & Sando, T. W. 1982: Origin of calc-alkaline series lavas at Medicine Lake volcano by fractionation, assimilation and mixing. *Contributions to Mineralogy and Petrology*, **80**, 160–82.

Groves, D. I., Goldfarb, R. J., Robert, F. & Hart, C. J. R. 2003: Gold deposits in metamorphic belts: overview of current understanding, outstanding problems, future research, and exploration significance. *Economic Geology*, **98**, 1–29.

Groves, D. I. & Phillips, G. N. 1987: The genesis and tectonic control on Archaean gold deposits of the Western Australian Shield – a metamorphic replacement model. *Ore Geology Reviews*, **2**, 287–322.

Grunder, A. L. 1995: Material and thermal roles of basalt in crustal magmatism: case study from eastern Nevada. *Geology*, **23**, 952–6.

Grundmann, G. & Morteani, G. 1989: Emerald mineralization during regional metamorphism: the Habachtal (Austria) and Leysdorp (Transvaal, South Africa) deposits. *Economic Geology*, **84**, 1835–49.

Guernina, S. & Sawyer, E. W. 2003: Large-scale melt-depletion in granulite terranes: an example from the Archean Ashuanipi Subprovince of Quebec. *Journal of Metamorphic Geology*, **21**, 181–201.

Guidotti, C. V. 2000: The classic high-temperature – low-pressure metamorphism of west-central Maine: is it post-tectonic or syntectonic? Evidence from porphyroblast–matrix relations: discussion. *Canadian Mineralogist*, **38**, 995–1006.

Guidotti, C. V. & Johnson, S. E. 2002: Pseudomorphs and associated microstructures of western Maine, USA. *Journal of Structural Geology*, **24**, 1139–56.

Guiraud, M., Powell, R. & Cottin, J. Y. 1996: Hydration of orthopyroxene-cordierite-bearing assemblages at Laouni, Central Hoggar, Algeria. *Journal of Metamorphic Geology*, **14**, 467–76.

Guiraud, M., Powell, R. & Rebay, G. 2001: H_2O in metamorphism and unexpected behaviour in the preservation of metamorphic mineral assemblages. *Journal of Metamorphic Geology*, **19**, 445–54.

Gustafson, J. K., Burrell, H. C. & Garretty, M. D. 1950: Geology of the Broken Hill ore deposit, Broken Hill, NSW, Australia. *Bulletin of the Geological Society of America*, **61**, 1369–438.

Gustafson, L. B. & Hunt, J. P. 1975: The porphyry copper deposits at El Salvador, Chile. *Economic Geology*, **70**, 857–912.

Guy, A. G. 1959: *Elements of Physical Metallurgy*. Reading, Massachusetts: Addison-Wesley.

Hälbich, I. W. & Altermann, W. 1991: The genesis of BIF in the Transvaal Supergroup, South Africa. *In*: Pagel, M. & Leroy, J. L. (eds.): *Source, Transport and Deposition of Metals*. Rotterdam: Balkema, pp. 287–90.

Halpin, J. A., Gerakiteys, C. L., Clarke, G. L., Belousova, E. A. & Griffin, W. L. 2005: In-situ U-Pb geochronology and Hf analyses of the Rayner Complex, east Antarctica. *Contributions to Mineralogy and Petrology*, **148**, 689–706.

Halpin, J. A., White, R. W., Clarke, G. L. & Kelsey, D. E. 2007: The Proterozoic P-T-t evolution of the Kemp Land coast, east Antarctica; constraints from Si-saturated and Si-undersaturated metapelites. *Journal of Petrology*, **48**, 1321–49.

Hammond, R. L. 1987: The influence of deformation partitioning on dissolution and solution transfer in low-grade tectonic mélange. *Journal of Metamorphic Geology*, **5**, 195–211.

Hand, M. & Dirks, P. H. G. M. 1992: The influence of deformation on the formation of axial-planar leucosomes and the segregation of small melt bodies within the migmatitic Napperby gneiss, central Australia. *Journal of Structural Geology*, **14**, 591–604.

Hand, M., Dirks, P. H. G. M., Powell, R. & Buick, I. S. 1992: How well established is isobaric cooling in Proterozoic orogenic belts? An example from the Arunta Inlier, central Australia. *Geology*, **20**, 649–52.

Hanel, M., Montenari, M. & Kalt, A. 1999: Determining sedimentation ages of high-grade metamorphic gneisses by their palynological record: a case study in the northern Schwarzwald (Variscan Belt, Germany). *International Journal of Earth Sciences*, **88**, 49–59.

Hanmer, S. 1979: The role of discrete heterogeneities and linear fabrics in the formation of crenulations. *Journal of Structural Geology*, **1**, 81–90.

Harker, A. 1893: On the migration of material during the metamorphism of rock-masses. *Journal of Geology*, **1**, 574–8.

Harker, A. 1909: *The Natural History of Igneous Rocks*. London: Methuen.

Harker, A. 1932: *Metamorphism*. London: Methuen.

Harley, S. L. 1984: An experimental study of the partitioning of Fe and Mg between garnet and orthopyroxene. *Contributions to Mineralogy and Petrology*, **86**, 359–73.

Harley, S. L. 1989: The origins of granulites: a metamorphic perspective. *Geological Magazine*, **126**, 215–47.

Harley, S. L. & Kelly, N. M. 2007: Zircon. Tiny but timely. *Elements*, **3**, 13–18.

Harley, S. L., Kelly, N. M. & Möller, A. 2007: Zircon behaviour and the thermal histories of mountain chains. *Elements*, **3**, 25–30.

Harley, S. L. & Santosh, M. 1995: Wollastonite at Nuliyam, Kerala, southern India: a reassessment of CO_2-infiltration at charnockite formation at a classic locality. *Contributions to Mineralogy and Petrology*, **120**, 83–94.

Harris, N. B. W., Ayres, M. & Massey, J. 1995: The geochemistry of granitic melts produced during the incongruent melting of muscovite: implications for the extraction of Himalayan leucogranite magmas. *Journal of Geophysical Research*, **100**, 15767–77.

Harris, N. B. W. & Massey, J. 1994: Decompression and anatexis of the Himalayan metapelites. *Tectonics*, **13**, 1537–46.

Harrison, T. M. 1981: Diffusion of ^{40}Ar in hornblende. *Contributions to Mineralogy and Petrology*, **78**, 324–331.

Harrison, T. M. 1983: Some observations on the interpretation of $^{40}Ar/^{39}Ar$ age spectra. *Chemical Geology (Isotope Geoscience Section)*, **1**, 319–38.

Harrison, T. M., Morgan, P. & Blackwell, D. P. 1986: Constraints on the age of heating of the Fenton Hill site, Valles Caldera, New Mexico. *Journal of Geophysical Research*, **91**, 1899–908.

Harte, B. & Hudson, N. F. C. 1979: Pelite facies series and the temperatures and pressures of Dalradian metamorphism in eastern Scotland. *In*: Harris, A. L., Holland, C. H. & Leake, B. E. (eds.): *The Caledonides of the British Isles. Geological Society of London Special Publication*, **8**, 323–37.

Hartel, T. H. D. & Pattison, D. R. M. 1996: Genesis of the Kapuskasing (Ontario) migmatitic mafic granulites by dehydration melting of amphibolite: the importance of quartz to reaction progress. *Journal of Metamorphic Geology*, **14**, 591–611.

Harvey, R. D. & Vitaliano, C. J. 1964: Wall rock alteration in the Goldfield district, Nevada. *Journal of Geology*, **72**, 564–79.

Hassler, S. W., Simonson, B. M., Sumner, D. Y. & Murphy, M. 2005: Neoarchaean impact spherule layers in the Fortescue and Hamersley Groups, Western Australia: stratigraphic and depositional implications of re-correlation. *Australian Journal of Earth Sciences*, **52**, 759–71.

Haydon, R. C. & McConachy, G. W. 1987: The stratigraphic setting of Pb-Zn-Ag mineralisation at Broken Hill. *Economic Geology*, **82**, 826–56.

Hayward, N. 1990: Determination of early fold axis orientations in multiply deformed rocks using porphyroblast inclusion trails. *Tectonophysics*, **179**, 353–69.

Hayward, N. 1992: Microstructural analysis of the classical spiral garnet porphyroblasts of south-east Vermont: evidence for non-rotation. *Journal of Metamorphic Geology*, **10**, 567–87.

Healy, B. C., Collins, W. J. & Richards, S. W. 2004: A hybrid origin for Lachlan S-type granites: the Murrumbidgee Batholith example. *Lithos*, **78**, 197–216.

Hedenquist, J. W. & Lowenstern, J. B. 1994: The role of magmas in the formation of hydrothermal ore deposits. *Nature*, **370**, 519–27.

Heinrich, C. A. 1986: Eclogite facies regional metamorphism of hydrous rocks in the central Adula nappe. *Journal of Petrology*, **27**, 123–54.

Heinrich, C. A., Driesner, T. D., Stefansson, A. & Seward, T. M. 2004: Magmatic vapor contraction and the transport of gold from the porphyry environment to epithermal ore deposits. *Geology*, **32**, 761–4.

Heinrich, C. A., Gunther, D., Audetat, A., Ulrich, T. & Frischknect, R. 1999: Metal fractionation between magmatic brine and vapor determined by microanalysis of fluid inclusions. *Geology*, **27**, 755–6.

Heinrich, E. W. 1980: *The Geology of Carbonatites*. Huntington, New York: Krieger.

Hemley, J. J. & Jones, W. R. 1964: Chemical aspects of hydrothermal alteration with emphasis on hydrogen metasomatism. *American Journal of Science*, **266**, 129–66.

Hemley, J. J., Montova, J. W., Marinenko, J. W. & Luce, R. W. 1980: Equilibria in the system Al_2O_3-SiO_2-H_2O and some general implications for alteration/mineralization processes. *Economic Geology*, **75**, 129–66.

Henry, C., Burkhard, M. & Goffé, B. 1996: Evolution of syn-metamorphic veins and their wallrocks through a Western Alps transect: no evidence for large-scale fluid flow. *Stable isotope, major and trace-element systematics. Chemical Geology*, **127**, 81–109.

Hensen, B. J. 1988: Chemical potential diagrams and chemographic projections: applications to the sapphirine granulites from Kiranur and Ganguvarpatti. Evidence for rapid uplift in the South Indian Shield? *Neues Jahrbuch für Mineralogie Abhandlungen*, **158**, 193–210.

Hewins, R. H. 1983: Dynamic crystallization experiments as constraints on chondrule genesis. *In*: King, E. A. (ed.): *Chondrules and Their Origin*. Houston: Lunar and Planetary Institute, pp. 122–33.

Hewins, R. H., Jones, R. H. & Scott, E. R. D. 1996: *Chondrules and the Protoplanetary Disk*. Cambridge: Cambridge University Press.

Hewitt, D. A. 1973: The metamorphism of micaceous limestones from south-central Connecticut. *American Journal of Science*, **273-A**, 444–69.

Hildreth, W. & Moorbath, S. 1988: Crustal contributions to arc magmatism in the Andes of central Chile. *Contributions to Mineralogy and Petrology*, **98**, 455–89.

Hill, M. L. 1985: Remarkable fossil locality: Crinoid stems from migmatites of the Coast Plutonic Complex, British Columbia. *Geology*, **13**, 825–6.

Hill, R. E., Barnes, S. J., Gole, M. J. & Dowling, S. E. 1990: Physical volcanology of komatiites. *Geological Society of Australia (Western Australian Division), Excursion Guide*, **1**.

Hill, R. E., Barnes, S. J., Gole, M. J. & Dowling, S. E. 1995: The volcanology of komatiites as deduced from field relationships in the Norseman-Wiluna greenstone belt, Western Australia. *Lithos*, **34**, 159–88.

Hills, E. S. 1963: *Elements of Structural Geology*. New York: Wiley.

Hinchey, A. M. & Carr, S. D. 2006: The S-type Ladybird leucogranite suite of southeastern British Columbia: geochemical and isotopic evidence for a genetic link with migmatite formation in the North American basement gneisses of the Monashee complex. *Lithos*, **90**, 223–48.

Hiroi, Y., Yokose, M., Oba, T., Kishi, S., Nohara, T. & Yao, A. 1987: Discovery of Jurassic radiolaria from acmite-rhodonite-bearing metachert of the Gosaisyo metamorphic rocks in the Abukuma terrane, northeastern Japan. *Journal of the Geological Society of Japan*, **93**, 445–8.

Hirsch, D. M., Ketcham, R. A. & Carlson, W. D. 2000: An evaluation of spatial correlation functions in textural analysis of metamorphic rocks. *Geological Materials Research*, **2/3**, 42 pp.

Hitzman, M. W., Oreskes, N. & Einaudi, M. 1992: Geological characteristics and tectonic setting of Proterozoic iron oxide (Cu-U-Au-REE) deposits. *Precambrian Research*, **58**, 241–87.

Hobbs, B. E. 1968: Recrystallization of single crystals of quartz. *Tectonophysics*, **6**, 353–402.

Hobbs, B. E., Archibald, N. J., Etheridge, M. A. & Wall, V. J. 1984: Tectonic history of the Broken Hill Block, Australia. *In*: Kröner, A. & Greiling, R. (eds.): *Precambrian Tectonics Illustrated*. Stuttgart: E. Schweizerbart'sche Verlagsbuchhandlung, pp. 353–68.

Hobbs, B. E., Means, W. D. & Williams, P. F. 1976: *An Outline of Structural Geology*. New York: Wiley.

Hobbs, B. E. & Talbot, J. L. 1966: The analysis of strain in deformed rocks. *Journal of Geology*, **74**, 500–13.

Hoersch, A. L. 1981: Progressive metamorphism of the chert-bearing Durness limestone in the Beinn an Dubhaich aureole, Isle of Skye, Scotland: A reexamination. *American Mineralogist*, **66**, 491–506.

Hoeve, J. 1978: Composition and volume changes accompanying soda metasomatic alterations, Vaestervik area, SE Sweden. *Geologische Rundschau*, **67**, 920–42.

Hoisch, T. D. 1987: Heat transport by fluids during Late Cretaceous regional metamorphism in the Big Maria Mountains, southeastern California. *Bulletin of the Geological Society of America*, **98**, 549–53.

Holdaway, M. J. & Lee, S. M. 1977: Fe–Mg cordierite stability in high-grade pelitic rocks based on experimental, theoretical and natural observations. *Contributions to Mineralogy and Petrology*, **63**, 175–98.

Holeywell, R. C. & Tullis, T. E. 1975: Mineral reorientation and slaty cleavage in the Martinsburg Formation, Lehigh Gap, Pennsylvania. *Bulletin of the Geological Society of America*, **86**, 1296–304.

Holland, T. J. B. 1980: The reaction albite = jadeite + quartz determined experimentally in the range 600–1200 °C. *American Mineralogist*, **65**, 129–34.

Holland, T. J. B. & Powell, R. 1985. An internally consistent thermodynamic dataset with uncertainties and correlations: 2. Data and results. *Journal of Metamorphic Geology*, **3**, 343–370.

Holland, T. J. B. & Powell, R. 1996a: Thermodynamics of order-disorder in minerals: I. Symmetric formalism applied to minerals of fixed composition. *American Mineralogist*, **81**, 1413–24.

Holland, T. J. B. & Powell, R. 1996b: Thermodynamics of order-disorder in minerals: II. Symmetric formulism applied to solid solutions. *American Mineralogist*, **81**, 1425–37.

Holland, T. J. B. & Powell, R. 1998. An internally-consistent thermodynamic data set for phases of petrological interest. *Journal of Metamorphic Geology*, **16**, 309–43.

Holland, T. J. B. & Richardson, S. W. 1979: Amphibole zonation in metabasites as a guide to the evolution of metamorphic conditions. *Contributions to Mineralogy and Petrology*, **70**, 143–8.

Hollis, J. A., Harley, S. L., White, R. W. & Clarke, G. L. 2006: Preservation of evidence for prograde metamorphism in UHT HP granulites, South Harris, Scotland. *Journal of Metamorphic Geology*, **24**, 263–79.

Hollister, L. S. 1969: Contact metamorphism in the Kwoiek area of British Columbia: an end-member of the metamorphic process. *Bulletin of the Geological Society of America*, **80**, 2465–94.

Hollister, L. S., Burruss, R. C., Henry, D. L. & Hendel, E.-M. 1979: Physical conditions during uplift of metamorphic terranes, as recorded by fluid inclusions. *Bulletin de Minéralogie*, **102**, 555–61.

Hollister, L. S. & Crawford, M. L. 1986: Melt-enhanced deformation: a major tectonic process. *Geology*, **14**, 558–61.

Hollister, V. F. 1975: An appraisal of the nature and source of porphyry copper deposits. *Minerals Science Engineering*, **7**, 225–33.

Holyoke, C. W. & Rushmer, T. 2002: An experimental study of grain scale melt segregations mechanisms in two common crustal rock types. *Journal of Metamorphic Geology*, **20**, 493–512.

Hopgood, A. M. & Bowes, D. R. 1978: Neosomes of polyphase agmatites as time-markers in complexly deformed migmatites. *Geologische Rundschau*, **67**, 313–30.

Hopgood, A. M., Bowes, D. R., Kouvo, O. & Halliday, A. N. 1983: U-Pb and Rb-Sr isotopic study of polyphase deformed migmatites in the Svecokarelides, southern Finland. *In*: Atherton, M. P. & Gribble, C. D. (eds.): *Migmatites, Melting and Metamorphism*. Nantwich: Shiva, pp. 80–92.

Hopwood, T. P. 1966: *The Relationship between Tectonic Style and Metamorphic Grade in the Cooma Complex, N.S.W.* Unpublished PhD thesis, University of Sydney.

Hopwood, T. P. 1976: Stratigraphy and structural summary of the Cooma metamorphic complex. *Journal of the Geological Society of Australia*, **23**, 345–60.

Horwitz, R. C. & Ramanaidou, E. R. 1993: Slumping in the Marra Supersequence Package in the southern Hamersley Province, Western Australia. *Australian Journal of Earth Sciences*, **40**, 339–44.

Huang, W. & Rubenach, M. J. 1995: Structural controls on syntectonic metasomatic tremolite and tremolite-plagioclase pods in the Molanite Valley, Mt. Isa, Australia. *Journal of Structural Geology*, **17**, 83–94.

Humphris, S. E. & Thompson, G. 1978: Hydrothermal alteration of oceanic basalts by seawater. *Geochimica et Cosmochimica Acta*, **42**, 107–26.

Hunahashi, M., Watanabe, J. & Woo Kim C. 1972: Zeolitic alteration of perlite complex in Tsuchihata mining district, northeast Japan. *Journal of the Faculty of Science, Hokkaido University, Series 4*, **15**, 41–80.

Huppert, H. E. & Sparks, R. S. J. 1988. The generation of granitic magmas by intrusion of basalt into continental crust. *Journal of Petrology*, **29**, 599–624.

Hutton, D. H. W. 1982: A tectonic model for the emplacement of the Main Donegal granite, NW Ireland. *Journal of the Geological Society of London*, **139**, 615–31.

Hutton, D. H. W. 1992: Granite sheeted complexes: evidence for the dyking ascent mechanism. *Transactions of the Royal Society of Edinburgh: Earth Sciences*, **83**, 377–82.

Hutton, D. H. W. & Ingram, G. M. 1992: The Great Tonalite Sill of southeastern Alaska and British Columbia: emplacement into an active contractional high angle reverse shear zone (extended abstract). *Transactions of the Royal Society of Edinburgh: Earth Sciences*, **83**, 383–6.

Hyndman, D. W. 1972: *Petrology of Igneous and Metamorphic Rocks*. New York: McGraw-Hill.

Indares, A. 1993: Eclogitized gabbros from the eastern Grenville Province: textures, metamorphic context and implications. *Canadian Journal of Earth Sciences*, **30**, 59–73.

Ingerson, E. 1938: Albite trends in some rocks of the Piedmont. *American Journal of Science*, **235A**, 127–41.

Ingerson, E. 1940: Fabric criteria for distinguishing pseudo-ripple marks from ripple marks. *Bulletin of the Geological Society of America*, **51**, 557–69.

Irvine, T. N. 1987: Layering and related structures in the Duke Island and Skaergaard intrusions: similarities, differences, and origins. *In*: Parsons, I. (ed.): *Origins of Igneous Layering*. Dordrecht: Reidel, pp. 185–245.

Irving, A. J. & Ashley, P. M. 1976: Anthophyllite-olivine-spinel, cordierite-anthophyllite and related hornfelses associated with metamorphosed serpentinites in the Goobarragandra district, neat Tumut, New South Wales. *Journal of the Geological Society of Australia*, **23**, 19–43.

Ishizuka, H. 1985: Prograde metamorphism of the Horokanai ophiolite in the Kamuikotan Zone, Hokkaido, Japan. *Japanese Journal of Petrology*, **26**, 391–417.

Ito, E. & Anderson, A. T. 1983: Submarine metamorphism of gabbros from the Mid-Cayman Rise: petrographic and mineralogic constraints on hydrothermal processes at slow-spreading ridges. *Contributions to Mineralogy and Petrology*, **82**, 371–88.

Jackson, D. H. & Santosh, M. 1992: Dehydration reaction and isotope front transport induced by CO_2 infiltration at Nuliyam, South India. *Journal of Metamorphic Geology*, **10**, 365–82.

Jackson, K. A. 1967: A review of the fundamental aspects of crystal growth. *In*: Peiser, H. S. (ed.): *Crystal Growth*. Oxford: Pergamon, pp. 17–24.

Jagoutz, E. 1994: Isotopic systematics of metamorphic rocks. *In*: Lanphere, M. A., Dalrymple, G. B. & Turrin, B. D. (eds.): *Abstracts of the Eighth International Conference on Geochronology, Cosmochronology, and Isotope Geology. United States Geological Survey Circular*, **1107**, 156.

James, H. L. 1992: Precambrian iron-formations: nature, origin, and mineralogic evolution from sedimentation to metamorphism. *In*: Wolf, K. H. & Chilingarian, G. V. (eds.): *Diagenesis, III. Developments in Sedimentology*, **47**, 543–89.

James, H. L. & Sims, P. K. 1973: Precambrian iron-formations of the world. Introduction. *Economic Geology*, **68**, 913–4.

James, P. R. 1976: Deformation of the Isua block, West Greenland: a remnant of the earliest stable continental crust. *Canadian Journal of Earth Sciences*, **13**, 816–23.

James, R. S., Grieve, R. A. F. & Pauk, L. (1978): The petrology of cordierite-anthophyllite gneisses and associated mafic and pelitic gneisses at Manitouwadge, Ontario. *American Journal of Science*, **278**, 41–63.

Jamieson, R. A. 1988: Textures, sequences of events, and assemblages in metamorphic rocks. *Mineralogical Association of Canada Short Course*, **14**, 189–212.

Jamieson, R. A. & Vernon, R. H. 1987: Timing of porphyroblast growth in the Fleur de Lys Supergroup, Newfoundland. *Journal of Metamorphic Geology*, **5**, 273–88.

Jamtveit, B., Bucher-Nurminen, K. & Austrheim, H. 1990: Fluid controlled eclogitization of granulites in deep crustal shear zones, Bergen Arcs, Western Norway. *Contributions to Mineralogy and Petrology*, **104**, 184–93.

Jamtveit, B., Wogelius, R. A. & Fraser, D. G. 1993: Zonation patterns of skarn garnets: records of hydrothermal system evolution. *Geology*, **21**, 113–16.

Janardhan, S., Newton, R. C. & Smith, J. V. 1979: Ancient crustal metamorphism at low P_{H_2O}: charnockite formation from Kappaldurga, south India. *Nature*, **278**, 511–14.

Jenkin, G. R. T., Craw, D. & Fallick, A. E. 1994: Stable isotopic and fluid inclusion evidence for meteoric fluid penetration into an active mountain belt; Alpine schist, New Zealand. *Journal of Metamorphic Geology*, **12**, 429–44.

Jenkin, G. R. T., Fallick, A. E. & Leake, B. E. 1992: A stable isotope study of retrograde alteration in S. W. Connemara, Ireland. *Contributions to Mineralogy and Petrology*, **110**, 269–88.

Jercinovic, M. J. & Williams, M. L. 2005: Analytical perils (and progress) in electron microprobe trace element analysis applied to geochronology: background acquisition, interferences, and beam irradiation effects. *American Mineralogist*, **90**, 526–46.

Joesten, R. 1974a: Local equilibrium and metasomatic growth of zoned calc-silicate nodules from a contact aureole, Christmas Mountains, Big Bend region, Texas. *American Journal of Science*, **274**, 876–901.

Joesten, R. 1974b: Pseudomorphic replacement of melilite by idocrase in a zoned calc-silicate skarn, Christmas Mountains, Big Bend region, Texas. *American Mineralogist*, **59**, 694–9.

Joesten, R. 1977: Evolution of mineral assemblage zoning in diffusion metasomatism. *Geochimica et Cosmochimica Acta*, **41**, 649–70.

Joesten, R. 1991: Local equilibrium in metasomatic processes revisited: diffusion-controlled growth of chert nodule reaction rims in dolomite. *American Mineralogist*, **76**, 743–56.

Joesten, R. & Fisher, G. 1988: Kinetics of diffusion-controlled mineral growth in the Christmas Mountains (Texas) contact aureole. *Bulletin of the Geological Society of America*, **100**, 714–32.

Johannes, W. 1983: On the origin of layered migmatites. *In*: Atherton, M. P. & Gribble, C. D. (eds.): *Migmatites, Melting and Metamorphism*. Nantwich: Shiva, pp. 234–48.

Johannes, W. 1985: The significance of experimental studies for the formation of migmatites. *In*: Ashworth, J. R. (ed.): *Migmatites*. Glasgow: Blackie, pp. 36–85.

Johannes, W. 1988: What controls partial melting in migmatites? *Journal of Metamorphic Geology*, **6**, 451–66.

Johannes, W. & Holtz, F. 1990: Formation and composition of H_2O-undersaturated granitic melts. *In*: Ashworth, J. R. & Brown, M. (eds.): *High Temperature Metamorphism and Crustal Anatexis*. London: Unwin Hyman, pp. 272–315.

Johnson, C. D. & Carlson, W. D. 1990: The origin of olivine-plagioclase coronas in metagabbros from the Adirondack Mountains, New York. *Journal of Metamorphic Geology*, **8**, 697–717.

Johnson, I. R. & Klinger, G. D. 1976: Broken Hill ore deposit and its environment. *In*: Knight, C. L. (ed.): *Economic Geology of Australia and Papua New Guinea. 1. Metals*. Melbourne: Australasian Institute of Mining and Metallurgy, pp. 476–91.

Johnson, M. R. W. 1960: The structural history of the Moine thrust zone at Lochcarron, wester Ross. *Transactions of the Royal Society of Edinburgh*, **64**, 139–68.

Johnson, M. R. W. 1962: Relationships of movement and metamorphism in the Dalradian of Banffshire. *Transactions of the Edinburgh Geological Society*, **19**, 29–64.

Johnson, M. R. W. 1967: Mylonite zones and mylonite banding. *Nature*, **213**, 246–47.

Johnson, S. E. 1990: Deformation history of the Otago schists, New Zealand, from progressively developed porphyroblast-matrix microstructures: uplift-collapse orogenesis and its implications. *Journal of Structural Geology*, **12**, 727–46.

Johnson, S. E. 1992: Sequential porphyroblast growth during progressive deformation and low-P high-T (LPHT) metamorphism, Cooma Complex, Australia: The use of microstructural analysis in better understanding deformation and metamorphic histories. *Tectonophysics*, **214**, 311–39.

Johnson, S. E. 1993a: Unravelling the spirals: a serial thin section study and three-dimensional computer-aided reconstruction of spiral-shaped inclusion trails in garnet porphyroblasts. *Journal of Metamorphic Geology*, **11**, 621–34.

Johnson, S. E. 1993b: Testing models for the development of spiral-shaped inclusion trails in garnet porphyroblasts; to rotate or not to rotate, that is the question. *Journal of Metamorphic Geology*, **11**, 635–59.

Johnson, S. E. 1999a: Near-orthogonal foliation development in orogens: meaningless complexity, or reflection of fundamental dynamic processes? *Journal of Structural Geology*, **21**, 1183–87.

Johnson, S. E. 1999b: Porphyroblast microstructures: a review of current and future trends. *American Mineralogist*, **84**, 1711–26.

Johnson, S. E. & Bell, T. H. 1996: How useful are 'millipede' and other similar porphyroblast microstructures for determining synmetamorphic deformation histories? *Journal of Metamorphic Geology*, **14**, 15–28.

Johnson, S. E. & Moore, R. R. 1996: De-bugging the 'millipede' porphyroblast microstructure: a serial thin-section study and 3-D computer animation. *Journal of Metamorphic Geology*, **14**, 3–14.

Johnson, S. E. & Vernon, R. H. 1995a: Stepping stones and pitfalls in the determination of an anticlockwise P-T-t-deformation path in the low-P, high-T Cooma Complex, Australia. *Journal of Metamorphic Geology*, **13**, 165–83.

Johnson, S. E. & Vernon, R. H. 1995b: Inferring the timing of porphyroblast growth in the absence of continuity between inclusion trails and matrix foliations: can it be reliably done? *Journal of Structural Geology*, **17**, 1203–6.

Johnson, S. E., Vernon, R. H. & Hobbs, B. E. 1994: Deformation and metamorphism of the Cooma Complex, southeastern Australia. *Geological Society of Australia Specialist Group in Tectonics and Structural Geology Field Guide*, **4**.

Johnson, T. E., Hudson, N. F. C. & Droop, G. T. R. 2001: Partial melting in the Inzie Head gneisses: the role of water and a petrogenetic grid in KFMASH applicable to anatectic pelitic migmatites. *Journal of Metamorphic Geology*, **19**, 99–118.

Jolly, W. T. & Smith, R. E. 1972: Degradation and metamorphic differentiation of the Keweenawan tholeiitic lavas of northern Michigan, U.S.A. *Journal of Petrology*, **13**, 273–309.

Joplin, G. A. 1968: *A Petrography of Australian Metamorphic Rocks*. Sydney: Angus & Robertson.

Jurewicz, S. R. & Watson, E. B. 1984: Distribution of partial melt in a felsic system: the importance of surface energy. *Contributions to Mineralogy and Petrology*, **85**, 25–9.

Jurewicz, S. R. & Watson, E. B. 1985: The distribution of partial melt in a granite system: the application of liquid phase sintering theory. *Geochimica et Cosmochimica Acta*, **49**, 1109–21.

Kamineni, D. C., Stone, D. & Johnston, P. J. 1991: Metamorphism and mineral chemistry of Quetico sedimentary rocks near Atikokan, Ontario, Canada. *Neues Jahrbuch für Mineralogie Abhandlungen*, **162**, 311–37.

Kanagawa, K. 1991: Change in dominant mechanisms for phyllosilicate preferred orientation during cleavage development in the Kitakami Slates of NE Japan. *Journal of Structural Geology*, **13**, 927–43.

Karlstrom, K. E., Miller, C. F., Kingsbury, J. A. & Wooden, J. L. 1993: Pluton emplacement along an active ductile thrust zone, Piute Mountains, southeastern California: Interaction between deformational and solidification processes. *Bulletin of the Geological Society of America*, **105**, 213–30.

Karlstrom, K. E. & Williams, M. L. 1995: The case for simultaneous deformation, metamorphism and plutonism: an example from Proterozoic rocks in central Arizona. *Journal of Structural Geology*, **17**, 59–81.

Keay, S., Collins, W. J. & McCulloch, M. T. 1997: A three-component Sr-Nd isotopic mixing model for granitoid genesis, Lachlan fold belt, eastern Australia. *Geology*, **25**, 307–10.

Kehlenbeck, M. M. 1972: Deformation textures in the Lac Rouvray anorthosite mass. *Canadian Journal of Earth Sciences*, **9**, 1087–98.

Keller, L. M., Abart, R., Stünitz, H. & de Capitani, C. 2004: Deformation, mass transfer and mineral reactions in an eclogites facies shear zone in a polymetamorphic metapelite (Monte Rosa nappe, western Alps). *Journal of Metamorphic Geology*, **22**, 97–118.

Kelly, N. M., Clarke, G. L. & Fanning, C. M. 2002: A two-stage evolution of the Neoproterozoic Rayner Structural Episode; new U-Pb sensitive high resolution ion microprobe constraints from the Oygarden Group, Kemp Land, East Antarctica. *Precambrian Research*, **116**, 307–30.

Kelly, N. M., Clarke, G. L. & Harley, S. L. 2006: Monazite behaviour and age significance in poly-metamorphic high-grade terrains: A case study from the western Musgrave Block, central Australia. *Lithos*, **88**, 100–34.

Kelly, N. M. & Harley, S. L. 2005: An integrated textural and chemical approach to zircon geochronology: refining the Archaean history of the Napier Complex, east Antarctica. *Contributions to Mineralogy and Petrology*, **149**, 57–84.

Kelsey, D. E., White, R. W., Powell, R., Wilson, C. J. L. & Quinn, C. D. 2003: New constraints on metamorphism in the Rauer Group, Prydz Bay, east Antarctica. *Journal of Metamorphic Geology*, **21**, 739–59.

Kemp, A. I. S. 2001: The role of anatectic processes in generating compositional diversity in metasediment-derived granitic rocks. *In*: Chappell, B. W. & Fleming, P. D. (eds.): *S-Type Granites and Related Rocks. Australian Geological Survey Organization Record*, **2001/02**, 63–4.

Kemp, A. I. S. 2004: Petrology of high-Mg, low-Ti igneous rocks of the Glenelg River Complex (SE Australia) and the nature of their interaction with crustal melts. *Lithos*, **78**, 119–56.

Kemp, A. I. S. & Gray, C. M. 1999: Geological context of crustal anatexis and granitic magmatism in the northeastern Glenelg River Complex, western Victoria. *Australian Journal of Earth Sciences*, **46**, 407–20.

Kenah, P. & Hollister, L. S. 1983: Anatexis in the Central Gneiss Complex, British Columbia. *In*: Atherton, M. P. & Gribble, C. D. (eds.): *Migmatites, Melting and Metamorphism*. Nantwich: Shiva, pp. 142–62.

Kenkmann, T. & Dresen, G. 2002: Dislocation microstructure and phase distribution in a lower crustal shear zone – an example from the Ivrea-Zone, Italy. *International Journal of Earth Sciences (Geologische Rundschau)*, **91**, 445–58.

Kennedy, W. Q. 1949: Zones of progressive regional metamorphism in the Moine Schists of the Western Highlands of Scotland. *Geological Magazine*, **86**, 43–56.

Kent, A. J. R., Ashley, P. M. & Fanning, C. M. 2000: Metasomatic alteration associated with regional metamorphism: an example from the Willyama Supergroup, South Australia. *Lithos*, **54**, 33–62.

Kerrich, R. 1986: Fluid infiltration into fault zones: chemical, isotopic, and mechanical effect. *Pure and Applied Geophysics*, **124**, 225–68.

Kerrich, R., Allison, I., Barnett, R. L., Moss, S. & Starkey, J. 1980: Microstructural and chemical transformations accompanying deformation of granite in a shear zone at Mieville, Switzerland: with implications for stress corrosion cracking and superplastic flow. *Contributions to Mineralogy and Petrology*, **73**, 221–42.

Kerrich, R., Fyfe, W. S., Gorman, B. E. & Allison, I. 1977: Local modification of rock chemistry by deformation. *Contributions to Mineralogy and Petrology*, **65**, 183–90.

Kerrick, D. M. 1972: Experimental determination of muscovite + quartz stability with $PH_2O < P_{total}$. *American Journal of Science*, **272**, 946–58.

Kerrick, D. M. 1977: The genesis of zoned skarns in the Sierra Nevada, California. *Journal of Petrology*, **18**, 144–81.

Killick, A. M. & Reimold, W. U. 1990: Review of the pseudotachylytes in and around the Vredefort Dome, South Africa. *South African Journal of Geology*, **93**, 350–65.

Kim, H. S. & Bell, T. H. 2005: Combining compositional zoning and foliation intersection axes (FIAs) in garnet to quantitatively determine early P-T-t paths in multiply deformed and metamorphosed schists: north-central Massachusetts, USA. *Contributions to Mineralogy and Petrology*, **149**, 141–63.

King, B. C. 1964: The nature of basic igneous rocks and their relations with associated acid rocks. Part IV. *Science Progress*, **52**, 282–92.

King, B. C. & Rast, N. 1955: Tectonic styles in the Dalradians and Moines of parts of the Central Highlands of Scotland. *Proceedings of the Geologists' Association*, **66**, 243–69.

King, B. C. & Rast, N. 1956: The small-scale structures of south-eastern Cowal, Argyllshire. *Geological Magazine*, **93**, 185–95.

King, D. T. & Petruny, L. W. 2004: Cretaceous–Tertiary boundary microtektite-bearing sands and tsunami beds, Alabama Gulf Coastal plain. *Lunar and Planetary Science Conference*, **35**, abstract.

King, E. A. (ed.) 1983: *Chondrules and their Origin*. Houston: Lunar and Planetary Institute.

King, H. F. & Thompson, B. P. 1953: The geology of the Broken Hill district. *In*: *Geology of Australian Ore Deposits*. Fifth Empire Mining and Metallurgical Congress, Australia and New Zealand, **1**, 533–77.

Kingsbury, J. A., Miller, C. F., Wooden, J. L. & Harrison, T. M. 1993: Monazite paragenesis and U–Pb systematics in rocks of the eastern Mojave Desert, California, U.S.A.: implications for thermochronometry. *Chemical Geology*, **110**, 147–67.

Kisters, A. F. M., Gibson, R. L., Charlesworth, E. G. & Anhaeusser, C. R. 1998: The role of strain localization in the segregation and ascent of anatectic melts, Namaqualand, South Africa. *Journal of Structural Geology*, **20**, 229–42.

Klein, C. 2005: Some Precambrian iron-formations (BIFs) from around the world: Their age, geologic setting, mineralogy, metamorphism, geochemistry, and origin. *American Mineralogist*, **90**, 1473–99.

Knill, J. L. 1960: The tectonic pattern in the Dalradian of the Craignish-Kimelfort district, Argyllshire. *Quarterly Journal of the Geological Society of London*, **115**, 339–64.

Knipe, R. J. 1981: The interaction of deformation and metamorphism in slates. *Tectonophysics*, **78**, 249–72.

Knipe, R. J. 1989: Deformation mechanisms – recognition from natural tectonites. *Journal of Structural Geology*, **11**, 127–46.

Knipe, R. J. & White, S. H. 1977: Microstructural variation of an axial plane cleavage around a fold – H.V.E.M. study. *Tectonophysics*, **39**, 355–80.

Knipe, R. J. & Wintsch, R. P. 1985: Heterogeneous deformation, foliation development and metamorphic processes in a polyphase mylonite. *In*: Thompson, A. B. & Rubie, D. C. (eds.): *Metamorphic Reactions. Kinetics, Texture and Deformation. Advances in Physical Chemistry*, **4**. New York: Springer, pp. 180–210.

Koeberl, C. 2006: Impact processes on the early Earth. *Elements*, **2**, 211–6.

Kohn, M. J. & Malloy, M. A. 2004: Formation of monazite via prograde metamorphic reactions among silicates: implications for age determinations. *Geochimica et Cosmochimica Acta*, **68**, 101–13.

Kohn, M. J. & Spear, F. S. 1989: Empirical calibration of geobarometers for the assemblage garnet+hornblende+plagiolcase+quartz. *American Mineralogist*, **74**, 77–84.

Kohn, M. J., Spear, F. S. & Valley, J. W. 1997: Dehydration-melting and fluid recycling during metamorphism: Rangeley Formation, New Hampshire, USA. *Journal of Petrology*, **38**, 1255–77.

Komiya, T., Maruyama, S., Masuda, T., Nohda, S. & Okamoto, K. 1999: Plate tectonics at 3.8–3.7 Ga; field evidence from the Isua accretionary complex, West Greenland. *Journal of Geology*, **107**, 515–54.

Koons, P. O. 1984: Implications to garnet-clinopyroxene geothermometry of non-ideal solid solution in jadeitic pyroxenes. *Contributions to Mineralogy and Petrology*, **79**, 258–67.

Koons, P. O. & Craw, D. 1991: Evolution of fluid driving forces and composition within collisional orogens. *Geophysical Research Letters*, **18**, 935–8.

Korzhinskii, D. S. 1950: Phase rule and geochemical mobility of elements, *Eighteenth International Geological Congress*, part 2, section A, 50–7.

Korzhinskii, D. S. 1959: *Physicochemical Basis of the Analysis of the Paragenesis of Minerals*. New York: Consultants Bureau.

Korzhinskii, D. S. 1966: On thermodynamics of open systems and the phase rule (A reply to D. F. Weill and W. S. Fyfe). *Geochimica et Cosmochimica Acta*, **30**, 829–36.

Korzhinskii, D. S. 1967: On thermodynamics of open systems and the phase rule (A reply to the second critical paper of D. F. Weill and W. S. Fyfe). *Geochimica et Cosmochimica Acta*, **31**, 1177–80.

Korzhinskii, D. S. 1970: *Theory of Metasomatic Zoning*. Oxford: Clarendon Press.

Koziol, A. M. & Newton, R. C. 1988: Redetermination of the garnet breakdown reaction and improvement of the plagioclase-garnet-Al_2SiO_5-quartz geobarometer. *American Mineralogist*, **73**, 216–23.

Krauskopf, K.1967: *Introduction to Geochemistry*. New York: McGraw-Hill.

Kresten, P. & Morogan, V. 1986: Fenitization at the Fen complex, southern Sweden. *Lithos*, **19**, 27–42.

Kretz, R. 1966a: Interpretation of the shape of mineral grains in metamorphic rocks. *Journal of Petrology*, **7**, 68–94.

Kretz, R. 1966b: Metamorphic differentiation at Einasleigh, north Queensland. *Journal of the Geological Society of Australia*, **13**, 561–82.

Kretz, R. 1983: Symbols for rock-forming minerals. *American Mineralogist*, **68**, 277–9.

Kretz, R. 1993: A garnet population in Yellowknife schist. Canada. *Journal of Metamorphic Geology*, **11**, 101–20.

Kretz, R. 1994: *Metamorphic Crystallization*. New York: Wiley.

Kretz, R. 2006: Shape, size, spatial distribution and composition of garnet crystals in highly deformed gneiss of the Otter Lake area, Québec, and a model for garnet crystallization. *Journal of Metamorphic Geology*, **24**, 431–49.

Kriegsman, L. M. 2001: Partial melting, partial melt extraction and partial back reaction in anatectic migmatites. *Lithos*, **56**, 75–96.

Kriegsman, L. M. & Hensen, B. J. 1998: Back reaction between restite and melt: implications for geothermobarometry and pressure-temperature paths. *Geology*, **26**, 1111–4.

Krogh, E. J. 1982: Metamorphic evolution of Norwegian country-rock eclogites, as deduced from mineral inclusions and compositional zoning in garnets. *Lithos*, **15**, 305–21.

Krogh, E. J. 1988: The garnet-clinopyroxene Fe-Mg geothermometer- a reinterpretation of existing experimental data. *Contributions to Mineralogy and Petrology*, **99**, 44–8.

Kröner, A., Williams, I. S., Compston, W., Bauer, N., Vitanage, P. W. & Perera, L. R. K. 1987: Zircon ion microprobe dating of high-grade rocks in Sri Lanka. *Journal of Geology*, **95**, 775–91.

Kruhl, J. H. & Vernon, R. H. 2005: Syndeformational emplacement of a tonalitic sheet-complex in a late-Variscan thrust regime: fabrics and mechanism of intrusion, Monte'e Senes, northeastern Sardinia. *Canadian Mineralogist*, **43**, 387–407.

Kukla, P. A., Kukla, C., Stanistreet, I. G. & Okrusch, M. 1990: Unusual preservation of sedimentary structures in sillimanite-bearing metaturbidites of the Damara Orogen, Namibia. *Journal of Geology*, **98**, 91–9.

Kuniyoshi, S. & Liou, J. G. 1974: Contact metamorphism of the Karmutsen volcanics, Vancouver Island, British Columbia. *Journal of Petrology*, **17**, 73–99.

Kwak, T. A. P. 1978: Mass balance relationships and skarn-forming processes in the King Island scheelite deposit, King Island, Tasmania, Australia. *American Journal of Science*, **278**, 943–68.

Kwak, T. A. P. 1987: W-Sn skarn deposits and related metamorphic skarns and granitoids. *Developments in Economic Geology*, **24**, 451 pp.

Kyte, F. T., Shukolyukov, A., Lumgmair, G. W., Lowe, D. R. & Byerly, G. R. 2003: Early Archean spherule beds: chromium isotopes confirm origin through multiple impacts of projectiles of carbonaceous chondrite type. *Geology*, **31**, 283–6.

LaBerge, G. L. 1966: Altered pyroclastic rocks in iron-formation in the Hamersley Range, Western Australia. *Economic Geology*, **61**, 147–61.

LaBerge, G. L. 1973: Pyroclastic rocks in South African iron-formation. *Economic Geology*, **68**, 1098–109.

Lafrance, B., John, B. E. & Frost, B. R. 1998: Ultra high-temperature and subsolidus shear zones: examples from the Poe Mountain anorthosite, Wyoming. *Journal of Structural Geology*, **20**, 945–55.

Lafrance, B., John, B. E. & Scoates, J. S. 1995: Syn-emplacement recrystallization and deformation microstructures in the Poe Mountain anorthosite, Wyoming. *Contributions to Mineralogy and Petrology*, **122**, 431–40.

Lafrance, B. & Vernon, R. H. 1993: Mass transfer and microfracturing in gabbroic mylonites of the Guadalupe Igneous Complex, California. *In*: Boland, J. N. & Fitz

Gerald, J. D. (eds.): *Defects and Processes in the Solid State: Geoscience Applications. The McLaren Volume*. Amsterdam: Elsevier, pp. 151–67.

Lafrance, B. & Vernon, R. H. 1999: Coupled mass transfer and microfracturing in gabbroic mylonites. *In*: Snoke, A. W., Tullis, J. & Todd, V. R. (eds.): *Fault-related Rocks: A Photographic Atlas*. Princeton: Princeton University Press, pp. 204–7.

Laing, W. P. 1980: Stratigraphic interpretation of the Broken Hill mines area. *In*: Stevens, B. P. J. (ed.): *A Guide to the Stratigraphy and Mineralization of the Broken Hill Block, New South Wales. Records of the Geological Survey of New South Wales*, **20**, 71–85.

Laing, W. P. 1996a: Volcanic-related origin of the Broken Hill Pb+Zn+Ag deposit, Australia. *In*: Pongratz, J. & Davidson, G. (eds.): *New Developments in Broken Hill Type Deposits*. CODES (University of Tasmania) Special Publication **1**, 53–66.

Laing, W. P. 1996b: Sedimentary structures in the Willyama Supergroup, Broken Hill domain: description and database. *Australian Geological Survey Organization Report*.

Lanzirotti, A. 1995: Yttrium zoning in metamorphic garnets. *Geochimica et Cosmochimica Acta*, **59**, 4105–10.

Laporte, D. 1994: Wetting behaviour of partial melts during crustal anatexis: the distribution of hydrous silicic melts in polycrystalline aggregates of quartz. *Contributions to Mineralogy and Petrology*, **116**, 486–99.

Lappin, A. R. & Hollister, L. S. 1980: Partial melting in the Central Gneiss Complex, near Prince Rupert, British Columbia. *American Journal of Science*, **280**, 518–45.

Lasaga, A. C. 1983: Geospeedometry: an extension of geothermometry. *In*: Saxena, S. K. (ed.): *Kinetics and Equilibrium in Mineral Reactions*. Berlin: Springer, pp. 82–114.

Lassey, K. R. & Blattner, P. 1988: Kinetically controlled oxygen isotope exchange between fluid and rock in one-dimensional advective flow. *Geochimica et Cosmochimica Acta*, **52**, 2169–76.

Lawrence, L. J. 1967: Sulphide neomagmas and highly metamorphosed sulphide deposits. *Mineralium Deposita*, **2**, 5–10.

Leake, B. E. & Skirrow, G. 1958: The pelitic hornfelses of the Cashel-Lough Wheelaun intrusion, County Galway, Eire. *Journal of Geology*, **68**, 23–40.

Le Breton, N. & Thompson, A. B. 1988: Fluid-absent (dehydration) melting of biotite in metapelites in the early stages of crustal anatexis. *Contributions to Mineralogy and Petrology*, **99**, 226–37.

Lee, J. H., Peacor, D. R., Lewis, D. D. & Wintsch, R. P. 1986: Evidence for syntectonic crystallization for the mudstone to slate transition at Lehigh Gap, Pennsylvania, U.S.A. *Journal of Structural Geology*, **8**, 767–80.

Le Fort, P. 1981: Mansalu leucogranite: A collision signature of the Himalayas. A model for its generation and emplacement. *Journal of Geophysical Research*, **86**, 10545–68.

Le Fort, P., Cuney, M., Deniel, C., France-Lanord, C., Sheppard, S. M. F., Upreti, B. N. & Vidal, P. 1987: Crustal generation of the Himalayan leucogranites. *Tectonophysics*, **134**, 39–57.

Leger, A. & Ferry, J. M. 1993: Fluid infiltration and regional metamorphism of the Waits River Formation, north-east Vermont. *Journal of Metamorphic Geology*, **11**, 3–29.

Legros, F., Cantagrel, J.-M. & Devouard, B. 2000: Pseudotachylyte (frictionite) at the base of the Arequipa volcanic landslide deposit (Peru); implications for emplacement mechanisms. *Journal of Geology*, **108**, 601–11.

Leith, C. K. & Mead, W. J. 1915: *Metamorphic Geology*. New York: Holt.

Le Maitre, R. W. 1976: The chemical variability of some common igneous rocks. *Journal of Petrology*, **17**, 589–98.

Lentz, D. R. & Goodfellow, W. D. 1993: Petrology and mass-balance constraints on the origin of quartz-augen schist associated with the Brunswick massive sulfide deposits, Bathurst, New Brunswick. *Canadian Mineralogist*, **31**, 877–903.

Li, X.-P., Zhang, L., Wei, C., Ai, Y. & Chen, J. 2007: Petrology of rodingite derived from eclogite in western Tianshan, China. *Journal of Metamorphic Geology*, **25**, 363–82.

Likhanov, I. I. & Reverdatto, V. V. 2002: Mass transfer during replacement of andalusite by kyanite in ferroaluminous metapelites in Yenesei Ridge. *Petrologiya*, **10**, 543–60.

Lindgren, W. 1925. Metasomatism. *Bulletin of the Geological Society of America*, **36**, 247–61.

Lindsay, D. H. 1983: Pyroxene thermometry. *American Mineralogist*, **68**, 477–493.

Lister, G. S. & Baldwin, S. L. 1996: Modelling the effect of arbitrary P-T-t histories on argon diffusion in minerals using the MacArgon program for the Apple Macintosh. *Tectonophysics*, **253**, 83–109.

Lister, G. S. & Williams, P. F. 1983: The partitioning of deformation in flowing rock masses. *Tectonophysics*, **92**, 1–33.

Lloyd, G. E. & Freeman, B. 1994: Dynamic recrystallization of quartz under greenschist conditions. *Journal of Structural Geology*, **16**, 867–81.

Lobato, L. M., Forman, J. M. A., Fyfe, W. S., Kerrich, R. & Barnett, R. L. 1983: Uranium enrichment in Archaean crustal basement associated with overthrusting. *Nature*, **303**, 235–7.

Lofgren, G. E. 1971: Devitrified glass fragments from Apollo 11 and 12 lunar samples. *In*: Levinson, A. A. (ed.): *Proceedings of the 2nd Lunar Science Conference; Volume 1, Mineralogy and Petrology. Geochimica et Cosmochimica Acta Supplement*, **2**, 949–55.

Loosveld, R. J. H. & Etheridge, M. A. 1990: A model for low-pressure facies metamorphism during crustal thickening. *Journal of Metamorphic Geology*, **8**, 257–67.

Lopez, S. & Castro, A. 2001: Determination of the fluid-absent solidus and supersolidus phase relationships of MORB-derived amphibolites in the range 4–14 kbar. *American Mineralogist*, **86**, 1396–403.

Losh, S. 1989: Fluid-rock interaction in an evolving ductile shear zone and across the brittle-ductile transition, central Pyrenees, France. *American Journal of Science*, **289**, 600–48.

Lowder, G. G. & Dow, J. A. S. 1978: Geology and exploration of porphyry copper deposits in North Sulawesi, Indonesia. *Economic Geology*, **73**, 628–44.

Lowe, D. R. 1999a: Petrology and sedimentology of cherts and related silicified sedimentary rocks in the Swaziland Supergroup. *In*: Lowe, D. R. & Byerly, G. R. (eds.): *Geologic Evolution of the Barberton Greenstone Belt, South Africa. Geological Society of South Africa Special Paper* **329**, 83–114.

Lowe, D. R. 1999b: Shallow-water sedimentation of accretionary lapilli-bearing strata of the Msauli Chert: evidence of explosive hydromagmatic komatiitic volcanism. *In*: Lowe, D. R. & Byerly, G. R. (eds.): *Geologic Evolution of the Barberton Greenstone Belt, South Africa. Geological Society of South Africa Special Paper* **329**, 213–32.

Lowe, D. R. & Byerly, G. R. 1986: Early Archean silicate spherules of probable impact origin, South Africa and Western Australia. *Geology*, **14**, 83–6.

Lowe, D. R., Byerly, G. R., Kyte, F. T., Shukolyukov, A., Asaro, F. & Krull, A. 2003: Spherule beds 3.47–3.24 billion years old in the Barberton Greenstone belt, South Africa: a record of large meteorite impacts and the influence on early crustal and biological evolution. *Astrobiology*, **3**, 7–48.

Lucas, S. B. & St-Onge, M. R. 1995: Syn-tectonic magmatism and the development of compositional layering, Ungava Orogen (northern Quebec, Canada), *Journal of Structural Geology*, **17**, 475–91.

Ludwig, K. R. 2003: User's manual for Isoplot/Ex, v3.0, a geochronological toolkit for Microsoft Excel. *Berkeley Geochronology Center Special Publication*, **4**, 70.

Lugmair, G. W. & Marti, K. 1978: Lunar initial $^{143}Nd/^{144}Nd$: differential evolution of the lunar crust and mantle. *Earth and Planetary Science Letters*, **39**, 349–57.

Maaløe, S. 1992: Melting and diffusion processes in closed-system migmatization. *Journal of Metamorphic Geology*, **10**, 503–16.

Maas, R., Nicholls, I. A. & Legg, C. 1997: Igneous and metamorphic enclaves in the S-type Deddick Granodiorite, Lachlan Fold Belt, SE Australia: petrographic, geochemical and Nd-Sr isotopic evidence for crustal melting and magma mixing. *Journal of Petrology*, **38**, 815–41.

Maddock, R. H. 1983: Melt origin of fault-generated pseudotachylytes demonstrated by textures. *Geology*, **11**, 105–8.

Maddock, R. H. 1986: Partial melting of lithic porphyroclasts in fault-generated pseudotachylytes. *Neues Jahrbuch für Mineralogie Abhandlungen*, **155**, 1–14.

Maddock, R. H. 1992: Effect of lithology, cataclasis and melting on the composition of fault-generated pseudotachylytes in Lewisian Gneiss, Scotland. *Tectonophysics*, **204**, 261–68.

Maddock, R. H., Grocott, J. & van Nes, M. 1987: Vesicles, amygdales and similar structures in fault-generated pseudotachylites. *Lithos*, **20**, 419–32.

Magloughlin, J. F. 1989: The nature and significance of pseudotachylite from the Nason terrane North Cascade Mountains, Washington. *Journal of Structural Geology*, **11**, 907–17.

Maltman, A. J. 1981: Primary bedding-parallel fabrics in structural geology. *Journal of the Geological Society of London*, **138**, 475–83.

Mancktelow, N. S. & Pennacchioni, G. 2003: The influence of grain boundary fluids on the microstructure of quartz-feldspar mylonites. *Journal of Structural Geology*, **26**, 47–69.

Manning, C. E. & Bohlen, S. R. 1991: The reaction titanite + kyanite = anorthite + rutile and titanite-rutile barometry in eclogites. *Contributions to Mineralogy and Petrology*, **109**, 1–9.

Marchildon, N. & Brown, M. 2001: Melt segregation in late syn-tectonic anatectic migmatites: an example from the Onawa contact aureole, Maine, USA. *Physics and Chemistry of the Earth (A)*, **26**, 225–9.

Marchildon, N. & Brown, M. 2002: Grain-scale melt distribution in two contact aureole rocks: implications for controls on melt localization and deformation. *Journal of Metamorphic Geology*, **20**, 381–96.

Marchildon, N. & Brown, M. 2003: Spatial distribution of melt-bearing structures in anatectic rocks from southern Brittany, France: implications for melt transfer at grain- to orogen-scale. *Tectonophysics*, **364**, 215–35.

Marcotte, S. B., Klepeis, K. A., Clarke, G. L., Gehrels, G. & Hollis, J. A. 2005: Intra-arc transpression in the lower crust and its relationship to magmatism in a Mesozoic magmatic arc. *Tectonophysics*, **407**, 135–63.

Mark, G. 1998: Albitite formation by selective pervasive sodic alteration of tonalite plutons in the Cloncurry district, Queensland. *Australian Journal of Earth Sciences*, **45**, 765–74.

Mark, G. & de Jong, G. 1996: Synchronous granitoid emplacement and episodic sodic-calcic alternation in the Cloncurry district: styles, timing and metallogenic significance. *James Cook University Economic Geology Research Unit (EGRU) Contribution*, **5**, 81–4.

Mark, G. & Foster, D. R. W. 2000: Magmatic-hydrothermal albite-actinolite-apatite-rich rocks from the Cloncurry district, NW Queensland. *Lithos*, **51**, 223–46.

Mark, G., Foster, D. R. W., Pollard, P. J., Williams, P. J., Tolman, J., Darvall, M. & Blake, K. L. 2004: Stable isotope evidence for magmatic fluid input during large-scale Na-Ca alteration in the Cloncurry Fe oxide Cu-Au district, NW Queensland, Australia. *Terra Nova*, **16**, 54–61.

Marlow, P. C. & Etheridge, M. A. 1977: Development of layered crenulation cleavage in mica schists of the Kanmantoo Group near Macclesfield, South Australia. *Bulletin of the Geological Society of America*, **88**, 873–82.

Marmo, B., Clarke, G. L. & Powell, R. 2002: Fractionation of bulk rock composition due to porphyroblast growth: effects on eclogite facies mineral equilibria, Pam Peninsula, New Caledonia. *Journal of Metamorphic Geology*, **20**, 151–66.

Marquer, D. & Burkhard, M. 1992: Fluid circulation, progressive deformation and mass-transfer processes in the upper crust: the example of basement-cover relationships in the External Crystalline Massifs, Switzerland. *Journal of Structural Geology*, **14**, 1047–57.

Martin, J. W. & Doherty, R. D. 1976: *Stability of Microstructure in Metallic Systems*. Cambridge: Cambridge University Press.

Martini, J. E. J. 1978: Coesite and stishovite in the Vredefort Dome, South Africa. *Nature*, **272**, 715–7.

Masch, L., Wenk, H.-R. & Preuss, E. 1985: Electron microscopy study of hyalomylonites – evidence for frictional melting in landslides. *Tectonophysics*, **115**, 131–60.

Mason, B. 1966: *Principles of Geochemistry*. Third edition. New York: Wiley.

Mason, R. 1978: *Petrology of the Metamorphic Rocks*. London: Allen and Unwin.

Massonne, H. J. & Schreyer, W. 1987: Phengite geobarometry based on the limiting assemblage with K-feldspar, phlogopite and quartz. *Contributions to Mineralogy and Petrology*, **96**, 212–24.

Masters, R. L. & Ague, J. J. 2005: Regional-scale fluid flow and element mobility in Barrow's metamorphic zones, Stonehaven, Scotland. *Contributions to Mineralogy and Petrology*, **150**, 1–18.

Mathison, C. I. 1987: Pyroxene oikocrysts in troctolitic cumulates – evidence for supercooled crystallisation and postcumulus modification. *Contributions to Mineralogy and Petrology*, **97**, 228–36.

Matland, M. L. & Kuenen, P. H. 1951: Sedimentary history of the Ventura basin, California, and the effects of turbidity currents. *Special Paper of the Society of Economic Paleontologists and Mineralogists*, **2**, 76–107.

Matthes, S. & Olesch, M. 1986: Polymetamorphic-metasomatic blackwall rocks of the Falkenberg granite contact near Erbendorf, Oberpfalz, Bavaria. *Neues Jahrbuch für Mineralogie Abhandlungen*, **153**, 325–62.

Mavrogenes, J. A., MacIntosh, I. W. & Ellis, D. J. 2001: Partial melting of the Broken Hill galena-sphalerite ore: experimental studies in the system PbS-FeS-ZnS-(Ag_2S). *Economic Geology*, **96**, 205–10.

Max, M. D. 1970: Metamorphism of Caledonian metadolerites in north-west County Mayo, Ireland. *Geological Magazine*, **107**, 539–47.

McBirney, A. R. & Hunter, R. H. 1995: The cumulate paradigm reconsidered. *Journal of Geology*, **103**, 114–22.

McBirney, A. R. & Noyes, R. M. 1979: Crystallization and layering in the Skaergaard intrusion. *Journal of Petrology*, **20**, 487–554.

McCaig, A. M. 1984: Fluid-rock interaction in some shear zones from the Pyrenees. *Journal of Metamorphic Geology*, **2**, 129–41.

McCaig, A. M. 1988: Deep fluid circulation in fault zones. *Geology*, **16**, 867–70.

McCaig, A. M. 1997: The geochemistry of volatile fluid flow in shear zones. *In*: Holness, M. B. (ed.): *Deformation-enhanced Fluid Transport in the Earth's Crust and Mantle*. London: Chapman & Hall, pp. 227–66.

McCaig, A. M. & Knipe, R. J. 1990: Mass-transport mechanisms in deforming rocks: recognition using microstructural and microchemical criteria. *Geology*, **18**, 824–27.

McCaig, A. M., Wayne, D. M., Marshall, J. D., Banks, D. & Henderson, I. 1995: Isotopic and fluid inclusion studies of fluid movement along the Gavarnie Thrust, Central Pyrenees – reaction fronts in carbonate mylonites. *American Journal of Science*, **295**, 309–43.

McCaig, A. M., Wickham, S. M. & Taylor, H. P. 1990: Deep fluid circulation in Alpine shear zones, Pyrenees, France – field and oxygen isotope studies. *Contributions to Mineralogy and Petrology*, **106**, 41–60.

McCall, G. J. H. 1971: Some ultra basic and basic igneous rock occurrences in the Archaean of Western Australia. *Special Publication of the Geological Society of Australia*, **3**, 429–42.

McCall, G. J. H. 2001: *Tektites in the Geological Record: Showers of Glass from the Sky*. London: Geological Society of London.

McCarthy, T. C. & Patiño-Douce, A. E. 1997: Experimental evidence for high-temperature felsic melts formed during basaltic intrusion of the deep crust. *Geology*, **15**, 463–6.

McClay, K. R. 1977: Pressure solution and Coble creep in rocks and minerals: a review. *Journal of the Geological Society of London*, **134**, 57–70.

McDougall, I. 1974 : The $^{40}Ar/^{39}Ar$ method of K-Ar age determination of rocks using HIFAR reactor. *Atomic Energy Australia, Australian Atomic Energy Commission*, **17(3)**, 3–12.

McKenna, L. W. & Hodges, K. V. 1988: Accuracy versus precision in locating reaction boundaries: Implications for the garnet-plagioclase-aluminium silicate-quartz geobarometer. *American Mineralogist*, **73**, 1205–8.

McKenzie, D. P. & Brune, J. P. 1972: Melting on fault planes during large earthquakes. *Geophysical Journal of the Royal Astronomical Society*, **29**, 65–78.

McLaren, A. C. 1991: *Transmission Electron Microscopy of Minerals and Rocks*. Cambridge: Cambridge University Press.

McLaren, A. C. & Pryer, L L. 2001: Microstructural investigation of the interaction and interdependence of cataclastic and plastic mechanisms in feldspar crystals deformed in the semi-brittle field. *In*: Boland, J. & Ord, A. (eds.): *Deformation Processes in the Earth's Crust. Tectonophysics*, **335**, 1–15.

McLellan, E. L. 1984: Deformational behaviour of migmatites and problems of structural analysis in migmatite terrains. *Geological Magazine*, **121**, 339–45.

McLellan, E. L. 1988: Migmatite structures in the Central Gneiss Complex, Boca de Quadra, Alaska. *Journal of Metamorphic Geology*, **6**, 517–42.

McLelland, J., Goldstein, A., Cunningham, B., Olson, C. & Orrell, S. 2002b: Structural evolution of a quartz-sillimanite vein and nodule complex in a late- to post-tectonic leucogranite, Western Adirondack Highlands, New York. *Journal of Structural Geology*, **24**, 1157–70.

McLelland, J., Morrison, J., Selleck, B., Cunningham, B., Olson, C. & Schmidt, K. 2002a: Hydrothermal alteration of late- to post-tectonic Lyon Mountain Granitic Gneiss, Adirondack Mountains, New York: Origin of quartz-sillimanite segregations, quartz-albite lithologies, and associated Kiruna-type low-Ti Fe-oxide deposits. *Journal of Metamorphic Geology*, **20**, 175–90.

McLelland, J. M. & Whitney, P. R. 1977: The origin of garnet in the anorthosite-charnockite suite of the Adirondacks. *Contributions to Mineralogy and Petrology*, **60**, 161–81.

McQueen, D. R., Kelly, W. C. & Clark, B. R. 1980: Kinematics of experimentally produced deformation bands in stibnite. *Tectonophysics*, **66**, 55–81.

Means, W. D. 1963: Mesoscopic structures and multiple deformation in the Otago schist, New Zealand. *New Zealand Journal of Geology and Geophysics*, **6**, 801–16.

Means, W. D. 1989: Synkinematic microscopy of transparent polycrystals. *Journal of Structural Geology*, **11**, 163–74.

Means, W. D. & Jessell, M. W. 1986: Accommodation migration of grain boundaries. *Tectonophysics*, **127**, 67–86.

Means, W. D. & Park, Y. 1994: New experimental approach to understanding igneous texture. *Geology*, **22**, 323–26.

Means, W. D. & Williams, P. F. 1974: Compositional differentiation in an experimentally deformed salt-mica specimen. *Geology*, **2**, 15–6.

Means, W. D. & Xia, Z. G. 1981: Deformation of crystalline materials in thin section. *Geology*, **9**, 538–43.

Mehnert, K. R. 1968: *Migmatites and the Origin of Granitic Rocks*. Amsterdam: Elsevier.

Mehnert, K. R. & Büsch, W. 1982: The initial stage of migmatite formation. *Neues Jahrbuch für Mineralogie Abhandlungen*, **145**, 211–38.

Meneilly, A. W. 1983: Development of early composite cleavage in pelites from west Donegal. *Journal of Structural Geology*, **5**, 83–97.

Merrihue, C. M. & Turner, G. 1966: Potassium argon dating by activation with fast neutrons. *Journal of Geophysical Research*, **71**, 2852–7.

Metz, P. & Trommsdorff, V. 1968: On phase equilibria in metamorphosed siliceous dolomites. *Contributions to Mineralogy and Petrology*, **18**, 305–9.

Meyer, C. & Hemley, J. J. 1967: Wall rock alteration. *In*: Barnes, H. L. (ed.): *Geochemistry of Hydrothermal Ore Deposits*. Second edition. New York: Wiley, pp. 166–236.

Meyer, C., Shea, E. P. & Goddard, C. C. 1968: Ore deposits at Butte, Montana. *In*: Ridge, J. D. (ed.): *Ore Deposits of the United States, 1933–1967 (Graton-Sale Volume). American institute of Mining, Metallurgical and Petroleum Engineers*, **2**, 1373–416.

Mikucki, E. J. & Ridley, J. R. 1993: The hydrothermal fluid of Archaean lode-gold deposits at different metamorphic grades: compositional constraints from ore and wallrock alteration assemblages. *Mineralium Deposita*, **28**, 469–81.

Miller, J. A. & Cartwright, I. 1997: Early meteoric fluid flow in high-grade, low-^{18}O gneisses from the Mallee Bore area, northern Harts Range, central Australia. *Journal of the Geological Society of London*, **154**, 839–48.

Miller, J. A. & Cartwright, I. 2000: Distinguishing between sea-floor alteration and fluid flow during subduction using stable isotope geochemistry: examples from Tethyan Ophiolites in the Western Alps. *Journal of Metamorphic Geology*, **18**, 467–82.

Miller, J. A. & Cartwright, I. 2006: Albite vein formation during exhumation of high-pressure terranes: a case study from alpine Corsica. *Journal of Metamorphic Geology*, **24**, 409–28.

Miller, R. B. & Paterson, S. R. 1999: In defense of magmatic diapirs. *Journal of Structural Geology*, **21**, 1161–73.

Miller, R. B. & Paterson, S. R. 2001: Construction of mid-crustal sheeted plutons: examples from the North Cascades, Washington. *Bulletin of the Geological Society of America*, **113**, 1423–42.

Miller, T., Baumgartner, L. P., Foster, C. T. & Venneman, T. W. 2004: Metastable prograde mineral reactions in contact aureoles. *Geology*, **32**, 821–24.

Milodowski, A. E. & Zalasiewicz, J. A. 1991: The origin and sedimentary, diagenetic and metamorphic evolution of chlorite-mica stacks in Llandovery sediments of central Wales, U.K. *Geological Magazine*, **128**, 63–278.

Milord, I. & Sawyer, E. W. 2003: Schlieren formation in diatexite migmatite: examples from the St Malo migmatite terrane, France. *Journal of Metamorphic Geology*, **21**, 347–62.

Milord, I., Sawyer, E. W. & Brown, M. 2001: Formation of diatexite migmatite and granite magma during anatexis of semi-pelitic metasedimentary rocks: an example from St Malo, France. *Journal of Petrology*, **42**, 487–505.

Misch, P. 1969: Paracrystalline microboudinage of zoned grains and other criteria for synkinematic growth of metamorphic minerals. *American Journal of Science*, **267**, 43–63.

Misch, P. 1970: Paracrystalline microboudinage in a metamorphic reaction sequence. *Bulletin of the Geological Society of America*, **81**, 2483–6.

Miura, Y., Uyedo, Y. & Kedves, M. 2001: Formation of Fe-Ni particles by impact processes. *Proceedings of Lunar and Planetary Science Conference XXXII*, Abstract **2140**.

Miyashiro, A. 1961: Evolution of metamorphic belts. *Journal of Petrology*, **2**, 277–311.

Mogk, D. W. 1990: A model for the granulite-migmatite association in the Archean basement of southwestern Montana, *In*: Vielzeuf, D. & Vidal, P. (eds.): *Granulites and Crustal Evolution*. Amsterdam: Kluwer, pp. 133–55.

Montel, J.-M., Foret, S., Veschambre, M., Nicollet, C. & Provost, A. 1996: Electron microprobe dating of monazite. *Chemical Geology*, **131**, 37–53.

Montel, J.-M., Kornprobst, J. & Vielzeuf, D. 2000: Preservation of old U-Th-Pb ages in shielded monazite: example from the Beni Bousera Hercynian kinzigites (Morocco). *Journal of Metamorphic Geology*, **18**, 335–34.

Moore, A. C. 1973: Studies of igneous and tectonic textures and layering in the rocks of the Gosse Pile intrusion, central Australia. *Journal of Petrology*, **14**, 49–79.

Moore, J. C. & Geigle, J. E. 1974: Slaty cleavage: incipient occurrences in the deep sea. *Science*, **183**, 509–10.

Moore, J. N. & Kerrick, D. M. 1976: Equilibria in siliceous dolomites of the Alta aureole, Utah. *American Journal of Science*, **276**, 502–24.

Morimoto, N., Fabries, J., Ferguson, A. K., Ginnzburg, I. V., Ross, M., Seifert, F. A., Zussmann, J., Aoki, K. & Gottardi, G. 1988: Nomenclature of pyroxenes. *Mineralogical Magazine*, **52**, 535–50.

Mørk, M. B. 1985: Incomplete high P-T metamorphic transitions within the Kvamsøy pyroxenite complex, west Norway: a case study of disequilibrium. *Journal of Metamorphic Geology*, **3**, 245–64.

Morris, R. C. 1985: Genesis of iron ore in banded iron-formation by supergene and supergene-metamorphic processes – a conceptual model. *In*: Wolf, K. H. (ed.): *Handbook of Strata-Bound and Stratiform Ore Deposits*. Volume 13. Amsterdam: Elsevier, pp. 73–235.

Morrison, J. 1994: Meteoric water-rock interaction in the lower plate of the Whipple Mountain metamorphic core complex, California. *Journal of Metamorphic Geology*, **12**, 827–40.

Morritt, R. F. C., Powell, C. M. & Vernon, R. H. 1982: Bedding-plane foliation. *In*: Borradaile, G. J., Bayly, M. B. & Powell, C. M. (eds.): *Atlas of Deformational and Metamorphic Rock Fabrics*. Berlin: Springer, pp. 46–9.

Mosher, S. 1980: Pressure solution deformation of conglomerates in shear zones, Narragansett basin, Rhode Island. *Journal of Structural Geology*, **2**, 219–26.

Moss, B. E., Haskin, L. A. & Dymek, R. F. 1995: Redetermination and reevaluation of compositional variations in metamorphosed sediments of the Littleton Formation, New Hampshire. *American Journal of Science*, **295**, 988–1019.

Moss, B. E., Haskin, L. A. & Dymek, R. F. 1996: Compositional variations in metamorphosed sediments of the Littleton Formation, New Hampshire, and the Carrabassett Formation, Maine, at sub-hand specimen, outcrop, and regional scales. *American Journal of Science*, **296**, 473–506.

Mottl, M. J. 1983: Metabasalts, axial hot springs, and the structure of hydrothermal systems at mid-ocean ridges. *Bulletin of the Geological Society of America*, **94**, 161–80.

Mueller, R. F. 1967: Mobility of the elements in metamorphism. *Journal of Geology*, **75**, 565–82.

Muffler, L. J. P. & White, D. E. 1969: Active metamorphism of Upper Cenozoic sediments in the Salton Sea geothermal field and the Salton Trough, southeastern California. *Bulletin of the Geological Society of America*, **80**, 157–82.

Murrell, S. A. F. 1985: Aspects of relationships between deformation and metamorphism that causes evolution of water. *In*: Thompson, A. B. & Rubie, D. C. (eds.): *Metamorphic Reactions. Kinetics, Textures, and Deformation. Advances in Physical Geochemistry*, **4**. New York: Springer, pp. 211–41.

Myers, J. D., Sinha, A. K. & Marsh, B. D. 1984: Assimilation of crustal material by basaltic magma: Strontium isotopic and trace element data for the Edgecumbe Volcanic Field, SE Alaska. *Journal of Petrology*, **25**, 1–26.

Myers, J. S. 1976: The early Precambrian gneiss complex of Greenland. *In*: Windley, B. F. (ed.): *The Early History of the Earth*. London: Wiley, pp. 165–76.

Myers, J. S. 1978: Formation of banded gneisses by deformation of igneous rocks. *Precambrian Research*, **6**, 43–64.

Myers, J. S. 1981: The Fiskenaesset anorthosite complex – a stratigraphic key to the tectonic evolution of the West Greenland gneiss complex 3000 – 2800 m.y. ago. *Special Publication of the Geological Society of Australia*, **7**, 351–60.

Nabelek, P. I. 1991: Stable isotope monitors. *In*: Kerrick, D. M. (ed.): *Contact Metamorphism. Mineralogical Society of America, Reviews in Mineralogy*, **26**, 395–436.

Nabelek, P. I. 2002: Calc-silicate reactions and bedding-controlled isotopic exchange in the Notch Peak aureole, Utah: implications for differential fluid fluxes with metamorphic grade. *Journal of Metamorphic Geology*, **20**, 429–40.

Nabelek, P. I. & Bartlett, C. D. 2000: Fertility of metapelites and metagraywackes during leucogranite generation: an example from the Black Hills, USA. *Transactions of the Royal Society of Edinburgh: Earth Sciences*, **91**, 1–14.

Nabelek, P. I. & Liu, M. 2004: Petrologic and thermal constraints on the origin of leucogranites in collisional orogens. *Transactions of the Royal Society of Edinburgh: Earth Sciences*, **95**, 73–85.

Nabelek, P. I., Russ-Nabelek, C. & Denison, J. R. 1992: The generation and crystallization conditions of the Proterozoic Harney Peak leucogranite, Black Hills, South Dakota, USA: petrologic and geochemical constraints. *Contributions to Mineralogy and Petrology*, **110**, 173–91.

Naggar, M. H. & Atherton, M. P. 1970: The composition and metamorphic history of some aluminium silicate-bearing rocks from the aureoles of the Donegal granite. *Journal of Petrology*, **11**, 549–89.

Naha, K. & Majumdar, A. 1971: Reinterpretation of the Aravalli basal conglomerate at Morchana, Udaipur district, Rajasthan, western India. *Geological Magazine*, **108**, 111–4.

Naldrett, A. J. & Gasparrini, E. L. 1971: Archaean nickel sulphide deposits in Canada: their classification, geological setting and genesis with some suggestions as to exploration. *Special Publication of the Geological Society of Australia*, **3**, 201–26.

Nell, J. 1985: The Bushveld metamorphic aureole in the Potgietersrus area: evidence for a two-stage metamorphic event. *Economic Geology*, **80**, 1129–52.

Nesbitt, R. W. 1971: Skeletal forms in the ultramafic rocks of the Yilgarn Block, Western Australia: evidence for an Archaean ultramafic liquid. *Special Publication of the Geological Society of Australia*, **3**, 331–47.

Newhouse, W. H. 1942: *Ore Deposits as Related to Structural Features*. Princeton: Princeton University Press.

Newton, R. C. 1990: Fluids and melting in the Archaean deep crust of southern India. *In*: Ashworth, J. R. & Brown, M. (eds.): *High-temperature Metamorphism and Crustal Anatexis*. London: Unwin Hyman, pp. 149–79.

Newton, R. C. 1992: Charnockite alteration evidence for CO_2 infiltration in granulite facies metamorphism. *Journal of Metamorphic Geology*, **10**, 383–400.

Newton, R. C. & Perkins, D. 1982: Thermodynamic calibration of geobarometers based on the assemblages garnet-plagioclase-orthopyroxene-clinopyroxene-quartz. *American Mineralogist*, **67**, 203–22.

Nicholson, R. 1966: Metamorphic differentiation in crenulated schists. *Nature*, **209**, 68–9.

Nicolas, A. & Poirier, J.-P. 1976: *Crystalline Plasticity and Solid State Flow in Metamorphic Rocks*. New York: Wiley-Interscience.

Niggli, P. 1954: *Rocks and Mineral Deposits* (translated by R. L. Parker). San Francisco: Freeman.

Norris, R. J. & Henley, R. W. 1976: Dewatering of a metamorphic pile. *Geology*, **4**, 333–6.

Norton, D. L. 1988: Metasomatism and permeability. *American Journal of Science*, **288**, 604–18.

Obata, M., Yoshimura, Y., Nagakawa, K., Odawara, S. & Osanai, Y. 1994: Crustal anatexis and melt migrations in the Higo metamorphic terrane, west-central Kyushu, Kumamoto, Japan. *Lithos*, **32**, 135–47.

O'Brien, P. J. & Rötzler, J. 2003: High-pressure granulites: formation, recovery of peak conditions and implications for tectonics. *Journal of Metamorphic Geology*, **21**, 3–20.

Oftedahl, C. 1967: A manganiferous chert in the Caledonian greenstone of Trondheim. *Det Kongelige Norske Videnskabers Selskabs Forhandlingar*, **40**, 48–54.

O'Hanley, D. S. 1996: *Serpentinites. Records of Tectonic and Petrological History*. Oxford: Oxford University Press.

O'Hara, K. 1988: Fluid flow and volume loss during mylonitization: an origin for phyllonite in an overthrust setting, North Carolina, U.S.A. *Tectonophysics*, **156**, 21–36.

O'Keefe, J. A. 1976: *Tektites and Their Origin*. Amsterdam: Elsevier.

O'Neill, H. St. C. & Wood, B. J. 1979: An experimental study of Fe-Mg partitioning between garnet and olivine and its calibration as a geothermometer. *Contributions to Mineralogy and Petrology*, **70**, 59–70.

Oliver, N. H. S. 1995: Hydrothermal history of the Mary Kathleen Fold Belt, Mt Isa Block, Queensland. *Australian Journal of Earth Sciences*, **42**, 267–80.

Oliver, N. H. S. 1996: Review and classification of structural controls on fluid flow during regional metamorphism. *Journal of Metamorphic Geology*, **14**, 477–92.

Oliver, N. H. S. 2001: Linking of regional and hydrothermal systems in the mid-crust by shearing and faulting. *In*: Boland, J. & Ord, A. (eds.): *Deformation Processes in the Earth's Crust. Tectonophysics*, **335**, 147–61.

Oliver, N. H. S., Cleverley, J. S., Mark, G., Pollard, P. J., Fu, B., Marshall, L. J., Rubenach, M. J., Williams, P. J. & Baker, T. 2004: Modeling the role of sodic alteration in the genesis of iron oxide-copper-gold deposits, Mount Isa block, Australia. *Economic Geology*, **99**, 1145–76.

Oliver, N. H. S., Valenta, R. K. & Wall, V. J. 1990: The effect of heterogeneous stress and strain on metamorphic fluid flow, Mary Kathleen, Australia, and a model for large-scale fluid circulation. *Journal of Metamorphic Geology*, **8**, 311–31.

Oliver, N. H. S. & Wall, V. J. 1987: Metamorphic plumbing system in Proterozoic calc-silicates, Queensland, Australia. *Geology*, **15**, 793–6.

Olsen, S. N. 1982: Open- and closed-system migmatites in the Front Range, Colorado. *American Journal of Science*, **282**, 1596–622.

Olsen, S. N. 1983: A quantitative approach to local mass balance in migmatites. *In*: Atherton, M. P. & Gribble, C. D. (eds.): *Migmatites, Melting and Metamorphism*. Nantwich: Shiva, pp. 201–33.

Olsen, S. N. 1985: Mass balance in migmatites. *In*: Ashworth, J. R. (ed.): *Migmatites*. Glasgow: Blackie, pp. 145–79.

Orville, P. M. 1969: A model for metamorphic differentiation origin of thin-layered amphibolites. *American Journal of Science*, **267**, 64–86.

Owada, M., Osanai, Y., Shimura, T., Toyoshima, T. & Katsui, Y. 2003: Crustal section and anatexis of lower crust due to mantle flux in the Hidaka metamorphic belt, Hokkaido, Japan. Hutton Symposium V Field Guidebook. *Geological Survey of Japan Interim Report*, **28**, 81–102.

Packham, G. H. & Crook, K. A. W. 1960: The principle of diagenetic facies and some of its implications. *Journal of Geology*, **68**, 392–407.

Page, R. W. & Laing, W. P. 1992: Felsic metavolcanic rocks related to the broken Hill Pb-Zn-Ag orebody, Australia: geology, depositional age and timing of high-grade metamorphism. *Economic Geology*, **87**, 2138–68.

Pan, Y. 1997: Zircon- and monazite-forming metamorphic reactions at Manitouwadge, Ontario. *Canadian Mineralogist*, **35**, 105–18.

Paria, P., Bhattachatya, A. & Sen, A. 1988: The reaction garnet + clinopyroxene + quartz = 2orthopyroxene + anorthite: a potential geobarometer for granulites. *Contributions to Mineralogy and Petrology*, **99**, 126–33.

Park, R. G. 1969: Structural correlation in metamorphic belts. *Tectonophysics*, **7**, 323–38.

Park, Y. & Means, W. D. 1996: Direct observation of deformation processes in crystal mushes. *Journal of Structural Geology*, **18**, 847–58.

Parr, J. M., Stevens, B. P. J., Carr, G. R. & Page, R. W. 2004: Subseafloor origin for the Broken Hill Pb-Zn-Ag mineralization, New South Wales, Australia. *Geology*, **32**, 589–92.

Parrish, R. R. 1990: U–Pb dating of monazite and its application to geological problems. *Canadian Journal of Earth Science*, **27**, 1431–50.

Parsons, I. (ed.) 1987: *Origins of Igneous Layering*. Dordrecht: Reidel.

Parsons, I. & Becker, S. M. 1987. Layering, compaction and post-magmatic processes in the Klokken intrusion. *In*: Parsons, I. (ed.): *Origins of Igneous Layering*. Dordrecht: Reidel, pp. 29–92.

Parsons, I. & Butterfield, A. W. 1981: Sedimentary features of the Nunarssuit and Klokken syenites, S. Greenland. *Journal of the Geological Society of London* **138**, 289–306.

Passchier, C. W., Myers, J. S. & Kröner, A. 1990: *Field Geology of High-grade Gneiss Terrains*. Berlin: Springer.

Passchier, C. W. & Trouw, R. A. J. 1996: *Microtectonics*. Berlin: Springer.

Patchett, P. J. 1980: Thermal effects of basalt on continental crust and crustal contamination. *Nature*, **283**, 559–61.

Paterson, M. S. 2001: A granular flow theory for the deformation of partially molten rock. *In*: Boland, J. & Ord, A. (eds.): *Deformation Processes in the Earth's Crust. Tectonophysics*, **335**, 51–61.

Paterson, S. R. & Miller, R. B. 1998: Mid-crustal magmatic sheets in the Cascade Mountains, Washington: implications for magma ascent. *Journal of Structural Geology*, **20**, 1345–63.

Paterson, S. R. & Tobisch, O. T. 1983: Pre-lithification structures, deformation mechanisms, and fabric ellipsoids in slumped turbidites from the Pigeon Point Formation, California. *Tectonophysics*, **222**, 135–49.

Paterson, S. R. & Tobisch, O. T. 1992: Rates of processes in magmatic arcs: implications for the timing and nature of pluton emplacement and wall rock deformation. *Journal of Structural Geology*, **14**, 291–300.

Paterson, S. R., Tobisch, O. T. & Morand, V. J. 1990: The influence of large ductile shear zones on the emplacement and deformation of the Wyangala Batholith, SE Australia. *Journal of Structural Geology*, **12**, 639–50.

Paterson, S. R. & Vernon, R. H. 1995: Bursting the bubble of ballooning plutons: a return to nested diapirs emplaced by multiple processes. *Bulletin of the Geological Society of America*, **107**, 1356–80.

Paterson, S. R. & Vernon, R. H. 2001: Inclusion trail patterns in porphyroblasts from the Foothills Terrane, California: a record of orogenesis or local strain heterogeneity? *Journal of Metamorphic Geology*, **19**, 351–72.

Paterson, S. R., Vernon, R. H. & Tobisch, O. T. 1989: A review of criteria for the identification of magmatic and tectonic foliations in granitoids. *Journal of Structural Geology*, **11**, 349–63.

Paterson, S. R., Yu, H. & Oertel, G. 1985: Primary and tectonic fabric intensities in mudrocks. *Tectonophysics*, **247**, 105–19.

Patiño-Douce, A. E. 1995: Experimental generation of hybrid silicic melts by reaction of high-Al basalt with metamorphic rocks. *Journal of Geophysical Research*, **100**, 15623–39.

Patiño-Douce, A. E. 1996: Effects of pressure and H_2O on the compositions of primary crustal melts. *Transactions of the Royal Society of Edinburgh: Earth Sciences*, **87**, 11–21.

Patiño-Douce, A. E. 1999: What do experiments tell us about the relative contributions of crust and mantle to the origin of granitic magmas? *In*: Castro, A., Fernández, C. & Vigneresse, J.-L. (eds.): *Understanding Granites; Integrating New and Classical Techniques. Geological Society of London Special Publication*, **168**, 55–75.

Patiño-Douce, A. E. & Johnston, A. D. 1991: Phase equilibria and melt productivity in the pelitic system: implications for the origin of peraluminous granitoids and aluminous granulites. *Contributions to Mineralogy and Petrology*, **107**, 202–18.

Pattison, D. R. M. 1991: Infiltration-driven dehydration and anatexis in granulite facies metagabbro, Grenville province, Ontario, Canada. *Journal of Metamorphic Geology*, **9**, 315–32.

Pattison, D. R. M. & Bégin, N. J. 1994: Zoning patterns in orthopyroxene and garnet in granulites: implications for geothermometry. *Journal of Metamorphic Geology*, **12**, 387–410.

Pattison, D. R. M. & Harte, B. 1985: A petrogenetic grid for pelites in the Ballachulish and other Scottish thermal aureoles. *Journal of the Geological Society of London*, **142**, 7–28.

Pattison, D. R. M. & Harte, B. 1988: Evolution of structurally contrasting anatectic migmatites in the 3-kbar Ballachulish aureole, Scotland. *Journal of Metamorphic Geology*, **6**, 475–94.

Pattison, D. R. M. & Harte, B., 1991. Petrography and mineral chemistry of metapelites. *In*: Voll, G., Topel, J., Pattison, D. R. M. & Seifert, F. (eds.): *Equilibrium and Kinetics in Contact Metamorphism: The Ballachulish Igneous Complex and its Thermal Aureole*. Heidelberg: Springer, pp. 135–79.

Pattison, D. R. M. & Newton, R. C. 1989: Revised experimental calibration of the garnet-clinopyroxene Fe-Mg exchange thermometer. *Contributions to Mineralogy and Petrology*, **101**, 87–103.

Pattison, D. R. M. & Tracy, R. J. 1991: Phase equilibria and thermobarometry of metapelites. *In*: Kerrick, D. M. (ed.): *Contact Metamorphism. Mineralogical Society of America, Reviews in Mineralogy*, **26**, 105–206.

Pearce, T. H. & Birkett, T. C. 1974: Archean metavolcanic rocks from Thackeray Township, Ontario. *Canadian Mineralogist*, **12**, 509–19.

Pepper, M. A. & Ashley, P. M. 1998: Volcanic textures in quartzofeldspathic gneiss of the Willyama Supergroup, Olary Domain, South Australia. *Australian Journal of Earth Sciences*, **45**, 971–8.

Perchuk, A. L., Burchard, M., Maresch, M. V. & Schertl, H.-P. 2005: Fluid-mediated modification of garnet interiors under ultrahigh-pressure conditions. *Terra Nova*, **17**, 545–53.

Percival, J. A. 1991: Granulite-facies metamorphism and crustal magmatism in the Ashuanipi complex, Quebec-Labrador, Canada. *Journal of Petrology*, **32**, 1261–97.

Perfit, M. R., Heezen, B. C., Rawson, M. & Donnelly, T. W. 1980: Chemistry, origin and tectonic significance of metamorphic rocks from the Puerto Rico Trench. *Marine Geology*, **34**, 125–56.

Perring, C. S., Pollard, P. J., Dong, G., Nunn, A. J. & Blake, K. L. 2000: The Lightning Creek sill complex, Cloncurry district, northwest Queensland: A source of fluids for the Fe-oxide–Cu–Au mineralization and sodic-calcic alteration. *Economic Geology*, **95**, 1067–89.

Peter, J. M. 2003: Ancient iron formations: their genesis and use in the exploration for stratiform base metal sulphide deposits, with examples from the Bathurst Mining Camp. *In*: Lentz, D. R. (ed.): *Geochemistry of Sediments and Sedimentary Rocks: Historical to Research Perspectives. Geological Association of Canada GEOtext*, **4**, 145–76.

Petford, N., Cruden, A. R., McCaffrey, K. J. W. & Vigneresse, J.-L. 2000: Granite magma formation, transport and emplacement in the Earth's crust. *Nature*, **408**, 669–73.

Petford, N., Kerr, R. C. & Lister, J. R. 1993: Dike transport of granitic magmas. *Geology*, **21**, 843–5.

Petford, N. & Koenders, M. A. 1998: Self-organization and fracture connectivity in rapidly heated continental crust. *Journal of Structural Geology*, **20**, 1425–34.

Pettijohn, F. J. 1949: *Sedimentary Rocks*. New York: Harper.

Pettijohn, F. J. & Potter, P. E. 1964: *Atlas and Glossary of Primary Sedimentary Structures*. New York: Springer.

Philippot, P. & Selverstone, J. 1991: Trace-element–rich brines in eclogitic veins: implications for fluid composition and transport during subduction. *Contributions to Mineralogy and Petrology*, **106**, 417–30.

Phillips, G. N. 1980: Water activity changes across an amphibolite-granulite facies transition, Broken Hill, Australia. *Contributions to Mineralogy and Petrology*, **75**, 377–86.

Phillips, G. N. & Groves, D. I. 1983: The nature of Archaean gold-bearing fluids as deduced from gold deposits of Western Australia. *Journal of the Geological Society of Australia*, **30**, 25–39.

Phillips, G. N. & Nooy, D. 1988: High grade metamorophic processes which influence Archaean gold deposits, with particular reference to Big Bell, Australia. *Journal of Metamorphic Geology*, **6**, 95–114.

Phillips, G. N., Wall, V. J. & Clemens, J. D. 1981: Petrology of the Strathbogie Batholith: a cordierite-bearing granite. *Canadian Mineralogist*, **19**, 47–63.

Philpotts, A. R. 1964: Origin of pseudotachylites. *American Journal of Science*, **262**, 1008–35.

Pitcher, W. S. & Berger, A. R. 1972: *The Geology of Donegal: A Study of Granite Emplacement and Unroofing*. New York: Wiley-Interscience.

Plessman, W. 1964: Gesteinslösung, ein Hauptfaktor beim Schieferungsprozess. *Geologische Mitteilungen*, **4**, 69–82.

Plimer, I. R. 1975: The geochemistry of amphibolite retrogression at Broken Hill, NSW. *Neues Jahrbuch für Mineralogie Monatshefte*, **10**, 471–81.

Plimer, I. R. 1977: The origin of the albite-rich rocks enclosing the cobaltian pyrite deposit at Thackaringa, NSW, Australia. *Mineralium Deposita*, **12**, 175–87.

Plimer, I. R. 1984: The mineralogical history of the Broken Hill Lode. *Australian Journal of Earth Sciences*, **31**, 379–402.

Poirier, J.-P. 1980: Shear localization and shear instability in materials in the ductile field. *Journal of Structural Geology*, **2**, 135–42.

Poirier, J.-P. & Guillopé, M. 1979: Deformation induced recrystallization of minerals. *Bulletin de Minéralogie*, **102**, 67–74.

Poitrasson, F., Chenery, S. & Bland, D. J. 1996. Contrasted monazite hydrothermal alteration mechanisms and their geochemical implications. *Earth and Planetary Science Letters*, **145**, 79–96.

Powell, C. M. 1969: Intrusive sandstone dykes in the Siamo Slate near Negaunee, Michigan. *Bulletin of the Geological Society of America*, **80**, 2585–94.

Powell, C. M. 1982: Overgrowths and mica beards on rounded quartz grains enclosed by cleavage folia. *In*: Borradaile, G. J., Bayly, M. B. & Powell, C. M. (eds.): *Atlas of Deformational and Metamorphic Rock Fabrics*. New York: Springer, pp. 300–1.

Powell, C. M. & Vernon, R. H. 1979: Growth and rotation history of garnet porphyroblasts with inclusion spirals in a Karakoram schist. *Tectonophysics*, **54**, 25–43.

Powell, R. 1978: *Equilibrium Thermodynamics in Petrology. An Introduction*. London: Harper & Row.

Powell, R. 1983: Processes in granulite-facies metamorphism. *In*: Atherton, M. P. & Gribble, C. D. (eds.): *Migmatites, Melting and Metamorphism*. Nantwich: Shiva, pp. 127–39.

Powell, R. 1985: Regression diagnostics and robust regression in geothermometer/geobarometer calibration: the garnet-clinopyroxene geothermometer revisited. *Journal of Metamorphic Geology*, **3**, 327–42.

Powell, R. & Downes, J. 1990: Garnet porphyroblast-bearing leucosomes in metapelites: mechanisms and an example from Broken Hill, Australia. In: Ashworth, J. R. & Brown, M. (eds.): *High Temperature Metamorphism and Crustal Anatexis*. London: Unwin Hyman, pp. 105–23.

Powell, R., Guiraud, M. & White, R. W. 2005: Truth and beauty in metamorphic phase equilibria: conjugate variables and phase diagrams. *Canadian Mineralogist*, **43**, 21–3.

Powell, R. & Holland, T. J. B. 1988: An internally consistent thermodynamic dataset with uncertainties and correlations: 3. Application methods, worked examples and a computer program. *Journal of Metamorphic Geology*, **6**, 173–204.

Powell, R. & Holland, T. J. B. 1993: On the formulation of simple mixing models for complex phases. *American Mineralogist*, **78**, 1174–80.

Powell, R., Holland, T. & Worley, B. 1998: Calculating phase diagrams involving solid solutions via non-linear equations, with examples using THERMOCALC. *Journal of Metamorphic Geology*, **16**, 577–88.

Powell, R., Will, T. M. & Phillips, G. N. 1991: Metamorphism in Archaean greenstone belts: calculated fluid compositions and implications for gold mineralization. *Journal of Metamorphic Geology*, **9**, 141–50.

Pressley, R. A. & Brown, M. 1999: The Phillips Pluton, Maine, USA: evidence of heterogeneous crustal sources, and implications for the origin of peraluminous granitoids and aluminous granulites. *Contributions to Mineralogy and Petrology*, **107**, 202–18.

Prestvik, T. 1980: The Caledonian ophiolite complex of Leka, north central Norway. Proceedings of the International Ophiolite Symposium. Cyprus Geological Survey Department, pp. 535–66.

Prior, D. J. 1987: Syntectonic porphyroblastic growth in phyllites: textures and processes. *Journal of Metamorphic Geology*, **5**, 27–39.

Proyer, A. 2003: Metamorphism of pelites in NKFMASH – a new petrogenetic grid with implications for the preservation of high-pressure mineral assemblages during exhumation. *Journal of Metamorphic Geology*, **21**, 493–509.

Putnis, A. 1992: *Introduction to Mineral Sciences*. Cambridge: Cambridge University Press.

Putnis, A. 2002: Mineral replacement reactions: from macroscopic observations to microscopic mechanisms. *Mineralogical Magazine*, **66**, 689–708.

Putnis, C. V., Tsukamoto, K. & Nishimura, Y. 2005: Direct observations of pseudomorphism: compositional and textural evolution at a fluid-solid interface. *American Mineralogist*, **90**, 1909–12.

Pyke, D. R., Naldrett, A. J. & Eckstrand, R. O. 1973: Archaean ultramafic flows in Munro Township, Ontario. *Bulletin of the Geological Society of America*, **84**, 955–78.

Pyle, J. M. & Spear, F. S. 1999: Yttrium zoning in garnet: coupling of major and accessory phases during metamorphic reactions. *Geological Materials Research*, **1**, 1–28.

Pyle, J. M. & Spear, F. S. 2003: Four generations of accessory-phase growth in low-pressure migmatites from SW New Hampshire. *American Mineralogist*, **88**, 338–51.

Pyle, J. M., Spear, F. S., Rudnick, R. L. & McDonough, W. F. 2001: Monazite–xenotime–garnet equilibrium in metapelites and a new monazite–garnet thermometer. *Journal of Petrology*, **42**, 2083–107.

Pyle, J. M., Spear, F. S., Wark, D. A., Daniel, C. G. & Storm, L. C. 2005: Contributions to the precision and accuracy of monazite electron microprobe ages. *American Mineralogist*, **90**, 547–77.

Raam, A. 1968: Petrology and diagenesis of Broughton Sandstone (Permian), Kiama district, New South Wales. *Journal of Sedimentary Petrology*, **38**, 319–31.

Råheim, A. & Green, D. H. 1974: Experimental determination of the temperature and pressure dependence of the Fe-Mg partitioning coefficient for coexisting garnet and clinopyroxene. *Contributions to Mineralogy and Petrology*, **48**, 179–203.

Raith, M. & Srikantappa, C. 1993: Arrested charnockite formation at Kottavattam, southern India. *Journal of Metamorphic Geology*, **11**, 815–32.

Ramsay, C. R. & Davidson, L. R. 1970: The origin of scapolite in the regionally metamorphosed rocks of Mary Kathleen, Queensland, Australia. *Contributions to Mineralogy and Petrology*, **25**, 41–51.

Ramsay, J. G. 1956: The supposed Moinian basal conglomerate at Glen Strathfarrar, Inverness-shire. *Geological Magazine*, **93**, 32–40.

Ramsay, J. G. 1962: The geometry and mechanics of formation of 'similar' type folds. *Journal of Geology*, **70**, 309–27.

Ramsay, J. G. 1967: *Folding and Fracturing of Rocks*. New York: McGraw-Hill.

Ramsay, J. G. & Graham, R. H. 1970: Strain variation in shear belts. *Canadian Journal of Earth Sciences*, **7**, 786–813.

Ramsay, J. G. & Huber, M. I. 1987: *The Techniques of Modern Structural Geology. Volume 2: Folds and Fractures*. London: Academic Press.

Ransom, B., Byerly, G. R. & Lowe, D. R. 1999: Subaqueous to subaerial Archean ultramafic phreatomagmatic volcanism, Kromberg Formation, Barberton Greenstone Belt, South Africa. *In*: Lowe, D. R. & Byerly, G. R. (eds.): *Geologic Evolution of the Barberton Greenstone Belt, South Africa. Geological Society of South Africa Special Paper*, **329**, 151–66.

Rasmussen, B., Fletcher, I. R. & Sheppard, S. 2003: Isotopic dating of the migration of a low-grade metamorphic front during orogenesis. *Geology*, **33**, 773–6.

Rasmussen, B. & Koeberl, C. 2004: Iridium anomalies and shocked quartz in a Late Archean spherule layer from the Pilbara craton: new evidence for a major asteroid impact at 2.63 Ga. *Geology*, **32**, 1029–32.

Read, C. M. & Cartwright, I. 2000: Meteoric fluid in filtration in the middle crust during shearing: examples from the Arunta Inlier, central Australia. *Journal of Geochemical Exploration*, **69–70**, 333–7.

Read, P. B. & Eisbacher, G. H. 1974: Regional zeolite alteration of the Sustut Group, north-central British Columbia. *Canadian Mineralogist*, **12**, 527–41.

Redden, J. A., Peterman, Z. E., Zartman, R. E. & Dewitt, E. 1990: U-Th-Pb zircon and monazite ages and preliminary interpretation of the tectonic development of Precambrian rocks in the Black Hills. *Geological Society of Canada Special Paper*, **37**, 229–51.

Reed, J. C. & Morgan, B. A. 1971: Chemical alteration and spilitization of the Catoctin greenstones, Shenandoah National Park, Virginia. *Journal of Geology*, **79**, 526–48.

Reid, J. B., Evans, O. C. & Fates, D. G. 1983: Magma mixing in granitic rocks of the central Sierra Nevada, California. *Earth and Planetary Science Letters*, **66**, 243–61.

Reimold, W. U. & Gibson, R. L. 1996: Geology and evolution of the Vredefort impact structure, South Africa. *Journal of African Earth Sciences*, **23**, 125–62.

Reimold, W. U. & Gibson, R. L. (eds.) 2006: *Processes of the Early Earth. Geological Society of America Special Paper*, **405**.

Reinhardt, J. & Rubenach, M. J. 1989: Growth of porphyroblasts relative to progressive deformation, temperature increase and time during prograde metamorphism. *Tectonophysics*, **158**, 141–61.

Reuter, A. & Dallmeyer, R. D. 1989: K-Ar and $^{40}Ar/^{39}Ar$ dating of cleavage formed during very low-grade metamorphism: a review. *In*: Daly, J. S., Cliff, R. A. & Yardley, B. W. D. (eds.): *Evolution of Metamorphic Belts. Geological Society of London Special Publications*, **43**, 161–71.

Reverdatto, V. V. 1970: Pyrometamorphism of limestones and the temperature of basaltic magmas. *Lithos*, **3**, 135–43.

Rice, J. M. 1977: Progressive metamorphism of impure dolomitic limestone in the Marysville aureole, Montana. *American Journal of Science*, **277**, 1–24.

Rice, J. M. & Ferry, J. M. 1982: Buffering, infiltration, and the control of intensive variables during metamorphism. *Reviews in Mineralogy*, **10**, 263–326.

Richards, S. M. 1966: The banded iron formations at Broken Hill, Australia, and their relationship to the lead-zinc orebodies. Parts I and II. *Economic Geology*, **61**, 72–96, 257–74.

Richards, S. W. & Collins, W. J. 2002: The Cooma Metamorphic Complex, a low-*P* high-*T* (*LPHT*) regional aureole beneath the Murrumbidgee Batholith. *Journal of Metamorphic Geology*, **20**, 119–34.

Richardson, S. W. 1970: The relation between a petrogenetic grid, facies series, and the geothermal gradient in metamorphism. *Fortschritte der Mineralogie*, **47**, 65–76.

Richter, F. M. & McKenzie, D. P. 1984: Dynamical models for melt segregation from a deformable matrix. *Journal of Geology*, **109**, 729–40.

Ridley, J. R. & Diamond, L. W. 2000: Fluid chemistry of orogenic lode gold deposits and implications for genetic models. *Reviews in Economic Geology*, **13**, 141–62.

Ridley, J. R., Groves, D. I. & Knight, J. T. 2000: Gold deposits in amphibolite and granulite facies terranes of the Archean Yilgarn craton, Western Australia: evidence and implications for synmetamorphic mineralization. *Reviews in Economic Geology*, **11**, 265–90.

Ridley, J. R. & Thompson, A. B. 1986: The role of mineral kinetics in the development of metamorphic microtextures. *In*: Walther, J. V. & Wood, B. J. (eds.): *Fluid-Rock Interactions during Metamorphism*. New York: Springer-Verlag, pp. 154–93.

Roache, T. J., Williams, P. J., Richmond, J. M. & Chapman, L. H. 2005: Vein and skarn formation at the Cannington Ag-Pb-Zn deposit, northeastern Australia. *Canadian Mineralogist*, **43**, 241–62.

Robbo, M., Borghi, A. & Compagnoni, R. 1999: Thermodynamic analysis of garnet growth zoning in eclogite facies granodiorite from M. Mucrone, Sesia Zone, western Italian Alps. *Contributions to Mineralogy and Petrology*, **137**, 289–303.

Robin, P.-Y. F. 1979: Theory of metamorphic segregation and related processes. *Geochimica et Cosmochimica Acta*, **43**, 1587–600.

Rock, N. M. S. 1976: Fenitisation around the Monchique alkaline complex, Portugal. *Lithos*, **9**, 263–79.

Roermund, H. L. M., Drury, M. R., Barnhoorn, A. & De Ronde, A. A. 2000: Super-silicic garnet microstructures from an orogenic peridotite, evidence for an ultra-deep (>6 GPa) origin. *Journal of Metamorphic Geology*, **18**, 135–47.

Rollinson, H. 1993: *Using Geochemical Data: Evaluation, Presentation, Interpretation*. Singapore: Longman.

Rosenberg, C. L. 2001: Deformation of partially molten granite: a review and comparison of experimental and natural case studies. *International Journal of Earth Sciences (Geologische Rundschau)*, **90**, 60–76.

Rosenberg, C. L. & Handy, M. R. 2000: Syntectonic melt pathways during simple shearing of a partially molten rock analogue (Norcamphor-Benzamide). *Journal of Geophysical Research*, **105**, 3135–49.

Rosenberg, C. L. & Handy, M. R. 2005: Experimental deformation of partially melted granite revisited: implications for the continental crust. *Journal of Metamorphic Geology*, **23**, 19–28.

Rosenfeld, J. L. 1968: Garnet rotations due to the major Paleozoic deformations in southeast Vermont. *In*: Zen, E-an, White, W. S., Hadley, J. B. & Thompson, J. B. (eds.): *Studies of Appalachian Geology: Northern and Maritime*. New York: Wiley, pp. 185–202.

Rosenfeld, J. L. 1970: Rotated garnets in metamorphic rocks. *Geological Society of America Special Paper*, **129**, 1–102.

Roser, B. P. & Nathan, S. 1997: An evaluation of elemental mobility during metamorphism of a turbidite sequence (Greenland Group, New Zealand). *Geological Magazine*, **134**, 219–34.

Rossoviskiy, L. N., Konovalenko, S. L. & Bovin, Y. P. 1981: Desilicified pegmatites with dravite and corundum, Southwestern Pamir. *International Geology Review*, **23**, 371–82.

Roy, A. B. 1978: Evolution of slaty cleavage in relation to diagenesis and metamorphism: a study from the Hunsrückschiefer. *Bulletin of the Geological Society of America*, **89**, 1775–85.

Rubatto, D. 2002: Zircon trace element geochemistry: distribution coefficients and the link between U-Pb ages and metamorphism. *Chemical Geology*, **184**, 123–38.

Rubatto, D. & Gebauer, D. 2000: Use of cathodoluminescence for U-Pb zircon dating by ion microprobe: some examples from the Western Alps. *In*: Pagel M., Barbin V., Blanc, P. & Ohnenstetter, D. (eds.): *Cathodoluminescence in Geosciences*. Berlin: Springer, pp. 373–400.

Rubatto, D., Gebauer, D. & Fanning, M. 1988: Jurassic formation and Eocene subduction of the Zermatt-Saas-Fee ophiolites: implications for the geodynamic evolution of the Central and Western Alps. *Contributions to Mineralogy and Petrology*, **132**, 269–87.

Rubatto, D. & Herman, J. 2007: Zircon behaviour in deeply subducted rocks. *Elements*, **3**, 31–6.

Rubenach, M. J. 2005: Relative timing of albitization and chlorine enrichment in biotite in Proterozoic schists, Snake Creek Anticline, Mount Isa Inlier, northeastern Australia. *Canadian Mineralogist*, **43**, 349–66.

Rubenach, M. J. & Barker, A. J. 1998: Metamorphic and metasomatic evolution of the Snake Creek Anticline, Eastern Succession, Mount Isa Inlier. *Australian Journal of Earth Sciences*, **45**, 363–72.

Rubenach, M. J. & Lewthwaite, K. A. 2002: Metasomatic albitites and related biotite-rich schists from a low-pressure polymetamorphic terrane, Snake Creek Anticline, Mount Isa Inlier, north-eastern Australia: microstructures and *P-T-d* paths. *Journal of Metamorphic Geology*, **20**, 191–202.

Rubie, D. C. 1986: The catalysis of mineral reactions by water and restrictions on the processes on the presence of aqueous fluid during metamorphism. *Mineralogical Magazine*, **50**, 399–415.

Rubie, D. C. 1990: Role of kinetics in the formation of eclogites. *In*: Carswell, D. A. (ed.): *Eclogite Facies Rocks*. Glasgow: Blackie, pp. 111–40.

Rubin, A. M. 1998: Dike ascent in partially molten rock. *Journal of Geophysical Research*, **103**, 20901–19.

Rumble, D. 1978: Mineralogy, petrology, and oxygen isotope geochemistry of the Clough Formation, Black Mountain, western New Hampshire, U.S.A. *Journal of Petrology*, **19**, 317–40.

Rumble, D. 1982a: The role of perfectly mobile components in metamorphism. *Annual Reviews of Earth and Planetary Science*, **10**, 221–33.

Rumble, D. 1982b: Stable isotope fractionation during metamorphic devolatilization reactions. *In*: Ferry, J. M. (ed.): *Characterization of Metamorphism Through Mineral Equilibria. Mineralogical Society of America, Reviews in Mineralogy*, **10**, 327–53.

Rumble, D., Ferry, J. M., Hoering, T. C. & Boucot, A. J. 1982: Fluid flow during metamorphism at the Beaver Brook fossil locality, New Hampshire. *American Journal of Science*, **282**, 886–919.

Rumble, D. & Hoering, T. C. 1986: Carbon isotopic geochemistry of graphite vein deposits from New Hampshire, U.S.A. *Geochimica et Cosmochimica Acta*, **50**, 1239–47.

Rumble, D., Oliver, N. H. S., Ferry, J. M. & Hoering, T. C. 1991: Carbon and oxygen isotope geochemistry of chlorite-zone rocks of the Waterville limestone, Maine, U.S.A. *American Mineralogist*, **76**, 857–66.

Rushmer, T. 1995: An experimental deformation study of partially molten amphibolite: application to low-melt fraction segregation. *Journal of Geophysical Research*, **100**, 15681–95.

Rushmer, T. 2001: Volume change during partial melting reactions: implications for melt extraction, melt geochemistry and crustal rheology. *Tectonophysics*, **342**, 389–405.

Rutherford, E. & Soddy, F. 1903: Radioactive change. *Philosophical Magazine*, **6**, 576–91.

Rutherford, M. J., Sigurdsson, H., Carey, S. & Davis, A. 1985: The May 18, 1980 eruption of Mount St. Helens. 1. Melt composition and experimental phase equilibria. *Journal of Geophysical Research*, **90**, 2929–47.

Rutter, E. H. 1972: The influence of interstitial water on the rheological behaviour of calcite rocks. *Tectonophysics*, **14**, 13–33.

Rutter, E. H. 1976: The kinetics of rock deformation by pressure solution. *Philosophical Transactions of the Royal Society of London*, **A283**, 203–19.

Rutter, E. H. 1983: Pressure solution in nature, theory and experiment. *Journal of the Geological Society of London*, **140**, 725–40.

Rutter, E. H. 1997: The influence of deformation on the extraction of crustal melts: a consideration of the role of melt-assisted granular flow. *In*: Holness, M. B. (ed.): *Deformation-enhanced Fluid Transport in the Earth's Crust and Mantle. Mineralogical Society Series*, **8**. London: Chapman & Hall, pp. 82–110.

Rutter, E. H. & Brodie, K. H. 1985: The permeation of water intro hydrated shear zones. *In*: Thompson, A. B. & Rubie, D. C. (eds.): *Kinetics, Textures and Deformation. Advances in Physical Geochemistry*, **4**. New York: Springer, pp. 242–50.

Rutter, E. H. & Neumann, D. H. K. 1995: Experimental deformation of partially molten Westerly granite under fluid-absent conditions with implications for the extraction of granite magmas. *Journal of Geophysical Research*, **100**, 15697–715.

Rye, R. O., Schuiling, R. D., Rye, D. M. & Jansen, J. B. H. 1976: Carbon, hydrogen and oxygen isotope studies of the regional metamorphic complex at Naxos, Greece. *Geochimica et Cosmochimica Acta*, **40**, 1031–49.

Sandiford, M. & Powell, R. 1986: Pyroxene exsolution in granulites from Fyfe Hills, Enderby Land, Antarctica: Evidence for 1000°C metamorphic temperatures in Archean continental crust. *American Mineralogist*, **71**, 946–54.

Sandiford, M. & Powell, R. 1991: Some remarks on high-temperature–low-pressure metamorphism in convergent orogens. *Journal of Metamorphic Geology*, **9**, 333–40.

Sandiford, M., Powell, R., Martin, S. F. & Perera, L. R. K. 1988: Thermal and baric evolution of garnet granulites from Sri Lanka. *Journal of Metamorphic Geology*, **6**, 351–64.

Sanford, R. F. 1982: Growth of ultramafic reaction zones in greenschist to amphibolite facies metamorphism. *American Journal of Science*, **282**, 543–616.

Sawyer, E. W. 1987: The role of partial melting and fractional crystallization in determining discordant migmatite leucosome compositions. *Journal of Petrology*, **28**, 445–73.

Sawyer, E. W. 1991: Disequilibrium melting and the rate of melt-residuum separation during migmatization of mafic rocks from the Grenville front, Quebec. *Journal of Petrology*, **32**, 701–38.

Sawyer, E. W. 1994: Melt segregation in the continental crust. *Geology*, **22**, 1019–22.

Sawyer, E. W. 1996: Melt segregation and magma flow in migmatites: implications for the generation of granite magmas. *Transactions of the Royal Society of Edinburgh: Earth Sciences*, **87**, 85–94.

Sawyer, E. W. 1998: Formation and evolution of granite magmas during crustal reworking: the significance of diatexites. *Journal of Petrology*, **39**, 1147–67.

Sawyer, E. W. 1999: Criteria for the recognition of partial melting. *Physics and Chemistry of the Earth (A)*, **24**, 269–79.

Sawyer, E. W. 2000: Melt distribution and movement in anatectic rocks. *Geophysical Research Abstracts*, **2**.

Sawyer, E. W. 2001: Melt segregation in the continental crust: distribution and movement of melt in anatectic rocks. *Journal of Metamorphic Geology*, **19**, 291–309.

Sawyer, E. W. & Barnes, S.-J. 1988: Temporal and compositional differences between subsolidus and anatectic migmatite leucosomes from the Quetico metasedimentary belt, Canada. *Journal of Metamorphic Geology*, **6**, 437–50.

Sawyer, E. W., Dombrowski, C. & Collins, W. J. 1999: Movement of melt during synchronous regional deformation and granulite facies anatexis, an example from the Wuluma Hills, central Australia. *In*: Castro, A., Fernandez, C. & Vigneresse, J.-L. (eds.): *Understanding Granites; Integrating New and Classical Techniques. Geological Society of London Special Publication* **158**, 221–37.

Sawyer, E. W. & Robin, P.-Y. F. 1986: The subsolidus segregation of layer-parallel quartz-feldspar veins in greenschist to upper amphibolite facies metasediments. *Journal of Metamorphic Geology*, **4**, 237–60.

Sayab, M. 2006: Decompression through clockwise *P-T* path: implications for early N-S shortening orogenesis in the Mesoproterozoic Mt Isa Inlier (NE Australia). *Journal of Metamorphic Geology*, **24**, 89–105.

Scally, A. & Simonson, B. M. 2005: Spherule textures in the Neoarchaean Wittenoom impact layer, Western Australia: consistency in diversity. *Australian Journal of Earth Sciences*, **52**, 773–83.

Scambelluri, M. & Phillipot, P. 2001. Deep fluids in subduction zones. *Lithos*, **55**, 213–27.

Schaefer, S. J. & Morton, P. 1991: Two komatiitic pyroclastic units, Superior Province, northwestern Ontario: their geology, petrography, and correlation. *Canadian Journal of Earth Sciences*, **28**, 1455–70.

Schaltegger, U., Fanning, C. M., Günther, D., Maurin, J. C., Schulmann, K. & Gebauer, D. 1999: Growth, annealing and recrystallization of zircon and preservation of monazite in high-grade metamorphism: conventional and in-situ U-Pb isotope, cathodoluminescence and microchemical evidence. *Contributions to Mineralogy and Petrology*, **134**, 186–201.

Scharbert, H. G. & Carswell, D. A. 1983: Petrology of garnet-clinopyroxene rocks in a granulite facies environment. Bohemian Massif of Lower Austria. *Bulletin minéralogique*, **106**, 761–74.

Scherer, E. E., Whitehouse, M. J. & Münker, C. 2007: Zircon as a monitor of crustal growth. *Elements*, **3**, 19–24.

Schiffman, D. & Smith, B. M. 1988: Petrology and oxygen isotope geochemistry of a fossil seawater hydrothermal system within the Soten Graben, northern Troodos Ophiolite, Cyprus. *Journal of Geophysical Research*, **93**, 4612–24.

Schmid, S. M. 1976: Rheological evidence for changes in the deformation mechanism of Solenhofen limestone towards low stress. *Tectonophysics*, **31**, 21–8.

Schmid, S. M., Boland, J. N. & Paterson, M. S. 1977: Superplastic flow in fine-grained limestone. *Tectonophysics*, **43**, 257–92.

Schmid, S. M., Panozzo, R. & Bauer, S. 1987: Simple shear experiments on calcite rocks: rheology and microfabric. *Journal of Structural Geology*, **9**, 747–78.

Schmid, S. M., Paterson, M. S. & Boland, J. N. 1980: High temperature flow and dynamic recrystallization in Carrara marble. *Tectonophysics*, **65**, 245–80.

Schoneveld, C. 1977: A study of typical inclusion patterns in strongly paracrystalline-rotated garnets. *Tectonophysics*, **39**, 453–71.

Schoneveld, C. 1979: *The Geometry and the Significance of Inclusion Patterns in Syntectonic Porphyroblasts*. Published PhD thesis, University of Leiden.

Schrijver, K. 1973: Bimetasomatic plagioclase-pyroxene reaction zones in granulite facies. *Neues Jahrbuch für Mineralogie Abhandlungen*, **119**, 1–19.

Schultz, R. W. 1966: Lower Carboniferous cherty ironstones at Tynagh, Ireland. *Economic Geology*, **61**, 311–42.

Schumacher, J. C., Hollocher, K. T., Robinson, P. & Tracy, R. J. 1990: Progressive reactions and melting in the Acadian metamorphic high of central Massachusetts and south-western New Hampshire, USA. *In*: Ashworth, J. R. & Brown, M. (eds.): *High-temperature Metamorphism and Crustal Anatexis*. London: Unwin Hyman, pp. 198–234.

Schumacher, R., Rötzler, K. & Maresch, W. V. 1999: Subtle oscillatory zoning in Erzgebirge, Germany. *Canadian Mineralogist*, **37**, 381–402.

Scott, J. S. & Drever, H. I. 1953: Frictional fusion along a Himalayan thrust. *Proceedings of the Royal Society of Edinburgh*, **65**, 121–42.

Seal, P. R., Clarke, A. H. & Morrissy, C. J. 1987: Stock-work tungsten (scheelite) – molybdenum mineralization, Lake George, southwestern New Brunswick. *Economic Geology*, **82**, 1259–82.

Searle, D. F. 1972: Mode of occurrence of cupriferous pyrite deposits of Cyprus. *Transactions of the Institution of Mining and Metallurgy*, **81B**, 189–97.

Sederholm, J. J. 1907: Om granit och gneiss. *Bulletin de la Commission géologique de Finlande*, **23**, 91–110.

Sederholm, J. J. 1967: *Selected Works: Granites and Migmatites*. Edinburgh: Oliver and Boyd.

Selleck, B. W., McLelland, J. M. & Bickford, M. E. 2005: Granite emplacement during tectonic exhumation: The Adirondack example. *Geology*, **33**, 781–4.

Selverstone, J. 1993: Micro- to macroscale interactions between deformational and metamorphic processes, Tauern Window, eastern Alps. *Schweizerische Mineralogische und Petrographische Mitteilungen*, **73**, 229–39.

Selverstone, J., Franz, G., Thomas, S. & Getty, S. 1992: Fluid variability in 2 GPa eclogites as an indicator of fluid behaviour during subduction. *Contributions to Mineralogy and Petrology*, **112**, 341–57.

Selverstone, J., Morteani, G. & Staude, J. M. 1991: Fluid channeling during ductile shearing; transformation of granodiorite into aluminous schist in the Tauern Window, Eastern Alps. *Journal of Metamorphic Geology*, **9**, 419–31.

Selverstone, J., Spear, F. S., Franz, G. & Morteani, G. 1984: High-pressure metamorphism in the SW Tauern Window, Austria: *P-T* paths from hornblende-kyanite-staurolite schists. *Journal of Petrology*, **25**, 501–31.

Seyfried, W. E., Berndt, M. E. & Seewald, J. S. 1988: Hydrothermal alteration processes at mid-ocean ridges: constraints from diabase alteration experiments, hot-spring fluids and composition of the oceanic crust. *Canadian Mineralogist*, **26**, 787–804.

Shand, S. J. 1916: The pseudotachylite of Parijs (Orange Free State), and its relation to "trop-shotten gneiss" and "flinty crush-rock." *Quarterly Journal of the Geological Society of London*, **72**, 198–221.

Shaw, D. M. 1956: Geochemistry of pelitic rocks. Part III: Major elements and general geochemistry. *Bulletin of the Geological Society of America*, **67**, 919–34.

Shaw, S. E. 1973: Preliminary report on the geochemistry of the mine sequence rocks Broken Hill, NSW. *In*: *Broken Hill Mines 1968. Australasian Institute of Mining and Metallurgy Monograph*, **3**, 185–98.

Shea, W. T. & Kronenberg, A. K. 1993: Strength and anisotropy of foliated rocks with varied mica contents. *Journal of Structural Geology*, **15**, 1097–121.

Shelley, D. M. 1993: *Igneous and Metamorphic Rocks under the Microscope*. London: Chapman & Hall.

Sheppard, S. M. F. 1986: Characterization and isotopic variations in natural waters. *In*: Valley, J. W., Taylor, H. P. & O'Neil, J. R. (eds.): *Stable Isotopes in High Temperature Geological Processes. Reviews in Mineralogy*, **16**, 165–83.

Sheppard, S. M. F. & Taylor, H. P. 1974: Hydrogen and oxygen isotope evidence for the origin of water in the Boulder batholith and the Butte ore deposits, Montana. *Economic Geology*, **69**, 926–46.

Shieh, Y.-N. 1974: Mobility of oxygen isotopes during metamorphism. *In*: Hofmann, A. W., Giletti, B. J., Yoder, H. S. & Yund, R. A. (eds.): *Geochemical Transport and Kinetics. Carnegie Institution of Washington Publication* **634**, 324–36.

Shieh, Y.-N. & Taylor, H. P. 1969: Oxygen and carbon isotope studies of contact metamorphism of carbonate rocks. *Journal of Petrology*, **10**, 307–31.

Shimura, T., Komatsu, M. & Iiyama, J. T. 1992: Genesis of the lower crustal garnet-orthopyroxene tonalites (S-type) of the Hidaka Metamorphic Belt, northern Japan. *Transactions of the Royal Society of Edinburgh: Earth Sciences* **83**, 259–68.

Shimura, T., Owada, M., Osanai, Y., Komatsu, M. & Kagami, H. 2004: Variety and genesis of the pyroxene-bearing S- and I-type granitoids from the Hidaka Metamorphic Belt, Hokkaido, northern Japan. *Transactions of the Royal Society of Edinburgh: Earth Sciences*, **95**, 161–79.

Shrock, R. R. 1948: *Sequences in Layered Rocks*. New York: McGraw-Hill.

Sibson, R. H. 1975: Generation of pseudotachylite by ancient seismic faulting. *Geophysical Journal of the Royal Astronomical Society*, **43**, 775–94.

Sibson, R. H. 1981: Control on low-stress hydrofracture dilatancy in thrust, wrench and normal fault terrains. *Nature*, **289**, 655–67.

Sibson, R. H. 1990: Faulting and fluid flow. *In*: Nesbitt, B. E. (ed.): *Fluids in Tectonically Active Regimes of the Continental Crust. Mineralogical Association of Canada Short Course*, **18**, 93–132.

Sibson, R. H. 1996: Structural permeability of fluid-driven fault-fracture meshes. *Journal of Structural Geology*, **8**, 1031–42.

Siddans, A. W. 1972: Slaty cleavage – a review of research since 1816. *Earth-Science Reviews*, **8**, 205–32.

Simonson, B. M. 1992: Geological evidence for a strewn field of impact spherules in the early Precambrian Hamersley Basin of Western Australia. *Bulletin of the Geological Society of America*, **104**, 829–39.

Simonson, B. M. 2003: Petrographic criteria for recognizing certain types of impact spherules in well-preserved Precambrian successions. *Astrobiology*, **3**, 49–65.

Simonson, B. M. & Glass, B. P. 2004: Spherule layers – Records of ancient impacts. *Annual Review of Earth and Planetary Sciences*, **32**, 329–61.

Simonson, B. M., Koeberl, C., McDonald, I. & Reimold, W. U. 2000: Geochemical evidence for an impact origin for a Late Archean spherule layer, Transvaal Supergroup, South Africa. *Geology*, **28**, 1103–6.

Simpson, C. 1986: Fabric development in brittle-to-ductile shear zones. *Pure and Applied Geophysics*, **124**, 269–88.

Simpson, C. & Wintsch, R. P. 1989: Evidence for deformation-induced K-feldspar replacement by myrmekite. *Journal of Metamorphic Geology*, **7**, 261–75.

Sinha, A. K., Hewitt, D. A. & Rimstidt, J. D. 1986: Fluid interaction and element mobility in the development of ultramylonites. *Geology*, **14**, 883–86.

Sivell, W. J. & Waterhouse, J. B. 1984: Oceanic ridge metamorphism of the Patuki Volcanics, D'Urville Island, New Zealand. *Lithos*, **17**, 19–36.

Slagstad, T., Jamieson, R. A. & Culshaw, N. G. 2005: Formation, crystallization, and migration of melt in the mid-orogenic crust: Muskoka Domain migmatites, Grenville Province, Ontario. *Journal of Petrology*, **46**, 893–919.

Slater, D. J., Yardley, B. W. D., Spiro, B. & Knipe, R. J. 1994: Incipient metamorphism and deformation in the Variscides of SW Dyfed, Wales: first steps towards isotopic equilibrium. *Journal of Metamorphic Geology*, **12**, 237–48.

Smith, B. R., Vogel, T. A. & Spence, W. H. 1969: Precambrian metaconglomerate mantling a granite dome at Plevna Lake in the Grenville Province of southeastern Ontario, Canada. *Bulletin of the Geological Society of America*, **80**, 297–302.

Smith, C. S. 1953: Microstructure. *Transactions of the American Society for Metals*, **45**, 533–75.

Smith, D. C. 1984: Coesite in clinopyroxene in the Caledonides and its implications for geodynamics. *Nature*, **310**, 641–44.

Smith, H. A. & Barreiro, B. 1990: Monazite U–Pb dating of staurolite grade metamorphism in pelitic schists. *Contributions to Mineralogy and Petrology*, **105**, 602–15.

Smith, R. E. 1968: Redistribution of elements in the alteration of some basic lavas during burial metamorphism. *Journal of Petrology*, **9**, 191–219.

Smith, R. E. 1974: The production of spilitic lithologies by burial metamorphism of flood basalts from the Canadian Keweenawan, Lake Superior. *In*: Amstutz, G. C. (ed.): *Spilites and Spilitic Rocks*. Berlin: Springer, pp. 403–26.

Smith, R. E., Perdrix, J. L. & Parks, T. C. 1982: Burial metamorphism in the Hamersley Basin, Western Australia. *Journal of Petrology*, **23**, 75–102.

Snoke, A. W., Tullis, J. A. & Todd, V. R. (eds.) 1999: *Fault-related Rocks: A Photographic Atlas*. Princeton: Princeton University Press.

Solar, G. S. & Brown, M. 2001: Petrogenesis of migmatites in Maine, USA: possible source of peraluminous leucogranite in plutons. *Journal of Petrology*, **42**, 789–823.

Solomatov, V. S. 2002: Constraints on the grain size in the mantles of terrestrial planets. *Lunar and Planetary Science*, **33**, abstract number 1446.

Sorby, H. C. 1853: On the origin of slaty cleavage. *Edinburgh New Philosophical Journal*, **55**, 137–48.

Sorby, H. C. 1908: On the application of quantitative methods to the structure and history of rocks. *Quarterly Journal of the Geological Society of London*, **64**, 171–232.

Spang, J. H., Oldershaw, A. E. & Stout, M. Z. 1979: Development of cleavage in the Banff Formation at Pigeon Mountain, Front Ranges, Canadian Rocky Mountains. *Canadian Journal of Earth Sciences*, **16**, 1108–15.

Sparks, H. A. & Mavrogenes, J. A. 2005: Sulfide melt inclusions as evidence for the existence of a sulfide partial melt at Broken Hill, Australia. *Economic Geology*, **100**, 773–9.

Spear, F. S. 1993: *Metamorphic Phase Equilibria and Pressure-Temperature-Time Paths*. Monograph. Washington, D.C.: Mineralogical Society of America.

Spear, F. S. & Cheney, J. T. 1989: A petrogenetic grid for pelitic schists in the system K_2O–FeO–MgO–Al_2O_3–SiO_2–H_2O. *Contributions to Mineralogy and Petrology*, **101**, 149–64.

Spear, F. S. & Kohn, M. J. 1996: Trace element zoning in garnet as a monitor of crustal melting. *Geology*, **24**, 1099–102.

Spear, F. S., Selverstone, J., Hickmott, D., Crowley, P. & Hodges, K. V. 1984: *P-T* paths from garnet zoning: A new technique for deciphering tectonic processes in crystalline terranes. *Geology*, **12**, 87–90.

Speer, J. A. 1981: The nature and magnetic expression of isograds in the contact aureole of the Liberty Hill pluton, South Carolina: Part II. *Bulletin of the Geological Society of America*, **92**, 1262–358.

Spray, J. G. 1987: Artificial generation of pseudotachylite using friction welding apparatus: simulation of melting on a fault plane. *Journal of Structural Geology*, **9**, 49–60.

Spray, J. G. 1995: Pseudotachylyte controversy: fact or friction? *Geology*, **23**, 1119–22.

Spray, J. G. & Thompson, L. M. 1994: Friction melt distribution in a multi-ring impact basin. *Nature*, **373**, 130–2.

Spry, A. 1969: *Metamorphic Textures*. Oxford: Pergamon.

Stallard, A. & Hickey, K. 2001: Fold mechanisms in the Canton Schist: constraints on the contributions of flexural flow. *Journal of Structural Geology*, **23**, 1865–81.

Stallard, A. & Hickey, K. 2002: A comparison of microstructural and chemical patterns in garnet from the Fleur de Lys Supergroup, Newfoundland. *Journal of Structural Geology*, **24**, 1109–23.

Stallard, A., Hickey, K. & Upton, G. J. 2003: Measurement and correlation of microstructures: the case of foliation intersection axes. *Journal of Metamorphic Geology*, **21**, 241–52.

Stanton, R. L. 1976: Petrochemical studies of the ore environment at Broken Hill, New South Wales: 1 – constitution of 'banded iron formations.' environmental synthesis. *Transactions of the Institution of Mining & Metallurgy*, **85**, B33–46.

Steiger, R. H. & Jäger, E. 1977: Subcommision on geochronology: convention of the use of decay constants in geo- and cosmochronology. *Earth and Planetary Science Letters*, **36**, 359–62.

Steinhardt, C. 1988: Lack of porphyroblast rotation in non-coaxially deformed schists from Petrel Cove, South Australia, and its implications. *Tectonophysics*, **158**, 127–40.

Stel, H. 1986: The effect of cyclic operation of brittle and ductile deformation on the metamorphic assemblage in catclasites and mylonites. *Pure and Applied Geophysics*, **124**, 289–307.

Stephens, W. E. 1992: Spatial, compositional and rheological constraints on the origin of zoning in the Criffell pluton, Scotland. *Transactions of the Royal Society of Edinburgh: Earth Sciences*, **83**, 191–9.

Stevens, B. P. J. 1998: A comparative study: calc-silicate ellipsoids from Broken Hill and diagenetic carbonate concretions from the Sydney Basin. *Quarterly Notes of the Geological Survey of New South Wales*, **106**, 11–15.

Stevens, B. P. J., Barnes, R. G., Brown, R. E., Stroud, W. J. & Willis, I. L. 1988: The Willyama Supergroup in the Broken Hill and Euriowie Blocks, NSW. *Precambrian Research*, **40/41**, 297–327.

Stevens, B. P. J. & Barron, L. M. 2002: Volcanic textures in the Palaeoproterozoic Hores Gneiss, Broken Hill, Australia. *Quarterly Notes of the Geological Survey of New South Wales*, no. **113**, 1–22.

Stevens, B. P. J. & Willis, I. L. 1983: Systematic classification of rock units: a key to mapping and interpretation of the Willyama Complex. *In*: Stevens, B. P. J. & Stroud, W. J. (eds.): *Rocks of the Broken Hill Block: Their Classification, Nature, Stratigraphic Distribution, and Origin. Records of the Geological Survey of New South Wales*, **21**, 1–56.

Stevens, G. 1997: Melting, carbonic fluids and water recycling in the deep crust: an example from the Limpopo Belt, South Africa. *Journal of Metamorphic Geology*, **15**, 141–54.

Stevens, G. 2007: Selective peritectic garnet entrainment as the origin of geochemical diversity in S-type granites. *Geology*, **35**, 9–12.

Stevens, G. & Clemens, J. D. 1993: Fluid-absent melting and the role of fluids in the lithosphere: a slanted summary? *Chemical Geology*, **108**, 1–17.

Stevens, G., Clemens, J. D. & Droop, G. T. R. 1997: Melt production during granulite-facies anatexis: experimental data from "primitive" metasedimentary protoliths. *Contributions to Mineralogy and Petrology*, **128**, 352–70.

Stevens, G., Prinz, S. & Rozendaal, A. 2005: Partial melting of the assemblage sphalerite + galena + chalcopyrite + S: implications for high-grade metamorphosed massive sulfide deposits. *Economic Geology*, **100**, 781–6.

Stewart, A. J. 1981: *Reynolds Range Region, Northern Territory. 1:100 000 Geological Map Commentary*. Canberra: Bureau of Mineral Resources, Geology and Geophysics.

Stillwell, F. L. 1918: The metamorphic rocks of Adelie Land. *Australasian Antarctic Expedition 1911–14 Scientific Report, Series A*, **3**.

Stillwell, F. L. 1922: The rocks in the immediate neighbourhood of the Broken Hill Lode and their bearing on its origin. *Memoirs of the Geological Survey of New South Wales*, **8**, 354–96.

Stipska, P. & Powell, R. 2005: Does ternary feldspar constrain the metamorphic conditions of high-grade meta-igneous rocks? Evidence from orthopyroxene granulites, Bohemian Massif. *Journal of Metamorphic Geology*, **23**, 627–47.

Stølen, L. K. 1994: The rift-related mafic dyke complex of the Rohkunborrri Nappe, Indre Troms, northern Norwegian Caledonides. *Norsk Geologisk Tidsskrift*, **74**, 35–47.

St-Onge, M. R. 1987. Zoned poikiloblastic garnets: P-T paths and syn-metamorphic uplift through 30 km of structural depth, Wopmay orogen, Canada. *Journal of Petrology*, **28**, 1–21.

Stowell, H. H., Taylor, D. L., Tinkham, D. K., Goldberg, S. A. & Ouderkirk, K. A. 2001: Contact metamorphic *P-T-t* paths from Sm-Nd garnet ages, phase equilibria modeling, and thermobarometry: Garnet Ledge, Southeastern Alaska. *Journal of Metamorphic Geology*, **19**, 645–60.

Streit, J. E. & Cox, S. F. 1998: Fluid infiltration and volume change during mid-crustal mylonitization of Proterozoic granite, King Island, Tasmania. *Journal of Metamorphic Geology*, **16**, 197–212.

Strong, D. F. & Hanmer, S. K. 1981: The leucogranites of southern Brittany: origin by faulting, frictional heating, fluid flux and fractional melting. *Canadian Mineralogist*, **19**, 163–76.

Stroud, W. J., Willis, I. L., Bradley, G. M., Brown, R. E., Stevens, B. P. J. & Barnes, R. G. 1983: Amphibole and/or pyroxene-bearing rocks. *In*: Stevens, B. P. J. & Stroud, W. J. (eds.): *Rocks of the Broken Hill Block: Their Classification, Nature, Stratigraphic Distribution, and Origin. Records of the Geological Survey of New South Wales*, **21**, 227–87.

Stünitz, H. 1993: Transition from fracturing to viscous flow in a naturally deformed metagabbro. *In*: Boland, J. N. & Fitz Gerald, J. D. (eds.): *Defects and Processes in the Solid State: Geoscience Applications. The McLaren Volume*. Amsterdam: Elsevier, pp. 121–50.

Stünitz, H. & Tullis, J. 2001: Weakening and strain localization produced by syn-deformational reaction of plagioclase. *International Journal of Earth Sciences (Geologische Rundschau)*, **90**, 136–48.

Stüwe, K. 1997: Effective bulk composition changes due to cooling: a model predicting complexities in retrograde reaction textures. *Contributions to Mineralogy and Petrology*, **129**, 43–52.

Stüwe, K. & Powell, R. 1989: Metamorphic segregations associated with garnet and orthopyroxene porphyroblast growth: two examples from the Larsemann Hills, east Antarctica. *Contributions to Mineralogy and Petrology*, **103**, 523–30.

Stüwe, K., Sandiford, M. & Powell, R. 1993: Episodic metamorphism and deformation in low-pressure, high-temperature terranes. *Geology*, **21**, 829–32.

Sutter, J. F. & Hunting, J. B. 1984: Laser microprobe $^{40}Ar/^{39}Ar$ dating of minerals in situ. *Proceedings of the Electron Microscopy Society of America*, **4**, 1525–9.

Sutton, S. J. 1991: Development of domainal slaty cleavage at Ococee Gorge, Tennessee. *Journal of Geology*, **99**, 789–800.

Suzuki, K. 1977: Local equilibrium during the contact metamorphism of siliceous dolomites in Kasuga-mur, Gifu-ken, Japan. *Contributions to Mineralogy and Petrology*, **61**, 79–89.

Suzuki, K., Adachi, M., Kato, T. & Yogo, S. 1999: CHIME dating method and its application to the analysis of evolutional history of orogenic belts. *Chikyukagaku (Geochemistry)*, **33**, 1–22.

Suzuki, K., Nasu, T. & Shibata, K. 1995: CHIME monazite ages of the Otagiri and Ichida Granites in the Komagane area, Nagano Prefecture. *Journal of Earth and Planetary Sciences, Nagoya University*, **47**, 17–30.

Sykes, M. L. & Moody, J. B. 1978: Pyrophyllite and metamorphism in the Caroline slate belt. *American Mineralogist*, **63**, 96–108.

Takagi, T., Kazuki, N., Collins, L. G. & Lizumi, S. 2007: Plagioclase-quartz rocks of metasomatic origin at the expense of granitic rocks of the Komaki district, southwestern Japan. *Canadian Mineralogist*, **45**, 559–80.

Talbot, J. L. 1964a: Crenulation cleavage in the Hunsrück-Schiefer of the middle Moselle region. *Geologische Rundschau*, **54**, 1026–43.

Talbot, J. L. 1964b: The structural geometry of rocks of the Torrens Group near Adelaide, South Australia. *Journal of the Geological Society of Australia*, **11**, 33–48.

Talbot, J. L. & Hobbs, B. E. 1968: The relationship of metamorphic differentiation to other structural features at three localities. *Journal of Geology*, **76**, 581–7.

Taylor, H. P. 1977: Water/rock interactions and the origin of H_2O in granitic batholiths. *Journal of the Geological Society of London*, **133**, 509–58.

Taylor, H. P. 1988: Oxygen, hydrogen and strontium isotope constraints on the origin of granites. *Philosophical Transactions of the Royal Society of London*, **A79**, 317–38.

Termier, H. & Termier, G. 1970: Sur une roche crystalline renfermant une écaille d'Echinderms (Massif du Tichka, Haut Atlas, Maroc). *Académie des Sciences Comptes Rendus*, D, **270**, 1092–5.

Teyssier, C. & Whitney, D. L. 2002: Gneiss domes and orogeny. *Geology*, **30**, 1139–42.

Thompson, A. B. 1975: Calc-silicate diffusion zones between marble and pelitic schist. *Journal of Petrology*, **16**, 314–46.

Thompson, A. B. 1982: Dehydration melting of pelitic rock and the generation of H_2O-undersaturated granitic liquids. *American Journal of Science*, **282**, 1567–95.

Thompson, A. B. 1999: Some time-space relationships for crustal melting and granitic intrusion at various depths. *In*: Castro, A., Fernández, C. & Vigneresse, J.-L. (eds.): *Understanding Granites; Integrating New and Classical Techniques. Geological Society of London Special Publication*, **168**, 7–25.

Thompson, A. B. 2000: P-T paths and H_2O recycling in migmatite terrains. *Geophysical Research Abstracts*, **2**.

Thompson, A. B., Tracy, R. J., Lyttle, P. & Thompson, J. B. 1977: Prograde reaction histories deduced from compositional zonation and mineral inclusions in garnet from the Gassetts Schist, Vermont. *American Journal of Science*, **277**, 1152–67.

Thompson, J. B. 1957: The graphical analysis of mineral assemblages in pelitic schists. *American Mineralogist*, **42**, 842–58.

Thompson, J. B. 1959: Local equilibrium in metasomatic processes. *In*: Abelson, P. H. (ed.): *Researches in Geochemistry, Volume 1*. New York: Wiley, pp. 427–57.

Thompson, J. B. 1970: Geochemical reaction and open systems. *Geochimica et Cosmochimica Acta*, **34**, 525–51.

Thompson, J. B. 1982: Reaction space – an algebraic and geometric approach. *Mineralogical Society of America Reviews in Mineralogy*, **13**, 33–52.

Thompson, P. H. 1978: Archean regional metamorphism in the Slave structural province – a new perspective on some old rocks. *In: Metamorphism of the Canadian Shield. Geological Survey of Canada Paper*, **78–10**, 85–102.

Thoni, M. & Jagoutz, E. 1992: Some aspects of dating eclogites in orogenic belts: Sm-Nd, Rb-Sr, and Pb-Pb isotopic results from the Austroalpine Saualpe and Koralpe type-locality (Carinthia/Styria, southeastern Austria). *Geochimica et Cosmochimica Acta*, **56**, 347–68.

Till, A. B. & Snee, L. W. 1995: $^{40}Ar/^{39}Ar$ evidence that formation of blueschists in continental crust was synchronous with foreland fold and thrust belt deformation, western Brooks Range, Alaska. *Journal of Metamorphic Geology*, **13**, 41–60.

Tilley, C. E. 1924: Contact metamorphism in the Comrie area of the Perthshire Highlands. *Quarterly Journal of the Geological Society of London*, **80**, 22–71.

Tilley, C. E. 1948: Earlier stages in the metamorphism of siliceous dolomites. *Mineralogical Magazine*, **28**, 272–76.

Tilley, C. E. 1951: The zoned contact-skarns of the Broadford area, Skye: a study of boron-fluorine metasomatism in dolomites. *Mineralogical Magazine*, **29**, 621–66.

Tilley, C. E. & Alderman, A. R. 1934: Progressive metasomatism in the flint nodules of the Scawt Hill contact zone. *Mineralogical Magazine*, **23**, 513–8.

Tinkham, D. K. & Ghent, E. D. 2005: XRMapAnal; a program for analysis of quantitative X-ray maps. *American Mineralogist*, **90**, 737–44.

Tobisch, O. T. 1955: Observations on primary deformed sedimentary structures in some metamorphic rocks from Scotland. *Journal of Sedimentary Petrology*, **35**, 415–9.

Tobisch, O. T., Barton, M. D., Vernon, R. H. & Paterson, S. R. 1991: Fluid-enhanced deformation: transformation of granitoids to banded mylonites, western Sierra Nevada, California, and southeastern Australia. *Journal of Structural Geology*, **13**, 1137–56.

Tobisch, O. T., McNulty, B. A. & Vernon, R. H. 1997: Microgranitoid enclave swarms in granitic plutons, central Sierra Nevada, California. *Lithos*, **40**, 321–39.

Tobisch, O. T. & Paterson, S. R. 1988: Analysis and interpretation of composite foliations in areas of progressive deformation. *Journal of Structural Geology*, **10**, 745–54.

Tomascak, P. B., Krogstad, E. J. & Walker, R. J. 1996: U-Pb monazite geochronology of granitic rocks from Maine: implications for Late Paleozoic tectonics in the Northern Appalachians. *Journal of Geology*, **104**, 185–95.

Tomkins, A. G. & Mavrogenes, J. A. 2001: Redistribution of gold within arsenopyrite and lollingite during pro- and retrograde metamorphism: application to timing of mineralization. *Economic Geology*, **96**, 525–34.

Tomkins, A. G. & Mavrogenes, J. A. 2002: Mobilization of gold as a polymetallic melt during pelite anatexis at the challenger deposit, South Australia: a metamorphosed Archean gold deposit. *Economic Geology*, **97**, 1249–71.

Tomkins, A. G., Pattison, D. R. M. & Frost, R. B. 2007: On the initiation of metal sulfide anatexis. *Journal of Petrology*, **48**, 511–35.

Tomkins, A. G., Pattison, D. R. M. & Zaleski, E. 2004: The Hemlo gold deposit, Ontario: a type example of melting and mobilization of a precious metal-sulfosalt assemblage during amphibolite facies metamorphism and deformation. *Economic Geology*, **99**, 1063–84.

Tracy, R. J. 1978: High grade metamorphic reactions and partial melting in pelitic gneiss, west-central Massachusetts. *American Journal of Science*, **278**, 150–78.

Tracy, R. J. 1985: Migmatite occurrences in New England. *In*: Ashworth, J. R. (ed.): *Migmatites*. Glasgow: Blackie, pp. 204–24.

Tracy, R. J. & Robinson, P. 1983: Acadian migmatite types in pelitic rocks of central Massachusetts. *In*: Atherton, M. P. & Gribble, C. D. (eds.): *Migmatites, Melting and Metamorphism*. Nantwich: Shiva, pp. 163–73.

Tranter, T. H. 1992: Underplating of an accretionary prism: an example from the LeMay Group of central Alexander Island, Antarctic Peninsula. *Journal of South American Earth Sciences*, **6**, 1–20.

Trendall, A. F. 1973a: Precambrian iron-formations of Australia. *Economic Geology*, **68**, 1023–34.

Trendall, A. F. 1973b: Varve cycles in the Weeli Wolli Formation of the Precambrian Hamersley Group, Western Australia. *Economic Geology*, **68**, 1089–97.

Trendall, A. F. 1983: The Hamersley Basin. *In*: Trendall, A. F. & Morris, R. C. (eds.): *Iron-Formation: Facts and Problems*. Amsterdam: Elsevier, pp. 69–129.

Trommsdorff, V. 1966: Progressive metamorphose kieseliger karbonatgesteine in den Zentralalpen zwishcen Bernin und Simplon. *Schweizerische Minerolgisches und Petrographisches Mitteilungen*, **46**, 431–60.

Tsujimori, T., Sisson, V. B., Liou, J. G., Harlow, G. E. & Sorenson, S. S. 2006: Very low-temperature record in subduction process: a review of worldwide lawsonite eclogites. *Lithos*, **92**, 609–24.

Tullis, J. 1983: Deformation of feldspars. *In*: Ribbe, P. H. (ed.): *Feldspar Mineralogy*. Second edition. *Mineralogical Society of America, Reviews in Mineralogy*, **2**, 297–323.

Tullis, J. & Yund, R. A. 1987: Transition from cataclastic flow to dislocation creep of feldspar; mechanisms and microstructures. *Geology*, **215**, 606–9.

Tullis, J. & Yund, R. A. 1991: Diffusion creep in feldspar aggregates: experimental evidence. *Journal Structural Geology*, **13**, 987–1000.

Tullis, J., Yund, R. A. & Farver, J. 1996: Deformation-enhanced fluid distribution in feldspar aggregates and implications for ductile shear zones. *Geology*, **24**, 63–6.

Turner, F. J. 1941: The development of pseudostratification by metamorphic differentiation in the schists of Otago, New Zealand. *American Journal of Science*, **239**, 1–16.

Turner, F. J. 1981: *Metamorphic Petrology. Mineralogical and Field Aspects*. Second edition. New York: McGraw-Hill.

Turner, F. J. & Weiss, L. E. 1963: *Structural Analysis of Metamorphic Tectonites*. New York: McGraw-Hill.

Upton, B. G. J. 1987: Gabbroic, syenogabbroic and syenitic cumulates of the Tugtutôq younger giant dyke complex, South Greenland. *In*: Parsons, I. (ed.): *Origins of Igneous Layering*. Dordrecht: Reidel, pp. 93–123.

Urai, J. L., Humphreys, F. J. & Burrows, S. E. 1980: *In situ* studies of the deformation and dynamic recrystallization of rhombohedral camphor. *Journal of Materials Science*, **15**, 1231–40.

Urai, J. L., Means, W. D. & Lister, G. S. 1986: Dynamic recrystallization of minerals. *In*: Hobbs, B. E. & Heard, H. C. (eds.): *Mineral and Rock Deformation: Laboratory Studies (The Paterson Volume). American Geophysical Union Geophysical Monograph*, **36**, 161–99.

Väisänen, M. & Hölttä, P. 1999: Structural and metamorphic evolution of the Turku migmatite complex, southwestern Finland. *Bulletin of the Geological Society of Finland*, **71**, 177–218.

Valiant, R. I., Barnett, R. L. & Hodder, R. W. 1983: Aluminum silicate-bearing rock and its relation to gold mineralization, Bousquet Mine, Bousquet Township, Quebec. *CIM (Canadian Institute of Mining and Metallurgy) Bulletin*, **76**, 81–90.

Vallance, T. G. 1960: Concerning spilites. *Proceedings of the Linnean Society of New South Wales*, **85**, 8–52.

Vallance, T. G. 1965: On the chemistry of pillow lavas and the origin of spilites. *Mineralogical Magazine*, **34**, 471–81.

Vallance, T. G. 1967: Mafic rock alteration and isochemical development of some cordierite-anthophyllite rocks. *Journal of Petrology*, **8**, 84–96.

Vallance, T. G. 1969: Spilites again: some consequences of the degradation of basalts. *Proceedings of the Linnean Society of New South Wales*, **94**, 8–51.

Vallance, T. G. 1974: Spilitic degradation of a tholeiitic basalt. *Journal of Petrology*, **15**, 76–96.

Valley, J. W., Bohlen, S. R., Essene, S. J. & Lamb, W. 1990: Metamorphism in the Adirondacks: II The role of fluids. *Journal of Petrology*, **31**, 555–96.

van der Molen, I. & Paterson, M. S. 1979: Experimental deformation of partially melted granite. *Contributions to Mineralogy and Petrology*, **70**, 299–318.

van der Pluijm, B. A. & Kaars-Sijpesteijn, C. H. 1984: Chlorite-mica aggregates: morphology, orientation, development and bearing on cleavage formation in very low-grade rocks. *Journal of Structural Geology*, **6**, 399–407.

Vance, J. A. 1968: Metamorphic aragonite in the prehnite-pumpellyite facies, northwest Washington. *American Journal of Science*, **266**, 299–315.

Vance, D. & Holland, T. 1993: A detailed isotopic and petrological study of a single garnet from the Gassett Schist, Vermont. *Contributions to Mineralogy and Petrology*, **114**, 101–18.

Vanderhaeghe, O. 1999: Pervasive melt migration from migmatites to leucogranite in the Shuswap metamorphic core complex, Canada: control of regional deformation. *Tectonophysics*, **312**, 35–55.

Vassallo, J. J. & Vernon, R. H. 2000: Origin of megacrystic felsic gneisses at Broken Hill. *Australian Journal of Earth Sciences*, **47**, 733–48.

Vavra, G., Gebauer, D., Schmid, R. & Compston, W. 1996: Multiple zircon growth and recrystallization during polyphase Late Carboniferous to Triassic metamorphism in granulites of the Ivrea Zone (Southern Alps): an ion microprobe (SHRIMP) study. *Contributions to Mineralogy and Petrology*, **122**, 337–58.

Vavra, G., Schmid, R. & Gebauer, D. 1999: Internal morphology, habit and U-Th-Pb microanalysis of amphibolite-to-granulite facies zircons: geochronology of the Ivrea Zone (Southern Alps). *Contributions to Mineralogy and Petrology*, **134**, 380–404.

Vernon, R. H. 1961: Banded albite-rich rocks of the Broken Hill district, New South Wales. *CSIRO Mineragraphic Investigations Technical Paper*, **3**, 59 pp.

Vernon, R. H. 1968: Microstructures of high-grade metamorphic rocks at Broken Hill, Australia. *Journal of Petrology*, **9**, 1–22.

Vernon, R. H. 1969: The Willyama Complex, Broken Hill area. *Journal of the Geological Society of Australia*, **16**, 20–56.

Vernon, R. H. 1970: Comparative grain-boundary studies in some basic and ultrabasic granulites, nodules and cumulates. *Scottish Journal of Geology*, **6**, 337–51.

Vernon, R. H. 1974: Controls of mylonitic compositional layering during non-cataclastic ductile deformation. *Geological Magazine*, **111**, 121–31.

Vernon, R. H. 1975: Deformation and recrystallization of a plagioclase grain. *American Mineralogist*, **60**, 884–88.

Vernon, R. H. 1976: *Metamorphic Processes*. London: Murby; New York: Wiley.

Vernon, R. H. 1977: Relationships between microstructures and metamorphic assemblages. *Tectonophysics*, **39**, 439–52.

Vernon, R. H. 1978a: Porphyroblast-matrix microstructural relationships in deformed metamorphic rocks. *Geologische Rundschau*, **67**, 288–305.

Vernon, R. H. 1978b: Pseudomorphous replacement of cordierite by symplectic intergrowths of andalusite, biotite and quartz. *Lithos*, **11**, 283–89.

Vernon, R. H. 1979: Formation of late sillimanite by hydrogen metasomatism (base-leaching) in some high-grade gneisses. *Lithos*, **12**, 143–52.

Vernon, R. H. 1982: Isobaric cooling of two regional metamorphic complexes related to igneous intrusions in southeastern Australia. *Geology*, **10**, 76–81.

Vernon, R. H. 1983: Restite, xenoliths and microgranitoid enclaves in granites. *Journal and Proceedings of the Royal Society of New South Wales*, **116**, 77–103.

Vernon, R. H. 1984: Microgranitoid enclaves in granites – globules of hybrid magma quenched in a plutonic environment. *Nature*, **309**, 438–39.

Vernon, R. H. 1986a: K-feldspar megacrysts in granites – phenocrysts, not porphyroblasts. *Earth-Science Reviews*, **23**, 1–63.

Vernon, R. H. 1986b: Evaluation of the "quartz-eye" hypothesis. *Economic Geology*, **81**, 1520–7.

Vernon, R. H. 1987a: Oriented growth of sillimanite in andalusite, Placitas-Juan Tabo area, New Mexico, U.S.A. *Canadian Journal of Earth Sciences*, **24**, 580–90.

Vernon, R. H. 1987b: Growth and concentration of fibrous sillimanite related to heterogeneous deformation in K-feldspar–sillimanite metapelites. *Journal of Metamorphic Geology*, **5**, 51–68.

Vernon, R. H. 1988a: Sequential growth of cordierite and andalusite porphyroblasts, Cooma Complex, Australia: microstructural evidence of a prograde reaction. *Journal of Metamorphic Geology*, **6**, 255–69.

Vernon, R. H. 1988b: Microstructural evidence of rotation and non-rotation of mica porphyroblasts. *Journal of Metamorphic Geology*, **6**, 595–601.

Vernon, R. H. 1989: Porphyroblast-matrix microstructural relationships – recent approaches and problems. *In*: Daly, J. S., Cliff, R. A. & Yardley, B. W. D. (eds.): *Evolution of Metamorphic Belts. Geological Society of London Special Publications*, **43**, 83–102.

Vernon, R. H. 1990a: Crystallization and hybridism in microgranitoid enclave magmas: microstructural evidence. *In*: Sawka, W. N. & Hildebrand, R. S. (eds.): Mafic inclusions in granites. *Journal of Geophysical Research*, **95**, 17849–59.

Vernon, R. H. 1990b: K-feldspar augen in felsic gneisses and mylonites – deformed phenocrysts or porphyroblasts? *Geologiska Foreningens i Stockholm Forhandlingar*, **112**, 157–67.

Vernon, R. H. 1991a: Interpretation of microstructures of microgranitoid enclaves. *In*: Didier, J. & Barbarin, B. (eds.): *Enclaves and Granite Petrology*. Amsterdam: Elsevier, pp. 277–91.

Vernon, R. H. 1991b: Questions about myrmekite in deformed rocks. *Journal of Structural Geology*, **13**, 979–85.

Vernon, R. H. 1996a: Problems with inferring *P-T-t* paths in low-*P* granulite facies rocks. *Journal of Metamorphic Geology*, **14**, 143–53.

Vernon, R. H. 1996b: Structural evidence of parent rocks in high-grade metamorphic areas – especially Broken Hill. *In*: Pongratz, J. & Davidson, G. (eds.): *New Developments in Broken Hill Type Deposits*. CODES (University of Tasmania) Special Publication, **1**, 17–20.

Vernon, R. H. 1998: Chemical and volume changes during deformation and prograde metamorphism of sediments. *In*: Treloar, P. J. & O'Brien, P. (eds.): *What Drives Metamorphism and Metamorphic Reactions? The Geological Society of London, Special Publications*, **138**, 215–46.

Vernon, R. H. 1999: Quartz and feldspar microstructures in metamorphic rocks. *Canadian Mineralogist*, **37**, 513–24.

Vernon, R. H. 2000a: Review of microstructural evidence of magmatic and solid-state flow. *Electronic Geosciences*, **5**:2.

Vernon, R. H. 2000b: *Beneath Our Feet: The Rocks of Planet Earth*. Cambridge: Cambridge University Press.

Vernon, R. H. 2004: *A Practical Guide to Rock Microstructure*. Cambridge: Cambridge University Press.

Vernon, R. H. 2005: Comment. *In*: Brown, M., Pitcher, W. S. & Vernon, R. H.: Invited comments on Clemens's 'Granites and granitic magmas'. *Proceedings of the Geologists' Association*, **116**, 23–25.

Vernon, R. H. 2007: Problems in identifying restite in S-type granites of southeastern Australia, with speculations on sources of magma and enclaves. *Canadian Mineralogist*, **45**, 147–78.

Vernon, R. H., Clarke, G. L. & Collins, W. J. 1990: Local mid-crustal granulite facies metamorphism and melting: an example in the Mount Stafford area, central Australia. *In*: Ashworth, J. R. & Brown, M. (eds.): *High-temperature Metamorphism and Crustal Anatexis*. London: Unwin Hyman, pp. 272–319.

Vernon, R. H. & Collins, W. J. 1988: Igneous microstructures in migmatites. *Geology*, **16**, 1126–9.

Vernon, R. H., Collins, W. J. & Paterson, S. R. 1993a: Pre-foliation metamorphism in low-pressure/high-temperature terrains. *Tectonophysics*, **219**, 241–56.

Vernon, R. H., Collins, W. J. & Richards, S. W. 2003: Contrasting magmas in metapelitic and metapsammitic migmatites in the Cooma Complex, Australia. *Visual Geosciences*, **8**, 45–54.

Vernon, R. H., Etheridge, M. A. & Wall, V. J. 1988: Shape and microstructure of microgranitoid enclaves: indicators of magma mingling and flow. *Lithos*, **22**, 1–11.

Vernon, R. H. & Flood, R. H. 1979: Microstructural evidence of time-relationships between metamorphism and deformation in the metasedimentary sequence of the northern Hill End Trough, New South Wales, Australia. *Tectonophysics*, **58**, 127–37.

Vernon, R. H., Flood, R. H. & D'Arcy, W. F. 1987: Sillimanite and andalusite produced by base-cation leaching and contact metamorphism of felsic igneous rocks. *Journal of Metamorphic Geology*, **5**, 439–50.

Vernon, R. H. & Johnson, S. E. 2000: Transition from gneiss to migmatite and the relationship of leucosome to peraluminous granite in the Cooma Complex, SE Australia. *In*: Jessell, M. W. & Urai, J. L. (eds.): *Stress, Strain and Structure. A Volume in Honour of WD Means. Journal of the Virtual Explorer*, **2** (print & CD).

Vernon, R. H. & Paterson, S. R. 1993: The Ardara pluton. Ireland: deflating an expanded intrusion. *Lithos*, **31**, 17–32.

Vernon, R. H. & Paterson, S. R. 2001: Axial-surface leucosomes in anatectic migmatites. In: Boland, J. N. & Ord, A. (eds.): *Deformation Processes in the Earth's Crust (Hobbs Volume), Tectonophysics*, **335**, 183–92.

Vernon, R. H. & Paterson, S. R. 2002: Igneous origin of K-feldspar megacrysts in deformed granites of the Papoose Flat pluton, California, USA. *Electronic Geosciences*, **7**, 31–39.

Vernon, R. H. & Paterson, S. R. In press: Mesoscopic structures resulting from crystal accumulation and melt movement in granites. *Transactions of the Royal Society of Edinburgh: Earth Sciences.*

Vernon, R. H., Paterson, S. R. & Foster, D. 1993b: Growth and deformation of porphyroblasts in the Foothills terrane, central Sierra Nevada, California: negotiating a microstructural minefield. *Journal of Metamorphic Geology*, 11, 203–22.

Vernon, R. H., Paterson, S. R. & Geary, E. E. 1989: Evidence for syntectonic intrusion of plutons in the Bear Mountains fault zone, California. *Geology*, **17**, 723–6.

Vernon, R. H. & Pooley, G. D. 1981: SEM/microprobe study of some symplectic intergrowths replacing cordierite. *Lithos*, **14**, 75–82.

Vernon, R. H. & Ransom, D. M. 1971: Retrograde schists of the amphibolite facies at Broken Hill, New South Wales. *Journal of the Geological Society of Australia*, **18**, 267–77.

Vernon, R. H., Richards, S. W. & Collins, W. J. 2000: Migmatite-granite relationships: origin of the Cooma Granodiorite magma, Lachlan Fold Belt, Australia. *Physics and Chemistry of the Earth (A)*, **26**, 267–71.

Vernon, R. H., White, R. W. & Clarke, G. L. In press: False metamorphic events inferred from misinterpretation of microstructural evidence and *P-T* data. *Journal of Metamorphic Geology.*

Vernon, R. H. & Williams, P. F. 1988: Distinction between intrusive and extrusive or sedimentary parentage of felsic gneisses: examples from the Broken Hill Block, NSW. *Australian Journal of Earth Sciences*, **35**, 379–88.

Vernon, R. H., Williams, V. A. & D'Arcy, W. F. 1983: Grainsize reduction and foliation development in a deformed granitoid batholith. *Tectonophysics*, **92**, 123–45.

Vidale, R. 1969: Metasomatism in a chemical gradient and the formation of calc-silicate bands. *American Journal of Science*, **267**, 857–74.

Vidale, R. 1974: Metamorphic differentiation layering in pelitic rocks of Dutchess County, New York. *In*: Hofmann, A. W., Giletti, B. J., Yoder, H. S. & Yund, R. A. (eds.): *Geochemical Transport and Kinetics. Carnegie Institution of Washington Publication* **634**, 273–86.

Vidale, R. A. & Hewitt, D. A. 1973: "Mobile" components in the formation of calc-silicate bands. *American Mineralogist*, **58**, 991–7.

Vielzeuf, D., Clemens, J. D., Pin, C. & Moinet, E. 1990: Granites, granulites and crustal differentiation. *In*: Vielzeuf, D. & Vidal, P. (eds.): *Granites and Crustal Evolution*. Dordecht: Kluwer, pp. 59–85.

Vielzeuf, D. & Holloway, J. R. 1988: Experimental determination of the fluid-absent melting relations in the pelitic system. *Contributions to Mineralogy and Petrology*, **98**, 257–76.

Vigneresse, J.-L. 2004: A new paradigm for granite generation. *Transactions of the Royal Society of Edinburgh: Earth Sciences*, **95**, 11–22.

Vigneresse, J.-L. In press: The building of granitic plutons from pervasive and continuous melting in the lower crust to discontinuous and spaced pluton building in the upper crust. *Transactions of the Royal Society of Edinburgh: Earth Sciences, Wally Pitcher Volume.*

Vigneresse, J.-L., Barbey, P. & Cuney, M. 1996: Rheological transitions during partial melting and crystallization with application to felsic magma segregation and transfer. *Journal of Petrology*, **37**, 1579–600.

Viljoen, M. J. & Viljoen, R. P. 1969: Evidence for the existence of a mobile extrusive peridotitic magma from the Komati Formation of the Onverwacht Group. *Special Publication of the Geological Society of South Africa*, **2**, 87–112.

Vogel, T. A. & Spence, W. H. 1969: Relict plagioclase phenocrysts from amphibolite grade metavolcanic rocks. *American Mineralogist*, **54**, 522–8.

Vogt, T. 1930: On the chronological order of deposition of the Highland schists. *Geological Magazine*, **67**, 68–73.

Voll, G. 1960: New work on petrofabrics. *Liverpool and Manchester Geological Journal*, **2**, 503–67.

Vollbrecht, A., Siegesmund, S. & Flaig, C. 1997: High-temperature deformation of a granitoid from the Zone of Erbendorf-Vohenstrauβ (ZEV). *Geologische Rundschau*, **86**, S141–54.

Wadsworth, W. J. 1961: The layered ultrabasic rocks of South-west Rhum, Inner Hebrides. *Philosophical Transactions of the Royal Society of London*, **A244**, 21–64.

Wager, L. R. 1961: A note on the origin of ophitic texture in the chilled olivine gabbro of the Skaergaard intrusion. *Geological Magazine*, **98**, 353–66.

Wager, L. R. & Brown, G. M. 1968: *Layered Igneous Rocks*. San Francisco: Freeman.

Wager, L. R., Vincent, E. A., Brown, G. M. & Bell, J. D. 1965: Marscoite and related rocks of the Western Red Hills Complex, Isle of Skye. *Philosophical Transactions of the Royal Society of London*, **A257**, 273–308.

Waight, T. E., Dean, A. A., Maas, R. & Nicholls, I. A. 2000a: Sr and Nd isotopic investigations towards the origin of feldspar megacrysts in microgranular enclaves in two I-type plutons of the Lachlan Fold Belt, south-east Australia. *Australian Journal of Earth Sciences*, **47**, 1105–12.

Waight, T. E., Maas, R. & Nicholls, I. A. 2000b: Fingerprinting feldspar phenocrysts using crystal isotopic composition stratigraphy: implications for crystal transfer and magma mingling in S-type granites. *Contributions to Mineralogy and Petrology*, **139**, 227–39.

Waight, T. E., Maas, R. & Nicholls, I. A. 2001: Geochemical investigations of microgranitoid enclaves in the S-type Cowra Granodiorite, Lachlan Fold Belt, SE Australia. *Lithos*, **56**, 165–86.

Wain, A. 1997: New evidence for coesite in eclogites and gneisses: defining an ultrahigh-pressure province in the western Gneiss Region of Norway. *Geology*, **25**, 927–30.

Wain, A. L., Waters, D. J. & Austrheim, H. 2001: Metastability of granulites and processes of eclogitisation in the UHP region of western Norway. *Journal of Metamorphic Geology*, **19**, 609–25.

Waldron, H. M. & Sandiford, M. 1988: Deformation volume and cleavage development in metasedimentary rocks from the Ballarat slate belt. *Journal of Structural Geology*, **10**, 53–62.

Walker, K. R., Joplin, G. A., Lovering, J. F. & Green, R. 1960: Metamorphic and metasomatic convergence of basic igneous rocks and lime-magnesia sediments of the Precambrian of north-western Queensland. *Journal of the Geological Society of Australia*, **6**, 149–78.

Walther, J. V. & Orville, P. M. 1982: Volatile production and transport in prograde metamorphism. *Contributions to Mineralogy and Petrology*, **79**, 252–7.

Walther, J. V. & Wood, B. J. 1984: Rate and mechanism in prograde metamorphism. *Contributions to Mineralogy and Petrology*, **88**, 246–59.

Ward, C. M. 1984: *Geology of the Dusky Sound Area, Fiordland, with Emphasis on the Structural-metamorphic Development of Some Porphyroblastic Staurolite Pelites*. Unpublished PhD thesis, University of Otago, Dunedin.

Ward, R. A., Stevens, G. & Kisters, A. F. M. 2007: Deformation-controlled, fluid-induced anatexis: a field and experimental study on the Damara Orogen, Namibia. Sixth Hutton Symposium, Stellenbosch, South Africa, Abstract Volume, pp. 222–3.

Wareham, C. D., Vaughan, A. P. M. & Miller, I. L. 1997: The Wiley Glacier complex, Antarctic Peninsula: pluton growth by pulsing of granitoid magma. *Chemical Geology*, **143**, 65–80.

Watanabe, J. & Hasegawa, K. 1986: Borosilicates (datolite, schorl) and aluminosilicates (andalusite, sillimanite) in the Oketo Rhyolite, Hokkaido. *Journal of the Faculty of Science, Hokkaido University*, Series IV, **21**, 583–98.

Watanabe, T., Yui, S. & Kato, A. 1970: Bedded manganese deposits in Japan, a review. *In*: Tatsumi, T. (ed.): *Volcanism and Ore Genesis*. Tokyo: University of Tokyo Press, pp. 119–42.

Waters, D. J. 1988: Partial melting and the formation of granulite facies assemblages in Namaqualand, South Africa. *Journal of Metamorphic Geology*, **6**, 387–404.

Waters, D. J. 2001: The significance of prograde and retrograde quartz-bearing intergrowth microstructures in partially melted granulite-facies rocks. *Lithos*, **56**, 97–110.

Waters, D. J. & Charnley, N. R. 2002: Local equilibrium in polymetamorphic gneiss and the titanium substitution in biotite. *American Mineralogist*, **87**, 383–96.

Waters, D. J. & Lovegrove, D. P. 2002: Assessing the extent of disequilibrium and overstepping of prograde metamorphic reactions in metapelites from the Bushveld Complex aureole, South Africa. *Journal of Metamorphic Geology*, **20**, 135–49.

Waters, D. J. & Whales, C. J. 1984: Dehydration melting and the granulite transition in metapelites from southern Namaqualand, S. Africa. *Contributions to Mineralogy and Petrology*, **88**, 269–75.

Watson, J. V. 1973: Effects of reworking on high-grade gneiss complexes. *Philosophical Transactions of the Royal Society of London*, **A273**, 443–55.

Watt, G. R., Oliver, N. H. S. & Griffin, B. J. 2002: Evidence for reaction-induced microfracturing in granulite facies migmatites. *Geology*, **28**, 327–30.

Wei, C. J., Powell, R. & Clarke, G. L. 2004: Calculated phase equilibria for low- and medium-pressure metapelites in the KFMASH and KMnFMASH systems. *Journal of Metamorphic Geology*, **22**, 495–508.

Weill, D. F. & Fyfe, W. S. 1964: A discussion of the Korzhinskii and Thompson treatment of thermodynamic equilibrium in open systems. *Geochimica et Cosmochimica Acta*, **28**, 565–76.

Weill, D. F. & Fyfe, W. S. 1967: On equilibrium thermodynamics of open systems and the phase rule (A reply to D. S. Korzhinskii). *Geochimica et Cosmochimica Acta*, **31**, 1167–76.

Weinberg, R. F. 1996: The ascent mechanism of felsic magmas: news and views. *Transactions of the Royal Society of Edinburgh: Earth Sciences*, **87**, 95–103.

Weinberg, R. F. 1999: Mesoscale pervasive felsic magma migration: alternatives to dyking. *Lithos*, **46**, 393–410.

Weinberg, R. F. & Podlachikov, Y. Y. 1994: Diapiric ascent of magmas through power-law crust and mantle. *Journal of Geophysical Research*, **99**, 9543–59.

Weinberg, R. F. & Searle, M. P. 1998: The Pangong Injection Complex, Indian Karakoram: a case of pervasive granite flow through hot viscous crust. *Journal of the Geological Society of London*, **155**, 883–91.

Weiss, L. E. 1972: *The Minor Structures of Deformed Rocks*. New York: Springer.

Weiss, L. E. & McIntyre, D. B. 1957: Structural geometry of Dalradian rocks at Loch Leven, Scottish Highlands. *Journal of Geology*, **65**, 575–602.

Wells, P. R. A. 1979: Chemical and thermal evolution of Archean sialic crust, southern West Greenland. *Journal of Petrology*, **20**, 187–226.

West, D. P. & Lux, D. R. 1993: Dating mylonitic deformation by the ^{40}Ar-^{39}Ar method: an example from the Norumbega fault zone, Maine. *Earth and Planetary Science Letters*, **120**, 221–37.

Wetherill, G. W. 1956: Discordant uranium lead ages. *Transactions of the American Geophysical Union*, **37**, 320–26.

Wheeler, J. 1992: Importance of pressure solution and Coble creep in the deformation of polymineralic rocks. *Journal of Geophysical Research*, **97**, 4579–86.

White, A. J. R. 1964: Clinopyroxenes from eclogites and basic granulites. *American Mineralogist*, **49**, 883–88.

White, A. J. R. 1966: Genesis of migmatites from the Palmer region of south Australia, Australia. *Chemical Geology*, **1**, 165–200.

White, A. J. R. & Chappell, B. W. 1977: Ultrametamorphism and granitoid genesis. *Tectonophysics*, **43**, 7–22.

White, A. J. R., Chappell, B. W. & Cleary, J. R. 1974: Geologic setting and emplacement of some Australian Paleozoic batholiths and implications for intrusion mechanisms. *Pacific Geology*, **8**, 159–171.

White, R. W. & Clarke, G. L. 1997: The role of deformation in aiding recrystallization: an example from a high pressure shear zone. *Journal of Petrology*, **38**, 1307–29.

White, R. W., Clarke, G. L. & Nelson, D. R. 1999: SHRIMP U-Pb zircon dating of Grenville-age events in the western part of the Musgrave Block, central Australia. *Journal of Metamorphic Geology*, **17**, 465–81.

White, R. W., Pomroy, N. E. & Powell, R. 2005: An *in situ* metatexite-diatexite transition in upper amphibolite facies rocks from Broken Hill, Australia. *Journal of Metamorphic Geology*, **23**, 579–602.

White, R. W. & Powell, R. 2002: Melt loss and the preservation of granulite facies mineral assemblages. *Journal of Metamorphic Geology*, **20**, 621–32.

White, R. W., Powell, R., & Clarke, G. L. 2002: The interpretation of reaction textures in Fe-rich metapelitic granulites of the Musgrave Block, central Australia: Constraints from mineral equilibria calculations in the system K_2O-FeO-MgO-Al_2O_3-SiO_2-H_2O-TiO_2-Fe_2O_3. *Journal of Metamorphic Geology*, **20**, 41–56.

White, R. W., Powell, R. & Clarke, G. L. 2003: Prograde metamorphic assemblage evolution during partial melting of metasedimentary rocks at low pressures: migmatites from Mt Stafford, central Australia. *Journal of Petrology*, **44**, 1937–960.

White, R. W., Powell, R. & Halpin, J. A. 2004: Spatially-focussed melt formation in aluminous metapelites from Broken Hill, Australia. *Journal of Metamorphic Geology*, **22**, 825–45.

White, R. W., Powell, R. & Holland, T. J. B. 2007: Progress relating to calculations of partial melting equilibria for metapelites. *Journal of Metamorphic Geology*, **25**, 511–27.

White, S. H. 1976: The effects of strain on the microstructures, fabrics and deformation mechanisms in quartzite. *Philosophical Transactions of the Royal Society of London*, **A283**, 69–85.

White, S. H., Burrows, S. E., Carreras, J., Shaw, N. D. & Humphreys, F. J. 1980: On mylonites in ductile shear zones. *Journal of Structural Geology*, **2**, 175–87.

White, S. H. & Johnston, D. C. 1981: A microstructural and microchemical study of cleavage lamellae in a slate. *Journal of Structural Geology*, **3**, 279–90.

White, S. H. & Knipe, R. J. 1978: Transformation- and reaction-enhanced ductility in rocks. *Journal of the Geological Society of London*, **135**, 513–6.

Whitney, J. A. 1988: The origin of granite: the role of and source of water in the evolution of granitic magmas. *Bulletin of the Geological Society of America*, **100**, 1886–97.

Whitney, P. R. & McLelland, J. M. 1973: Origin of coronas in metagabbros of the Adirondack Mountains, New York. *Contributions to Mineralogy and Petrology*, **39**, 81–98.

Wickham, S. M. 1987a: Crustal anatexis and granite petrogenesis during low-pressure regional metamorphism: the Trois Seigneurs Massif, Pyrenees, France. *Journal of Petrology*, **28**, 127–69.

Wickham, S. M. 1987b: The segregation and emplacement of granitic magmas. *Journal of the Geological Society of London*, **144**, 281–97.

Wickham, S. M., Janardhan, A. S. & Stern, R. J. 1994: Regional carbonate alteration of the crust by mantle-derived magmatic fluids, Tamil Nadu, South India. *Journal of Geology*, **102**, 379–98.

Wickham, S. M. & Taylor, H. P. 1985: Stable isotopic evidence for large-scale seawater infiltration in a regional metamorphic terrane: the Trois Seigneurs Massif, Pyrenees, France. *Contributions to Mineralogy and Petrology*, **91**, 122–37.

Wickham, S. M. & Taylor, H. P. 1987: Stable isotope constraints on the origin and depth of penetration of hydrothermal fluids associated with Hercynian regional metamorphism and crustal anatexis in the Pyrenees. *Contributions to Mineralogy and Petrology*, **95**, 255–68.

Widmer, T. & Thompson, A. B. 2001: Local origin of high pressure vein material in eclogite facies rocks of the Zermatt-Sass Zone, Switzerland. *American Journal of Science*, **301**, 627–56.

Wiebe, R. A. 1994: Silicic magma chambers as traps for basaltic magmas: the Cadillac Mountain intrusive complex, Mount Desert Island, Maine. *Journal of Geology*, **102**, 423–37.

Wiebe, R. A. 1996: Mafic-silicic layered intrusions: the role of basaltic injections on magmatic processes and the evolution of silicic magma chambers. *Transactions of the Royal Society of Edinburgh: Earth Sciences*, **87**, 233–42.

Wijbrans, J. R. & McDougall, I. 1986: $^{40}Ar/^{39}Ar$ dating of white micas from an Alpine high-pressure metamorphic belt on Naxos (Greece): the resetting of the argon isotopic system. *Contributions to Mineralogy and Petrology*, **93**, 187–94.

Wilcox, R. E. 1944: Rhyolite-basalt complex on Gardiner River, Yellowstone Park, Wyoming. *Bulletin of the Geological Society of America*, **55**, 1047–80.

Willemse, J. & Viljoen, E. A. 1970. The fate of argillaceous material in the gabbroic magma of the Bushveld Complex. *In*: Visser, D. J. L. & von Gruenewaldt, G. (eds.): *Symposium on the Bushveld and other Layered Intrusives. Geological Society of South Africa, Special Publication*, **1**, 336–66.

Williams, E. 1959: Discussion on sedimentary structures in the metamorphic rocks and orebodies of Broken Hill by M. A. Condon. *Proceedings of the Australasian Institute of Mining and Metallurgy*, **189**, 77–9.

Williams, M. L. 1991: Heterogeneous deformation in a ductile fold-thrust belt: the Proterozoic structural history of the Tusas Mountains, New Mexico. *Bulletin of the Geological Society of America*, **103**, 171–88.

Williams, M. L. & Burr, J. L. 1994: Preservation and evolution of quartz phenocrysts in deformed rhyolites from the Proterozoic of southwestern North America. *Journal of Structural Geology*, **16**, 203–21.

Williams, M. L., Hanmer, S., Kopf, C. & Darrach, M. 1995: Syntectonic generation and segregation of tonalitic melts from amphibolite dikes in the lower crust, Striding-Athabasca mylonite zone, northern Saskatchewan. *Journal of Geophysical Research*, **100**, 15717–34.

Williams, M. L. & Jercinovic, M. J. 2002: Microprobe monazite geochronology: putting absolute time into microstructural analysis. *Journal of Structural Geology*, **24**, 1013–28.

Williams, M. L., Jercinovic, M. J. & Terry, M. P. 1999: Age mapping and dating of monazite on the electron microprobe: deconvoluting multistage tectonic histories. *Geology*, **27**, 1023–6.

Williams, M. L. & Karlstrom, K. E. 1996: Looping *P-T* paths and high-*T*, low-*P* middle crustal metamorphism: Proterozoic evolution of the southwestern United States. *Geology*, **24**, 1119–22.

Williams, M. L., Melis, E. A., Kopf, C. F. & Hanmer, S. 2000: Microstructural tectonometamorphic processes in the development of gneissic layering: a mechanism for metamorphic segregation. *Journal of Metamorphic Geology*, **18**, 41–57.

Williams, M. L., Scheltema, K. E. & Jercinovic, M. J. 2001: High-resolution compositional mapping of matrix phases: implications for mass transfer during crenulation cleavage development in the Moretown Formation, western Massachusetts. *Journal of Structural Geology*, **23**, 923–39.

Williams, P. F. 1967: Structural analysis of the Little Broken Hill area of New South Wales. *Journal of the Geological Society of Australia*, **14**, 317–32.

Williams, P. F. 1972: Development of metamorphic layering and cleavage in low grade metamorphic rocks at Bermagui, Australia. *American Journal of Science*, **272**, 1–47.

Williams, P. F. 1983a: Large scale transposition by folding in northern Norway. *Geologische Rundschau*, **72**, 589–604.

Williams, P. F. 1983b: Timing of deformation and the mechanism of cleavage development in a Newfoundland mélange. *Maritime Sediments and Atlantic Geology*, **19**, 31–48.

Williams, P. F. 1985: Multiply deformed terrains – problems of correlation. *Journal of Structural Geology*, **7**, 269–80.

Williams, P. F. 1990: Differentiated layering in metamorphic rocks. *Earth-Science Reviews*, **29**, 267–81.

Williams, P. F., Collins, A. R. & Wiltshire, R. G. 1969: Cleavage and penecontemporaneous deformation structures in sedimentary rocks. *Journal of Geology*, **77**, 415–25.

Williams, P. F., Means, W. D. & Hobbs, B. E. 1977: Development of axial-plane slaty cleavage and schistosity in experimental and natural materials. *Tectonophysics*, **42**, 139–58.

Williams, P. J. 1983: The genesis and metamorphism of the Arinteiro-Bama Cu deposits, Santiago de Compostela, northwestern Spain. *Economic Geology*, **78**, 1689–700.

Williams, P. J. 1988: Metalliferous economic geology of the Mt Isa Eastern Succession, Queensland. *Australian Journal of Earth Sciences*, **45**, 329–41.

Williams, P. J. 1994: Iron mobility during synmetamorphic alteration in the Selwyn Range area, NW Queensland: implications for the origin of ironstone-hosted Au-Cu deposits. *Mineralium Deposita*, **29**, 250–60.

Williams, P. J. & Baker, T. 1995: Regional-scale association of skarn alteration and base-metal deposits: the Cloncurry district, Mount Isa Inlier, Queensland, Australia. *Transactions of the Institution of Mining and Metallurgy*, **104**, B187–96.

Williams, P. J. & Heinemann, M. 1993: Maramungee: a Proterozoic Zn skarn in the Cloncurry district, Mount Isa Inlier, Queensland, Australia. *Economic Geology*, **88**, 1114–34.

Willis, I. L., Stevens, B. P. J., Stroud, W. J., Brown, R. E., Bradley, G. M. & Barnes, R. G. 1983: Metasediments, composite gneisses, and migmatites. *In*: Stevens, B. P. J. & Stroud, W. J. (eds.): *Rocks of the Broken Hill Block: Their Classification, Nature, Stratigraphic Distribution, and Origin. Records of the Geological Survey of New South Wales*, **21**, 57–125.

Willner, A. P., Herve, F. & Massonne, H. J. 2000: Mineral chemistry and pressure-temperature evolution of two contrasting high-pressure-low-temperature belts in the Chonos Archipelago, Southern Chile. *Journal of Petrology*, **41**, 309–30.

Wilshire, H. G. 1961: Layered diatremes near Sydney, New South Wales. *Journal of Geology*, **69**, 473–84.

Wilson, C. J. L. 1984: Shear bands, crenulation and differentiated layering in ice-mica models. *Journal of Structural Geology*, **6**, 303–19.

Wilson, C. J. L. 1986: Deformation-induced recrystallization of ice: the application of *in situ* experiments. *In*: Heard, H. C. & Hobbs, B. E. (eds.): *Mineral and Rock Deformation: Laboratory Studies (The Paterson Volume). American Geophysical Union Geophysical Monograph*, **36**, 213–32.

Wilson, G. 1952: Ptygmatic structures and their formation. *Geological Magazine* **89**, 1–21.

Wilson, G., Watson, J. & Sutton, J. 1953: Current-bedding in the Moine series of north-west Scotland. *Geological Magazine*, **90**, 377–87.

Wilson, M. 1989: *Igneous Petrogenesis. A Global Tectonic Approach*. London: Chapman & Hall.

Windley, B. F. & Bridgewater, D. 1971: The evolution of Archaean low- and high-grade terrains. *Special Publication of the Geological Society of Australia*, **3**, 33–46.

Winkler, H. G. F. 1967: *Petrogenesis of Metamorphic Rocks*. Second edition. Berlin: Springer.

Winkler, H. G. F. 1979: *Petrogenesis of Metamorphic Rocks*. Fifth edition. Berlin: Springer.

Winkler, H. G. F., Boese, M. & Marcpoulos, T. 1975: Low temperature granitic melts. *Neues Jahrbuch für Mineralogie Monatshefte*, 245–68.

Winkler, H. G. F. & von Platen, H. 1961: Experimentelle gesteinmetmorphose V. Experimentelle anatektische Schmelzen und ihre petrogenetische Bedeutung. *Geochimica et Cosmochimica Acta*, **24**, 250–9.

Wintsch, R. P., Webster, J. R., Bernitz, J. A. & Fout, J. S. 1990: Geochemical and geological criteria for the discrimination of high-grade gneisses of intrusive and extrusive origin, eastern Connecticut. *Special Paper of the Geological Society of America*, **245**, 187–208.

Wintsch, R. P. & Yi, K. 2002: Dissolution and replacement creep: a significant deformation mechanism in mid-crustal rocks. *Journal of Structural Geology*, **24**, 1179–93.

Wolf, M. B. & Wyllie, P. J. 1991: Dehydration melting of solid amphibolite at 10 kbar: Textural development, liquid interconnectivity and applications to the segregation of magmas. *Contributions to Mineralogy and Petrology*, **44**, 151–79.

Wood, B. J. & Fraser, D. G. 1976: *Elementary Thermodynamics for Geologists*. Oxford: Oxford University Press.

Wood, B. J. & Walther, J. V. 1986: Fluid flow during metamorphism and its implications for fluid-rock ratios. *In*: Walther, J. V. & Wood, B. J. (eds.): *Fluid-Rock Interaction during Metamorphism*. New York: Springer-Verlag, pp. 89–108.

Wood, D. A., Gibson, I. L. & Thompson, R. N. 1976: Element mobility during zeolite facies metamorphism of the Tertiary basalts of eastern Iceland. *Contributions to Mineralogy and Petrology*, **55**, 241–54.

Wood, D. S. 1974: Current views of the development of slaty cleavage. *Annual Reviews of Earth and Planetary Sciences*, **2**, 369–401.

Woodland, B. G. 1985: Relationship of concretions and chlorite-muscovite porphyroblasts to the development of domainal cleavage in low-grade metamorphic deformed rocks from north-central Wales, Great Britain. *Journal of Structural Geology*, **7**, 205–15.

Woodsworth, G. J. 1977: Homogenization of zoned garnets from pelitic schists. *Canadian Mineralogist*, **15**, 230–42.

Worden, R. H., Droop, G. T. R. & Champness, P. E. 1991: The reaction antigorite → olivine + talc H_2O in the Bergell aureole, N. Italy. *Mineralogical Magazine*, **55**, 367–77.

Worley, B. & Powell, R. 1998a: Making movies: phase diagrams in temperature, pressure, composition and time. *In*: Treloar, P. J. & O'Brien, P. (eds.): *What Drives Metamorphism and Metamorphic Reactions? Geological Society of London, Special Publications*, **138**, 263–74.

Worley, B. & Powell, R. 1998b: Singularities in NCKFMASH (Na_2O–CaO–K_2O–MgO–FeO–Al_2O_3–SiO_2–H_2O). *Journal of Metamorphic Geology*, **16**, 169–88.

Worley, B., Powell, R. & Wilson, C. J. L. 1997: Crenulation cleavage formation: evolving diffusion, deformation and equilibration mechanisms with increasing metamorphic grade. *Journal of Structural Geology*, **19**, 1121–36.

Wright, T. O. & Henderson, J. R. 1992: Volume loss during cleavage formation in the Meguma Group, Nova Scotia, Canada. *Journal of Structural Geology*, **14**, 281–90.

Wyborn, D., Chappell, B. W. & Johnston, R. M. 1981: Three S-type volcanic suites from the Lachlan Fold belt, southeast Australia. *Journal of Geophysical Research*, **86**, 10335–48.

Wyllie, P. J. 1962: The petrogenetic model, an extension of Bowen's petrogenetic grid. *Geological Magazine*, **99**, 558–69.

Wyllie, P. J. 1977: Crustal anatexis: an experimental review. *Tectonophysics*, **43**, 41–71.

Wyllie, P. J. 1983: Experimental studies on biotite- and muscovite-granites and some crustal magmatic sources. *In*: Atherton, M. P. & Gribble, C. D. (eds.): *Migmatites, Melting and Metamorphism*. Nantwich: Shiva, pp. 12–26.

Wyllie, P. J., Huang, W. L., Stern, C. R. & Maaløe, S. 1976: Granitic magmas: possible and impossible sources, water contents and crystallization sequences. *Canadian Journal of Earth Sciences*, **13**, 1007–19.

Wynne-Edwards, H. R. 1969: Tectonic overprinting in the Grenville province, southwestern Quebec. *Special Paper of the Geological Association of Canada*, **5**, 163–82.

Yang, P. & Pattison, D. R. M. 2006: Genesis of monazite and Y zoning in garnet from the Black Hills, South Dakota. *Lithos*, **88**, 233–253.

Yang, P. & Rivers, T. 2002: The origin of Mn and Y annuli in garnet and thermal dependence of P in garnet and Y in apatite in calc-pelite and pelite, Gagnon terrane, western Labrador. *Geological Materials Research*, **4**, 1–35.

Yardley, B. W. D. 1977a: The nature and significance of the mechanism of sillimanite growth in the Connemara Schist, Ireland. *Contributions to Mineralogy and Petrology*, **10**, 235–42.

Yardley, B. W. D. 1977b: An empirical study of diffusion in garnet. *American Mineralogist*, **62**, 793–800.

Yardley, B. W. D. 1977c: Petrogenesis of migmatites in the Huntly–Portsoy area, northeast Scotland – a discussion. *Mineralogical Magazine*, **41**, 292–4.

Yardley, B. W. D. 1977d: Relationships between the chemical and modal compositions of metapelites from Connemara, Ireland. *Lithos*, **65**, 53–8.

Yardley, B. W. D. 1978: Genesis of the Skagit Gneiss migmatites, Washington, and the distinction between possible mechanisms of migmatization. *Bulletin of the Geological Society of America*, **89**, 941–51.

Yardley, B. W. D. 1989: *An Introduction to Metamorphic Petrology*. Essex: Longman.

Yardley, B. W. D. 2005: Metal concentrations in crustal fluids and their relationship to ore formation. *Economic Geology*, **100**, 613–32.

Yardley, B. W. D. & Barber, J. P. 1991: Melting reactions in the Connemara Schists: the role of water infiltration in the formation of amphibolite facies migmatites. *American Mineralogist*, **76**, 848–56.

Yardley, B. W. D. & Bottrell, S. H. 1992: Silica mobility and fluid movement during metamorphism of the Connemara schists, Ireland. *Journal of Metamorphic Geology*, **10**, 453–64.

Yardley, B. W. D., Bottrell, S. H. & Cliff, R. A. 1991a: Evidence for a regional-scale fluid loss event during mid-crustal metamorphism. *Nature*, **349**, 151–4.

Yardley, B. W. D., Gleeson, S., Bruce, S. & Banks, D. 2000: Origin of retrograde fluids in metamorphic rocks. *Journal of Geochemical Exploration*, **69–70**, 281–86.

Yardley, B. W. D. & Lloyd, G. E. 1995: Why metasomatic fronts are really metasomatic sides. *Geology*, **23**, 53–6.

Yardley, B. W. D., Rochelle, C. A., Barnicoat, A. & Lloyd, G. E. 1991b: Oscillatory zoning in metamorphic minerals: an indicator of infiltration metasomatism. *Mineralogical Magazine*, **55**, 357–65.

Yardley, B. W. D. & Valley, J. W. 1997: The petrologic case for a dry lower crust. *Journal of Geophysical Research*, **102**, 12173–85.

Yardley, B. W. D. & Valley, J. W. 2000: Comment on 'The petrologic case for a dry lower crust' by Yardley, B. D. W. & Valley, J. W. – Reply. *Journal of Geophysical Research*, **105**, 6065–8.

York, D., Hall., C. M., Yanase, Y., Hanes, J. A. & Kenyon, W. J. 1981: $^{40}Ar/^{39}Ar$ dating of terrestrial minerals with a continuous laser. *Geophysical Research Letters*, **8**, 1136–8.

Yund, R. A. & Tullis, J. 1991: Compositional changes of minerals associated with dynamic crystallization. *Contributions to Mineralogy and Petrology*, **108**, 346–55.

Zack, T., Rivers, T., Brumm, R. & Kronz, A. 2004: Cold subduction of oceanic crust: Implications for a lawsonite eclogite from Dominican Republic. *European Journal of Mineralogy*, **16**, 909–16.

Zeck, H. P. 1970: An erupted migmatite from Cerro del Hoyazo, SE Spain. *Contributions to Mineralogy and Petrology*, **26**, 225–46.

Zeitler, P. K. 1989: The geochronology of metamorphic processes. *In*: Daly, J. S., Cliff, R. A. & Yardley, B. W. D. (eds.): *Evolution of Metamorphic Belts. Geological Society of London Special Publications*, **43**, 131–47.

Zeitler, P. K. & Chamberlain, C. P. 1991: Petrogenetic and tectonic significance of young leucogranites from the northwestern Himalaya, Pakistan. *Tectonics*, **10**, 729–41.

Zen, E-an. 1966: Construction of pressure-temperature diagrams for multi-component systems after the method of Schreinemakers – a geometric approach. *United States Geological Survey Bulletin*, **1225**.

Zeng, L., Saleeby, J. B. & Asimov, P. 2005: Nd isotope disequilibrium during crustal anatexis: A record from the Goat Ranch migmatite complex, southern Sierra Nevada batholith, California. *Geology*, **33**, 53–6.

Zharikov, A. 1970: Skarns. *International Geology Review*, **12**, 541–59, 619–47, 760–76.

Zhu, X. K. & O'Nions, R. K. 1999: Zonation of monazite in metamorphic rocks and its implications for high temperature thermochronology: a case study from the Lewisian terrain. *Earth and Planetary Science Letters*, **171**, 209–20.

Zwart, H. J. 1960a: The chronological succession of folding and metamorphism in the central Pyrenees. *Geologische Rundschau*, **50**, 203–18.

Zwart, H. J. 1960b: Relations between folding and metamorphism in the central Pyrenees, and their chronological succession. *Geologie en Mijnbouw*, **39**, 163–80.

Zwart, H. J. 1962: On the determination of polymetamorphic mineral associations, and its application to the Bosost area (Central Pyrenees). *Geologische Rundschau*, **52**, 38–65.

Zwart, H. J. & Calon, T. J. 1977: Chloritoid crystals from Curaglia; growth during flattening or pushing aside? *Tectonophysics*, **39**, 477–86.

Index